Recent Advances in Biotechnology

(Volume 1)

Microbial Biopolyester Production, Performance and Processing Microbiology, Feedstocks, and Metabolism

Edited By:

Martin Koller

Office of Research Management and Service,
c/o Institute of Chemistry,
University of Graz
NAWI Graz, Heinrichstrasse 28/III, 8010 Graz,
Austria

Bentham Science Publishers Ltd.
Executive Suite Y - 2
PO Box 7917, Saif Zone
Sharjah, U.A.E.
Email: subscriptions@benthamscience.org

BENTHAM SCIENCE

CONTENTS

FOREWORD i

PREFACE iv
THE CONTRIBUTING SCIENTISTS v
CHAPTER 1 vii
CHAPTER 2 viii
CHAPTER 3 viii
CHAPTER 4 ix
CHAPTER 5 ix
CHAPTER 6 x
CHAPTER 7 xi
CHAPTER 8 xi

LIST OF CONTRIBUTORS xiv

CHAPTER 1 NOVEL INEXPENSIVE FEEDSTOCKS FROM AGRICULTURE AND INDUSTRY FOR MICROBIAL POLYESTER PRODUCTION 3
GENERAL ASPECTS OF POLYHYDROXYALKANOATE PRODUCTION 3
Motivation for Utilization of Inexpensive Substrates 4
PHA Producing Strains and their Characterization 5
PHA Biosynthetic Pathways and their Regulation 6
Enhancement of PHA yields by Application of Exogenous Stress 11
PRODUCTION OF PHA USING LIPID SUBSTRATES 13
Production of PHAs from Lipid Substrates Employing Bacterium Cupriavidus necator 16
Production of PHAs from Lipid Substrates by Other Bacterial Strains 21
Production of PHA from Lignocellulose-Based Substrates 23
Characterization of Lignocellulose Biomass 24
Pretreatment of Lignocellulose Materials from PHA Production 25
Hydrolysis of Cellulose and Hemicellulose 25
Composition of Hydrolysates 27
Detoxification of Hydrolysates 29
Production of PHA on Hydrolysated Lignocellulose Materials 34
Lignocellulose Biomass as Substrate for PHA Production 34
Utilization of Wood Biomass for PHA Production 35
MOLASSES 43
PHA Production on Molasses 45
CHEESE WHEY 48
PHA Production by Naturally Producing Bacteria 52
PHA Production by Genetically Engineered Bacteria 54
GLYCEROL AS A CHEAP SUBSTRATE FOR PHA PRODUCTION 58
Glycerol as Waste from Biodiesel Production 58
Production of PHA on Waste Glycerol Based Substrate 59
STARCH-BASED SUBSTRATES 64
Chemical Composition of Starch 65
STARCHY MATERIALS 66
Hydrolysis of Starchy Materials 68
Productions of PHA on Starch-Based Substrates 70
Utilization of Starchy Materials by PHA Producing Bacteria 70

Untreated Starchy Substrates 72
PETROCHEMICAL PLASTIC WASTE TRANSFORMATION TO PHA 74
PRODUCTION OF PHA FROM VOLATILE FATTY ACIDS 77
CONCLUDING REMARKS 78
CONFLICT OF INTEREST 79
ACKNOWLEDGEMENTS 79
REFERENCES 80

CHAPTER 2 PHA BIOPOLYESTER PRODUCTION FROM SURPLUS WHEY: MICROBIOLOGICAL AND ENGINEERING ASPECTS 100
THE NEED FOR VALUE-ADDED AND SUSTAINABLE UPGRADING OF WHEY: HOW TO TURN A SURPLUS PRODUCT INTO A USEFUL RAW MATERIAL 101
WHAT CHARACTERIZES PHASS AS SO CALLED "GREEN PLASTICS"? 107
Talking about "Green Plastics": Clarifying some Inconsistencies 107
Significance and Particularities of PHAs 110
Composition and Production of PHAs 112
PHAs: The Broad Range of their Applications 114
Factors Impacting the Economic Viability of PHA Production 117
PHA Research: Quo Vadis? 119
Raw Materials 119
New Powerful Production Strains and Genetic Engineering 120
Mixed Culture Approaches 123
Downstream Processing 125
Improved Process Design 126
Follow-up Materials: Advanced Blends and Composites Based on PHA 128
THE ENTIRE PROCESS TO CONVERT THE RAW MATERIAL (MILK) TO PHA BIOPOLYMERS 131
General Aspects 131
Significance of Whey´s Protein Fraction 132
Case Examples for PHA Production Starting from Whey by Diverse Microbes 134
Gram-Negative Wild Type Organisms as Whey-Based PHA-Producers 135
Gram-Positive Wild Type Organisms as Whey-Based PHA-Producers 144
Genetically Engineered Organisms as Whey-Based PHA-Producers 145
Mathematical Modelling of Whey-Based PHA Production 148
CONCLUDING REMARKS 153
CONFLICT OF INTEREST 155
ACKNOWLEDGEMENTS 155
REFERENCES 156

CHAPTER 3 BIOPOLYESTERS AND RELATED VALUABLE PRODUCTS IN PHOTOTROPHIC MICROBES 175
INTRODUCTION 176
PHA Production and Metabolism in Anoxygenic Phototrophic Microbes 178
PHA Production by Anaerobic Photosynthetic Bacteria 178
PHA Production by Aerobic Anoxygenic Phototrophic Bacteria 180
Biosynthetic Routes 181
Polyester Synthases 183
Factors Controlling the Formation of PHA 186
In Vivo Formation of Polyester Particles, Self-Assembly and Structure 188
Catabolism of Polyester in Purple Bacteria 189
PHA Depolymerases 191

Cyanobacterial Production of Polyhydroxyalkanoates 192
Cyanobacterial Strains Capable of Synthesizing PHA 193
Screening for PHA Synthesizing Capability 194
Production of PHB from CO2 in Cyanobacteria 195
Efforts toward Improved PHA Production 195
Optimization of Culture Conditions 195
Genetic Modification for Higher Yield 199
Recovery of PHA from Cyanobacteria 204
Practical Applications of Cyanobacterial PHA 206
CONCLUDING REMARKS 208
CONFLICT OF INTEREST 208
ACKNOWLEDGEMENTS 209
NOTES 209
REFERENCES 209

CHAPTER 4 BIOPOLYMER PRODUCTION BY MIXED MICROBIAL CULTURES: INTEGRATING REMEDIATION WITH VALORIZATION 226
INTRODUCTION 227
Mixed Culture Definition 227
Historical Background 228
Mixed Cultures versus Pure Cultures 229
METABOLISM 231
STRATEGIES OF PHA PRODUCTION BY MMC 235
First Attempts 235
Three-Step Process 238
Operation Modes 242
Anaerobic/Aerobic System 243
Aerobic Dynamic Feeding System 245
Other Possibilities 253
APPLICATIONS 254
CONCLUSION 256
CONFLICT OF INTEREST 256
ACKNOWLEDGEMENTS 256
REFERENCES 256

CHAPTER 5 RALSTONIA EUTROPHA AND THE PRODUCTION OF VALUE ADDED PRODUCTS: METABOLIC BACKGROUND OF THE WILD-TYPE STRAIN AND ITS ROLE AS A DIVERSE, GENETICALLY-ENGINEERED BIOCATALYST ORGANISM 265
INTRODUCTION 266
A Brief History of R. eutropha (C. necator): What's in a Name? 266
R. eutropha Research, the Early Years 268
POLYHDROXYALKANOATE HOMEOSTASIS 270
PHA Homeostasis - Key Proteins 272
β-Ketothiolase (PhaA, BktB) 272
Acetoacetyl-CoA Reductase (PhaB) 273
PHA Synthases (PhaC) 274
Phasins (PhaP) 277
Phasin Regulation by PhaR 278
Multifunctional Granule Associated Protein (PhaM) 279
Depolymerases (phaZ) 280
R-Specific Enoyl-CoA Hydratase (PhaJ) 283

3-Hydroxybutyrate Oligomer Hydrolase (PhaY) 283
THE BUSINESS OF PHA: R. EUTROPHA AS A PLATFORM ORGANISM FOR INDUSTRIAL CULTIVATIONS 284
R. eutropha Productions: Fermentative PHA Production 284
Metabolic Pathways and Genetics Involved in Production of PHBHV 287
Engineered R. eutropha and P(HB-co-HHx) Synthesis 289
Metabolic Pathways and Genetics Involved in Production of P(HB-co-HHx) 289
Fermentation Strategies 292
PHA Biosynthesis Strategy - Nitrogen vs. Phosphorus Limitation 294
Waste Streams as Carbon Feedstocks 295
MORE THAN JUST A GOOD IDEA: UTILITY OF PHA BIOPOLYMERS 298
An Outlook of R. eutropha and the PHA Industry 299
Biosynthesis of Other Unique Biopolymer Compounds by R. eutropha 300
Autotrophy 301
Growth in the Presence of Carbon Monoxide [CO] 306
Ralstonia eutropha JMP134 – Master Bioremediator 307
Aromatic Degradation Pathways 308
β-Ketoadipate Pathway 308
2,4-dichlorophenoxyacetate 309
4-Methylcatechol Ortho Ring-Cleavage Pathway 310
Gentisate Ring-Cleavage Pathway 310
Phenylacetyl-CoA Ring-Cleavage Pathway 312
Homogentisate Ring-Cleavage Pathway 313
2,3-dihydroxyphenylpropionate Meta Ring-Cleavage Pathway 314
Catechol Meta Ring-Cleavage Pathway 315
Chlorohydroxyquinol Ortho Ring-Cleavage Pathway 315
Aminohydroquinone Ring-Cleavage Pathway 316
2-Aminobenzoyl-CoA Ring-Cleavage Pathway 317
R. eutropha JMP134 – A Question 317
The Dawning of a New Era – Biofuels Production in R. eutropha 318
Short-Chained Alcohol Production 319
Utilization of CO2 and Electricity 321
Medium-Chain Length Alcohol Production 322
Eutrophanol? Future Directions for Fuel Production 323
CONCLUDING REMARKS: RALSTONIA WRAP UP 324
CONFLICT OF INTEREST 324
ACKNOWLEDGEMENTS 324
NOTES 325
REFERENCES 325

CHAPTER 6 PHBHV BIOSYNTHESIS BY HALOFERAX MEDITERRANEI: FROM GENETICS, METABOLISM, AND ENGINEERING TO ECONOMICAL PRODUCTION 348
INTRODUCTION 349
THE GENOME OF H. MEDITERRANEI 350
THE KEY ENZYMES FOR PHBHV BIOSYNTHESIS 352
PHA Synthase (PhaEC) 352
PHA Biosynthetic Ketothiolases (PhaA and BktB) 354
PHA-Specific Acetoacetyl-CoA Reductase (PhaB) 356
PHA Granule-Associated PhaP and PhaR 356
MULTIPLE PATHWAYS FOR PROPIONYL-COA SUPPLYING 360
ENZYMES AND PATHWAYS FOR PHBHV MOBILIZATION 362

PhaZh1-Drived PHBHV Degradation 362
PhaJ-Mediated PHA Mobilization 364
PRODUCTION OF PHBHV FROM INEXPENSIVE CARBON RESOURCES 366
Rice Bran and other Agricultural By-Products 366
Whey Hydrolysates from Dairy Industry 367
Vinasse and Stillage from Ethanol Industry 368
Surplus Glycerol from the Biodiesel Industry and Chitin 369
ENGINEERING AND TAILOR-MADE PRODUCTION OF PHBHV 369
Genetic Engineering of H. mediterranei for PHA Production 369
Production of Tailor-Made PHA 371
CONCLUDING REMARKS 373
NOTES 374
CONFLICT OF INTEREST 374
ACKNOWLEDGEMENTS 374
REFERENCES 374

CHAPTER 7 BACTERIAL GENETIC MODIFICATIONS FOR IMPROVING POLYHYDROXYALKANOATES PRODUCTION FROM INEXPENSIVE CARBON SOURCES ... 380
TRADITIONAL AND DNA-BASED METHODS TO LOOK FOR PHA PRODUCERS 381
Testing New Isolates for the Ability to Produce PHA 382
Traditional Methods 382
DNA-Based Methods 383
Molecular and Phenotypic Classification: The Case of C. necator 385
The Metagenomic Perspective 386
MAKING PHA PRODUCING STRAINS ABLE TO UTILIZE DIFFERENT CARBON SOURCES 389
PHAs from Whey 391
PHAs from Molasses 393
PHAs from Waste Lipids 395
PHAs from Agricultural Residues 400
CLONING PHA GENES IN NON PRODUCING BACTERIA 406
CONCLUDING REMARKS 414
CONFLICT OF INTEREST 416
AKNOWLEDGEMENTS 416
REFERENCES 417

CHAPTER 8 MEDIUM-CHAIN-LENGTH POLY-3-HYDROXYALKANOATES 436
INTRODUCTION 436
DISCOVERY AND POTENTIAL APPLICATIONS 437
METABOLIC ROUTES TO MCL-PHA 439
Structurally Related Carbon Sources 440
Structurally Unrelated Compounds 441
Chain Elongation Reactions, in which Acetyl-Coa Moieties are condensed to 3-Hydroxyacyl-CoA 442
mcl-PHA Synthesis 442
Key Enzymes/Genes 445
NATURALLY OCCURRING MCL-PHA PRODUCING MICROORGANISMS 445
MCL-PHA MATERIAL PROPERTIES 447
Tensile, Thermal and Crystallization Properties 447
Block Copolymers 448
Processing and Blending 449
PRODUCTION METHODS AND ECONOMIC CONSIDERATIONS 450

Fermentation Methods 450
Chemostat Studies 451
Fed-Batch Studies 452
Oxygen Supply 455
SEPARATION METHODS 456
Solvent Extraction 456
Differential Digestion of Biomass and PHA 458
mcl-PHA Latex 460
THE FUTURE 462
CONFLICT OF INTEREST 463
ACKNOWLEDGEMENTS 463
REFERENCES 464

SUBJECT INDEX 474

FOREWORD

I am very happy to receive an invitation from Dr. Martin Koller to write something on my motivation for PHA research, how the knowledge about PHA and PHA-related R&D has changed in the time since I started with this topic, and what my ideas are for the future of PHA. I take this invitation as an initiative to think the past and future of these fascinating materials.

I started my PhD thesis in 1986 under the guidance of Professor Robert M Lafferty at the Graz University of Technology (TU Graz), Austria. At that time, as an organic chemist working on chemical synthesis of Nylon-6 under explosive or flammable conditions in organic solvents, I was very much attracted by the idea of using gentle conditions like aqueous phase, room temperature and normal pressure to make plastics. My initial research was on fermentation process development using *Alcaligenes latus* DSM 1122, 1123 and 1124 for production of copolymers PHBHV consisting of 3-hydroxybutyrate (3HB) and 3-hydroxyvalerate (3HV). At that time, ICI successfully launched the PHBHV product trademarked as BIOPOL, the whole world was enthusiastic for bioplastics as they seemed to solve the environmental and petroleum resource crisis. At the Institute of Biotechnology (today: Institute of Biotechnology and Biochemical Engineering) of TU Graz, I was able to interact not only with my supervisor Professor Lafferty but also with Dr. Gerhart Braunegg, Dr. Walter Steiner and Dr. Helmut Schwab as well as many highly talented PhD students. Dr. Gerhart Braunegg is the inventor of the gas chromatography (GC) method for qualitative and quantitative PHA determination. Although this GC method was invented and established in the 1970s, it is still the most popular method for PHA studies, as manifested by its numerous citations in the recent literature. For that all PHA researchers should pay respects to Gerhart. In addition, Gerhart has done some excellent works on using waste substrates for PHA production. He also actively promotes the concept of continuous fermentation for PHA production. This concept has an in-depth impact on my research career, which I will describe shortly.

I left Graz in 1990 to conduct my postdoc studies in University of Nottingham in UK. My postdoc study aimed to use PHA as a matrix to encapsulate drug for sustainable release. Very soon we found that PHA *in vivo* degradation was too slow to meet the sustainable release goal compared with other polymers like polylactide (PLA), poly(lactide-*co*-glycolide) (PLGA) and human serum albumin. However, the flexible mechanical property combined with biodegradability and biocompatibility makes PHA a suitable material for implant tissue engineering applications, and this forms part of my career interest in my research conducted at Tsinghua University/Beijing where I become an independent principle investigator.

PHA research has been changed dramatically since I started my PhD thesis 30 years ago. The changes can be summarized in the following areas:

1. Much wider PHA diversity which evolves into "PHAome" similar to "Genome" or "Proteome";
2. Much more "Synthetic Biology" approaches for engineering PHA production strains in a more effective way;
3. Open (unsterile) and continuous fermentation processes for high cell density and high PHA contents growth;
4. High-value applications of PHA in areas of medical implant applications and smart material developments.

Since the breakthrough in weakening beta-oxidation in *Pseudomonas* spp. by our lab in 2010, it has become possible to precisely control the structures of PHA either in the form of homopolymers, random copolymers or block copolymers. This will really achieve tailor-made PHA based on property requirements. Since *Pseudomonas* spp. are the strains used to generate the "PHAome", the economy to obtain diverse PHA is very important depending on the ability to reach high cell density growth for *Pseudomonas* spp. In this area, Dr. Bruce A Ramsay has done some excellent researches to realize high cell density growth of *Pseudomonas putida* containing over 50% PHA. The contribution of Bruce also contains the efficient extraction of PHA from *Pseudomonas* spp. In the future, PHAome produced by high cell density growth of *Pseudomonas* spp. will allow diverse PHA be produced in economically competitive way.

With the rapid developments of synthetic biology, many possibilities can be exploited to improve PHA production strains, including promoter engineering and ribosome-binding-site (RBS) optimization to enhance PHA content, minimized genome to, *inter alia*, increase substrate to PHA conversion efficiency, reprogramming cell growth pattern to accelerate cell growth, new pathway assembly for PHAome from glucose only, mixed substrate fermentation for single PHA molecules, morphology engineering the bacterial cell shapes for enhanced PHA accumulation and reduction on downstream purification cost. All of those efforts should lead to the significant reduction on PHA production costs.

Since PHA production cannot compete with the petrochemical plastic industries as it suffers from high consumptions on energy and freshwater, discontinuous processing, low product concentration and thus high product recovery cost as well as low substrate to product conversion efficiency. Therefore, to make PHA production competitive, we need to develop fermentation technology which is energy and fresh water saving and run in a continuous way instead of a batch way. To meet this need, a seawater technology based on *Halomonas* spp.

grown in seawater in an unsterile and continuous process has been developed. Since seawater contains around 35 g/L NaCl and *Halomonas* spp. prefer higher pH, the process allows *Halomonas* spp. to grow in an open and continuous way for at least two months infection free. This seawater biotechnology achieves fresh water saving, energy saving and reduces process complexity combined with use of low cost fermentors, it significantly reduces the production cost of PHA and possibly other bioproducts. The *Halomonas* spp. have been engineered to allow them to become a platform for low cost production of biofuels, biomaterials and chemicals.

This vast diversity constituting the "PHAome" is waiting for exploitation. Except for very few PHAs that are commercially available from the market for application developments, such as PHB, PHBHV, P3HB4HB and PHBHHx, all other PHAs are produced by individual labs across the world in very small amounts for academic curiosity. Moving advanced materials like PHAs from the laboratory to the marketplace to address critical challenges in environment, energy, transportation, healthcare, and other areas of application concerns takes far too long, and at too high a cost. How to accelerate the pace of discovery and deployment of advanced PHA materials has been a central question for all researchers and stakeholders in this field. All these depend on the availability of the diverse PHAs in sufficient quantities and with consistent properties for studying their thermal and mechanical properties as well as other application potentials. It should be a global effort to establish platforms to supply diverse PHA based on PHAome information in sufficient quantities and material property consistency for various application developments. The understanding of the PHAome concept should allow us to construct a suitable bacterial platform that is able to provide consistent PHA molecular structures with less molecular weight variations for constant material properties. This is very important for commercial developments.

I hope this information is helpful for PHA researchers.

Thank you for your attention!

Dr. George Guo-Qiang Chen
Department of Microbiology and Biomaterials
Center for Synthetic and Systems Biology; School of Life Sciences
Tsinghua University
Beijing 100084, China
Tel: +86-10-62783844
Fax: +86-10-62794217
E-mail: chengq@mail.tsinghua.edu.cn

PREFACE

This *e*Book volume sheds light on contemporary R&D endeavors managed in various research institutions all over the world; these endeavors have one aim in common: *Replacing established plastics usually stemming from the conversion of fossil feed stocks by "bio-inspired" alternatives provided by nature!* The utmost significance of these endeavors shall be visualized by the annual pile of plastics of about 330 Mt produced worldwide; the predominant part of these plastics is highly resistant against (bio-)degradation, and only an evanescent share of about 20% of spent plastics is subjected to recycling procedures. Sustainable alternatives, so called "environmentally degradable plastics", exist, but do not even cover 5% of the global plastic market. In addition, although very common today, the term "environmentally degradable plastic" is used highly ambiguously and often explicitly incorrectly; in many cases, so called "bio-plastics" do not conform to the attributes "biobased", "biodegradable", or "biocompatible".

Microbial polyhydroxyalkanoates (PHA), a versatile group of polyesters produced by nature as prokaryotic storage materials, are contemporarily attracting increasing attention as polymeric basic materials to be implemented in various fields of the plastic market. This fact is manifested by the nowadays exponentially growing number of scientific publications and patent applications related to this topic. Not only does the biobased nature of PHAs underline their significance for sustainable future developments, but also the high versatility, flexibility and adjustability of their properties, their biodegradability and, to a steadily increasing extent, their biocompatibility. To develop overall sustainable and efficient production processes, one has to take into account all single process steps, starting from microbiology and enzymology, selection of suitable inexpensive raw materials, optimizing process engineering and process regime, until enhancement of product recovery in terms of time-, energy-, and material input.

My personal motivation to edit this *e*Book volume started already more than one decade ago, when I was doing my Master thesis on novel, ecologically benign methods for extraction of PHA from microbial biomass in the group of Professor Gerhart Braunegg at Graz University of Technology. Since then, I got more and more fascinated by these unique biopolyesters. Not only the multiple and complex metabolic roles, the multifaceted functions of PHA in divers ecosystems, the high diversity of PHA accumulating microorganism and the inestimable variety of different "plastic like" properties of different types of PHA attracted my attention, but also the variety of cultivation strategies in discontinuous and continuous mode which can be implemented to produce PHA in an efficient way at pre-defined composition and quality. Most of all, raw material aspects which need to be addressed to make PHA production cost-efficient for me always constituted the central field of interest; in this context, I had the honor to perform my doctoral thesis, guided by G. Braunegg, on production of PHA from whey as an abundant raw material in the frame of the EU-FP5 project WHEYPOL. After this project, several other research groups worldwide investigated whey-based production of PHA and other valued bio-products; this gave reason for a chapter dedicated to this topic in the volume at hand (see Chapter 2). Years later, we upgraded another industrial surplus material to a feedstock for PHA biosynthesis, namely waste lipids from the rendering and animal processing industry. Here, enabled by the EU-FP7 funded project ANIMPOL, our multidisciplinary and international consortium demonstrated how the utilization of this waste stream on the one hand contributes to minimizing waste disposal issues, and, on the other hand, provides for cost-efficient production of PHA biopolyesters and even positively contributes to biodiesel industry. As a common feature of all PHA-related research activities I carried out together with G. Braunegg, one should emphasize the strong industrial embedding in all projects, either as direct project sponsors, advisors, or as active project partners. At this point, I would like to take the chance to express my gratitude to all our industrial partners and financiers who enabled our research and, by cooperating with us, underlined the high industrial significance of this novel class of biomaterials.

Coming to the core part of the two volumes of this *e*Book issue, all single steps, starting from the intracellular procedures on the genetic and enzymatic level, the selection of the adequate raw materials, the biotechnological production process including kinetic analysis and mathematical modelling, process engineering aspects regarding fermentation modes, feeding regimes and bioreactor selection, until the downstream processing for product recovery and refining have to be understood and optimized; this goes in parallel with market issues and holistic consideration of sustainability aspects. Fig. (**1**) illustrates the seven pillars which have to be addressed to come to efficient PHA production processes. The volume at hand concentrates on microbiological, genetic, enzymatic, substrate-related and metabolic issues, whereas aspects of engineering (fermentation technology and downstream processing), kinetic modelling, and polymer characterization, as well as sustainability- and market issues will be the focus of the forthcoming volume 2.

THE CONTRIBUTING SCIENTISTS

Although an increasing number of high-quality monographs and reviews are contemporarily published, none of these addresses all aspects of the holistic PHA production process in a comprehensive way. The volumes *"Microbial biopolyesters – Production, Performance and Processing aspects 1 and 2"* of the *e*Book series "*Recent Advances in Biotechnology*" address this shortcoming in sixteen articles written by carefully selected leading scientists and their associates, performing their work on three different continents. Of course, it was connected to considerable efforts to select the most outstanding groups for all special chapters. These research groups are since years or sometimes even since decades active in different scientific areas involved in this fascinating and strongly emerging field of research, as demonstrated by their tremendous number of related publications and patents.

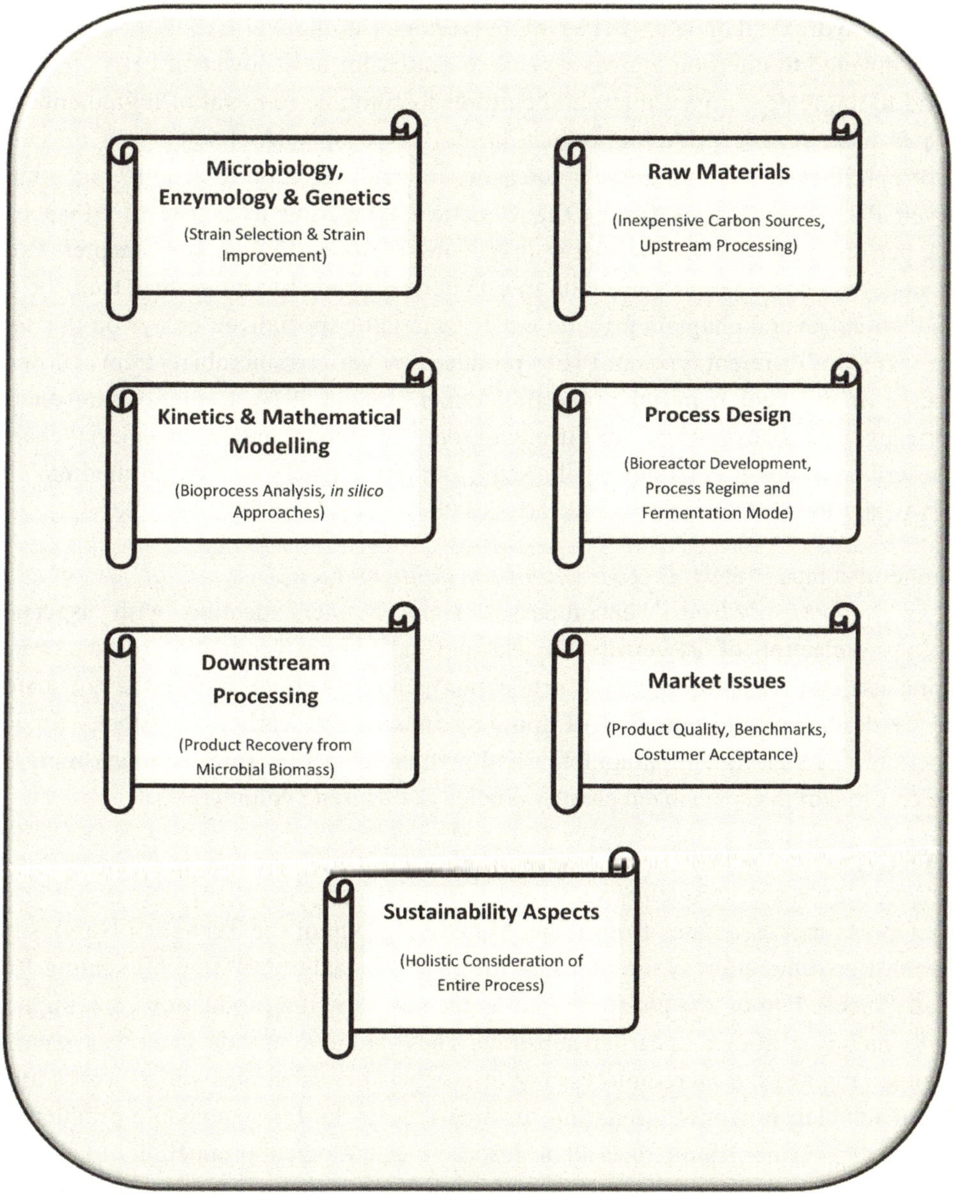

Fig. (1). The seven pillars of efficient PHA bioproduction.

I am outstandingly happy and thankful to all these scientists for their willingness to spend their precious time in contributing to the work at hand. As a result, I am convinced that these two volumes are outstanding composites of comprehensive articles covering all single aspects which are nowadays recognized to be decisive for sustainable and efficient production of microbial polyhydroxyalkanoate (PHA) biopolyesters. In the volume 1 at hand, major attention is devoted to microbiological aspects to search for new powerful PHA production strains, and to implement mixed microbial cultures to combine removal of ingredients from waste water bodies with PHA bioproduction; further a special chapter is dedicated to photoautotroph microbes which attract increasing interest based on their high potential for solar-driven PHA biosynthesis from CO_2. Genetic engineering as a tool to enhance the metabolic performance of PHA producing organisms is covered by another chapter. Due to the high importance of the carbonaceous raw materials needed as microbial feedstock for PHA production, several chapters provide either generic or specialized essays on this topic. The huge variety of different types of PHAs produced by various microbial strains allows for a vast range of different material properties, dependent on PHA´s composition on their monomeric level. This gave reason for different groups of authors to dedicate special chapters to well-described or novel PHA homopolyesters, and to the emerging field of medium chain length PHA (*mcl*-PHA), a fascinating new class of "biolatexes" and "bioelastomers".

The volume at hands, *"Part 1: The Microbe's Point of View (Aspects of Microbiology, Feedstocks, and Metabolism)"* encompasses eight chapters dealing with aspects of microbiology (selection of powerful archaeal and eubacterial production strains, genetic engineering as a tool for optimized PHA production), and puts emphasize on the raw material selection by describing the potential of applying inexpensive carbon feedstocks for PHA bioproduction. The subsequent paragraphs shall give the respected readership an impression about the contents of the specialized chapter articles of this issue volume:

CHAPTER 1

Stanislav Obruca and associates from Ivana Marova´s group at the Technical University of Brno, Czech Republic, address the fact that the main obstacle until now preventing PHAs from finally succeeding on the plastic market is the still too high production cost, important portion of which is attributed to carbon substrate. These authors provide an extensive and up-to-date compilation of the incredible variety of inexpensive substrates which were globally investigated for their potential upgrading towards feedstocks for sustainable production of PHA. Hence, this chapter summarizes advances in biotechnological production of PHAs from various carbon-rich waste products stemming from different fields of industry, especially food industry and agriculture. Special attention is devoted to PHA production from inexpensive lipids, side streams from dairy-, sugar-, and biodiesel production, lignocellulosics, starch-based materials, diverse volatile organic acids, and exotic, precarious substrates such as petrochemical plastic waste. The authors underline that substrates for PHAs production should meet several basic criteria such as availability in constant amount and quality during whole year, stability, low cost of collection, transportation and processing, and emphasize that, for ethical reasons, application of raw materials for biopolymer production must be in no conflict with human food chain.

CHAPTER 2

The volumes of generated surplus whey almost equal to those of the whole milk processed for cheese production, and by far surmount the needs of its restricted classical markets such as animal fodder, pharmaceutical matrices, or additives in food production. As a consequence, enormous quantities of whey are contemporarily disposed in aquatic environments, provoking prevailing ecological concern due to its considerable biochemical oxygen demand, and high expenses for the disposal. This was the motivation for our article describing efforts in the past and nowadays to apply whey as a substrate for biotechnological PHA production. We show that, using different production scales, production strains and production modes, surplus whey can be upgraded to the role of a carbon feedstock for the production of PHB homopolyester or, by switching to advanced feeding strategies and application of powerful novel production strains, to tailor-made PHA co- and terpolyesters with customer-oriented material properties. Microbiological, engineering, and economic aspects of whey-based PHA production by meso- and extremophile wild-type or genetically engineered prokarytotic production strains are covered. The route needed to make efforts in this field competitive with current PHA-production on prized feedstocks of nutritional significance is presented. This competitiveness is addressed both in environmental and economic aspects. In addition, the chapter addresses also alternative routes for sustainable and value-added conversion of surplus whey, *e.g.*, for generation of green energy carriers or various organic bulk- and fine chemicals.

CHAPTER 3

Gabor Rakhely and his colleagues from the Hungarian Academy of Sciences compiled a comprehensive review on the production of biopolyesters and related valued products from photoautotrophic microorganisms such as different purple non-sulfur-, sulfur photosynthetic- and cyanobacteria. The authors address the metabolic functions and particularities of anabolism and catabolism of these biomaterials in photoautotrophs and clearly visualize that these intracellular ongoing differ significantly from observations in heterotrophic PHA producers. Moreover, the authors present photoautotrophic PHA production by cyanobacteria and related organisms from simple, abundant carbon sources like CO_2 or syngas as one of the most promising routes to enable efficient PHA production, and discuss the impact of addition of organic substrates to improve the PHA yield. With this chapter, the authors clearly show that application of phototrophic microbes definitely can be regarded as a serious strategy for efficient production of PHA homo- and heteropolyesters. PHA composition- and yield might be varied by the growth conditions and especially the illumination regime. Although at the moment the efficacy of phototrophic PHA production processes is significantly lower than for established technologies applying chemoheterotrophic microbes, it might be further improved by genetically modified strains and/or bioprocess optimization. Alternatively, waste utilization combined with CO_2 fixation might also constitute a viable approach.

CHAPTER 4

PHA production by mixed microbial cultures (MMC) as an alternative to application of pure microbial cultures became the central point of many research works. This motivated Luisa Serafim and colleagues from Portugal, to contribute the first comprehensive review on this topic. Firstly, they discovered wastewater treatment systems, namely those related with biological phosphorus removal, application of MMC became a process to be considered for polymer production intended to replace petrochemical derived plastics. The ability to store PHA provides microbial communities a competitive advantage for survival under transient conditions typical of waste treatment systems. By this way, MMC can continuously adapt to the operational conditions, increasing the share of PHA-storing organisms in the microbial community. As further benefit, this approach is connected to minimum requirements of sterility. MMC can produce PHA using organic waste or industrial by-products as substrates, allowing for their valorization. The use of waste, together with the lower requirements in sterility and process control, could signify a decrease on PHA production costs. This chapter intends to contextualize the use of MMC for PHA production, by describing the different approaches and comparing them with the traditional processes employing pure or genetically modified organisms. As a resume, the authors conclude that, although being challenging, PHA production by MMC constitutes a way of waste treatment through its valorization. Being PHA production a competitive advantage for microbial cultures, the enrichment in PHA-storing bacteria depends more on the operational conditions imposed to the process rather than the existence of sterility. PHA producing processes using MMC and waste as substrates could help to reduce the polymer costs but at the same time using the existing facilities of waste/wastewater treatment systems and residues on location could contribute to the overall profit of industries and waste water treatment plants.

CHAPTER 5

Christopher Brigham, Abdulrahman A. Kehail, and Jacob D. Palmer from the University of Massachusetts, USA, address the high versatility of the best investigated PHA producing microbe: *Ralstonia eutropha* (a.k.a., *Cupriavidus necator*). The authors point out two principal reasons for the popularity of *R. eutropha* for biotechnology: 1) the organism´s versatile metabolism and acceptance of a wide variety of carbon sources from sugars and lipids to carbon dioxide and aromatic compounds for growth and bioproductions, 2) its capability to store a large amount of carbon as intracellular polymer. The article further underlines that this strain is not only known as the model system for PHA biosynthesis and mobilization, but, moreover, *R. eutropha* is being used as a host organism for heterologous bioconversions, such as the synthesis of "Eutrophanol" as a novel biofuel. This chapter chronicles the key discoveries of *R. eutropha* biology and presents the biomanufacturing context that inspired many research groups to tailor the organism as a novel biocatalyst. The authors emphasize that there are still many challenges to be faced and to be overcome to develop bioprocesses with *R. eutropha* as the biocatalyst. Firstly, the costs of PHA production must be lowered significantly to compete with the cost of petroleum-based plastic as raw material. Secondly, to engineer heterologous biosynthesis pathways in *R. eutropha*, the path forward to high yield productions must be clear.

CHAPTER 6

Another fascinating PHA producer is described by the contribution of Hua Xiang from the Chinese Academy of Science: *Haloferax mediterranei*, an extremely halophilic archaeon has shown promise for the production of the copolyester poly(3-hydroxybutyrate-*co*--hydroxyvalerate) (PHBHV) from many low-cost carbon sources of industrial origin, and can be considered as one of the rising stars on the sky of biotechnology. This extremophile organism, originally isolated from Mediterranean brines, can be cultivated in highly saline cultivation media at almost zero sterility precautions and produces high-quality PHBHV copolyesters from unrelated carbon sources which, after biomass harvest, can easily be recovered from the bioactive cell material. To obtain a complete understanding of PHBHV biosynthesis and to provide novel strategies for PHA production by haloarchaea, the complete genome of *H. mediterranei* has been deciphered, and a genome-wide investigation of the genetics and metabolism of this ancient haloarchaeon involved in PHBHV biosynthesis was performed only recently in the past few years. This exploration covers the identification of key genes for PHBHV biosynthesis, the specific precursor supplying pathways, the PHA granule modulation, and PHBHV mobilization, as well as its linkage to haloarchaeal-specific acetyl-CoA and propionyl-CoA assimilation. These advances have not only revealed many haloarchaeal-specific enzymes and pathways for PHBHV biosynthesis, such as the archaeal-type PHA synthases, the haloarchaeal-type β-ketothiolases with two distinct subunits, and the novel 3-hydroxypropionate pathway coupling CO_2 fixation into PHBHV, but have also provided strategies for bioengineering haloarchaea for economical production of PHBHV or for tailor-made PHAs of even higher value. In addition to PHBHV production, the chapter also reveals other valued compounds produced by this strain, such as carotenoid pigments, hydrolytic enzymes, xanthan-like polysaccharides, or antimicrobial halocins.

CHAPTER 7

Sergio Casella, Lorenzo Favaro, and Marina Basaglia from the University of Padova, Italy, emphasize that several phases of the PHA production chain can be identified where genetic tools can efficiently be implemented in view of possible industrial applications, starting from the microbial strain isolation and characterization, through all the biotechnological steps, until the downstream processing. Leaving aside the huge number of genetic modifications accomplished on many bacterial species to clarify basic biochemical, genetic and metabolic aspects of PHA metabolism, the chapter reveals that several intriguing attempts have been reported aiming at improving the performance of the microorganisms to be used for a possible PHA production process. Most of natural microorganisms can synthesize PHAs with varied monomers only at lab scale from structurally-related precursors, which are frequently expensive, not miscible in water, toxic even at low concentrations and high oxygen demanding. That is why researchers have addressed their attention to inexpensive and renewable unrelated carbon sources. Moreover, the exploitation of the metabolic potential of bacteria for the production of PHAs with tailor-made monomer composition from unrelated carbon sources rapidly became one of the main objectives. At the same time, the improvement of yield has to be faced not only with the modification of growth conditions, but possibly by re-engineering metabolic flux to drive carbon utilization toward PHA biosynthesis pathway.Most of the available information deals with the use of genetic tools for (*i*) characterization and identification of new isolates, (*ii*) testing new isolates for the presence of relevant genes involved in PHA production/degradation, (*iii*) metagenomic approach in complex environments for searching relevant genes (and strains), (*iv*) assessment of PHA genes expression and control of depolymerase genes, (*v*) metabolic engineering of producing and not producing bacterial strains, (*vi*) improving downstream processes, (*vii*) making PHA producing strains able to utilize different inexpensive carbon sources, (*viii*) cloning PHA genes in other recipient bacteria. The authors conclude that a wide range of bacteria capable of naturally producing PHA or those obtained through recombinant DNA techniques are not limited to those already described, but that many other organisms with desirable properties could be exploited in the future for producing recombinant PHAs.

CHAPTER 8

Bruce Ramsay contributes a chapter on a special class of PHA biopolyesters which attracts increasing attention due to numerous anticipated applications, namely on *mcl*-PHA. Accumulated only by a manageable group of related Pseudomonas species, these special polyesters are remarkable in their physic-chemical properties (especially low crystallinity and low glass transition temperature), structural diversity, chemical functionality, and areas of applications; the reader will learn that investigation of *mcl*-PHA properties and exploration of special fields of their application are still in their infancy. *Mcl*-PHA is highly water resistant, yet readily susceptible to enzymatic degradation. They can, *inter alia*, be used in packaging, "biolatex", paints, toners, "smart adhesives" and in many biomedical applications but have yet to be produced commercially. *Mcl*-PHAs are most efficiently produced from carboxylic acids but can also be produced from many non-related substrates. Despite their numerous potential applications, *mcl*-PHAs cannot yet compete with the polymers currently used due to their high cost of production. The author describes recent advances in the biotechnological production and downstream processing for product recovery, as well as strain development in order to improve polymer quality and finally to reduce PHA production cost by increasing productivity, which should soon lead to the adoption of *mcl*-PHAs by industry. The author also emphasizes the need to implement new, unprecedented engineering strategies to accelerate the market break-through of *mcl*-PHA, such as multistage continuous approaches to tailor product composition, or optimized oxygen transfer by pressurized bioreactor facilities in order to generate high cell densities which increases the overall volumetric productivity.

With this issue volume, we wish to address scientists active in the field of biopolymer science, microbiologists, genetic engineers, and biotechnologists in general. The *e*Book is also dedicated to students of higher level which are involved in the fields of polymer chemistry, environmental science, biotechnology, microbiology, and overall life sciences; we hope that this *e*Book is helpful for you, and can give you some insight into the fascinating world of these multifaceted biopolymers! In addition, we strongly believe that the issue volume at hands also attracts the attention of representatives of polymer industry, agroindustry, and biofuel industry. Do you have a carbonaceous waste stream to be upgraded to a feedstock for PHA production? Do you want to get to feet on the ground of innovative polymeric products? Is it conceivable to construe your production lines as a "biorefinery" system, where material stream now undergoing cost-intensive disposal could act as feedstocks, hence, suddenly get a value? This might provide for the ignition sparks for a broader implementation of PHA biopolyesters on industrial scale.

I supremely hope that the investigative and scientific efforts collected and summarized in this *e*Book will inspire researchers all over the world to intensify their activities in this area, and to attract the interest of undergraduates, graduates, as well as of innovative representatives from relevant industrial sectors. Most of all, these accomplishments shall contribute to accelerate the impatiently aspired market break-through of real "bio-plastics" which honestly deserve this designation. This *e*Book underlines that "bio-inspired" solutions for prevailing ecological problems are already developed by experts from the fields of engineering and life sciences, or at least in an advanced status of development; these solutions are waiting for their broad implementation in "White Biotechnology"!

Again, I would like to cordially thank all contributors of this issue for their supreme work.

Dr. Martin Koller
Office of Research Management and Service,
c/o Institute of Chemistry,
University of Graz
NAWI Graz, Heinrichstrasse 28/III, 8010 Graz,
Austria

List of Contributors

Abdulrahman A. Kehail	Department of Bioengineering, University of Massachusetts Dartmouth, 285 Old Westport Road, North Dartmouth, MA 02747, USA
András Fülöp	Bay Zoltán Nonprofit Ltd. for Applied Research, Division for Biotechnology, Közép fasor 52, H-6726 Szeged, Hungary
Bruce A. Ramsay	Department of Chemical Engineering, Queen's University, Kingston, ON K7L 3N6, Canada
Christopher J. Brigham	Department of Bioengineering, University of Massachusetts Dartmouth, 285 Old Westport Road, North Dartmouth, MA 02747, USA
Dan Kucera	Materials Research Centre, Faculty of Chemistry, Brno University of Technology, Purkynova , Brno, Czech Republic
Diogo Queirós	CICECO-Aveiro Institute of Materials, Chemistry Department, University of Aveiro, Campus Universitário de Santiago, 3810-193 Aveiro, Portugal
Gábor Rákhely	Department of Biotechnology, University of Szeged; Institute of Biophysics, Biological Research Center, Hungarian Academy of Sciences, Közép fasor 52, H-6726 Szeged, Hungary
Gerhart Braunegg	ARENA Arbeitsgemeinschaft für Ressourcenschonende & Nachhaltige Technologien, Inffeldgasse 21b, 8010 Graz, Austria
Hua Xiang	State Key Laboratory of Microbial Resources, Institute of Microbiology, Chinese Academy of Sciences, Beijing, People's Republic of China
Ivana Marova	Materials Research Centre, Faculty of Chemistry, Brno University of Technology, Purkynova , Brno, Czech Republic
Jacob D Palmer	Department of Bioengineering, University of Massachusetts Dartmouth, 285 Old Westport Road, North Dartmouth, MA 02747, USA
Lorenzo Favaro	Department of Agronomy Food Natural resources Animals and Environment, University of Padova, AGRIPOLIS, Italy
Luísa Seuanes Serafim	CICECO-Aveiro Institute of Materials, Chemistry Department, University of Aveiro, Campus Universitário de Santiago, 3810-193 Aveiro, Portugal
Lukas Marsalek	Department of Biotechnology, Institute of Applied Microbiology, University of Natural Resources and Life Sciences, Gregor-Mendel-Strasse 33, 1180 Wien, Austria
Marina Basaglia	Department of Agronomy Food Natural resources Animals and Environment, University of Padova, AGRIPOLIS, Italy

Martin Koller	Office of Research Management and Service, c/o Institute of Chemistry, University of Graz, NAWI Graz, Heinrichstrasse 28/III, 8010 Graz, Austria ARENA Arbeitsgemeinschaft für Ressourcenschonende & Nachhaltige Technologien, Inffeldgasse 21b, 8010 Graz, Austria
Paulo Costa Lemos	LAQV, REQUIMTE, Chemistry Department, Faculty of Science and Technology, University Nova of Lisbon, 2829-516 Caparica, Portugal
Pavla Benesova	Materials Research Centre, Faculty of Chemistry, Brno University of Technology, Purkynova , Brno, Czech Republic
Péter B. Kós	Department of Biotechnology, University of Szeged, Institute of Plant Biology, Biological Research Center, Hungarian Academy of Sciences, Közép fasor 52, H-6726 Szeged, Hungary
Sergio Casella	Department of Agronomy Food Natural resources Animals and Environment, University of Padova, AGRIPOLIS, Italy
Simona Rossetti	Water Research Institute, C.N.R., *Via* Salaria Km 29,300, 00015 Monterotondo, Italy
Stanislav Obruca	Materials Research Centre, Faculty of Chemistry, Brno University of Technology, Purkynova , Brno, Czech Republic

Recent Advances in Biotechnology

(Volume 1)

Microbial Biopolyester Production, Performance and Processing

Microbiology, Feedstocks, and Metabolism

CHAPTER 1

Novel Inexpensive Feedstocks from Agriculture and Industry for Microbial Polyester Production

Stanislav Obruca[*], **Pavla Benesova, Dan Kucera** and **Ivana Marova**

Materials Research Centre, Faculty of Chemistry, Brno University of Technology, Purkynova 118, 612 00 Brno, Czech Republic

Abstract: Polyhydroxyalkanoates (PHAs) are very promising materials which might serve as environmentally-friendly alternative to petrochemical plastics. The main obstacle preventing PHAs from entering the market massively is the final cost of the polymeric material, important portion of which is attributed to carbon substrate. Hence, the researchers have been intensively looking for cheap substrates enabling sustainable production of PHAs. Hence, this chapter summarizes the advances in biotechnological production of PHAs from various waste products originating from industries of different fields, especially from food industry and agriculture.

Keywords: Agricultural waste, Biodiesel industry, Biotechnological production, Carbon-rich waste, Cheese whey, Glycerol, Inexpensive substrates, Lignocelluloses, Lipid substrates, Molasses, Polyhydroxyalkanoates (PHA), Sugar industry, Upstream processing.

GENERAL ASPECTS OF POLYHYDROXYALKANOATE PRODUCTION

Polyhydroxyalkanoates (PHAs) are polyesters of hydroxyacids (usually 3-hydroxyacids) which are synthesized by various microorganisms under the conditions of nutrient limitation in the presence of an excessive carbon and energy source. Due to their low solubility and high molecular weight, PHAs do not cause the increase of intracellular osmotic pressure and they are, therefore, ideal storage

[*] **Corresponding author Stanislav Obruca:** Materials Research Centre, Faculty of Chemistry, Brno University of Technology, Purkynova 118, 612 00 Brno, Czech Republic; Tel: +420 776 788 818; Email: Stana.O@seznam.cz.

Martin Koller (Ed.)

materials. In this role they are much more common storage materials in microorganisms than is glycogen, polyphosphates, or fats, for example [1].

PHAs were discovered by Lemoigne who isolated and characterized poly(3-hydroxybutyrate) (PHB), a member of PHAs family, from *Bacillus megaterium* in 1926. Since this first report the PHAs accumulation has been discovered in many microorganisms including Gram-negative and Gram-positive species (*i.e.*, autotrophs, heterotrophs, phototrophs, aerobes, anaerobes) and archaebacteria. In 1976, Imperial Chemical Industries (ICI Ltd., UK) recognized the potential applicability of PHAs to replace some of the commodity plastics produced from petrochemical resources. Despite the fact that the bacterially produced PHAs were expensive as compared to oil-derived plastics, it was expected that the price of crude oils would rise to high levels, making the production of PHB economically feasible. Because no rise in oil prices occurred, PHB originated as a biodegradable and biocompatible plastic material [2].

Currently, PHAs still attract the attention of many researchers and, therefore, many aspects of their biological function, metabolism, biotechnological production, but also the processing and applications were understood. This chapter aims at providing a detailed insight into the current state of the art PHAs production from inexpensive and/or waste substrates.

Motivation for Utilization of Inexpensive Substrates

The main obstacle preventing the PHAs market from massive expansion is their cost. Generally, PHAs are several times more expensive than their competitors - polymers of petrochemical origin. Principally, currently known technologies of PHA production have three essential disadvantages: i. the use of relatively expensive carbon substrates; ii. the use of aseptic culture of selected or genetically modified strains which require high expenses for the sterilization of equipment and medium, as well as for the maintenance of aseptic conditions during the biosynthesis of bioplastics iii. The use of expensive, often flammable and toxic organic solvents and/or energy- and reagent-consuming methods for the extraction of PHAs from bacterial cells [3].The aim of this chapter is to face the first and the most important problem of industrial PHA production - the cost of entering substrate and the utilization of inexpensive carbon substrates for biotechnological production of PHAs.

The cost of common substrate (pure sugars or fatty acids) is known to account for almost 50% of the total production costs of PHAs [4]. Therefore, many researchers have been looking for inexpensive substrates enabling economically feasible PHAs production process. Numerous potential substrates, including but not limited to waste products and side products of food industry and agriculture, have been investigated in this perspective. Generally, inexpensive substrates for biotechnological production of PHAs should meet some basic requirement:

a. they should be available in sufficient amount and quality during whole year
b. they should not undergo rapid spoilage by the action of naturally occurring microbes
c. their cost should be sufficiently low even when taking into account the economic aspects of their collection and transportation
d. there should not be other, more profitable technology of their utilization
e. the substrates for PHAs production should not compete with human food chain.

Usually the utilization of inexpensive substrates provides lower PHA yields as compared to processes using pure sugars of fatty acids. Additionally, low overall product yields contribute to the cost, as less efficient bioconversion would require more substrate for the production of the same amount of product. The problem is compounded by the fact that, in most cases, PHA production gives low product concentrations and low overall volumetric productivities. The latter leads to higher capital and operating costs, especially regarding the application of large bioreactor facilities [5].

Hence, despite the fact that the production of PHAs from inexpensive substrates is widely accepted as correct strategy, there are many obstacles which prevent it from large scale industrial applications.

PHA Producing Strains and their Characterization

Polyhydroxyalkanoates are complex class of polyoxoesters which are synthesized by most genera of eubacteria and members of the *Halobacteriaceae* family of the

Archea [6]. Only recently, a PHA synthase gene was also identified in the genome of an uncultivated *Crenarchaecota* [7]. Most of these prokaryotes synthesize poly(3-hydroxybutyric acid) and other PHAs as a complex storage material deposited in insoluble cytoplasmic inclusions. Meanwhile, approximately 150 different polyhydroxyalkanoic acids are now known to occur as constituents of PHAs. Among these there are 3-hyroxyacids of 6 to 14 carbon atoms with saturated, unsaturated, straight, or branched aliphatic side chains. PHAs can be classified into two groups depending on the number of carbon atoms in the monomer units: short-chain-length (*scl-*) PHAs, which consist of 3-5 carbon atoms, and medium-chain-length (*mcl-*) PHAs, which consist of 6-14 carbon atoms [2]. The monomer units in these microbial polyesters are all in the R-configuration because of the stereospecificity of biosynthetic enzymes. The molecular weights of PHAs range from hundreds thousands to millions depending on the specific PHA, the producing microorganism, and the growth conditions. Recently it was found that low-molecular weight oligomer PHB (120 -200 monomers) is widely distributed in microorganisms, plants and animals, including humans [8]. This form of PHB occurs as PHB- calcium phosphate complex in biomembranes as an ion channel, in *E.coli* this complex has been found to occur in large amounts in the membrane of cells being competent for genetic transformations [9]. Nevertheless, for large scale production as bioplastics high molecular weight polymer should be produced.

To realize sustainable industrial process based on the use of PHA polyester, high manufacturing costs could be reduced by establishing an efficient recombinant production system of PHA with desired properties. Recombinant *E.coli* strains belong to the most important overproducers. Besides other bacteria, PHB biosynthesis was successfully conferred to plant *Arabidopsis thaliana* and yeast *Saccharomyces cerevisiae* [2].

PHA Biosynthetic Pathways and their Regulation

The biochemistry of PHB biosynthesis is very well known. There are at least four different pathways for natural PHA synthesis, which consist of three enzymatic reactions catalyzed by successive action of 3-ketothiolase (*phbA*), acetoacetyl CoA reductase (*phbB*) and PHB synthase (*phbC*). These pathways differ with respect to microorganisms in which they are found [10].

In *Cupriavidus necator (*former *Ralstonia eutropha, Wautersia eutropha and Alcaligenes eutrophus*) and most of other PHB producing bacteria PHB biosynthesis results from the metabolism of carbohydrates. The first reaction involves the condensation of two acetyl CoA molecules into acetoacetyl CoA catalyzed by 3-ketothiolase. In the second step the product is stereo-selectively reduced to R-3-hydroxybutyrylCoA in a reaction catalyzed by NADPH-dependent acetoacetyl-CoA reductase. The final reaction involves the polymerization of R-3-hydroxybutyryl CoA enantiomer into PHB catalyzed by PHB synthase.

A few exceptions of this general pathway are known. In *Rhodopseudomonas rubrum* (*Rhodospirillum rubrum*), the oxidation of fatty acids leads to the biosynthesis of PHA. The first reaction remains the same as that for *C. necator.* Acetoacetyl CoA is then reduced to (*S*)-3-hydrobutyryl CoA and subsequently converted to (*R*)-3-hydroxybutyryl CoA by two stereospecific enoyl-CoA hydratases [2]. The third type of PHA biosynthetic pathway is found in most of *Pseudomonas* species, belonging to rDNA homology group I (*P. oleovorans*). In these organisms the oxidation of alkanes, alkanols or alkanoic acids leads to the synthesis of *mcl*-PHAs. The fourth type of PHA biosynthetic pathway occurs in *Pseudomonas* belonging to rDNA homology group II. The synthesis of PHA results from *de novo* fatty acid synthesis pathway which involves the synthesis of copolymers of *mcl*-PHA from acetyl CoA [10].

The PHB biosynthesis and accumulation occur in the most of bacteria if carbon source is provided in excess and if another nutrient, such as nitrogen, sulfur, phosphate, iron, magnesium, potassium or oxygen is limiting [11]. A copolymer of 3-hydroxybutyrate and 3-hydroxyvalerate PHBHV can be synthesized by *C. necator* and many other microorganisms from either mixed substrate of glucose and propionic acid or direct precursor of 3-hydroxyvalerate. If propionate is fed, essentially the same biochemical pathway as for PHB synthesis is used, but propionyl CoA and acetyl CoA are condensed by 3-ketothiolase to give 3-ketovaleryl CoA, which leads to the incorporation of 3-hydroxyvalerate monomers into the polymer. Several microorgansims are capable of accumulating PHBHV from substrates unrelated to 3-hydroxyvalerate. These organisms include

Corynebacterium, *Nocardia* and *Rhodococcus* strains and mutants of *R.eutropha* (24). The terpolymers containing 3-hydroxybutyrate, 3-hydroxyvalertae and 3-hydroxyhexanoate monomers were produced by *Rhodococcus* sp. grown on the medium supplied by hexanoic acid. Using some specific precursors, the copolymers of 3HB and 4HB, terpolymers of 3HB, 4HV and 5HV and polymers containing 4HV and 3-HP can be synthesized by different *C. necator* strains. Many other variations of PHA composition can be found in *Alcaligenes*, *Pseudomonas* and *Burkholderia* strains.

The PHB synthesis is controlled at the enzymatic level as well as at the transcription level [12]. At the enzymatic level intracellular concentration of acetyl CoA and free coenzyme play a central role. It has been shown that PHB biosynthesis is stimulated by both intracellular concentration of NAD(P)H and high ratios of NAD(P)H/NAD(P). Under balanced growth conditions acetyl CoA is oxidized *via* the TCA cycle and NAD(P)H is generated and used for biosynthetic purposes. When the growth ceases the NAD(P)H concentration increases, which reduces the activity of TCA cycle enzymes citrate synthase and isocitrate dehydrogenase. As a result acetyl CoA cannot be oxidized *via* the TCA cycle and enters the PHB synthetic pathway. Citrate synthase is potentially important control point in the whole PHB synthesis process based on its ability to control the availability of free CoA, which inhibits the activity of 3-ketothiolase. The work with recombinant strains demonstrated that the biosynthesis rate of PHB is controlled by 3-ketothiolase and acetoacetyl CoA reductase, whereas the content of PHB is controlled by PHB synthase [2, 13].

PHB synthase is another key regulation enzyme in the PHA formation. It is proposed that PHB synthase is post-translationally activated by acetyl phosphate. The enzyme converting acetyl CoA to acetyl phosphate is controlled by acetyl CoA concentration and C/N ratio. Therefore, acetyl phosphate could act as a signal of C/N balance affecting the PHB metabolism. It has been found in the cyanobacterium *Synechoccoccus sp.* MA19 that the PHB synthase activity is exclusively found in isolated cell membrane fractions [12]. More detailed analysis of PHB synthases revealed that these soluble enzymes change the conformation and turn into amphipathic molecules upon covalent catalysis of polyester-chain formation. A self-assembly process is initiated resulting in the formation of insoluble cytoplasmic inclusions with a phospholipid monolayer and covalently attached polyester synthases on the surface [2]. The surface-attached enzymes show a remarkable increase in their activity. Detailed structural, genetic and biochemical characterizations of PHA synthases from 45 different bacteria were performed [2]. Primary structures of these synthases show an overall identity of 8 - 96% with only eight strictly conserved amino acid residues and six conserved blocks. Moreover, all PHA synthases seem to contain the lipase box in which the essential active site Ser is replaced by Cys. PHA synthases comprise a new family of enzymes with unique features, particularly considering the functional role in biogenesis of water-insoluble intracellular granules, and the association with a phospholipid monolayer. These enzymes can be assigned to four classes based on their substrate specificity and subunit composition. Classes I and II consist of 1 subunit (phaC). Class I preferentially utilizes CoA thioesters of various R---hydroxy fatty acids with 3-5 C atoms, whereas class II preferentially utilizes thioesters with 6-14 C atoms. Classes III and IV comprise PHA synthases consisting of two types of subunits (*phaC*, *phaE*/*phaR*) and both prefer 3-5C thioester chains [2, 14].

The conserved-domain homology search strongly suggested that PHB synthases probably belong to the α/β-hydrolase superfamily, with conserved cysteine residue as catalytic nucleophile [2]. In Class I conserved catalytic triad of Cys-319, Asp-480 and His-508 was found, which are directly involved in covalent catalysis. In other classes of PHA synthase similar conserved triads residues were found. The α/β-hydrolase-based catalytic mechanism, particularly considering lipases and cysteine proteases, provides a good model for classes I-III of PHA synthases as indicated by mutational analysis of the *C. necator* class I PHA synthase [2], the *A.vinosum* class III [14] and the *P.aeruginosa* class II PHA synthase [2].

During the course of evolution, the changes in *phb* genes resulted in many rearrangements in the *phb* loci [15], *e.g.*, the clustering of *phb* genes in an operon with the same or separate transcriptional units, the fusion or splicing, the changes in orientation and many others. Detailed genome analysis including the regulation mechanisms could be important mainly in potential industrially producing strains. In various microorganisms the promoters have been identified based on the sequence upstream of *pha* and *phb* genes [12]. Among those a phosphate starvation inducible promoter was identified in *Acinetobacter sp.* which seems to be under the *pho* regulon control. It was shown that the LuxR regulatory protein (quorum sensing) controls also the synthesis of PHB in *Vibrio harveyi* [16]. Furthermore, it was demonstrated that the *ntrB* and *ntrC* genes involved in nitrogen regulation in various bacteria are also involved in the regulation of PHB synthesis by ammonia in *Azospirillum brasilense* SP7 [12].

Many other factors involved in regulatory mechanisms at the transcription level have been described for *Pseudomonas*. These factors influence the composition of copolymer by the formation of either PHB homopolymer or copolymer containing monomeric units from 4 to 12 C atoms. For *P.aeruginosa* PAO1 it has been shown that PHA accumulation from gluconate requires a functional RpoN sigma factor, whereas PHA accumulation was only reduced in the absence of RpoN when the cells were grown on fatty acids [12]. RpoN is an alternative sigma factor of RNA polymerase which is involved in the growth phase dependent activation of certain inducible promoters.

In addition, an open reading frame of described putative translational products exhibits the similarity to AlgP, a member of the prokaryotic histone H1-like family of regulators. The gene product of the ORF homologue of *P.oleovorans*, called *phaF* is involved in transcriptional regulation of *pha* gene expression and, in addition, the PhaF protein has also a structural function, *i.e.*, the binding to PHA granules presumably imbedded in phospholipids, forming a layer between the hydrophobic polymer and the hydrophillic cytoplasm. In addition, it is known that granule associated proteins of *C. necator* and other bacteria are only synthesized when the cells accumulate PHB. Only those amounts of proteins are synthesized which can be bound to the granules, since soluble proteins were never detected [17].

In general, it has been assumed that not only the synthesis of PHA but also the PHB degradation by PHB depolymerases is highly regulated. Most PHA degrading bacteria repress PHA depolymerase gene expression in the presence of a soluble carbon source that favors high growth rates. However, the findings in *P.oleovorans* [18] and *C. necator* [19] confirmed that PHA depolymerase is always present and active, and that PHA is synthesized and degraded simultaneously. Although the regulation of PHA metabolism must occur in many if not in all bacteria, just little is known on this subject. However, the examples listed above indicate large diversity of regulatory mechanisms including global and specific systems which are required for efficient PHA accumulation in response to specific environmental and intracellular signals [2].

Enhancement of PHA yields by Application of Exogenous Stress

Generally, it is assumed that the ability of PHAs synthesis and intracellular degradation improves bacterial stress resistance [20]. For instance Ayub *et al.* studied cold resistance of Antarctic bacterium *Pseudomonas* sp. 14-3 [21, 22]. Its stress tolerance was analyzed under PHA accumulating and non-accumulating conditions and significantly higher levels of stress resistance were observed when PHAs were produced. In recent study, the same authors tested the PHA synthase negative mutants of *Psedomonas*sp. 14-3 in order to investigate the mechanism of involvement of PHA in stress resistance [23]. The mutant strain was unable to grow at 10° C and was more susceptible to freezing than its parental strain. PHAs were necessary for the development of oxidative stress response induced by cold treatment. The $NADH/NAD^+$ ratio and NADPH content decreased strongly in the mutant while only minor changes were observed in parental strain. Authors proposed that the PHA metabolism modulated the availability of reducing equivalents, contributing to the alleviation of oxidative stress produced by low temperature.

However, the interconnection of PHAs metabolism and stress response of bacteria can also provide interesting practical outcome. Based on the results of various researchers, the application of appropriate dose of exogenous stress may be used as an interesting and cheap strategy to enhance the PHAs yields. Obruca *et al.* studied the influence of various stress factors on the PHA production and observed that an approximate 30% increase in PHAs production by *C. necator* H16 was obtained when hydrogen peroxide or ethanol were applied at the beginning of stationary phase [24]. Hydrogen peroxide enhanced the activity of pentosephosphate pathway that consequently increased the intracellular $NADPH/NADP^+$ ratio. During the ethanol metabolization, in which alcohol dehydrogenase is involved, acetyl-CoA and reduced coenzymes NAD(P)H were formed. It stimulated the flux of acetyl-CoA into PHB biosynthetics rather than to TCA cycle. Furthermore, the application of both stress factors increased the enzymatic activities of β-ketothiolase and acetoacetyl-CoA reductase which probably supported the PHB accumulation in bacterial cells. Moreover, the molecular weight of PHB produced under stress conditions was significantly higher as compared to control cultivation [25]. Positive effect of ethanol stress on PHB production was also observed in genetically modified *Pichia pastoris*. Ethanol formed during initial phase of oxygen limited growth was further utilized for the PHB production [26].

There are several reports that appropriate osmotic pressure might induce the PHAs accumulation in several bacterial strains. NaCl stress in *Rhizobium* had been previously investigated and it was found that the changes in the osmolarity of peripheral environment could induce a new set of protein expressions which are different from those brought on by the heat stress. This stress response has helped the cells overcome the osmotic stress and aided the acclimatization at high osmotic pressure. Furthermore, NaCl stress resulted in the accumulation of PHAs in *Rhizobium* DDSS-69 cultures grown under unbalanced growth conditions [27]. The effect of different NaCl concentrations on the PHA productivity using *Cupriavidus necator* has been investigated by Passanha *et al.* The highest PHA production was obtained with the addition of 9 g/L NaCl, which yielded 30% higher PHA than in the control culture. Increasing the addition of NaCl to 15 g/L it was found that the production of PHA was inhibited. Hence; the addition of NaCl can; therefore, be used as a simple, low-cost, sustainable, non-toxic and non-reactive external stress strategy to increase the PHA productivity [28]. Similarly, the bacterial strain *Zobellella denitrificans* had been isolated which was suitable for the PHAs production using contaminated glycerol and the maximum PHAs accumulation in the presence of 20 g/L of NaCl was achieved [29]. The osmotic stress induced by both NaCl and PEG was also observed to increase the *mcl*-PHA production in *Pseudomonas fluorescens* [30]. Furthermore, numerous halophilic bacteria strains such as *Haloferax mediterranei* or *Halomonas boliviensis* are considered to be the candidates for biotechnological production of PHAs since these bacteria are able to convert various carbon sources to PHAs with high efficiency. Their ability to prosper in high salinity represents an important advantage since the biotechnological process can be conducted under semi-sterile conditions which decrease its overall cost [31]. Further, another advantage stemming from the utilization of halophiles for PHAs production is the fact that cheaper sea water can be used instead of expensive fresh water, which also reduces the cost of production process [32].

There are also reports that the exposition of bacterial culture to heavy metals might have positive impact on the PHAs productivity of several bacterial strains. In non-endophyte strain *Azospirillum brasilense* Sp7 heavy metals induced an enhanced accumulation of PHB. In contrast, the response of the endophytic strain *Azospirillum brasilense* Sp245 to heavy metal uptake was found to be much less pronounced [33]. Similarly, the addition of nickel was observed to increase the PHAs production in *Cupriavidus pauculus* KPS 201 and *Bacillus firmus* AND 408 [34] and cadmium supported the PHAs accumulation in *Cupriavidus taiwanensis* EJ02 which is known for its resistance to heavy metals [35]. Nevertheless, the application of heavy metals as inducers of PHAs production does not seem to be very promising strategy due to their toxicity and potential pollution of waste waters coming out of the biotechnological process.

An interesting strategy how to increase the productivity of the process was reported by Follonier *et al.* for the *mcl*-PHAs production employing *Pseudomonas putida* KT2440. Their study revealed that elevated pressure is an adequate and efficient way to enhance the oxygen transfer rate and *mcl*-PHA productivity. The authors estimated that the oxygen provided to the culture broth under elevated pressure would be sufficient to triple the *mcl*-PHA productivity in chemostat system. Hence, even putting cells under the pressure seems to be simple and efficient way to enhance the productivity of the biotechnological process [36].

PRODUCTION OF PHA USING LIPID SUBSTRATES

Generally, lipids and fat-rich materials are considered being very promising substrates for biotechnological production of PHAs, because they are metabolized *via* the β-oxidation pathway leading to the formation of acetyl-CoA which is the crucial substrate of PHAs biosynthetic pathway. Furthermore, unlike other carbon sources, the theoretical yield coefficient of PHB production from oils is over 1 g PHB per g of plant oil, since lipids comprise higher number of carbon atoms per weight than for instance carbohydrates [37]. Moreover, lipid based substrates are also very attractive because they contain low amount of water which makes them more stable against microbial spoilage and the cost of their transportation is lower than for water rich substrates such as cheese whey. In particular, plant oils represent very promising group of lipid substrates which attract the attention of numerous researchers dealing with the production of PHAs. Apart from noble plant oils, there are also attempts to utilize waste oils such as waste frying oils for the production of PHA. These potential sources provide numerous advantages over noble oils-they do not compete with human food chain, their utilization does not place demands on agricultural land and; moreover, potentially problematic waste is transformed into the high value product [38].

Recently lipid stocks have mostly been channeled into the biodiesel production. Biodiesel is a possible alternative to conventional petrol-based diesel and can be produced by the transesterification of triacylglycerols (TAGs) from vegetable oil and animal fats. Depending on the climate and soil conditions, different kinds of edible vegetable oils are being used as the main feedstocks for biodiesel production, such as sunflower oil in Europe, soybean oil in the USA, rapeseed oil in Canada and palm oil in Southeast Asia. Although the commercial use of biodiesel is still limited by its high price, it can be expected that global demand for vegetable oils will increase [39].

Generally, unlike noble oils, waste frying oils and other waste lipids, such as oil extracted from spent coffee grounds, are not optimal substrates for biodiesel production due to high peroxide and acid value. High content of free fatty acids (FFA) in oil negatively influences the transesterification process and; thus, the biodiesel production. High content of FFA in oil tends to form large amounts of unwanted soap by-products during alkali catalyzed transesterification process, which promotes the formation of stable emulsions. This increases the viscosity of the reactant mixture and creates serious problems at the biodiesel separation from glycerol during the processing and ultimately lowers the yield substantially [40, 41]. This may be partially solved by the application of acid-catalyzed trans-esterification; however, high content of FFA affects also the physical properties of oil because it increases the kinematic viscosity of oil and high viscosity of oil results in high power consumption when mixing the reactant in the esterification process and, moreover, high viscosity of the reactants mixture results in lower biodiesel yields [42].

On the other side, we have observed that even highly devalued frying oils are better carbon sources for the PHAs production than noble ones. It is probably caused by the changes which happened during frying process — fatty acid composition usually changed. The concentration of free fatty acids is enhanced and also some components of fried food such as proteins, lipids or carbohydrates migrate to oil. It is likely that these changes are beneficial for the bacterial growth and PHB accumulation [38]. For instance, we observed good correlation between the acid value of oils and the PHB yields ($R^2 = 0.9058$). The inducible enzymatic system of PHAs bacterial producers involves extracellular lipase and its lipase-specific foldase. For instance, LipA of *Cupriavidus necator* which is widely employed for the PHAs production from lipids was determined to be the non-specific extracellular lipase. Because it is able to completely hydrolyze TAG into fatty acids and glycerol [41]. Oppositely, free fatty acids (FFA) are simple substrates which may be easily absorbed and utilized by bacteria cells without previous enzymatic treatment, which may be important during initial stage of cultivation when lipolytic enzymes are about to be excreted. Furthermore, FFA act as natural surfactants and; therefore, enable the emulsification of oil and make TAG susceptible to lipases action. Moreover, FFA are likely to serve as inducers of lipase expression. Therefore, it seems logical that higher content of free fatty acids in oil increases the growth and the metabolic activity of used bacterial strain. Hence, the properties of waste oils, which make them non-suitable for biodiesel production, are useful in their utilization for the production of PHAs.

On the contrary there are also several drawback connected to the PHAs production from waste oils containing high amount of FFA and high peroxide value. Some waste oils can contain the products of fatty acids decomposition which might be toxic for bacterial strains and; therefore, they decrease the productivity of PHAs process. The investigations are also needed to better understand and overcome the technical difficulties in the use of oils for producing PHAs. One major drawback is the foam formation, which decreases the concentration of dissolved oxygen. This can be overcome by the addition of antifoam reagents, but it is not known how these reagents reduce or increase the bacteria growth rates. Oils also cause problems to downstream processes to obtain bacterial cells and purified PHAs, which means that in addition there is a need for developing the cell breakage procedures, more investigations are needed on how to separate PHAs from cultures with oily substrates. Thus, there is a need to develop and optimize new fermentation strategies and novel methods for the product recovery [39].

Although vegetable oils in both pure and spent forms seem to be attractive carbon sources for the PHAs production, only little has been reported on the costs of their production from oily substrates. One source reports that the costs of carbon sources constitute 40% of the final cost of biopolymers produced by bacterial fermentation. If pure oil is used, the price of PHAs will still be higher than that of synthetic plastic. However, using waste substrates reduces the total financial outlay, which means that the final cost of PHAs could be competitive with the price of polypropylene. As previously reported, the price of rapeseed is around 3.51 Euro/kg [38] whereas its waste equivalent is about half of that price [43]. Also the entire integrated process of PHAs production should be considered because other factors such as up-stream and down-stream processes can influence the final price of PHAs. It is estimated that approximately 30% of the overall production costs come from the recovery of PHAs from bacterial cells [44]. Some separation methods are especially expensive, such as solvent extraction, involving extra costs for the disposal of used solvent.

Production of PHAs from Lipid Substrates Employing Bacterium *Cupriavidus necator*

The Gram-negative facultative chemolithoautotrophic hydrogen-oxidazing bacterium *Cupriavidus necator* (formerly *Ralstonia eutropha*, *Wautersia eutropha* and *Alcaligenes eutrophus*) represents one of the model organisms for PHA metabolism. Moreover, due to its capability of utilizing fatty acids, triacyl-glycerols and other substrates while accumulating high amounts of PHAs, *C. necator* is considered as a candidate for industrial production of PHA from lipid-based substrates [45]. This bacterium is capable of accumulation of homopolymer

PHB and copolymer PHBHV; however, for the accumulation of the later valerate, propionate or other propionyl-CoA generating substrate must be introduced into the fermentation media [46]. Further, this bacterium was genetically engineered in order to incorporate 3-hydroxyhexanoate (3HHx) monomer units into the polymer structure. The copolymer of 3-hydroxybutyrate and 3-hydroxyvalkerate (P3HB-*co*-3HHx) reveals superior mechanical and technological properties to most of other members of PHAs family. In order to force *C. necator* to produce P(3HB-*co*-3HHx) directly from intermediates of fatty acid metabolism, a strain containing specific heterologous genes must be engineered. As native PHA synthase (PhaC) in *C. necator* is specific only for *scl* monomers, it is typically replaced with a PHA synthase gene the product of which exhibits broader substrate specificity, with the ability to polymerize both scl and *mcl* monomers. For instance, there are reports on the preparation of two recombinant strains capable of high polymer accumulation per cell dry mass (CDM) (64-73%) and high 3HHx monomer content (up to 30 mol-%) in resulting PHA when these strains were grown on palm oil as the sole carbon source. These strains contained a broad-specificity PHA synthase, isolated from *Rhodococcus aetheriovarans,* and the *phaJ* gene from *P. aeruginosa,* resulting in the increase of intracellular 3HHx-CoA concentrations for the PHA biosynthesis. In both strains, these genes are present on the PHA production plasmid pCB113 [47].

The enzymatic inducible system of *C. necator* responsible for the utilization of TAG was recently described. It involves extracellular lipase (LipA) and its lipase-specific foldase (LipB). LipA was determined to be non-specific extracellular lipase, because it is able to completely hydrolyze TAG into fatty acids and glycerol. *C. necator* mutant strains with a Δ*lipA* or Δ*lipB* single deletion mutation completely eradicated the strains' ability to hydrolyze and utilize TAGs from palm oil. No extracellular lipase activity was detected in the supernatant of these single deletion strains. LipA was determined to be the only extracellular lipase present in *C. necator* H16. Importantly, the *lipA* overexpression under a constitutively expressed promoter not only resulted in more secreted LipA enzyme and subsequent TAG hydrolysis but also shortened the culture lag phase to 5 hrs as compared to the wild type (12 h), through faster emulsification of lipids during the shake flask experiments [48]. It should be pointed out that lipases are not only necessary enzymes for the utilization of TAG and, thus, for the PHAs production on lipids, but they also represent very important biotechnological products since many industrial processes can be catalyzed by lipases owing to their multifaceted chemo-, regio-, and enantioselectivities. It is estimated that lipases account for 21% of the world's enzyme industry and contribute to an estimated annual sale of 1.5 billion US-$ [47]. Hence, lipase can be considered as very attractive side-product of PHAs production on plant oils and other lipid substrates. Since PHAs are intracellular storage compounds and lipase is excreted into the cultivation media, both products can be easily separated for instance by the removal of cells by centrifugation or microfiltration. After that, lipase can be isolated from the cultivation media.

Table 1. Production of PHA on various lipid substrates employing *C. necator*.

Substrate	**Cultivation**	**Polymer**	**CDM [g/L]**	**PHA [wt.-%]**	**PHA [g/L]**	**$Y_{P/S}$ [g/g]**	**Ref.**
Noble Oils							
Palm kernel oil	Batch. Flaks	PHBHV	3.0	73.0	2.2	-	[49]
Palm kernel oil	Batch. Flaks	P(3HB-*co*-4HB)	3.3	68.0	2.2	-	[49]
Sesame oil	Batch. Flaks	PHB	7.4	62.0	4.6	0.46	[50]
Olive oil	Batch. Flaks	P(3HB-*co*-3HHx)	3.5	76.0	2.7	-	[51]
Corn oil	Batch. Flaks	P(3HB-*co*-3HHx)	3.6	77.0	2.8	-	[51]
Palm oil	Batch. Flaks	P(3HB-*co*-3HHx)	3.6	81.0	2.9	-	[51]
Soybean oil	Fed-Batch. Fermentor	P(3HB-*co*-3HHx)	126.0	74.0	102.1	0.72	[37]
Palm oil	Extended batch	P(3HB-*co*-3HHx)	99.0	70.0	69.1	-	[52]
Coconut oil	Batch. Flaks	P(3HB-*co*-3HHx)	2.6	48.0	1.3	-	[53]
Jathropa oil	Batch. Fermentor	PHB	20.1	77.1	15.5	0.99	[54]
Waste Lipids							
Waste frying rapeseed oil	Batch. Flasks	PHB	-	-	1.2	-	[55]
Waste frying rapeseed oil	Fed-Batch. Fermentor	PHBHV	138.5	75.9	105.1	0.83	[38]
Waste frying rapeseed oil	Batch. Fermentor	PHBHV	18.7	85.5	16.0	-	[56]

(Table 1) contd.....

Substrate	Cultivation	Polymer	CDM [g/L]	PHA [wt.-%]	PHA [g/L]	$Y_{P/S}$ [g/g]	Ref.
Waste frying rapeseed oil	Batch. Fermentor. hydrolyzate of whey proteins was used as a nitrogen source	PHB	31.0	90.5	28.1	0.94	[57]
Spent coffee ground oil	Batch. Fermentor	PHB	29.4	90.1	26.5	0.88	[41]
Spent coffee ground oil	Fed-Batch. Fermentor	PHB	55.4	89.1	49.4	0.82	[41]
Spent coffee ground oil	Batch. Fermentor	PHB	16.7	78.4	13.1	0.77	[58]
Fat-containing waste produced from the margarine manufacturing	Batch. Fermentor	PHB	11.2	56.0	6.3	-	[59]
Used-palm olein	Batch. Fermentor	PHB	14.0	83.0	11.6		[60]

Wide range of noble plant oils such as sesame oil, corn oil, palm oil, coconut oil soybean oil or jatropha oil were used for the biotechnological production of various PHAs including PHB, PHBHV or P(3HB-*co*-3HHx) employing *Cupriavidus necator* as a producing strain. Furthermore, also waste lipid substrates such as waste frying oils, waste after margarine production, oil extracted from spent coffee grounds or used palm olein can be utilized for the PHA production with *C. necator* as shown in Table **1**. Generally, the PHAs content per cell dry mass varied between 48 and 90% which proves great productivity potential of the bacterium. Besides efficient PHA accumulation per CDM as described above, high total PHA production per liter is desired for an economical PHA production process. Efficient process was described by Kahar *et al.* who produced 95.5 g/L P(3HB-*co*-3HHx) from soybean oil under the phosphate limitation. The fermentation was started with initial concentration of soybean oil of 20 g/L and 0.4% ammonium chloride. Soybean oil and ammonium chloride were fed in pulses, shortly before limitation back to initial concentrations [37]. Further, Riedel *et al.* also reported high cell density fermentation using palm oil as the sole carbon source for the P(3HB-co-3HHx) copolymer production. The authors reached high concentration of the product at the end of cultivation - 102 g/L and the data suggest that the fermentation process is scalable and profitable [52]. Also waste frying rapeseed oil was used as a substrate for high cell density cultivation and PHAs production. Obruca *et al.* described the fed-batch cultivation of *C. necator* on waste frying oil. Propanol at 1 vol.-% enhanced both the PHA and biomass formation significantly and, furthermore, resulted in the incorporation of 3-hydroxyvalerate units into the PHA structure; hence, propanol was used as a precursor of 3HV. In the fed-batch cultivation performed in laboratory bioreactor, the concentration of oil was maintained at 20 g/L, propanol was added at 18th hr of cultivation and after that its concentration was controlled at 1%, $(NH_4)_2SO_4$ was maintained at 3 g/L until 30 hrs when the nitrogen feeding was stopped to induce the nitrogen limitation. After 72 hrs the CDM was 138 g/L. Cells contained 76% of PHA with 8% content of 3HV. Total yield of PHA was 105 g/L. The volumetric productivity was 1.46 g/(Lh) and the yield coefficient was 0.83 g of PHA per g of oil [38].

Spent coffee grounds (SCG), an important waste product from coffee industry, contain approximately 15 wt.-% of coffee oil. Therefore; the utilization of oil extracted from SCG as a feedstock for the PHA production exhibits several advantages. The coffee industry keeps steadily growing; the annual worldwide production of green coffee beans exceeded 8 million tons in 2010. Therefore, the coffee by-products and especially spent coffee grounds (SCG) are available in millions of tons which represent a sustainable carbon source for the PHA production and the PHAs production from oil extracted from spent coffee grounds has been recently independently described by Cruz *et al.* [58] and Obruca *et al.* [41]. Based on the results of the latter group oil extracted from spent coffee grounds represented the most suitable carbon source for the PHAs production when compared to other inexpensive/waste oils. Employing the fed-batch mode of cultivation, the PHB yields, the PHB content in biomass, the volumetric productivity and the $Y_{P/S}$ yields coefficient reached 49.4 g/L, 89.1%, 1.33 g/(Lh) and 0.82 g per g of oil, respectively. Moreover, since the oil extraction decreased the calorific value of SCG by only about 9% (from 19.61 MJ/kg to 17.86 MJ/kg) residual SCG after the oil extraction can be used as a fuel to, at least partially, cover the heat and energetic demands of the fermentation. Considering the data obtained in laboratory bioreactor in the fed-batch mode of cultivation scaled-up to common industrial bioreactor of the volume of 100 m^3, the annual production of PHB is estimated to make 668 tons utilizing 5,573 tons of SCG. Since the coffee oil content in SCG is about 15%, 4737 tons of SCG after the oil extraction can be combusted generating theoretically approx. 85 TJ (2.3 107 kWh) of energy, the efficiency of industrial biomass boilers presents the value between 65 and 85%. The generation of such a high amount of energy would significantly improve the economic aspects of the PHAs production process [41].

Production of PHAs from Lipid Substrates by Other Bacterial Strains

Apart from *C. necator*, there are also several other strains capable of production of scl PHAs from lipid substrates. Kimura *et al.* reported the biosynthesis of PHB from various plant oils such as rice oil, safflower oil, soya oil or olive oil employing bacterium *Chromobacterium* sp. which was isolated from soil at Hottate River. Authors obtained rather high biomass as well as PHB concentrations, the PHB content in biomass varied between 35-56% [61]. Similarly, Allias and Tan isolated palm oil-utilizing bacteria from the palm oil mill effluent, they identified it as a strain closely related to *Burkholderia cepacia.* The bacterium was able to produce PHB using palm olein as the sole carbon source and PHBHV when propionate or valerate were added to the cultivation media [62]. Further, *Burkholderia* sp. USM (JCM15050) isolated from oil-polluted wastewater is capable of utilizing palm oil products and glycerol to synthesize PHB. To gain the ability to produce the polymer containing 3HHx, a plasmid harbouring the PHA synthase gene of *Aeromonas caviae* was successfully transformed into the strain. The resulting transformant incorporated approximately 1 mol-% of 3HHx in the polymer when crude palm kernel oil or palm kernel acid oil was used as the sole carbon source [63]. Another lipid substrate which seems to have great potential in the PHAs production is the seed oil of *Jatropha curcas* plant. Among other favorable properties, *J. curcas* seed oil is non-edible, widely available, and can be cheaply produced. In the study of Allen *et al.*, *Pseudomonas oleovorans* was grown in a mineral salt medium supplemented with saponified *J. curcas* seed oil as the only carbon source under the batch fermentation. Optimum PHA yield of 26.06% cell dry mass was achieved after 72 hrs of cultivation [64].

Furthermore, numerous bacterial strains belonging to *Pseudomonas* are capable of accumulation of *mcl*-PHAs when cultivated on lipid substrates. The problem is that some strains which were reported to be capable of the PHAs production from plant oils and other lipid substrates such as *Pseudomonas aeruginosa* belong among opportunistic pathogen and; therefore, they are not suitable for the PHAs production at an industrial level [65 - 67]. Further, *Pseudomonas chlororaphis* HS21 was also previously used to produce *mcl*-PHAs from palm oil [68]. Also Walsh *et al.* employed *P. chlororaphis* for the biotechnological production of *mcl*-PHAs from rapeseed oil and acid hydrolyzate of rapeseed oil. The *mcl*-PHAs volumetric productivity with rapeseed oil as the substrate was 0.53 g/(Lh) after 25 hrs with the yield of 0.22 g of PHA g per g of oil, while the volumetric productivity with hydrolyzate of rapeseed oil as the substrate was 0.54 g/(Lh) after 25 hrs with lower yield of 0.15 g per g of substrate. Thus, under the fermentation conditions tested, the hydrolyzate of rapeseed oil was an inferior substrate for the PHAs production when compared to non-pretreated rapeseed oil [65]. Further, *Comamonas testosteroni* was able to produce *mcl*-PHA from a range of oils including castor seed oil, coconut oil, mustard oil, cottonseed oil, groundnut oil, olive oil, and sesame oil [69]. *Pseudomonas resinovorans* B2649 was shown to utilize lard, butter oil, olive oil, sunflower oil, coconut oil and soybean oil to produce *mcl*-PHA16 [70], and *Pseudomonas saccharophilia* was cultured using soybean oil and sunflower oil to produce *mcl*-PHA [71]. Follonier *et al.* used *Pseudomonas resinovorans* for the biotechnological production of *mcl*-PHA in 2-step fermentation process, hydrolyzed pomace as a growth substrate and waste frying oil as *mcl*-PHA precursor. The hydrolyzate of Solaris grapes proved to be a very promising growth substrate as compared to that obtained from apricots [72]. The synthesis of *mcl*-PHA by *Pseudomonas* sp. Gl01 using saponified waste palm oil as a sole carbon source was investigated. It was shown that the analyzed strain accumulated the biopolymers during the growth phase. Up to 43% of *mcl*-PHAs were produced at 17 hr, when *Pseudomonas* sp. Gl01 was grown for 48 hrs in the bioreactor containing 15 g/L of saponified waste palm oil [73]. Table **2** provides an overview on the literature reports of *scl*- and *mcl*-PHA production based on lipid substrates employing various bacterial strains.

Table 2. Production of *scl*- and *mcl*-PHA on lipid substrates employing various bacterial strains.

Bacterial Strain	Substrate	Cultivation	Polymer	CDM [g/L]	PHA [wt.-%]	PHA [g/L]	$Y_{P/S}$ [g/g]	Ref.
***scl*-PHA**								
Chromobacterium sp.	Rice oil	Batch, E. flasks	PHB	12.9	55.8	7.2	0.72	[61]

(Table 2) contd.....

Bacterial Strain	Substrate	Cultivation	Polymer	CDM [g/L]	PHA [wt.-%]	PHA [g/L]	$Y_{P/S}$ [g/g]	Ref.
	Safflower oil	Batch, E. flasks	PHB	17.6	49.8	8.8	0.88	[61]
	Soya oil	Batch, E. flasks	PHB	13.8	42.4	5.9	0.59	[61]
	Olive oil	Batch, E. flasks	PHB	12.3	45.6	5.6	0.56	[61]
	Linseed oil	Batch, E. flasks	PHB	9.6	48.7	4.7	0.47	[25]
	Peanut oil	Batch, E. flasks	PHB	12.6	35.1	4.4	0.44	[61]
	Corn oil	Batch, E. flasks	PHB	10.1	47.0	4.7	0.47	[61]
Burkholderia cepacia	Palm olein	Batch, E. flasks	PHB	5.1	43.1	2.2	-	[62]
Burkholderia sp.	Palm oil	Batch, E. flasks	P(3HB-*co*-3HV-*co*-3HHx)	1.5	76	1.1	-	[63]
Pseudomonas oleovorans	Jathropa curcas oil	Batch, E. flasks	PHBHV	-	26	-	-	[64]
***mcl*-PHA**								
Pseudomonas aeruginosa	Palm oil	Batch, Ferm.	*mcl*-PHA	2.2	36.0	0.8	0.11	[66]
Pseudomonas aeruginosa	Peanut oil	Batch, E. flasks	*mcl*-PHA	4.0	3.8	0.2	-	[67]
	Olive oil	Batch, E. flasks	*mcl*-PHA	4.5	6.0	0.3	-	[67]
	Waste oil	Batch, E. flasks	*mcl*-PHA	5.4	5.7	0.3	-	[67]
Pseudomonas chlororaphis	Hydrolyzed rapeseed oil	Batch, Fermentor	*mcl*-PHA	76.0	17.0	12.9		[65]
Pseudomonas chlororaphis	Palm kernel oil	Batch, Fermentor	*mcl*-PHA	3.3	45.0	1.5		[68]
Comamonas testosteroni	various oils	Batch, E. flasks	*mcl*-PHA	-	79 - 88	-	-	[69]
Pseudomonas resinovorans	coconut oil	Batch, E. flasks	*mcl*-PHA	-	57	-	-	[70]
Pseudomonas saccharophila	soybean and sunflower oil	Batch, E. flasks	*mcl*-PHA	-	-	-	-	[71]
Pseudomonas resinovorans	Waste frying oil	Batch, Fermentor	*mcl*-PHA	10.2	12.4	1.3	-	[72]
Pseudomonas sp. Gl01	Saponified waste palm oil	Batch, Fermentor	*mcl*-PHA	-	43.0	-	-	[73]

Production of PHA from Lignocellulose-Based Substrates

Lignocellulosic biomass includes agricultural and forestry residues, portions of municipal solid waste as well as herbaceous and woody crops. With the annual generation of 80 billion tons, lignocellulosic materials have a great potential for the production of wide variety of industrial and commodity products including paper, lumber, bioethanol, biodegradable polymers (*e.g.*, PHAs) and a range of fine chemicals. Such materials are abundant and competitive in price with petroleum, and, therefore, lignocellulosic biomass can provide a sustainable resource also for fermentative production of PHAs. On the other hand, the utilization of lignocellulosic materials as substrates for biotechnological purposes is accompanied by many obstacles stemming from nature of these materials such as relatively low amount of fermentable sugars, the cost of hydrolysis process, the presence of inhibitors in hydrolysates and the cost of their removal in large volumes. Moreover, also the issues related to substrate accessibility and transportation should be taken into account. Due to these problems, the utilization of lignocellulose based substrates is challenging and only limited number of processes seem to be profitable at the moment. The most of industrial units dealing with lignocellulose biomass utilization work in the experimental mode. The fact is that many technological as well as scientific problems need to be solved and, therefore, the aim of this chapter is to focus on lignocellulosic biomass to summarize recent advances in biotechnological production of PHAs from various lignocellulose-based materials.

Characterization of Lignocellulose Biomass

Lignocellulosics, comprised of cellulose, pectin, hemicellulose, and lignin, are the most abundant raw materials on Earth; therefore, they are considered as potential substrates for the production of various chemical, fuels and materials. Generally, the composition of plant cell wall varies in cellulose (40-80%), hemicellulose (10-40%), and lignin (5-25%) content depending on the type of biomass used [74].

Cellulose is the most widespread organic material in the world. The dominant structure of this polysaccharide of plant cell walls is linear β-(1→4)-D-glucopyranoside polymer. Due to the fact that cellulose possesses a substantial degree of crystallinity, it functions as rigid, loadbearing component of the cell wall. Furthermore, the rigidity of the cellulose microfibril is strengthened within a matrix of hemicelluloses and pectins [75].

Pectins are acidic cell wall polysaccharides functioning as a sol-like matrix, providing water and ion retention, support and facilitation of cell wall modifying enzymes, cell wall porosity, cell-to-cell adhesion, cell expansion, cell signaling, developmental regulation, and defense [75, 76].

Hemicelluloses are highly versatile materials. Generally, they are classified according to the main sugar residue in the backbone, *e.g.*, xylans, mannans, and glucans, with xylans and mannans being the most prevalent. The average degree of polymerization of hemicelluloses is in the range of 80-200. They are usually linked with various other cell-wall components such as cellulose, cell-wall proteins, lignin, and other phenolic compounds by covalent and hydrogen bonding, and by ionic and hydrophobic interactions. Depending on the plant species, developmental stage, and tissue type, hemicelluloses can be divided into several subclasses which can be grouped into two general categories based on the hydration of fibers. Low hydration polysaccharides include the arabinoxylan, glucuronoxylans, xyloglucans and linear mannans. With the exception of linear mannans, which serve mainly as a seed storage compound, this class of hemicelluloses functions primarily to stabilize the cell wall through hydrogen-bonding interactions with cellulose and covalent interaction with lignin and are generally water-soluble in their native state due to their branched structure. The second group of hemicelluloses, composed mainly of hydrocolloids, is used primarily as an extracellular energy and raw materials storage system and as a water retention mechanism in seeds. Galactoglucomannans, glucomannans, galactomannans, and β-glucans tend to be heavily hydrated and have fewer, if any, ester-linked side chains [75, 77, 78].

Pretreatment of Lignocellulose Materials from PHA Production

Hydrolysis of Cellulose and Hemicellulose

Cellulose and hemicellulose are polysaccharides which can be hydrolyzed into sugars and fermented or chemically altered to valuable fuels and chemicals [75]. The hydrolysis technologies may involve psychical treatment, chemical methods such as hydrolysis by concentrated or diluted acid, as well as enzymatic methods whereby chemical and enzymatic hydrolysis are often combined in consecutive steps. The composite formed by cellulose, hemicellulose and lignin is responsible for remarkable resistance against hydrolysis and enzymatic attack [79]. Out of the three components, lignin is the most recalcitrant to degradation whereas cellulose is more resistant to hydrolysis as compared to hemicelluloses [75, 80].

Generally, the pre-treatment of cellulosic prior to its enzymatic hydrolysis by cellulases significantly improves the glucose yields. The goal of the pre-treatment process is at least partial removal of lignin and hemicelluloses, the reduction of cellulose crystallinity as well as the increase of porosity of the lignocellulosic materials. The pre-treatment must meet the following requirements:

a) improve the formation of sugars or the ability to subsequently form sugars by hydrolysis

b) avoid the formation of by products which are inhibitory to the subsequent hydrolysis and fermentation process

c) avoid the degradation or loss of carbohydrate, and

d) be cost-effective. Various types of pre-treatment methods are known, *e.g.*, physical, physicochemical, chemical, biological, and electrical or their combination [81].

Cellulase refers to a group of enzymes which, acting together, hydrolyse cellulose. Glucose represents the final product of enzymatically driven hydrolysis of cellulose, and its yields can approach 100%. The cellulose-hydrolysing enzymes are divided into three major groups: endoglucanases, cellobiohydrolases and β-glucosidases. The endoglucanases act randomly on internal bonds of cellulose chain, while cellobiohydrolases attack the chain ends, releasing cellobiose. β-glucosidases are only active on cello-oligosaccharides and cellobiose and release glucose monomers units [75, 80]. Cellulases are distributed throughout the bio-sphere; they are mostly manifested in fungi and microbial sources. Although a large number of microorganisms are capable of cellulose degradation, only a few of these microorganisms produce significant quantities of cell-free enzymes capable of complete hydrolysis of crystalline cellulose *in vitro* [82]. Fungi are the main cellulase-producing microorganisms, though a few bacteria and actinomycetes have also been recently reported to the yield cellulase activity. Microorganisms of the genera *Trichoderma* and *Aspergillus* are thought to be the cellulase producers, and crude enzymes produced by these microorganisms are commercially available for industrial and agricultural use [81, 82].

Hemicellulose is the second most abundant renewable biomass formed of heterogenous polymers built up by pentoses, hexoses and sugar acids. Anyway, in the case of enzymatic hydrolysis of hemicellulose, there are various enzymes responsible for their degradation. Hemicellulases are produced by the species of bacteria and fungi, as well as by some plants [75, 80].

Acid hydrolysis of lignocellulosic biomass consisting of hemicelluloses and cellulose is performed by concentrated or diluted acids: acid catalyzes the breakdown of long carbohydrate chains to form shorter chain oligomers and then to sugar monomers. Because of amorphous form of hemicelluloses, less severe conditions are required for its hydrolysis in comparison with crystalline cellulose. The advantages of acid hydrolysis are that acid can penetrate lignin without the pre-treatment and the rate of acid hydrolysis is faster than that of enzyme hydrolysis [73]. Sulphuric and hydrochloric acids are the most commonly used catalysts for the hydrolysis of lignocellulosic biomass. This process is carried out at high temperatures to achieve acceptable rates of biomass conversion. On the other hand, high temperature increases the rate of lignocellulose biomass derived sugars decomposition thus causing the formation of toxic compounds which further decreases the yields of fermentable sugars [83, 84].

Composition of Hydrolysates

Lignocellulosic materials are converted by hydrolysis to fermentable sugars, which subsequently can be fermented to a variety of products such as ethanol, other alcohols, organic acids, single cell proteins and lipids [85]. Cellulose can be hydrolysed to yield glucose molecules, while hemicelluloses can be broken up *i.e.*, hydrolysed to yield simpler molecules such as arabinose, mannose, glucose, galactose, xylose and uronic acid [75, 77].

The hydrolysis of lignocellulosic waste by chemical or enzymatic treatments or a combination of both, releases several other degradation products which prove to be harmful to microbial fermentation. For example, acid hydrolysis of lignocellulosic materials produces several inhibitory compounds which inhibit cells, affect specific growth rates and thus considerably decrease the yield of the biotechnological process. Based on the origin, inhibitors are usually divided into three major groups, i. weak acids, ii. furan derivatives, iii. phenolic compounds. Phenolic and aromatic compounds are released especially from lignin [75, 82]. Some phenolic compounds include the extractives which are also derived from released sugars after the hydrolysis. The effect of phenolic and other aromatic compounds, which may inhibit both microbial growth and product yield, are very variable, and can be related to specific functional groups. One possible mechanism is that phenolic compound interferes with the cell membrane by influencing its function and changing its protein-to-lipid ratio. Phenolic compounds are also investigated with regard to the inhibition of enzymatic hydrolysis of cellulose; the experiments with phenols suggest that one way in which they affect proteins is inducing the precipitation [86].

Lignocellulose hydrolysates contain aliphatic acid, such as acetic acid, formic and levulinic acid. Acetic acid is formed primarily by the hydrolysis of acetyl group of hemicellulose, while formic acid and levulinic acid arise as acid-catalyzed thermochemical degradation products from polysaccharides. The last mentioned acid-formic acid is the degradation product of furfural and hydroxymethylfurfural (HMF). Undissociated acids enter the cell through the diffusion over the cell membrane and then dissociate due to the neutral cytosolic pH-value. The dissociation of acid leads to the decrease in the intracellular pH-value, which may cause the cell death. The last main group of inhibitors present in lignocellulosic hydrolysates is a group of furan derivatives. Furan aldehydes furfural and HMF are formed by the dehydration of pentose and hexose sugars. These furan aldehydes are known as the most potent inhibitors of microbial cell growth and can inhibit ethalogenic microorganisms such as yeasts [85 - 87].

Lignin is a complex polymer which belongs to the group of aromatic alcohols consisting of phenylpropane units. Because of its structural complexity, lignin provides a structural function in plants as well as the stiffness and rigidity to the plant cell wall. In order to make the lignocellulose more accessible for acids or enzymes in subsequent hydrolysis, lignin needs to be broken down in the pre-treatment step. However, as in the case of hemicellulose and cellulose, lignin degradation leads to the generation of inhibitory by-products [85, 88].

Besides furans and phenolics, there are other products generated in the pre-treatment of lignocellulose. They are extracted from wood by the use of various solvents and play a role in the defense system of a plant as well as the precursors of certain chemicals [89]. These products include tannic and terpene acids such as caproic acid, caprylic acid, pelargonic acid or palmitic acid. The type and the concentration of acids discharged into the hydrolysates depends strongly on the feedstock, hydrolysis temperature, acid concentration and overall severity of the pretreatment. For example, feedstocks with high content of acetylated carbohydrates such as the agricultural residues and hardwood, contain higher amounts of aliphatic acids than softwood [13]. These raw material extractives cause higher cell membrane permeation, disturb the proton gradient over the inner mitochondrial membrane which inhibits the regeneration of ATP in mitochondria and eventually can lead to cellular death [90].

The hydrolysis of lignocellulosic feedstocks might also release inorganic ions (K^+, Na^+,Cl^-) which originate from the raw material and chemicals added during the pretreatment and possibly from the process equipment [86, 88, 89]. Inorganic ions also come from the addition of salts which in turn lead to higher osmotic pressure. If ions cannot cross the cell membrane, the cell will shrink as a result of hypertonic environment [88, 79].

Detoxification of Hydrolysates

Several types of detoxification strategies can be used to detoxify slurries and hydrolysates. It is always important to distinguish what substances should be removed prior to the cultivation and which on the other hand do not have to be removed from the culture media due to their low degree of inhibition or microbial tolerance and select detoxification procedure accordingly.

There are several methods based on the physical treatment of the hydrolysates.

Evaporation

Evaporation is a simple method to remove volatile compounds such as acetic acid and furans from lignocellulosic hydrolysates. The overall decrease of some of these compounds can reach up to 100% as was demonstrated previously by Wilson *et al.* [91]. In this study, the decrease in volatile inhibitory compounds enhanced the ethanol productivity compared to the reference fermentation. On the other hand, although lowering the amounts of volatile inhibitory products through the evaporation can be beneficiary in some cases in terms of product formation and biomass concentration, the same process can as well cause the opposite effect on the microorganisms by increasing the concentration of non-volatile compounds [92].

Membrane Separation

Membrane separation can be a suitable and less expensive alternative to ion exchange resin. Functional groups attached to adsorptive micro porous membranes may be used to capture the cell wall derived inhibitors from the lignocellulosic hydrolysates. After the pre-treatment, the content of hydrolysates can be fed through the membrane pores which predominantly bind the inhibitory products and thus greatly reduce the purification and processing time. Another great advantage of membrane extraction is the separation of organic phase containing the inhibitory cell wall derived products from the aqueous phase. The extraction of sulfuric acid, levulinic acid or acetic acid from organic phase solvents can be applied for higher productivity as was previously observed by Grzenia *et al.* [93]. Even more efficient extraction of acetic acid was confirmed with anion exchange membranes (Sartobind Q) than with ion exchange resin (Amberlyst A21). Related to the membrane separation systems are the molecular sieves which contain small pores of uniform size and can therefore capture inhibitory compounds of bigger sizes. The molecular sieves have been shown to perform well when capturing acetic acid (40%) and furfurals of up to 82% capture efficiency [94].

Chemical Methods

Other possible way to detoxify hydrolysates is with the help of chemicals. There are several chemical methods commonly used for removing the toxic compounds as by-products of hydrolysis step.

Neutralization

Before lignocellulosic lysates are used for the microbial fermentation, the neutralization of the acidic environment is essential. In the process of neutralization, added base reacts with acid to form water and salt. Through this process, furfurals and phenolics can be extracted by subsequent neutralization. The most common compounds used for the neutralization are sodium hydroxide and calcium hydroxide. The latter is widely used in over-liming process of removing the volatile inhibitory compounds [85].

Calcium Hydroxide Over-Liming

In some of the studies, the over-liming has been claimed to be the most efficient and economical method of the lignocellulose hydrolysate detoxification. The process has been shown in which volatile inhibitory compounds such as furfural and hydroxymethyl furfural are removed [85]. The method uses calcium hydroxide which reacts with acids in acidic hydrolysates from the lignocellulose pretreatment. The reaction of alkali with acid results in the precipitation of calcium sulfate (gypsum) lowering the soluble salt concentration in the culture media. Because the soluble salt concentration is decreased, the microbial growth is no longer limited by the hyperosmotic environment that would otherwise occur.

Nevertheless, in all cases, the over-liming as well as other alkali detoxification methods have one problem in common. The alkali detoxification products not only react with inhibitors but also with sugars. As a result, the sugar concentration is decreased through the formation of reactive enolate intermediates by calcium ions (using calcium hydroxide) or any other alkali ions [95]. Furthermore, it is very difficult to remove the precipitated particles completely, which may further complicate the isolation of intracellular products (*e.g.*, PHAs) of following fermentation.

Activated Charcoal Treatment

The treatment with activated charcoal is known to be an effective method with high absorbance capacity to remove lignin-derived inhibitors and acetic acid without affecting the levels of sugar in the hydrolysate. Through this process, the fermentation with high product yield comparable to the inhibitor free fermentation can be achieved due to low affinity of sugars towards charcoal [85, 88]. Furthermore, activated charcoal does not require the pH-value adjustments as in the case of anion exchange resins or alkali treatment. However, the application of this method is quite expensive since activated carbon cannot be recycled back into the process after use and around 10% of charcoal granules become degraded during each thermal reactivation cycle. Moreover, the method efficiency depends on the contact time, temperature and the amount of activated carbon applied [88]. The increase of contact time has been shown to affect the clarification process while elevated temperatures enhance the absorption due to faster rate of diffusion from the solution to the activated charcoal [96].

Ion Exchange Resins

The ion exchange detoxification is not a cost effective method but on the other hand it yields great results in terms of detoxification efficiency of hemicellulose and cellulose derived inhibitors. The ion exchange process can therefore drastically enhance the fermentability of the hydrolysate [78], especially when combined with other detoxification procedures such as overliming, activated charcoal or evaporation. There are several types of resins nowadays which can be applied for the purification such as anion exchangers, cation exchangers and also a resin without charged groups [97].

Biological Methods

Biological methods take advantage of living microorganisms and/or their microbial enzymatic capabilities to detoxify lignocellulose hydrolysates [88]. The biological methods keep constantly improving in terms of enzyme engineering, microbial acclimatization or microbial genetic engineering. Nowadays, these methods are in demand because they are simple to carry out, environmentally friendly, have fewer side reactions and require milder work conditions and less energy. However, they have several limitations including slow reaction time, inhibitor specificity and loss of fermentable sugars [88, 90].

Microbial Pretreatment of Lignocellulose

A number of microorganisms such as yeast, bacteria and fungi can be applied as a pretreatment step to detoxify the inhibitory compounds and degrade lignin before actual fermentation. It has been reported earlier that bacterial species such as *Ureibacillus thermosphaericus* or filamentous soft-rot fungus *Trichoderma reesei* can degrade furfurals, HMF, aromatic compounds and weak acids at comparable efficiency as overliming or alkali treatment [88]. Even though the microorganisms consume some amounts of fermentable sugars, the benefit of the pretreatment surpasses the negatives leading to higher product yields, sugar-to-product transformation rate and more importantly, the detoxification step is avoided [88, 90].

Another method related to microbial treatment is *in situ* microbial detoxification. Unlike the pre-treatment of lignocellulosic material with microorganisms in a separate step before actual fermentation, the microorganisms which help detoxify the inhibitory products are incubated along with the production strains in one vessel. A variety of microorganisms (yeast, fungi and bacteria) as well as genetically engineered strains can be co-inoculated. These "helpers" can deploy enzymes such as laccases and peroxidases which can oxidise the inhibitors or partly digest lignocellulose by lignin-degrading enzymes [90].

Microbial Acclimatization

The adaptation of microorganisms to the lignocellulitic hydrolysate is an effective and inexpensive method to improve the product yields by reducing the inhibitory effect of toxic compounds as well as by decreasing the loss of fermentable sugars [89]. The microorganisms can adapt to the presence of inhibitory compounds by gradual concentration increase of inhibitory compounds in the lysate. Throughout rounds of incubations they develop defense mechanisms which protect them from the toxic compounds in the form of changes in enzymatic specificity and/or cell physiology. Long term adaptation has been successfully developed by Liu *et al.* who adopted *Pichia stipitis* in the prolonged lag phase to reduce HMF to less toxic compound, 2,5-bis-hydroxymethylfuran. The yeast was able to reduce the aldehyde group on the furan ring of HMF into an alcohol [98]. Another group managed to develop *S. cerevisiae* strain with high tolerance to furfural and HMF by cultivating the strain for over 300 generations in the media with higher furfural concentrations. The strain not only survived in higher furfural concentrations but its lag phase was significantly shortened from 90 h at the beginning of the adaption phase to 17 h at the end [99].

Enzymatic Detoxification

Enzymatic detoxification is widely used and promising method to remove toxic compounds from lignocellulosic hydrolysate. One of the main advantages of using enzymes is low utility cost compared to other methods due to mild conditions applied as well as the substrate specificity and high catalytic activity allowing fast detoxification process. However, enzymes require prolonged incubation and their price is usually very high [88, 90]. Enzymes used in the industry are laccases and peroxidases derived from basidiomycetous white rot fungi (*Trametes versicolor, Phenorochete chrysosporium, Cythus bulleri, C. stercoreous, and Pycnoporous cinnabarinus*) [85].

Production of PHA on Hydrolysated Lignocellulose Materials

Lignocellulose Biomass as Substrate for PHA Production

As was mentioned above, lignocellulosics represent highly abundant but also rather versatile group of materials. The properties and sustainability of these materials with respect to their utilization for biotechnological production of PHAs as well as its hydrolysis and further treatment should be always performed with regards to particular biomass stream and fermentation process.

Apart from the substrate selection and preparation, the key importance is usually ascribed to the selection of PHAs producing bacterial strain. Generally, many bacteria have been screened to produce PHAs, however, only a few have been considered being candidates for the production of these polyesters on a large scale. The suitability of a bacterium for PHAs production from such a complex and challenging substrates depends on many different factors *e.g.*, the ability to utilize fermentable sugars present in hydrolysates (including pentoses), the capability of tolerating or even eliminating potential microbial inhibitors, the stability and safety of the organism, the growth and accumulation rates, achievable cell densities and PHAs contents, the extractability of the polymer and also the molecular weights of accumulated PHAs [2].

Since PHAs are intracellular products, the total PHAs yields are dependent on reached concentration of biomass. Therefore, reaching high cell density is the first condition of reasonable process of PHAs production. The fed-batch cultivation strategy, which is often in use to reach high cell densities, can be in the case of utilization of lignocellulose hydrolysates complicated by the phenomenon called carbon catabolite repression (CCR). In the presence of sugar mixtures, a preferential consumption of one of the sugars (usually glucose or other hexoses) is observed while other sugars (usually pentoses) remain unutilized in the medium. Employing the fed-batch cultivation, their concentration in medium can reach inhibiting concentration which significantly decreases the productivity of the process [100]. In some bacterial strains CCR is mediated by the proteins of phosphotransferase system (PTS). Lopes *et al.* [101] studied the catabolite repression in PTS mutants of *Burholderia sacchari* IPT101 to improve the total carbon up-take in sugar mixtures. The wild strain only started consuming xylose after glucose was completely depleted, while one UV mutant was able to consume glucose and xylose simultaneously [101].

Utilization of Wood Biomass for PHA Production

Among various types of lignocellulosics forest biomass represents an enormous reservoir of renewable carbon-rich materials. Globally, approximately 80 billion tons of woody biomass is generated per annum, with the production of total plant matter estimated at roughly 180 billion tons annually. While the bulk of the cellulosic component is efficiently exploited by the paper/pulp industry, the hemicellulosic and lignin fractions are vastly underutilized process streams, which hold the potential as platform intermediates in the production of value-added, bio-based PHAs [102]. Hence, maple hemicellulosic hydrolysate was utilized as a renewable feedstock for the PHAs production employing *Burkholderia cepacia* by Pan *et al.* [103]. To increase the fermentability of wood hydrolysate, several detoxification methods were tested. Overliming combined with low-temperature sterilization resulted in the highest removal of total inhibitory phenolics (65%). The fed-batch fermentation exhibited maximum PHB production after 96 hrs - 8.72 g/L broth and 51.4% of cell dry mass) [104]. In their further work the authors estimated the contribution of individual inhibitors of wood hydrolysate to total antimicrobial effect. The results indicated that syringic acid was the most important inhibitor among three phenolics analysed and the synergistic inhibition was observed for the combinations of vanillin/syringic acid and vanillic acid/syringic acid. Additionally, strong synergistic effects were observed for the combinations of acetate/phenolics and levulinic acid/furfural [104].

Bowers *et al.* also focused on the utilization of wood biomass for the PHAs production. In their study, *Pinus radiata* wood chips were subjected to high-temperature mechanical pre-treatment or steam explosion in the presence of sulphur dioxide before being enzymatically treated to produce corresponding hydrolysates. Two PHB-producing bacteria *Novosphingobium nitrogenifigens* and *Sphingobium scionense* were grown on these hydrolysates. The highest PHB yields (biomass, PHB content and PHB yield - 1.23 g/L; 32% and 0.4 g/L, respectively) were observed in *Sphingobium scionense* [105]. Further, Silva *et al.* tested the PHAs production employing *Brevundimonas Vesicularis* and *Sphingopyxis macrogoltabida* using acid-hydrolysed sawdust as a carbon source. The bacterial strains were capable to accumulate terpolymer consisting of 3-hydroxybutyrate, 3-hydroxyvalerate and lactic acid (3-hydroxypropionic acid), the polymer content in cells reached 72% of cell dry mass, however, the total PHAs yields were limited by low biomass growth [106]. Similar copolymer consisting of 3-hydroxybutyrate and lactic acid was also produced by transgenic *Escherichia coli*. In particular, β-xylosidase and endoxylanase were engineered into the bacteria which was additionally designed to express the PHA synthase from *Pseudomonas sp.* 61-3 harboring a Ser325Th/Gln481Lys mutation [PhaC1Ps (ST/QK)], propionyl- CoA transferase (PCT) from *Megasphaera elsdenii*, and β-ketothiolase (PhaA) and NADPH-dependent acetoacetyl-CoA reductase (PhaB) from *Cupriavidus necator* under a *C. necator* constitutive promoter. The PHAs production yields using xylan as a sole carbon source were minimal; however, when the xylan-based media was supplemented with a single sugar (xylose or arabinose) to permit the accumulation of xylan derived xylose in the media, the PHA production yields increased. The highest biomass and PHAs yields obtained within this study were 8.9 and 3.6 g/L, respectively [107].

Lignocellulose-Based Agricultural Waste for PHA Production

Apart from wood biomass, there is another source of lignocellulose-based materials which can be used as a substrate for the PHA production - the waste streams of agriculture and food industry. These materials are generated in enormous amounts during processing of agricultural plants. Despite the fact that some of these waste- and by-products can be used for various purposes such as animal feeding or fertilization, the production of PHAs represents very interesting and promising strategy for their valorization yielding high-value products. Hence, there are numerous papers dealing with the conversion of these waste materials into PHAs.

Bagasse is a fibrous matter that remains after sugarcane or sorghum stalks are crushed to extract their juice. Since bagasse is highly abundant and available waste material, there were several attempts to produce PHAs from acid sugarcane bagasse hydrolysate. Silva *et al.* employed *Burkholderia cepacia* and *Burkholderia sacchari*. To improve the fermentability of the hydrolysate, the three-step detoxification process (consisting of i. concentration in water bath ii. overliming iii. activated charcoal treatment) was applied. After the treatment, the highest biomass and PHAs yields were observed in *B. sacchari* 4.4 and 2.7 g/L, respectively [108]. Even higher PHAs and biomass yields (11.1 g/L biomass, 6.3 g/L PHB) on sugarcane bagasse hydrolysate prepared by diluted acid treatment were obtained employing *Cupriavidus necator* [109] and also *Bacillus thuringiensis* was reported to be capable of sugarcane bagasse hydrolysate utilization and PHB production yielding 10.6 g/L of biomass and 4.2 g/L PHB [110]. Finally, another type of bagasse - stemming from manufacturing of tequila was used for the PHAs production employing *Saccharophagus degradans*, which can readily attach to cellulosic fibers, degrade cellulose, and utilize this as the primary carbon source while producing PHB. This approach does not require any hydrolysis prior to the cultivation, which in turns theoretically reduces the cost of up-stream processing. Unfortunately, the study by Munoz and Riley did not reveal the PHB yields, thus it is very difficult to estimate the efficiency of such a process [111].

Another very promising waste substrate for the PHAs production is spent coffee grounds. Coffee is one of the world's most popular beverages and has grown steadily in commercial importance during the last 150 years. Nowadays, coffee is, after petroleum, the second largest traded commodity in the world. In 2010 the worldwide annual production of coffee beans exceeded 8 million tons. During the preparation of coffee beverage or the manufacturing of instant coffee, raw coffee powder is mixed with hot water or steam under the conditions favoring the release of aroma compounds and other coffee-bean constituents into the liquid, which generates solid residues known as spent coffee grounds. Spent coffee grounds can be considered as very promising substrate for the PHAs production. At first, SCG contain approximately 15% of oil, which can be simply extracted and converted into PHB by *Cupriavidus necator* [41, 58]. The residual solids after the oil extraction contain significant portion of hemicelluloses and cellulose. Therefore, they were hydrolysed (chemical hydrolysis by diluted acid was followed by the cellulose treatment) and converted into PHAs employing *Burkholderia cepacia.* Hexoses (predominantly mannose and galactose) were substantially dominating sugars of the hydrolysate which may be an important factor positively influencing the PHAs production. Moreover, the hydrolysate contained levulinic acid (the product of hexose degradation), which served as a precursor of 3-hydroxyvalerate resulting in the accumulation of PHBHV copolymer [112].

Coir pitch, the byproduct of coconut fibers and waste material from the coir industry, is stable and not easily degradable due to its high lignin content. Coir pith takes a decade to decompose thereby posing environmental hazard and disposal problem. Thus, Prabu and Murugesan investigated the PHB production from coir pitch by *Azotobacter beijerinickii.* The waste material was at first partially delignified and further enzymatically hydrolysed by cellulase. The amount of PHB produced by *A. beijerinickii* from the coir pitch hydrolysate was 2.4 g/L [113].

Also the rice straw hydrolysate obtained by chemical hydrolysis using diluted acid was converted into PHA by *Bacillus firmus.* Acid pretreated black liquor contained sugars and sugar degradation products such as formic acid, acetic acid, furfural and hydroxymethyl furfural. The bacterium grew in the hydrolysate medium without any detoxification and it could accumulate 1.9 g/L biomass with 1.7 g/L PHB and the PHB content in the cell was 89% [114].

Oil palm is the most abundant industrial crop in Southeast Asia. Each year, more than 15 million tonnes of oil palm empty fruit bunch are generated by palm oil industries in Malaysia. Oil palm empty fruit bunch consists of cellulose (50.4%), hemicellulose (21.9%), lignin (10%), and ash (17.7%). Zhang *et al.* investigated the PHAs production from oil palm empty fruit bunch. The material was first chemically pretreated and enzymatically hydrolysed by a cellulase cocktail and further used as a substrate for the PHAs production employing *Bacillus megaterium*. Tryptone was identified as its best nitrogen source, the PHB content and the production reached 51.6% and 12.48 g/L, respectively, the productivity reached 0.260 g/(Lh) [115].

Furthermore, also the residues of wheat processing such as wheat straw and wheat bran can be considered as very promising potential substrates for the PHAs production. The world wheat production estimated for the period of 2012-2013 is about 660 million tonnes, of which about 15-20% is straw. Asia and Europe are the primary production regions, with about 43% and 32%, respectively, while North America is the third largest production region with 15% of global wheat production [116]. Van-Thuoc *et al.*studied the PHB production from the wheat bran enzymatic hydrolysate using *Halomonas boliviensis*. The PHB yields were improved by the addition of butyric acid and sodium acetate, which can be obtained by anaerobic digestion of solid potato waste. The biomass and PHB yields obtained in this study were 8.0 and 4.0 g/L, respectively [117]. In addition, wheat straw was also converted into PHB by *Burkholderia sacchari*. The lignocellulosic hydrolysates were prepared from ground wheat straw using the AFEX process as a pre-treatment followed by an enzymatic hydrolysis of cellulose and hemicellulose fractions. Resulting hydrolysate contained glucose, xylose and arabinose as the most predominant sugars. The PHB production was performed in the fed-batch cultivation mode and the feeding strategy was set to overcome CCR. Very high maximum polymer concentration of 105 g/L was reached after 61 hours of cultivation corresponding to the accumulation of 72% of CDM. The polymer yield and the productivity were 0.22 g/g total sugar consumed and 1.6 g/(Lh), respectively [100].

Hydrolysed wheat straw was also used for the PHA production in the study of Cesario *et al.* The lignocellulosic hydrolysate was prepared using the AFEX process as the pre-treatment followed by an enzymatic hydrolysis of cellulose and hemicellulose fractions [100]. Hydrolysed wheat straw was utilized by the PHA producing strain *Bulkholderia sacchari*, the efficiency of production on waste substrate was compared to the PHA production where glucose served as a carbon source. γ-butyrolactone was used in this study as a precursor of the 4HB units. Hydrolysed wheat straw as a substrate containing γ-butyrolactone revealed the maximum polymer concentration, the polymer content and the composition of 24 g/L, 0.5 g/L, 27% and 5 mol-% 4HB, respectively. On hydrolysed wheat straw, a similar copolymer composition was achieved compared to the glucose containing media [118].

Furthermore, also other types of straws were used for the PHA production by Saratale *et al.* Rice paddy straw, wood straw and soybean straw were used as waste lignocellulosic substrate for the PHA production by *Cupriavidus necator*. For better saccharification, it is necessary to include a pre-treatment step to degrade lignin layers of straw in order to increase the enzymatic accessibility of cellulose and hemicellulose. In this study, the alkali pre-treatment using NaOH, KOH or NaOCl was applied to all biomass samples. The pre-treatment (2% NaOH, 121°C, 30 min) of rice paddy straw resulted in the maximum yield of 703 mg of reducing sugars per gram of paddy straw with 84.19% hydrolysis yield after two enzymatic hydrolysis processes, which is higher than in the case of soybean straw (72.0%) and wood straw (68.0%). The enzymatic step was provided by cellulose from *Trichoderma reesei.* Greater PHA accumulation with maximal PHB production (10.1 g/L) was observed with rice paddy straw hydrolysate. Wood straw and soybean straw have significantly lower PHB yields in comparison with rice paddy straw (Table **3**) [119].

Degraded corn stover is another possible waste substrate which can be converted to PHAs by bacteria species. Shailesh *et al.* reported the possibility of hydrolysed corn stover utilization by *Paracoccus sp.* isolated from Lonar Lake. The hydrolysis of corn stover was provided by one-site produced cellulase from the co-culture of *T. reesei* and *A. niger*. The PHA content and the production reached up to 72.4% and 9.71 g/L respectively from 40 g/L corn stover hydrolysed sugars [120].

Around 3.4 billion hectares of total land area comprise of grass lands which cover 69% of global agricultural area. The grass lands also play an important role in Europe's agriculture, *i.e.*, the total agricultural area constitutes 164 million hectares of cultivated land and 76 million hectares of that are permanent grass lands. Grass lands do not require the addition of fertilizers, annual ploughing of soil, and are a carbon sink which generate very low net greenhouse gas emissions. It was estimated that about 19 million hectares of pastures could become available for bioenergy feedstocks without compromising Europe's food and feed sectors by 2030 [121]. According to Davis *et al.*, grass biomass can be turned into *mcl*-PHAs employing *Pseudomonas* strains. Grass biomass was pre-treated (2% NaOH 120°C or hot water 120°C) and delignified and further enzymatically digested. Tested *Pseudomonas* strains accumulated 20-34% of *mcl*-PHA per biomass when these hydrolysates were used as sole carbon and energy source [122]. Further, Koller *et al.* used green grass juice as an additional complex nitrogen and phosphate substrate to improve the PHB production by *Cupriavidus necator* [123]. Also saccharified water hyacinth hydrolysates were used for the PHB production by *C. necator*. The bacteria preferred enzymatic hydrolysate to acid hydrolysate and the PHB production was optimized using the response surface methodology. After 72 hrs of cultivation, 35 g/L of reducing sugar containing water hyacinth hydrolysate supplemented with 1.5 g/L $(NH_4)_2SO_4$ in laboratory scale fermentor gave 12 g/L of cell dry mass and 7 g/L of PHB [124].

The PHAs production was also reported on the cellulose hydrolysates. For instance Zhang *et al.* published the PHB production on enzymatically hydrolysed pulp fiber sludge employing *C. necator*. By controlling the concentrations of the inorganic salts in the growth medium, almost 78% of the cell mass was converted to pure PHB. Efforts were made to find the conditions for bacterial growth in the form of a biofilm on cheap and reusable carrier. A number of positively charged carriers were tested, and the anion exchanger DEAE-Sephadex A-25 was chosen for the packed-bed biofilm culture [125]. Chemically hydrolysed cellulose and propionate as a precursor of 3-hydroxyvalerate were used as substrates for the PHBHV production by transgenic *Escherichia coli* with superior resistance to 5-hydroxymethylfurfural. The yields for the cell growth and the polymers were almost the same as those performed for the case of analytical grade glucose [126].

Table 3. Production of PHA on lignocellulose-based substrates employing various bacterial strains.

Production Strain	Carbon Source	PHA Type	Cultivation	Biomass [g/L]	PHA [wt.-%]	PHA [g/L]	$Y_{P/S}$ [g/g]	Ref.
Bulcholderia sacchari	wheat straw hydrolysate	PHB	Fed-batch	88.0	27.0	24.0	n.a	[118]
Paracoccus sp. LL1	corn stover hydrolysate	PHB	Batch	14.8	72.4	9.7	0.25	[120]
Rastolnia eutropha	rice paddy straw	PHB	Batch	15.5	70.1	10.8	n.a	[119]
	wood straw	PHB	Batch	11.4	59.5	6.1	n.a	
	soybean straw	PHB	Batch	12.1	62.2	6.6	n.a	
Burkholderia cepacia	xylose	PHB	Batch	2.6	60.0	1.6	0.11	[127]
Pseudomonas strains	grass biomass	*mcl*-PHA	Batch	0.9	33.0	0.3	n.a	[122]
Azotobacter beijerinickii	coir pitch	PHB	Batch	5.0	48.0	2.4	n.a	[123]
Burkholderia sacchari	wheat straw hydrolysate	PHB	Fed-batch	145.8	72.0	105	0.22	[100]
Recombinant E. coli	beech wood xylan and xylose	P(LA-*co*-3HB)	Batch	8.9	40.4	3.6	n.a	[107]
Burkholderia cepacia	wood hydrolysate	PHB	Batch	n.a	n.a	n.a	n.a	[111]
Ralstonia eutropha	bagasse hydrolysate	PHB	Batch	11.1	56.5	6.3	n.a	[109]
Bacillus firmus	rice straw hydrolysate	PHB	Batch	1.9	89.0	1.7	n.a	[114]
Burkholderia cepacia	wood hydrolysate	PHBHV	Batch	n.a	40.0	2.0	n.a	[101]
Recombinant E. coli	cellulose hydrolysate + propionate	PHBHV	Batch	4.2	50.0	2.1	n.a	[126]
Bacillus megaterium	oil palm empty fruit bunch	PHB	Batch	24.2	51.6	12.5	0.23	[115]
Burkholderia cepacia	wood hydrolysate	PHB	Fed batch	16.9	51.4	8.7	0.19	[110]
Sacharophagus degradans	waste from tequila bagasse	PHA	Batch	n.a	n.a	n.a	n.a	[116]
Burkholderia cepacia	bagasse hydrolysate	PHB	Batch	4.4	53.0	2.3	0.29	[108]

(Table 3) contd.....

Production Strain	Carbon Source	PHA Type	Cultivation	Biomass [g/L]	PHA [wt.-%]	PHA [g/L]	$Y_{P/S}$ [g/g]	Ref.
Burkholderia sacchari	bagasse hydrolysate	PHB	Batch	4.4	62.0	2.7	0.39	[108]
Halomonas boliviensis	wheat bran + potato waste hydrolysate	PHB	Batch	8.0	50.0	4.0	0.2	[117]
Ralstonia eutropha	water hyacinth hydrolysates	PHB	Batch	12.0	58.3	7.0	0.13	[124]
Burkholderia sacchari	xylose	PHB	Batch	5.5	58.0	3.2	0.26	[128]
Burkholderia sacchari	xylose	PHB	Batch	5.3	50.0	2.7	0.17	[101]
Ralstonia eutropha	enzymatically hydrolysed pulp fibre	PHB	Batch	8.8	31.9	2.8	n.a	[125]
Sphingobium scionense	enzymatically hydrolysed softwood	PHB	Batch	1.2	32.0	0.4	0.22	[105]
Burkholderia cepacia	spent coffee grounds hydrolysate	PHBHV	Batch	5.5	56.0	3.1	0.24	[112]
Brevundimonas Vesicularis	acid hydrolysed sawdust	P(LA-*co*-3HB-*co*-HV)	Batch	0.3	78.0	0.3	n.a	[106]
Bacillus thuringiensis	bagasse hydrolysate	PHB	Batch	10.6	39.6	4.2	n.a	[110]

MOLASSES

Molasses is the main by-product of sugar industry. The term molasses is applied to the final effluent obtained in the preparation of sugar by repeated crystallization [129]. The major sources for sugar molasses are sugar beet and sugar cane that contains high proportion of sucrose [130]. The quality and amount of obtained molasses depends on local conditions of growth and the effects of weather. Besides the natural factors, the processing of crops in the sugar factory such as the efficiency of the juice clarification, the method of crystallization during boiling and the separation of the sugar crystals from the low-grade massecuite is also highly responsible for obtained molasses quality [129]. Table **4** provides a compilation of typical composition of beet sugar molasses and cane sugar molasses.

Table 4. Typical composition of beet sugar molasses and cane sugar molasses [117].

Constituent		Beet Molasses [%]		Cane Molasses [%]	
Water		16.5		20.0	
Organic Constituents					
Sugars:	Sucrose	51		32	
	Glucose	-		14	
	Fructose	-	53	16	62
	Invert sugar	1		-	
	Raffinose	1		-	
Non-sugars: Nitrogenous material, free and bound acids, soluble gummy substances		19		10	
Inorganic Constituents (ash)					
SiO_2		0.1		0.5	
K_2O		3.9		3.5	
CaO		0.26		1.5	
MgO		0.16		0.1	
P_2O_5		0.06		0.2	
Na_2O		1.3	11.5	-	8.0
Fe_2O_3		0.02		0.2	
Al_2O_3		0.07		-	
Carbonate residue		3.5		-	
Sulfate residue		0.55		1.6	
Chlorides		1.6		0.4	

Sugar molasses cannot be further used in foods and is, thus, ideal for the consideration as an inexpensive carbon source for fermentative processes [130]. In white sugar factories the yield of molasses is 4% on beets, corresponding to up to 25% on sugar. The average sugar content in the sugar beets is 16-18% and only 13-14% of sugar can be recovered as a commercial product. As an average, 2.2-2.6% of sugar on beets is transferred into molasses when raw sugar is produced. The yield of molasses is affected by various factors and differs from batch to batch [129]. In 2004, the global sugar production generated about 39 million tons of cane molasses and 12 million tons of beet molasses [131].

The objective of the sugar industry is to produce molasses the purity of which is as low as possible. Commercial molasses has ordinarily a quotient of around 60, *i.e.*, approximately 48% sugar is present in molasses the solids content of which is 80% [129].

There are a few ways of using of molasses. The most promising way of utilization of molasses is its employment as a carbon source in the biotechnological production. It is technically and economically highly important [129].

Molasses from sugar cane and sugar beets is biotechnologically utilized for more than half a century. Conversely, soy molasses in recent years meets increasing use in biotechnology ever. Soy molasses, a by-product from soybean processing to soy proteins, is another co-product stream that could benefit from new uses to alleviate the disposal problems. It contains high content (30%) of soluble carbohydrates, especially sucrose, stachyose, and raffinose [130].

Beet molasses is moderately alkaline with the pH-value of about 8.0 and has a characteristic odor. Cane molasses is acid and has more acidulous or fruity-aromatic odor. Highly viscous molasses is not a homogeneous liquid or true solution of sugar in a non-sugar solution; molasses always contains suspended components of varying composition and in varying amounts [129].

PHA Production on Molasses

The PHA production from cane molasses was studied by Gomaa who studied the PHB production employing *Bacillus subtilis* and *Escherichia coli* isolated from industrial contamination. The amount of accumulated PHA followed a similar pattern to its growth for each treatment indicating a growth-related production, yielding the maximum PHA production of 54.1 and 47.16% (5.3 and 2.8 g/L) for *B. subtilis* and *E. coli*, respectively after 96 hrs of cultivation in the medium containing 6 and 8% of molasses, respectively and decreased thereafter. The growth and PHA yields were improved by introducing 1% of ethanol into the molasses medium. Ammonium sulfate and ammonium nitrate were used as nitrogen sources at the concentration of 1 g/L. The optimum environmental conditions which influenced the PHA production by the two strains were the inoculum concentration of 8%, the pH-value of 7.0 and the temperature of 35°C [132].

Another bacterium which is capable of utilization of sugarcane molasses is *Bacillus flexus*. El-Sheekh *et al.* [133] employed *Bacillus flexus* ME-77. The highest PHB production (4.5 g/L) was achieved using 30 g/L of sugarcane molasses (SCM) at the pH-value of 7.5 and applying 1, 0.2 and 0.03 g/L of phosphate, $MgSO_4$ $7H_2O$ and $CaCl_2$, respectively, at 27°C incubation temperature and 200 rpm agitation speed. The optimization of production method led to the highest PHB yield (4.2 g/L) and CDM 5.4 g/L.

Page [134] reported the production of copolymer PHBHV at 19-22 g/L by *Azotobacter vinelandii* UWD grown on beet molasses. The hydroxyvalerate content in the polymer was increased by the addition of 10 to 40 mM valerate as a supplement.

Cane molasses was found to be an excellent substrate for the growth of *Bacillus megaterium* for subsequent PHA production by Kulpreecha *et al.* [135]. The total biomass and the PHA content were obtained 72.7 g/L and 42%, respectively. Albuquerque *et al.* [136] demonstrated the production of 61% of PHA in CDM using molasses as a complex feedstock in the 2-stage CSTR cultivation. Omar *et al.* [137] reported the production of PHB at 3 g/L and the PHB content of 50% of CDM by *Bacillus megaterium* grown on beet molasses. The cultivation of *B. megaterium* at 30 l scale gave a lower yield - the PHB content of 25% of cells and the cell density of 3.4 g L^{-1} after 14 hrs of growth.

Sugarcane molasses and corn steep liquor as a cheap nitrogen source was chosen for the PHB production by *Bacillus megaterium* in order to reduce the cost of PHB production by Gouda *et al.* [138]. The best growth was obtained with 3% molasses, while the maximum yield of PHB (46.2% of CDM) was obtained with 2% molasses.

The PHAs accumulation was also reported in the media derived from soy molasses with high sucrose content. Solaiman *et al.* [139] tested *Pseudomonas corrugata* on inexpensive soy molasses (5%) as a carbon substrate for the production of *mcl*-PHA and achieved the cell dry mass concentration of 3.4 g/L with the PHB content of 5-17% in the biomass. The most prominent repeat-unit monomers in PHAs were 3-hydroxydodecanoate, 3-hydroxyoctanoate, 3-hydroxydodecanoate, and 3-hydroxytetradecenoate. The conversion of these unusual saccharides in soy molasses into PHA seems to be possible by *Bacillus* sp. CL1. PHAs accumulated up to 90% of CDM in *Bacillus* sp. CL1 [140]. Law *et al.* [141] also approved that two isolated *Bacillus* strains, HF-1 and HF-2, synthesized PHB from hydrolyzed soy and malt waste.

Also a lot of efforts were targeted on genetic modification of particular bacterial strains in order to reach highly efficient production of PHA from molasses. Liu *et al.* [142] successfully replaced glucose as sole carbon source by beet molasses. The recombinant *Escherichia coli* strain (HMS174/pTZ18u-PHB) was employed for the PHB production. The final cell dry mass, the PHB content and the PHB productivity were 39.5 g/L, 80 wt.-% and 1 g/Lh, respectively, in the 5 l stirred tank fermenter after 31.5 hrs of fed-batch fermentation with constant pH-value and dissolved O_2 (pO_2) content. Zhang *et al.* [143] employed the recombinant *Escherichia coli, Klebsiella oxytoca* and *Klebsiella aerogenes*. The authors describe the accumulation of about 3 g/L of PHB after 37 hrs of cultivation in shaking flask. The concentration of PHB in cell was approximately 50%.

Table **5** provides a summary of literature reports on molasses-based PHB production processes both on shaking flask and bioreactor scale.

Table 5. Summary of PHB production on molasses published in literature.

Production Strain	Molasses	Fermentation Mode	Biomass [g/L]	PHA [wt.-%]	PHA [g/L]	$Y_{p/s}$ [g/g]	Ref.
Bacillus flexus ME-77	Sugarcane	Flasks, Batch	5.4	77,7	4.2	*n.a.*	[133]
Bacillus sp. 112A	Sugarcane	Flasks, Batch	*n.a.*	*n.a.*	1.9	*n.a.*	[144]
recombinant *E. coli* strain	Beet	Fermentor, Fed- batch	39.5	80.0	31.6	*n.a.*	[142]
Bacillus subtilis	Sugarcane	Flasks, Batch	11.0	54.1	5.3	*n.a.*	[132]
Escherichia coli	Sugarcane	Flasks, Batch	7.6	47.2	2.8	*n.a.*	[132]
Bacillus megaterium	Sugarcane	Flasks, Batch	3.6	59.4	2.2	*n.a.*	[138]
Pseudomonas corrugata	Soy	Flasks, Batch	1.5-3.6	5-17	*n.a.*	*n.a.*	[139]
Ralstonia eutropha	Sugarcane	Flasks, Batch	*n.a.*	39	*n.a.*	*n.a.*	[145]

(Table 5) contd.....

Production Strain	Molasses	Fermentation Mode	Biomass [g/L]	PHA [wt.-%]	PHA [g/L]	$Y_{p/s}$ [g/g]	Ref.
Bacillus megaterium	Sugarcane	Flasks, Batch	8.78	61.6	*n.a.*	0.39	[135]
Mixed microbial cultures	Sugarcane	Flasks, Batch	72.7	42.0	*n.a.*	*n.a.*	[136]
Bacillus megaterium	Beet	Flasks, Batch	3.7	50.0	*n.a.*	*n.a.*	[137]

CHEESE WHEY

Cheese whey represents an important side product of cheese manufacturing. Whey contains most of water-soluble components (lactose, whey proteins, milk salts, *etc.*) of transformed milk and, of course, water. The precise whey composition is dependent on the method of cheese or caseins manufacturing and it also varies during the season of milk production. Generally, whey is often classified as "sweet" if derived from the precipitation of caseins by rennet, or as "acid", when casein coagulation was carried out by the acidification of milk either by partial fermentation of lactose to lactic acid or by the addition of mineral acids (HCl, H_2SO_4, *etc.*) [146]. In general, whey produced from rennet-coagulated cheeses is low in acidity (pH about 6.5), while the production of acid cheeses (such as ricotta or cottage cheese) or caseins yields acidic whey (a.k.a. fermented whey) with the pH-values lower than 5. Cheese-whey salts comprise NaCl and KCl (more than 50% of total salts), calcium salts (primarily phosphate) and others. Cheese whey also contains appreciable quantities of other components, such as lactic (0.05% [wt./vol.]) and citric acids, non-protein nitrogen compounds (urea and uric acid), some vitamins belonging to B group, free amino acids, *etc;* [147]. Table **6** summarizes the typical composition of sweet and acidic whey.

Whey proteins represent about 20% of milk proteins. The most abundant whey proteins are β-lactoglobulin (50%), α-lactoalbumin (12%), immunoglobulins (10%), serum albumin (5%) and proteose peptones (0.23%). The Protein Efficiency Ratio value and the essential amino acid content of whey proteins are higher than those of caseins. Moreover; their biological value exceeds even that of whole egg proteins [148].

The main differences between the two types of whey are the mineral content, the pH-value and the composition of whey protein fraction. The acid coagulation approach results in the decrease of pH to the value of about 4.5, which is necessary for the casein precipitation. At this low pH-value, calcium present in casein micelles is partially solubilized and released into the solution. This causes higher calcium content in acidic whey.

Table 6. Typical composition of sweet and acid cheese whey [149].

Component	Sweet Whey [g/L]	Acidic (Fermented) Whey [g/L]
Total Solids	63-70	63-70
Lactose	46-52	44-46
Protein	6-10	6-8
Calcium	0.4-0.6	1.2-1.6
Phosphate	1.3	2.4
Lactate	2	6.4
Chloride	1.1	1.1

On the other hand, the precipitation of casein *via* the action of rennet results in the release of specific fragment of κ-casein molecule. This fragment is termed glycomacropeptide (GMP) and due to its solubility it ends up in whey. GMP is an important protein of sweet whey, because it constitutes approximately 20% of its protein content, however, it is not present in acidic whey [150].

Dairy industry all over the world generates huge amount of whey per litter of processed milk. About 50% of total world cheese whey production is treated and transformed into various food products (of which 45% is used directly in liquid form, 30% in the form of powdered cheese whey, 15% as lactose and byproducts from its removal, and the rest as cheese-whey-protein concentrates) the remaining 50% is disposed as waste [151]. This potentially dangerous waste can be used for the production of various biotechnological products by microbial fermentation. Generally, the processes based on microbial cultures on cheese whey are considered as the most profitable alternatives for the transformation of cheese whey surplus [147]. However, there are many potential facts and risks which should be taken into account prior to the utilization of cheese whey as a fermentation medium.

Whey is foremost a diluted solution and many difficulties encountered in whey handling arise from this fact. First of all, it complicates the transportation of whey and, secondly, cheese whey tends to spoil by the action of naturally occurring microorganisms which significantly devaluates its potential application as the fermentation medium. Therefore, it is worthy to decrease the concentration of water resulting in the preparation of cheese whey concentrate. During the procedure of pre-concentration, the cheese whey is, at least, partially sterilized. Moreover, due to high osmotic pressure, this form of cheese whey is much more microbially stable and, in addition, because of the decrease of whey weight it also simplifies the process of whey transportation.

Since most biotechnological productions employ the microbial monoculture, it is necessary to overcome any possible contaminations of fermentation medium by undesired microorganisms. Hence, the medium must be sterilized prior to the cultivation. The sterilization of whole cheese whey is problematic, because heating of whey to as high temperature as 121°C results in the precipitation of whey proteins [152]. The protein precipitate seriously complicates the fermentation as well as the recovery of the biomass or the down-stream processing of fermentations products. Therefore, the removal of excessive proteins prior to the sterilization is usually needed. This can be done by decreasing the pH-value by mineral acids and boiling [153]. On the other hand, the whey proteins have high nutritive value which would be reduced by any improper treatment. It is reasonable to remove the proteins by more gentle method which allows their subsequent application in human diet or in animal feeding. The most commonly used methods nowadays are the ultrafiltration and the diafiltration because of their advantages of cost reduction, high processes speed, absence of denaturation or protein structure modification and the fact that the protein concentrate is free of salts, thereby making it suitable for all kinds of human foods [154]. The process of protein removal ends up with two products. The first is called the whey protein concentrate and it contains almost exclusively the protein fraction of whole whey. The second products, termed the whey permeate, contains lactose and all other soluble substances. The protein recovery can be highly profitable and reduces the effluent load of whey to about 10 000 ppm of COD. However; the whey permeate still contains about 50 000 - 60 000 ppm of COD, mostly represented by lactose [146].

Another option how to overcome the precipitation of cheese whey proteins during the sterilization is their hydrolysis *in situ* by proteolytic enzymes or microbes [155]. Whey protein hydrolysis yields the peptide mixture with increased solubility and altered foaming characteristics. In addition, peptides can be utilized by microbes during the fermentation as carbon, nitrogen and essential amino acids source and, thus, they could represent very important factor supporting the growth of microorganisms and the fermentation process [156, 157].

Some microbial strains suitable for the production of various substances do not show the β-galactosidase activity, in other words, these microbes are not capable of utilizing lactose directly. On the other hand, they can utilize glucose and/or galactose and, hence, they can be used for the fermentation using whey with pre-hydrolyzed lactose. Lactose can be cleaved by acids or by simple application of β-galactosidase into the cheese whey prior or after the sterilization [158, 159] or by using immobilized enzyme or whole cell system [160]. The question is whether the process of biotechnological production remains profitable even after the inclusion of relatively expensive process of whey lactose hydrolysis.

Whole whey or whey permeate are usually not sufficient medium for optimal microbial growth and for the production of desired metabolites. Hence, the whey medium is supplemented with various substances serving as additional nitrogen source and with specific growth factors required by particular microorganism. Generally, the yeast extract and the yeast autolysate are used as complex nitrogen sources [153, 161, 162] or nitrogen can be also added in the form of mineral salts such as $(NH_4)_2SO_4$, NH_4NO_3, NH_4Cl, *etc.* [153, 163, 164], depending on the preferences of particular microbe. The introduction of inorganic phosphate was also found to be beneficial for some biotechnological productions using whey as a substrate [165 - 167].

Besides mentioned supplements, the addition of other nutrients (molasses, corn steep molasses), vitamins, minerals, amino acids and inorganic supplements to whey or whey permeate were also investigated [153, 165 - 168]. Of course, it is crucial to take into account the demands of microbes to be employed.

PHA Production by Naturally Producing Bacteria

The PHA production from cheese whey by naturally PHA producing and lactose utilizing strain was studied by Yellore and Desai [169] who studied the PHB production employing *Methylobacterium sp.* ZP24, isolated from a local pond. The isolate was able to utilize deproteinized whey and produce 1.1 g/L of polymer. The introduction of $(NH_4)_2SO_4$ enhanced the PHB yields 2.5-fold, so the bacterium produced 2.6 g/L of PHB which represented 44% of CDM. Further, the production process was scaled-up by Nath *et al.* [164]. Since *Methylobacterium sp.* ZP24 showed the growth associated PHB production, an intermittent feeding strategy having lactose and ammonium sulfate at varying concentration was used in order to reach higher yield of the polymer. About 1.5-fold increase in the PHB production was obtained by this feeding strategy. Further increase in the PHB production by 0.8-fold could be achieved by limiting the dissolved oxygen levels in the fermentor. Decreased oxygen level is thought to increase the flux of acetyl-CoA towards the PHB accumulation over the TCA cycle. The maximum yields reached 3.91 g/L when *Methylobacterium sp.* ZP24 was able to convert 1 g of lactose into 0.32 g of PHB.

Another bacterium which is capable of direct transformation of lactose into PHB is *Bacillus megaterium*. Obruca *et al.* [153] employed *B. megaterium* CCM 2037. The optimization of medium composition in terms of whey dilution, Mg^{2+}, $(NH_4)_2SO_4$ and PO_4^{2-} supplementation improved the PHB yields about 50-fold (the biomass and the PHB yields 2.82 and 1.05 g/L, respectively) as compared to whey itself. Furthermore, the PHB yields were improved by about 40% by introducing 1% ethanol into the medium at the beginning of the stationary phase of growth (the biomass 2.87 g/l, the PHB yields 1.48 g/L). Pandian *et al.* [170] converted whey into PHB using *Bacillus megaterium* isolated from brackish water. The medium was supplemented with rice bran and see water and also the fed-batch fermentation strategy was developed so the maximum production of PHB (11.32 g/L) occurred at the 36th hour of cultivation.

Azotobacter vinelandii was used for the PHB production on whey using bacterial peptone as a nitrogen source. It was observed that by increasing the inoculum size the time of incubation can be reduced considerably [171]. Further, *Pseudomonas hydrogenovora* produced PHA on enzymatically cleaved lactose in whey permeate. The application of hydrolyzed whey permeate turned out to be advantageous as compared to the utilization of pure sugars. Sodium valerate was introduced into the fermentation media to incorporate 3HV units into the PHA structure and to form PHBHV copolymer. The process yielded 1.44 g/L of copolymer containing 21% of 3HV [158]. Pantazaki *et al.* [172] focused on the *mcl*-PHA production from cheese whey employing *Thermus thermophilus* HB8 under the nitrogen limitation. PHA was accumulated in up to 35% of CDM after 24 hrs of cultivation. The heteropolymer consisted of short chain length 3-hydroxyvalerate and medium chain length 3-hydroxyheptanoate, 3-hydroxynanoate and 3-hydroxyundecanoate. Further, Povolo *et al.* [173] reported that *Hydrogenophaga pseudoflava* is capable of accumulation of terpolymer consisting of 3-hydroxybutyrate, 3-hydroxyvalerate and 4-hydroxybutyrate from structurally unrelated carbon sources such as lactose and cheese whey. The terpolymer reveals superior mechanical properties as compared to PHB or PHBHV. On the other hand, the PHA content was rather low (in the range of 3-10% of CDM).

Very interesting fermentation strategy of the PHA production from cheese whey was suggested by Koller *et al.* [174]. The authors of the study developed a viable strategy to cope with potential contamination during the PHA production from cheese whey employing *Hydrogenophaga psedoflava*. Using whey as a substrate the risk of microbial contaminations rises as whey itself often (over 75% of the whey samples) contains infectants; *e.g.*, the spore forming, gram positive bacterium *Bacillus cereus*. During the PHA production, the growth of *B. cereus* occurs at the expense of the production strain, resulting in the loss of the fermentation batch. The thermal processing, which is frequently used to prevent the microbial contamination is expensive, time- and energy-consuming and does not necessarily have the desired effect on thermal resistant spore forming bacteria. The authors observed that the application of vancomycin inhibited the growth of contaminant *B. cereus* but not the PHA producing strain. Moreover, the specific PHA production rates of *H. pseudoflava* decreased with increasing antibiotics in a minimal medium but in the media supplemented with yeast extract the specific PHB production rates increased as the antibiotics concentrations increased.

Koller *et al.* [175] also investigated the extremely halophile archaeon *Haloferax mediterranei* for the production of two different high-performance polyhydroxyalkanoates (PHAs). A copolyester containing 3-hydroxyvalerate was produced from whey sugars as sole carbon source. The maximum specific growth rate (m_{max}) and the maximum specific PHA production rate ($q_{p\ max}$) were determined to be 0.10 L/h and 0.15 L/h, respectively. The concentration of PHBHV in cells reached 72.8%. Unfortunately, after the fermentative production of PHA-rich biomass and the subsequent cell harvest and downstream processing for PHA recovery, environmentally hazardous, highly saline residues, namely spent fermentation broth and cell debris, remain as the residues [176]. Nevertheless, these waste streams were used for recycling experiments to assess their recyclability in subsequent production processes. It was demonstrated that spent fermentation broth can be used to replace a considerable part of fresh saline fermentation medium in subsequent production processes. In addition, 29% of the expensive yeast extract, needed as nitrogen and phosphate source for efficient cultivation of the microorganism, can be replaced by cell debris from prior cultivations.

Furthermore, cheese whey might also serve as a very promising nitrogen source for the PHA producing bacteria. Protease hydrolyzed whole cheese whey applied as a complex nitrogen source improved the PHA production by *Cupriavidus necator* on waste frying oils by about 40%, the product yield coefficient reached 0.94 g of PHA per g of oil and the final PHA content in cells exceeded 90% of CDM. Positive effects of whole whey hydrolysate on the PHA production were attributed to the amino acids composition. Whole whey hydrolysate was rich in proline, leucine, isoleucine and methionine, which were observed to promote the PHA accumulation in bacterial cells. Therefore, the hydrolysate of cheese whey seems to be superior to more expensive complex nitrogen sources such as soya peptone, yeast extract or casein hydrolysate [57].

PHA Production by Genetically Engineered Bacteria

Apart from the microbes mentioned in the previous text, there is a great number of bacterial strains which are either not able to utilize lactose or they do not accumulate PHAs. Therefore, a lot of efforts were targeted on genetic modification of particular bacterial strain in order to reach highly efficient production of PHAs from cheese whey.

Cupriavidus necator is a well-known PHA producer, but unable to grow on lactose. Thus, Povolo at al [177]. choose one of intracellular PHB depolymerases, the enzyme degrading intracellular stocks of PHA, to insert *lac* genes of *E. coli*. This would have the effect of allowing the polymer production on lactose and, in the same time, of removing a part of the PHA intracellular degradation system. The disruption of PHA depolymerase was achieved by a gene replacement after isolating a fragment of this gene and interrupting it with a cartridge containing the *lac* genes and a synthetic promoter. The growth and polymer production studies of genetically modified strain mRePTin whey permeate and hydrolyzed whey permeate as carbon sources, were performed. Lower PHA degradation and higher yields were obtained as compared to the wild-type strain.

Finally, many researchers reported very efficient PHA production using *Escherichia coli* harboring the PHA biosynthetic genes of various natural PHA producers. The fermentation strategies for the production of PHB from whey by recombinant *Escherichia coli* strain CGSC 4401 harboring the *Alcaligenes latus* PHA biosynthesis genes were developed by Ahn *et al.* [178]. The pH-stat fed-batch cultures were carried out with concentrated whey solution containing 280 g of lactose equivalent per liter. The cell and PHB concentrations of 119.5 and 96.2 g/L, respectively, were obtained at the 37.5th hour of cultivation. In the second study, Ahn *et al.* [179] even improved the fermentation strategy. The same *E. coli* strain was employed for the PHB production using the cell recycle membrane system. The final cell concentration, the PHB concentration and the PHB content of 194 g/L, 168 g/L and 87%, respectively, were obtained at the 36.5th hr by the pH-stat cell recycle fed-batch culture using the whey solution concentrated to contain 280 g of lactose per liter as a feeding solution.

Different recombinant *E. coli* strains were used for the PHB accumulation in whey-based broth. The timing of PHB biosynthesis in recombinant *E. coli* was controlled using the agitation speed of stirred tank fermentor. The PHB content of 80% could be obtained with oxygen limitation by decreasing the agitation speed to only 500 rpm. The final yields reached were 31 g/L of biomass containing 80%

of PHB [180]. Nikel *et al.* [167] aimed at statistical optimization of culture medium using powdered cheese whey by recombinant *E. coli* strain harboring the biosynthetic geneses from a natural isolate *Azotobacter sp.* The study demonstrated that the optimal concentrations of the three components, defined as those yielding the maximal biomass and PHB production in shaken flasks, were 37.96 g/L of deproteinated milk whey powder, 29.39 g/L of corn steep liquor and 23.76 g/L of phosphates. The cultivation in the optimized medium yielded 9.41 g/L of biomass and 6.12 g/L of PHB in the culture broth.

In another study Nikel *et al.* [181] employed the same recombinant strain for the PHB production fed-batch mode using concentrated and deproteinated whey solution containing 25% of lactose as a feeding solution.

Table 7 provides a summary of PHA production processes from cheese whey reported in literature. Despite the legislative restrictions arising from the application of genetically modified strains, it seems that recombinant *E. coli* is the most promising candidate for industrial production of PHA using whey. Employing these recombinant strains, it is possible to achieve very high cell densities in relatively short time. Because the PHA production is, in these cases, growth associated, the PHA contents in cells reach up to 87% of CDM.

That is the most of all the bacterial strains tested so far. This is very important factor influencing the economics of PHA production, because the PHA content of produced biomass strongly affects the efficiency and the price of the down-stream process.

For example, Lee and Choi [182] reported that relatively low PHB content in cells - 50% results in high recovery cost of 4.8 US-$/kg of PHB. On the other hand, the recovery cost for a process with 88% of PHB cell content was only 0.92 US-$/kg of PHB. Lower PHB content clearly resulted in high recovery cost. This is mainly due to the requirement of large amounts of digesting agents for breaking the cell walls and to increased cost of waste disposal.

Table 7. Summary of PHB productions on cheese whey published in literature.

Production Strain	Substrate	Fermentation Mode	Biomass [g/L]	PHA [wt.-%]	PHA [g/L]	$Y_{p/s}$[1] [g/g]	Ref.
Methylobacterium sp. ZP24	Deproteinized whey $(NH_4)_2SO_4$	Erlenmeyer flask, Batch	5.9	44	2.6	*n.a.*[2]	[169]
Methylobacterium sp. ZP24	Deproteinized whey $(NH_4)_2SO_4$	Fermentor, Fed-batch	*n.a.*	*n.a.*	3.9	0.32	[164]
Hydrogenophaga psedoflava	Cheese whey	Erlenmeyer flask	1.46	10.1	0.15	*n.a.*	[173]
Bacillus megaterium CCM 2037	Deproteinized whey $(NH_4)_2SO_4$, phosphate	Erlenmeyer flask, Batch	2.87	52	1.5	0.08	[153]
Bacillus megaterium SRKP-3	Whey Rice Bran, Sea Water	Fermentor, Fed-batch	*n.a.*	*n.a.*	11.3	*n.a.*	[170]
Azotobacter vinelandii MTCC 124	Whey Bacterial peptone	Erlenmeyer flask, Batch	*n.a.*	*n.a.*	1.4	*n.a.*	[171]
Pseudomonas hydrogenovora	Hydrolyzed whey permeate	Fermentor, Fed-batch	*n.a.*	12	1.4	*n.a.*	[175]
Thermus thermophilus HB8	Deproteinized whey	Erlenmeyer flask, Batch	1.6	35	0.5	*n.a.*	[172]
Cupriavidus necator (*E. coli* lac genes)	Whey	Erlenmeyer flask, Batch	8.8	23	2.0	*n.a.*	[177]
Escherichia coli (*A. latus* PHA biosynthetic genes)	Concentrated whey	Fermentor, Fed-batch	119.5	81	96.2	*n.a.*	[178]
Escherichia coli (*A. latus* PHA biosynthetic genes)	Concentrated whey	Fermentor, Fed-batch, Cell recycle system	194	87	168.0	*n.a.*	[179]
Escherichia coli (*C. necator*PHA biosynthetic genes)	Whey	Fermentor, Fed-batch, Oxygen limitation	31	80	25.0	*n.a.*	[180]
Escherichia coli (*Azotobacter* sp. biosynthetic genes)	Deproteinized whey powder Corn steep liquor	Erlenmeyer flask, Batch	9.4	65	6.1	*n.a.*	[167]
Escherichia coli (*Azotobacter* sp. biosynthetic genes)	Deproteinized whey concentrate	Fermentor, Fed-batch	70	73	51.1	*n.a.*	[181]

Notes: [1] Yield in terms of g of PHA per g of substrate; [2]*n.a.* stands for: data not available.

GLYCEROL AS A CHEAP SUBSTRATE FOR PHA PRODUCTION

The increasing price of natural resources year by year and rapid depletion of petroleum reserves have motivated the government, industry and community to consider for an economical replacement made from renewable sources [183]. The increase in global biodiesel production makes the development of sustainable processes for the utilization of its main by-product, crude glycerol, imperative. Biodiesel is produced through the transesterification of triglycerides with a monovalent alcohol (methanol, ethanol, *etc.*), generally catalysed by NaOH and KOH or acid, to fatty acid alkyl esters [184, 185]. The production of biodiesel provides relatively large amount of co-products and waste. Besides glycerol as the main by-product, it is gluten meal, gluten feed, oilcake, or wastewater [186].

Glycerol is an organic compound and it is synonymous to glycerine, propan-1,2,--triol, 1,2,3-propanetriol, 1,2,3-trihydroxypropane, glycerol, and glycol alcohol. Glycerol is virtually nontoxic to both human and environment [187]. The global production of glycerol from biodiesel production has climbed from 200,000 tonnes in 2003 to more than 2 million tonnes in 2013. Anyway, just 12% of worldwide production is made synthetically; the rest is obtained as a side-product from the biodiesel production. The industrial application of crude glycerol in food, pharmaceutical and cosmetics industries, its main markets, requires a costly refining process in order to achieve a necessary high purity. In addition to the applications in food industry and polymer industry, glycerol has been considered as the feedstock for new industrial fermentations. Especially attractive is the microbial conversion of raw glycerol into 1,3-propanediol, H_2, ethanol or citric acid. Equally interesting could be the conversion of glycerol into polyhydroxyalkanoates [180].

Glycerol as Waste from Biodiesel Production

As was mentioned at the beginning, glycerol can be generated from the transesterification of fat and oils in biodiesel plant. The transesterification is a chemical reaction in which fat and triglycerides react with alcohol such as methanol in the presence of catalyst to produce fatty acid methyl esters with glycerol as a by-product. As the transesterification reaction progresses, the reaction stream is separated into two phases, which are biodiesel-rich phase (top layer) and glycerol-rich phase (bottom phase). Glycerol produced from oleochemical or biodiesel plant is in crude form that contains various impurities depending on the process and the type of materials processed [188].

The production of 100 kg of biodiesel yields approximately 10 kg of impure (crude) glycerol, with the purity of 50-55%. Approximately 12-16% is represented by alkalis, especially in the form of alkali soaps and hydroxides, 15 to 18% are methyl esters, next there are 8 to 12% of methanol, and 3% of water. In addition to methanol and soaps, crude glycerol also contains a variety of elements such as Ca, Mg, P, or S [189, 190]. To produce biodiesel, a feedstock that contains fat or oils can be used in the transesterification process. In general, the biodiesel feedstock can be categorized into three groups: i) the first generation contains *e.g.*, palm oil, rapeseed, soybean oil, sunflower oil or peanut oil, ii) the second generation is represented by jatropha oil, sea mango, tobacco seed oil, salmon oil, waste cooking oils or tallow iii) microalgae are the third generation of biodiesel feedstock [187].

Crude glycerol is produced immediately from the reaction and the impurities such as used catalyst and the excess of alcohol can be found in the glycerol phase. Crude glycerol has high salt and free fatty acid content as well as substantial colour. Further, pure glycerol has numerous industrial uses and is an important industrial feedstock in food, drug and cosmetics industries. On the other hand, the applications of crude glycerol are highly limited due to the presence of salt and impurities and its fuel value is also marginal [187, 190].

Production of PHA on Waste Glycerol Based Substrate

In recent years, the research has focused on the development of alternative PHAs production processes, including the use of waste/surplus based feedstock. This approach permits for a lower investment and operating costs for the global process.

However, rapid growth of biodiesel industry has created a surplus glycerol that has resulted in decreasing prices. Many microorganisms are able to metabolize glycerol, therefore many publications are focused on the possibility of using crude glycerol as potential carbon source for the production of polyhydroxylakanoates. Moreover, not only PHAs can be synthetized by the utilization of glycerol, but also organic acids which can serve as precursors for the production of PHAs copolymers with better mechanical properties, *e.g.*, PHBHV. Bories *et al.* reported the production of propionic acid by *Propionibacterium* utilizing glycerol as carbon source [191].

Nowadays, many researches are focused on the usage of glycerol, as waste product from the biodiesel production, for the conversion to valuable product. One of possible applications of glycerol is the conversion to polyhydroxyalkanoates. There are several studies outlining the options of PHA production (Table **8**).

Table 8. Production of PHA on waste glycerol based substrate.

Production Strain	Carbon Source	PHA Type	Cultivation	Biomass [g/L]	PHA [wt.-%]	PHA [g/L]	Ref.
Mixed microbial cultures	crude glycerol	PHB	Batch	n.a	47.0	n.a	[188]
Cupriavidus necator	glycerol-rich - phase	P(3HB-*co*-4HB)	Fed-batch	n.a	36.1	10.6	[192]
Pseudomonas mediterranean	cruise glycerol	*mcl*-PHA	Batch fermentor	3.4	n.a.	1.7	[198]
pseudomonas corugata	crude glycerol	*mcl*-PHA		4.2	n.a.	1.2	
Halomonas sp.	glycerol	PHA	Batch	4.5	56.5	2.5	[197]
Pseudomonas putida	glycerin	PHA	Fed-batch	30.0	31.0	n.a	[199]
Pseudomonas mediterranea	partially refined glycerol	*mcl*-PHA	Batch	3.3	21.0	n.a	[200]
	reagent grade glycerol	*mcl*-PHA		3.4	23.0		
Haloferax mediterranei	crude glycerol	P(3HB-*co*-HV)	Batch	n.a	76.0	16.2	[196]
Cupriavidus necator	crude glycerol	PHB	Fed-batch fermentor	37.0	72.9	27.0	[193]
		PHBHV	Fed-batch	35.2	66.4	23.4	
Bulkholderia cepacia	crude glycerol	PHB	Batch fermentor	23.6	31.4	7.4	[195]
Pseudomonas oleovorans	crude glycerol	PHB	Batch	3.0	38.0	1.1	[201]

(Table 8) contd.....

Production Strain	Carbon Source	PHA Type	Cultivation	Biomass [g/L]	PHA [wt.-%]	PHA [g/L]	Ref.
Pseudomonas putida	crude glycerol	*mcl*-PHA	Fed-batch fermentor	19.1	35.6	6.2	[202]
Cupriavidus necator	crude glycerol	PHBHV	Fed-batch fermentor	19.6	55.6	10.9	[194]

The resulting composition of PHAs is influenced by selected bacterial strain. There are several studies reporting the production of PHB, PHBHV, respectively. The most frequent bacterial strains which are found to be able to utilize glycerol, as reported in presented studies, were *C. necator*, *B. cepacia*, or archeon *Haloferax mediterranei*.

Cavallieiro *et al.* reported the utilization of untreated glycerol by *C. necator* producing P(3HB-*co*-4HB) and its copolymer with 3-hydroxyvalerate in the presence of γ-butyrolactone as a precursor [192]. The experiments were carried out using waste glycerol (pH 6.1, containing 1174.5 g of glycerol per litre). This liquid phase rich in glycerol was generated in a pilot plant producing biodiesel from refined vegetable oils with sodium methoxide as catalyst. The final PHA concentration and the PHA content were 10.9 g/L and 36.1%, respectively. The content of 4HB in the presence of γ-butyrolactone was 17.1%.

Kachrimanidou *et al.* presented the study about the production of polyhydroxybutyrate (PHB) and poly(3HB-*co*-3HV) using the strain *Cupriavidus necator* cultivated on crude glycerol, sunflower meal (SFM) hydrolysates and levulinic acid as the sole fermentation feedstocks [193]. The fed-batch bioreactor fermentations led to the production of 27 g/L of PHB with an intracellular content of 72.9%. Continuous feeding of levulinic acid led to the production of up to 23.4 g/L of PHBHV with an intracellular content of 66.4% and 3HV content of 22.5 mol-%.

The potential of rapeseed hydrolysates in the mixture with glycerol as nutrient supplement for the production of PHBHV by *C. necator* was presented by Garcia *et al.* [194]. The highest PHA concentration (10.9 g/L) and content (55.6%) were achieved in the fed-batch fermentation. For the purposes of fed-batch cultivation, the concentrations of glycerol of 21 g/L and free amino nitrogen of 0.4 g/L were used. The rapeseed meal hydrolysate was added at 48 hr, while crude glycerol was added at 58 and 72.5 hr. The 3HV content was 9.8 mol-% and 3.3 mol-% at 10 and 58 hr, respectively. The 3HV content remained constant after 58 hrs of fed-batch fermentation.

Zhu *et al.* proved the ability of *Burkholderia cepacia* to synthesize PHB from glycerol present in the concentrations ranging from 3% to 9% [195]. Increasing of the glycerol concentration resulted in a gradual reduction of biomass, PHA yield, and molecular mass of PHB. At 24 hr or 48 hr, 2% or 5% glycerol was added to the medium, and the cultures were harvested at 72 h. When biodiesel-glycerol was used as a sole carbon source in shake flasks, *B. cepacia* produced 5.8 g/L of dry biomass and contained up to 81.9% of PHB of total dry biomass when using 3% biodiesel-glycerol as a feedstock over 96 hrs of growth. However, increasing of the glycerol concentration from 3% to 9% resulted in a decrease of microbial biomass of *B. cepacia* of almost 50%.

Archaeon *Haloferax mediterranei* was selected for the production of PHA co- and terpolyesters using inexpensive crude glycerol phase from the biodiesel production as a carbon source by Hermann-Krauss *et al.* [196]. Crude glycerol was assessed by the comparison with the application of pure glycerol. Applying pure glycerol in the concentration of 10 g/L, a copolyester with the molar fraction of 3-hydroxybutyrate of 0.90 mol/mol and 3-hydroxyvalerate of 0.10 mol/mol, was produced, an intracellular PHA content was 75.4% in the biomass. Supplying γ-butyrolactone as 4-hydroxybutyrate precursor resulted in a poly(3HB-*co*-3HV-*co*-4HB) terpolyester containing 3HV (0.12 mol/mol) and 4HB (0.05 mol/mol) in the poly[(*R*)-3-hydroxybutyrate] (PHB) matrix. The comparison of crude glycerol to pure glycerol phase did not reveal any negative effect in terms of productivity or polyester properties.

There are several studies providing the suitability of glycerol as a substrate for the production of *mcl*-PHAs by *Pseudomonas sp* [197]. To cut down the cost of production and optimize the PHA/dried cell weight, different concentrations (1, 2 and 5%) of refined glycerol and crude glycerol from the biodiesel process were used as a carbon source in the batch fermentation process by Palmeri *et al.* [198]. Commercial refined glycerol (purity 99.8%, pH-value 7.0) or crude glycerol (pH

8.6), a mixture of glycerol and free and methyl esters fatty acids, were tested. *Pseudomonas mediterranea* as well as *Pseudomonas corrugata*, grew and synthesized a *mcl*-PHA elastomer both on crude and refined glycerol carbon sources. The best PHA/dried cells ratios were obtained when crude glycerol was supplied at 2% of glycerol, giving more than 60% of PHA. Diversely, both strains of *P. corrugata* produced very similar *mcl*-PHA elastomer regardless of carbon source.

Castro *et al.* studied the possibilities of *mcl*-PHA production by *Pseudomonas putida* strains using raw glycerol as the only carbon/energy source [197]. The experiments were carried out using commercial refined glycerol (purity 99%) or crude waste glycerol (adjusted to pH-value 7.0), a mixture of glycerol and fatty acids. With 2% of pure glycerol, the maximum yield of PHA was obtained after 96 hrs of cultivation, with the biomass reaching 4.14 g/L and PHA accumulating to 1.78 g/L ~ 43%. At higher glycerol concentrations (3%), higher biomass and PHA yields were observed (4.53 g/L of biomass and 2.56 g/L of PHA ~ 56.5%). In the case of waste glycerol, probably the cell growth and the PHA accumulation were inhibited due to the substances present in the composition of material. The best PHA production seemed to be at the glycerol concentration of 2% with the biomass yield of 2.88 g/L and the PHA accumulation of 0.95 g/L ~ 39%; and 33% of PHA at the concentration of 3% of waste glycerol.

Pseudomonas putida bacterial strain was also tested for the PHA production from glycerol by Javers *et al.* [199]. *Pseudomonas putida* was grown on a complex medium composed of co-products of ethanol and biodiesel industries to assess the organism's capability to produce *mcl*-PHA. When glycerol was added to the bioreactor (total addition of 75 g/L), the final cell dry mass and PHA content were 30 g/L and 31%, respectively. At 28 hr, the first of four glycerol additions was made; the glycerol utilization rate peaked at 2.87 g/(Lhr).

P. mediterranea synthesizes amorphous *mcl*-PHA when grown on crude glycerol, whereas on both, the reagent grade and partially refined glycerol, it produces two very similar distinctive *mcl*-PHAs. The results show, that, thanks to a different metabolic pathway, *P. mediterranea* synthesizes *mcl*-PHAs on a substrate containing 1-2% of refined glycerol or partially refined glycerol [200].

The report of Ashby *et al.* was focused on the utilization of refined glycerol and two different crude glycerol samples with widely varying compositions to demonstrate the possibilities of using glycerol in the PHB synthesis [201]. One refined and two crude glycerol (from the biodiesel production) samples were utilized to produce PHB by *Pseudomonas oleovorans*. The concentration of 2% of refined glycerol was chosen to supply the bacterium carbon source for the growth and production of PHB. Under the batch culture conditions, the cell dry mass and the PHB yields reached a maximum of 3.0 g/L and 1.14 g/L, respectively, giving the maximum cell productivity of 38% after 72 hrs. However, only 45% of glycerol was utilized to achieve these values leaving 55% of glycerol as unused and ultimately wasted.

Sodium terephthalate (TA) produced from the PET pyrolysis product and waste glycerol from the biodiesel manufacture were supplied to *Pseudomonas putida* in a fed-batch bioreactor by Kenny *et al.* [202]. *P. putida* reached the cell dry mass of 8.70 g/L and PHA of 2.61 g/L in 48 hr when grown on solely TA. When TA and waste glycerol were supplied in the combination, the biomass productivity [g/(Lhr)] increased between 1.3- and 1.7-fold and the PHA productivity [g/(Lhr)] increased 1.8- to 2.2-fold compared to solely TA supplied. The monomer of PHA accumulated from TA or waste glycerol was predominantly composed of 3-hydroxydecanoic acid. The PHA monomers 3-hydroxytetradeeanoic acid and 3-hydroxytetradecenoic acid were not present in PHA accumulated from solely TA but were present when waste glycerol was supplied to the fermentation.

STARCH-BASED SUBSTRATES

Starch is a principal component of most seeds (maize, rice or wheat), tubers (potato) and roots (arrowroot and tapioca). Most commercial starch is produced from maize, followed by wheat, tapioca and potato [203]. Next to cellulose and chitin, starch is one of the most important carbohydrates in the plant system [204]. Within the plant, cell starch is stored in the form of granules located in amyloplasts, intracellular organelles surrounded by lipoprotein membrane. Starch granules are highly variable in size and shape depending on the plant material. Granules contain both, amorphous and crystalline internal regions in respective proportions of about 30/70 [205]. Many microorganisms can hydrolyse starch, especially fungi which are hence suitable for the solid substrate fermentation application involving starchy substrates. Glucoamylase, α-amylase, β-amylase, pullulanase and isoamylase are involved in the processes of starch degradation [206].

Starch is the most abundant carbohydrate reserve in plants and is found in leaves, flowers, fruits, seeds, and in different types of stems and roots, or plants. It serves as a source of carbon and energy. Starch is stored and cyclically mobilized during the seed germination, fruit maturation and the sprouting of tubers. In plants, starch accumulates in many different photosynthetic or non-photosynthetic tissues. One of the most abundant and universally distributed forms of polysaccharide storage concerns transient starch, which is accumulated in plant leaf cells during the day and broken down at night to achieve more or less constant supply of sucrose to non-photosynthetic tissues. Major attention was devoted to the long term form of storage present in endosperm of cereals, parenchyma of tubers, cotyledons of legume seeds. Major starch sources are cereals (40-90%), roots (30-70%), tubers (65-85%), legumes (25-50%) and some immature fruits like bananas or mangos, which contain approximately 70% of starch by dry weight. The accumulation pattern of starch granules in each plant tissue, the shape, size, structure and composition are unique to each botanical species [207, 208].

Chemical Composition of Starch

Starch is mainly composed of two homopolymers of D-glucose: amylase, mostly linear α-D(1,4)-glucan and branched amylopectin, having the same backbone structure as amylose but with many α-1,6-linked branch points. Starch has different proportions of amylose and amylopectin ranging from about 10-20% of amylose and 80-90% of amylopectin depending on the source [205]. The molecular weight of amylopectin is about 1000 times as high as the molecular weight of amylose. In addition to α-D-1,4 bonds which are present in amylose and linear segments of amylopectin, amylopectin molecule has α-D-1,6 bonds which occur every 20 to 30 anhydroglucose units. Glucose units with α-1,6 glucosidic linkages are the branching points which cause the interlinkages of glucose residue which give rise to a ramified or bush-like structure of amylopectin molecule. Starch is totally biodegradable in a wide variety of environments. It can be hydrolysed into glucose by enzymatic activity of microorganisms. The starch polymer, because of its complex structure, requires a combination of enzymes (endoamylases, and exoamylases) for its depolymerization into oligosaccharides and smaller sugars, or to be transformed by transferring oligoglucosidic linkages and residues by the creation of new bonds (debranching enzymes and glycosyl-transferases) [209].

Minor components associated with starch correspond to three categories of materials: (i) particulate material, composed mainly of cell-wall fragments; (ii) surface components, removable by extraction procedures; and (iii) internal components. Lipids represent the most important fraction associated with starch granules. High contents are generally observed for cereal starches: 0.8-1.2 and 0.6-0.8% for wheat and normal maize, respectively. The main constituents of surface components are proteins, enzymes, amino acids and nucleic acids. Some components can be extracted without the granule disruption: approximately 10% of proteins and 10-15% of lipids [210].

STARCHY MATERIALS

The harvested parts of our staple crop plants are starch-storing organs. Cereal seeds (rice, maize, wheat, barley, sorghum, and others) are the most important group, followed by tubers (*e.g.*, potato, sweet potato, yam) and storage roots (*e.g.*, cassava, taro), and seeds of beans and peas. Much of our agricultural land is devoted to their growth. Approximately 2500 million tonnes of starch crops are harvested annually [211].

Amongst various starchy materials available throughout the world, corn, sugarcane, wheat, potato, corn stover, molasses or purified starch are being processed in many biotechnological productions. As the demand and the cost of these starchy crop materials keeps raising day by day it has become indispensable to use substitute resources [212].

Cassava is a tropical root crop which is an economically available fermentable source and is produced by numerous countries. It is incorporated into animal feed (20%) and about a similar proportion is converted into starch for industrial purposes whereas; some of the portion is also used as a food source in several developing countries. About 50-70% of starch content is recovered from the cassava root, which can be detailed as follows: 95% of organic matter, 3% of protein, and 2% of lipids. Due to low ash content and rich organic nature it can be used as a substrate for many biotechnological productions. In addition, it can also be easily hydrolyzed by various techniques. As cassava starch does not have much industrial application in food industries compared to corn starch, it also lacks the competition in terms of price and is available throughout the year due to its flexibility in terms of planting and harvesting [212, 213]. Some of important characteristics of the cassava plant are the endurance, the resistance to disease and drought and it can grow in relatively low-quality soils [214].

Another important and well known source of starch are potatoes. Amyloses isolated from potato starches have higher degree of polymerization than those isolated from cereals. Non-carbohydrate constituents are present in starchy granules, and both, the actual constituents and their relative amounts vary. Potato starch granules have very low concentrations of lipids, whereas the lipid content of wheat starch granules can be in excess of 10 g/kg. Potato starches as well as starches from other tuberous plants are unique due having covalently bound phosphate esterified to approximately 0.5% of the glucose residues. The potato worldwide production was almost 400 millions of tonnes in 2013 [215].

Starch is the major constituent of wheat endosperm as well. Starch generally accounts for about 80% of the dry mass of the starchy endosperm and comprises a mixture of two polymers, amylose and amylopectin, in the ratio of about 1:3 [216]. Wheat is produced in most of the countries and accounts for about 19% of the world's caloric supplies. It is used primarily as flour for making bread, pastry, pasta and noodles *etc*. In addition, the by-products from milling wheat into flour are used as feed. The annual global production of dry wheat is about 529 Tg. Asia (43%) and Europe (32%) are the primary production regions [217].

Corn is significant global feedstock of starch. Maize contains about 72% of starch, 10% of protein, and 4% of fat, but has lower protein content then rice or wheat. Maize provides lots of B vitamins and essential minerals along with the fibre [218]. On the basis of amylose and amylopectin ratio, corn can be divided into normal, waxy and high amylose. Waxy maize is a starch variant of normal maize which contains 100% of amylopectin whereas normal maize contains 75% of amylopectin and 25% of amylose [219].

Hydrolysis of Starchy Materials

Due to the fact that after the starch hydrolysis large amount of fermentable sugars are released, both ways of hydrolysis - chemical and enzymatic are available. The enzymatic hydrolysis of starch is traditionally performed as a two-step procedure, including the liquefaction and saccharification. In the first step, a water suspension of starch is partially hydrolysed with α-amylase to maltodextrins. In the second step, liquefied starch is transformed into low molecular weight oligosaccharides such as glucose or maltose. Each step requires a specific enzyme that must be separated and inactivated after each run. The cost of enzymes is one of the main disadvantages of the enzymatic hydrolysis [220]. On the other hand, enzymatic hydrolysis allows greater control over the amylolysis, the specificity of the reaction, and the stability of generated products [221].

The other possible way how to hydrolyse starch to obtain fermentable sugars is by the acid hydrolysis which is quite a high energy demanding process and requires more energy for heating. The acid hydrolysis of starch to produce sugars has been a commercial method since 1914, when the first plant producing the glucose syrup was built in France. In addition to D-glucose, other products are formed, such as oligosaccharides, and the products of decomposition and dehydration of D-glucose, such as 5-hydroxymethylfurfural, levulinic acid, formic acid, and colored products [222]. This phenomenon can be beneficial for the PHA production; the product of starch decomposition, such as levulinic acid, could serve as a precursor for the production of PHA copolymers with better mechanical properties. Unfortunately, high concentration can lead to the inhibition of biotechnological process [112].

The enzymes commonly used for the starch processing are generally classified as amylases. Basically, there are four types of starch-converting enzymes: (i) endoamylases; (ii) exoamylases; (iii) debranching enzymes; (iv) transferases. Endo-acting enzymes or endoamylases are able to cleave α-1,4-glycosidic bonds present in the inner part (endo-) of amylose or amylopectin chain. Enzyme α-amylase is a well-known endoamylase that hydrolyses α-1,4-linkages in the interior part of starch polymer in a random fashion, which leads to the formation of linear and branched oligosaccharides. These enzymes are found in a wide variety of microorganisms, belonging to *Bacteria* and *Archaea*. Exoamylases such as glucoamylases and α-glucosidases cleave both α-1,4 and α-1,6 bonds on the external glucose residues of amylose or amylopectin from the non-reducing end and thus produce only glucose (glucoamylase and α-glucosidases). Transferases are another group of starch-converting enzymes which cleave an α-1,4-glycosidic bond of the donor molecule and transfer a part of donor to a glycosidic acceptor with the formation of a new glycosidic bond. The debranching enzymes catalyse the hydrolysis of α-1,6-glycosidic bonds in amylopectin and/or glycogen and related polymers. The affinity of debranching enzymes towards α-1,6-bond distinguishes these enzymes from other amylases which show primary affinity towards α-1,4-glycosidic linkages. The debranching enzymes can be classified into two major groups, that is, direct and indirect [209].

As was mentioned above, the two-stage hydrolysis of starch with subsequent liquefaction and saccharification is commonly used for the production of starch hydrolysates. The liquefaction stage converts insoluble starch granules into soluble polymer fragments (maltodextrins). For the purpose of liquefaction process, thermostable bacterial alpha-amylase is used. The enzyme treatment is initially run at about 105°C, followed by the incubation at 90-95°C. The liquefaction is usually a continuous process and takes approximately 1-3 hrs. The second step is the saccharification process - breaks polymer fragments into desired starch sugars. This is performed by maltose-producing enzymes, such as microbial β-amylase or fungal α-amylase, or by glucose-producing enzymes, *i.e.*, glucoamylase. The saccharification of liquefied starch to glucose usually runs for up to 72 hrs. Shorter time for the hydrolysis reaction, *e.g.*, 30-48 hrs, can be applied if a simultaneous application of glucoamylase and pullulanase or glucoamylase and isoamylase is provided. In recent years new procedures of saccharification and liquefaction are being developed, such as unconventional single-stage method for starch hydrolysis, which would facilitate a simplification of the hydrolysis process, more efficient utilization of enzymes, shortened hydrolysis time and higher efficiency of starch conversion [223 - 225].

Productions of PHA on Starch-Based Substrates

Starchy materials from starch crops or wastewater from starch manufacturing processes are the substrates which are in the point of interest for the PHA production because they are cheap, renewable and locally available. Nowadays, the companies are successful in reducing the costs of amylase production. Inexpensive efficient amylase enzymes are released to the markets. Therefore the use of enzymatic hydrolysate of starchy feedstocks as a substrate for biotechnological productions might be technically and economically feasible [214]. The PHA production based on hydrolysed starch has been reported in many studies, the overview of those productions is shown in Table **9**.

Utilization of Starchy Materials by PHA Producing Bacteria

Potato starch processing waste from the potato starch industry, which includes effluents and potato residues, causes serious environmental problems. A potato starch factory with annual starch production of 10 kt discharges 720 t of effluents and 192 t of potato residues daily [226]. One possible application of potato waste can be the utilization by PHA producing microorganisms.

Haas *et al.* used waste potato starch as a low cost carbon substrate for the production of PHB [227]. The liquefaction and saccharification steps were provided by use of $CaCl_2$ solution (95°C), for the rehydration the starch liquozyme supra was added. For the saccharification process AMG300 was used at 60°C for 18 hrs. Hydrolysed waste potato starch contained 570 g/L of glucose and 58 g/L of maltose. Using the liquefaction and saccharification method described above, about 9% of starch was lost to maltose, which couldn´t be utilized by PHA producing bacteria *Rastolnia eutropha*. The culture displayed the biomass and PHB accumulation rates similar to the fed-batch reactor conducted with pure glucose. The final results were given by 179 g/L of biomass and 94 g/L of PHB. Hence, waste potato starch can replace glucose in the PHB production.

Thanks to high content of starch present in wheat grain, it can be considered as a potential substrate for many biotechnological productions. Wheat hydrolysates were tested as a carbon source for the conversion to PHAs by *C. necator* in the study of Kouinas *et al.* [228]. The conversion efficiency from starch to glucose

was from 78% to 94% depending on the part of wheat grain. The hydrolysis was carried out at 68°C for 5 min to gelatinise the starch and then the reaction temperature was reduced to 60°C which was an optimal temperature for the application of glucoamylase. The medium with the highest nitrogen concentration gave the highest microbial biomass concentration (29.9 g/L), growth yield (0.28 g residual microbial biomass per g glucose), and PHB yield (0.43 g per g glucose).

Huang *et al.* used extruded rice bran and extruded cornstarch as carbon sources to produce PHA by an archaea *Haloferax mediterranei* which cannot use native rice bran or cornstarch as a carbon source. Extruded materials were mixed with α-amylase and dried at 50°C for 8 hrs. Extruded starchy materials reduced the cost of the process, moreover the cell concentration and the PHA accumulation were higher. In this study, the authors obtained the cell concentration of 140 g/L, the PHA concentration of 77.8 g/L, and the PHA content of 55.6% by using extruded rice bran and extruded cornstarch in the ratio of 1:8 as the carbon materials in repeated fed-batch fermentation process [229].

Cassava is other very important starchy material which was used in many studies as potential carbon source for the PHA production. Cassava is widely cultivated in the world, which implies that a considerable amount of by-products is produced, and appropriate utilization for these by-products must be found to minimize the environmental impacts. Cassava starch wastewater was also used as very interesting starch-based waste substrate for the PHA production [230].

Kreuger *et al.* tested the bacterial strains capable of bioconverting the cassava starch by-product into PHAs. PHA producing bacterial strains were selected according to their ability to produce amylase to degrade the substrate. From a total of tested 72 strains, four isolates of *B. megaterium* were selected based on their capacity to bioconvert hydrolysed cassava starch by-product into PHB. After 48 hrs of cultivation, 4.97 g of dry biomass/l containing 29.7% of PHB were obtained [213].

Another study dealing with cassava starch hydrolysate as a substrate for PHA producing bacteria is reported by Poomipuk *et al.* The hydrolysis of cassava starch was performed by the liquefaction at 90°C, pH-value 6.5 for 45 min with commercial α-amylases derived from *Bacillus licheniformis* followed by the saccharification at 60°C, pH-value 4 for 60 min with glucoamylase obtained from *Aspergillus niger*. The resulting glucose concentration of cassava starch hydrolysate after the treatments described above is 159 g/L and 198 g/L of total sugar concentration. The bacterium capable of accumulating PHAs, *Cupriavidus sp.*, which was isolated from cassava starch wastewater was able to utilize hydrolysed cassava starch. The cultivation of isolated *Cupriavidus sp.* under an N-limited condition provided fairly high biomass and PHA concentrations with the maximum value of 5.97 and 2.43 g/L respectively. The greatest maximum PHA content (62%) was also obtained in the cultivation [214].

Halophile bacteria *Halomonas boliviensis* was also used for the testing of PHB production using the starch hydrolysate as a substrate by Quillaguamam *et al.* [231]. Starch was hydrolysed by use of amylase from *Bacillus halodurants*. Different amounts of α-amylase were utilized for the starch hydrolysis, resulting in the production of mixtures of maltooligosaccharides (G1-G6) at varying ratios. Maltose (G2) and maltotetrose (G4) were the main oligosacharides formed when relatively high amount of amylase was used for the starch hydrolysis. The highest PHB accumulation (56%) by *H. boliviensis* was obtained when 6.4 U per mL of amylase at 55°C for 1 hr was used for the starch hydrolysis.

Untreated Starchy Substrates

Depending on the ability of producing strain to amylolytically hydrolyse starch, starch can be supplied to its oligomers (maltrotetraose, maltohexaose), dimers (maltise) and monomers (glucose) as a substrate, either directly or after enzymatic, or chemical hydrolysis [110]. Few studies were reported regarding the production of PHA from raw starch.

For the production of PHB engineered *Corynebacterium glutamicum* was used, expressing the PHB biosynthetic genes with α-amylase on its cell surface in the study of Song *et al.* [232]. PHB was produced from soluble starch using α-amylase on *C. glutamicum* cell-surface originating from *Streptococcus bovis*. The biomass and the PHB yield obtained in this study were 6.1 and 0.39 g/L respectively. Moreover, the use of starch gave higher polymer yield compared to the utilization of glucose.

Other study reported by Sheu *at* al. deals with the copolymer production of PHBHV, which was produced by thermophilic bacterium *Caldimonas taiwanensis* utilizing starch as a carbon source and valerate as a precursor of 3-hydroxyvalerate units [233]. *C. taiwanensis* is able to produce starch-degrading enzymes into medium, which efficiently digests soluble starch into usable reducing sugars. In this study major resources of starch production and consumption, cassava, wheat, corn and potatoes, were respectively mixed with valerate as the carbon source. Wheat starch yielded the highest biomass, but the lowest PHA content among used starches, which can be caused by the presence of gluten providing the nitrogen source for bacterial growth. The highest PHA content was obtained in case of cassava starch (67%) which contained 0.05% of valerate, leading to the production of copolymer with the monomer composition: 3HB 87 mol-% and 3HV 13 mol-%.

Table 9. Productions of PHA on starch-based substrates.

Production Strain	Carbon Source	PHA Type	Cultivation	Biomass [g/L]	PHA [wt.-%]	PHA [g/L]	Ref.
Rastolnia eutropha	waste potato starch	PHB	Fed-batch Fermentor	179.0	55.0	94.0	[227]
Bacillus megaterium	cassava starch	PHB	Batch	4.9	29.7	1.4	[213]
Haloferax mediterranei	extruded cornstarch, rice bran	PHB	Fed-batch Fermentor	140.0	55.6	77.8	[229]
Rastolnia eutropha	cassava starch	PHB	Batch	5.9	61.6	2.4	[214]
Corynebacterium glutamicum	starch	PHB	Bathc	6.1	6.4	0.4	[232]
Caldimonas taiwanensis	cassava starch	PHA	Batch	2.8	67.0	n.a.	[233]
	corna starch			3.3	65.0		
	potato			2.6	55.0		
	sweet potato			1.6	52.0		
	wheat starch			4.1	42.0		
Bacillus thuringiensis	starch	PHB	Batch	6.8	41.5	2.8	[110]
Halomonas boliviensis	starch hydrolysate	PHB	Batch	n.a	56.0	n.a	[231]
Bacillus cereus	starch	PHB	Batch	1.0	48.0	0.5	[234]

(Table 9) contd.....

Production Strain	Carbon Source	PHA Type	Cultivation	Biomass [g/L]	PHA [wt.-%]	PHA [g/L]	Ref.
Cupriavidus necator	wheat hydrolysate	PHB	Batch	29.9	n.a	51.1	[228]

Gowda *et al.* reported the possibility of PHA production using hydrolytic potential of *Bacillus thuringiensis* from agro-waste biomass [110]. Comparative PHAs production by *B. thuringiensis* IAM 12077 in biphasic growth conditions using glucose and starch showed appreciable levels of growth (5.7 and 6.8 g/L) and PHA production (58.5 and 41.5%) with the PHA yield of 3.3 and 2.8 g/L, respectively. The nitrogen deficiency supported the maximum PHA yield (2.46 g/L) and accumulation (53.3%). The maximum growth (3.6 g/L), PHB yield (2.6 g/L) and PHA accumulation (72.8%) were obtained with C:N ratio of 8:1 using starch in the concentration of 10 g/L as the carbon source.

Direct utilization of starch by PHA producing *Bacillus* species was tested in the study of Halami *et al.* [234]. Among eight cultures tested, one of native cultures was found to be a potential PHA producer, accumulating about 50% of PHA of CDM. The marginal growth was seen if soluble starch was used as a carbon source at 1 and 2%, the PHA production was 45.5 and 48%, respectively, after 72 hrs of fermentation.

PETROCHEMICAL PLASTIC WASTE TRANSFORMATION TO PHA

Apart from renewable resources, there are also reports on the utilization of completely different raw materials for the biotechnological production of PHAs - the petrochemical plastic waste. Petrochemical based plastics, produced annually on the 100 million ton scale, pervade modern society as a result of their versatile and highly desirable properties. However, once disposed of, many of these plastics pose major waste management problems due to their recalcitrance. In the U.S. alone, over 3 million tons of polystyrene are produced annually, 2.3 million tons of which end up in a landfill. Therefore; the group of Professor Kevin O´Connor processed the conversion of polystyrene, a major petrochemical plastic waste, to *mcl*-PHA using a two-step chemo-biotechnological process. The chemical step involved the pyrolysis of polystyrene (520°C) in a fluidized bed reactor to generate an oil composed predominantly of styrene (82%) and low levels of other aromatic compounds. The distillation of this oil increased the styrene content to 90%. Distilled and non-distilled styrene oils were used as fermentable substrates for *Pseudomonas putida* CA-3 which in turn accumulated *mcl*-PHA to 30% of the cell dry mass. A single pyrolysis run and four fermentation runs resulted in the conversion of 64 g of polystyrene to 6.4 g of PHA [235]. A higher cell density was observed in cells supplied with nitrogen and thus further increases in the overall production of PHA were observed through nitrogen feeding. The highest yield of PHA was 0.28 g of PHA per g of styrene supplied with the nitrogen feeding rate of 1.5 mg/(Lh) [236]. Similarly, also Tan *et al.* aimed at the utilization of styrene. They tested twelve newly-isolated styrene-degrading *Pseudomonads* which were able to produce *mcl*-PHA. The best-performing isolate was identified as *Pseudomonas putida* NBUS12. This bacterium achieved a final biomass of 1.28±0.10 g/L and PHA content of 32.49% of biomass. The extracted polymer was mainly comprised of 3-hydroxyhexanoate, 3-hydroxyoctanoate, 3-hydroxydecanoate, 3-hydroxydodecanoate, and 3-hydroxy-tetradecanoate monomers at the ratio of 2:42:1257:17:1 [237].

Beside polystyrene, the researchers also focused on the utilization of poly(ethylene terephthalate) (PET). PET is one of many petrochemical based plastics which contribute greatly to the convenience of everyday life. Best known for its use in plastic bottles it is produced on a multimillion tonne scale worldwide. Like other petrochemical plastics, the success of PET as a convenience bulk commodity polymer has led to post consumer PET products becoming a major waste problem. The current technologies recycle PET to a low value product and thus the factors such as high relative cost of sorting and low value of the downstream product contribute to poor recycling rates for PET. The conversion of PET to a high value product (upcycling) should lead to higher levels of PET recycling. Hence, professor O´Connor and his group developed similar process of PET utilization as described above for polystyrene. The thermal treatment of PET in the absence of air (pyrolysis) generates terephthalic acid (TA) as the major product. Generated TA is potentially a feedstock for the microbial synthesis of the value added biodegradable polymer *mcl*-PHA. PET was pyrolyzed at 450°C resulting in the production of a solid, liquid, and gaseous fraction. Liquid and gaseous fractions were burnt for energy recovery, whereas the solid fraction constituting mainly of TA was used as the feedstock for the bacterial production of PHA. The strains previously reported to grow on TA were unable to accumulate PHA. The researchers; therefore, isolated bacteria from soil exposed to PET granules at a PET bottle processing plant. From the 32 strains isolated, three strains capable of the accumulation of medium chain length *mcl*-PHA from TA as a sole source of carbon and energy were selected for further study. These isolates were identified using 16S rDNA techniques as *P. putida* (GO16), *P. putida* (GO19), and *P. frederiksbergensis* (GO23). *P. putida* GO16 and GO19 accumulated PHA composed predominantly of a 3-hydroxydecanoic acid monomer while *P. frederiksbergensis* GO23 accumulated 3-hydroxydecanoic acid as the predominant monomer with increased amounts of 3-hydroxydodecanoic acid and 3-hydroxydodecenoic acid compared to the other two strains [238].

Finally, the same research group also worked on the utilization of polyethylene (PE). This polymer is produced more than any other polymer type, making up 29.1% of worldwide polymer production. PE has a broad range of physico-chemical properties, which enable its use in a variety of products. These include heavy-duty commodities, such as water pipes, food containers, toys and chemical storage containers. PE is also widely used in the film applications such as plastic bags, agricultural films, bubble wrap or multilayer and composite films. Like other petrochemical plastics, the success of PE as a convenience bulk commodity polymer has led to post consumer PE products becoming a major waste problem. The thermal treatment of PE in the absence of air generated a complex mixture of low molecular weight paraffins with the carbon chain lengths from C8 to C32 (PE pyrolysis wax). Several bacterial strains were able to grow and produce PHA from this PE pyrolysis wax. The addition of biosurfactant (rhamnolipids) allowed for greater bacterial growth and PHA accumulation of tested strains. Some strains were only capable of growth and PHA accumulation in the presence of the biosurfactant. *Pseudomonas aeruginosa* PAO-1 accumulated the highest level of PHA with almost 25% of CDM when supplied with the PE pyrolysis wax in the presence of rhamnolipids. The change of nitrogen source from ammonium chloride to ammonium nitrate resulted in faster bacterial growth and earlier onset of PHA accumulation [239].

PRODUCTION OF PHA FROM VOLATILE FATTY ACIDS

Numerous microbial species are known to ferment dissolved organic compounds into organic acids such as acetic, propionic, butyric, valeric and lactic acid and other fermentation products under the anaerobic conditions. This so called acidogenic fermentation is an important step in anaerobic digestion of organic compounds to methane and carbon dioxide [240, 241].

Bengtsson *et al.* reported that acidogenic fermentation of cheese whey permeate and three pulp and paper mill effluents can serve as the first step in the process of PHAs production employing mixed bacterial culture. From the batch experiments, it was found that in case of whey and one of the paper mill effluents nearly 100% of the soluble COD was readily fermentable. In continuous chemostat experiments with these two effluents, varying the retention time and pH-value of the reactors exhibited significant impact on the amount and composition of volatile fatty acids produced. The main fermentation products were acetate, propionate and butyrate [240]. In another study the same authors described the production of PHAs by activated sludge from paper mill wastewater. The investigated three-stage process consisted of acidogenic fermentation which converted the wastewater organic matter to volatile fatty acids, an activated sludge system operating under feast/famine conditions to enrich for PHAs producing organisms and the accumulation of PHAs in the batch experiments. After the fermentation of wastewater, 74% of the soluble COD was present as volatile fatty acids (in particular acetate, propionate, butyrate and valerate). The maximum PHA content achieved was 48% of the sludge dry weight and the three stage process exhibited a potential to produce 0.11 kg of PHA per kg of influent COD treated. The resulting materials consisted of 31-47 mol-% of 3-hydroxybutyrate and 53-69 mol-% of 3-hydroxyvalerate [242]. Further, the same group of authors employed the same approach to study the production of PHAs also from fermented sugar cane molasses. The produced polymers contained five types of monomers, namely 3-hydroxybutyrate, 3-hydroxyvalerate, 3-hydroxy-2-methyl-butyrate, 3-hydroxy-2-methylvalerate and the medium chain length monomer 3-hydroxyhexanoate. The mixed bacterial culture employed for the PHAs production was enriched in glycogen accumulating organisms. The high polymer yields and production rates obtained in the study suggest that the enrichment of mixed culture by glycogen accumulating microbes can be a fruitful strategy for mixed culture production of PHA from waste substrates [243].

Furthermore, volatile fatty acid rich effluent is also formed during the biotechnological production of hydrogen. Actually, the treatment of resulting acid-rich material is considered being one of the most problematic parts of the biohydrogen production process [241]. On the other hand, since bacterial monocultures as well as mixed cultures can turn volatile fatty acids into PHAs, the coupling of hydrogen production with subsequent PHAs production from volatile fatty acids seems to be a reasonable solution of above mentioned problems yielding two valuable products - bio-hydrogen and PHAs. Such a strategy was studied by Mohan *et al.* who used acid-rich effluents from the hydrogen bioreactor for the PHAs production. The authors reported the PHB production productivity of 25% especially at lower substrate loading conditions. The decreases in PHB production up to 6% which were observed with the increase in substrate load might be due to the presence of high concentration of residual carbon along with acid metabolites. The terminal integration of bioplastics production with dark-fermentative H_2 production process appears to be a promising strategy to enhance the treatment efficiency and to reduce the overall process cost [241].

CONCLUDING REMARKS

To sum up, the fact that the PHA producing bacteria are capable of utilizing wide range of waste substrates coming from various industrial fields opens many opportunities for potentially interesting industrial processes of PHA production. Nevertheless, there are several drawbacks which should be taken into account and which principally reduce the number of sustainable and economically viable processes of PHA production. Firstly, the substrates for the PHAs production should meet several basic criteria such as the availability (in constant amount and quality during whole year), stability, low cost of collection, transportation and processing or no competition with human food chain. The fact is that many substrates, the utilization of which is described in the literature, do not meet these requirements. Secondly, the utilization of waste substrates is usually connected with the decrease of productivity in comparison with pure substrates. Therefore, appropriate selection of producing strain and substrate is crucial from the point of view of sustainability and economic feasibility of the process. Based on literature data, it seems that, for instance, the utilization of waste and inexpensive lipids for the biotechnological production of PHAs employing *Cupriavidus necator* could be industrially interesting. Furthermore, also the bioconversion of cheese whey by genetically modified *E. coli* provides very interesting PHAs yields. Nonetheless, we have to admit that it is still questionable whether the biotechnological process producing PHAs as the only product can be economically feasible at all. On the other hand, the concept of industrial bio-refineries becomes attractive and it can be expected that its importance will increase in the following decades. These facilities are intended to completely convert the biomass including various waste products and by-products of food industry, *etc.* into multiple valuable products such as biofuels, biopolymers and other valuable biomolecules but also into heat and energy. Since the substrates are utilized completely in the cascade of processes and the whole system is flexible because it can shift among several products based on actual market demands, *etc.*, the bio-refineries might be more economically viable than traditional single-product manufacturing facilities. In the concept of bio-refinery, PHAs might be important or even crucial members of product portfolio which might further reduce the production cost of PHAs and help these valuable environmentally-friendly polymers compete with petrochemical based plastics and; therefore, partially replace them in appropriate applications.

CONFLICT OF INTEREST

The authors confirm that they have no conflict of interest to declare for this publication.

ACKNOWLEDGEMENTS

This work was supported by the project Materials Research Centre at FCH BUT—Sustainability and Development no.LO1211 and by national COST project LD15031 of the Ministry of Education, Youth and Sports of the Czech Republic and by the project GA15-20645S of the Czech Science Foundation (GACR).

REFERENCES

[1] Müller HM, Seebach D. Poly(hydroxyalkanoates): A fifth class of physiologically important organic biopolymers. Angew Chem Int Ed Engl 1993; 32(4): 477-502. [http://dx.doi.org/10.1002/anie.199304771]

[2] Kessler B, Wilholt B. Poly(3-hydroxyalkanoatesg. In: Flickinger MC, Drew SW, Eds. Encyclopedia of Bioprocess Technology - Fermentation, Biocatalysis and Bioseparation. New York: John Wiley 1999; pp. 2024-40.

[3] Ivanov V, Stabnikov V, Ahmed Z, Dobrenko S, Saliuk A. Production and applications of crude polyhydroxyalkanoate containing bioplastic from the organic fraction of municipal solid waste. Int J Environ Sci Technol 2015; 12(2): 725-38. [http://dx.doi.org/10.1007/s13762-014-0505-3]

[4] Choi JI, Lee SY. Process analysis and economical evaluation for poly(3-hydroxybutyrate) production by fermentation. Bioprocess Eng 199717(6): 335-42.

[5] Kaur G, Roy I. Strategies for large-scale production of polyhydroxyalkanoates. Chem Biochem Eng Q 2015; 29(2): 157-72. [http://dx.doi.org/10.15255/CABEQ.2014.2255]

[6] Rehm BH. Polyester synthases: natural catalysts for plastics. Biochem J 2003; 376(Pt 1): 15-33. [http://dx.doi.org/10.1042/bj20031254] [PMID: 12954080]

[7] Quaiser A, Ochsenreiter T, Klenk H-P, *et al.* First insight into the genome of an uncultivated crenarchaeote from soil. Environ Microbiol 2002; 4(10): 603-11. [http://dx.doi.org/10.1046/j.1462-2920.2002.00345.x] [PMID: 12366755]

[8] Reusch RN, Sadoff HL. Putative structure and functions of a poly-beta-hydroxybutyrate/calcium polyphosphate channel in bacterial plasma membranes. Proc Natl Acad Sci USA 1988; 85(12): 4176-80. [http://dx.doi.org/10.1073/pnas.85.12.4176] [PMID: 2454464]

[9] Reusch RN, Sadoff HL. D-(-)-poly-beta-hydroxybutyrate in membranes of genetically competent bacteria. J Bacteriol 1983; 156(2): 778-88. [PMID: 6415039]

[10] Naik S, Venu Gopal SK, Somal P. Bioproduction of polyhydroxyalkanoates from bacteria: a metabolic approach. World J Microbiol Biotechnol 2008; 24(10): 2307-14. [http://dx.doi.org/10.1007/s11274-008-9745-z]

[11] Schlegel HG, Gottschalk G, Von Bartha R. Formation and utilization of poly-β-hydroxybutyric acid by Knallgas bacteria (Hydrogenomonas). Nature 1961; 191(4787): 463-5. [http://dx.doi.org/10.1038/191463a0] [PMID: 13747776]

[12] Kessler B, Witholt B. Factors involved in the regulatory network of polyhydroxyalkanoate metabolism. J Biotechnol 2001; 86(2): 97-104. [http://dx.doi.org/10.1016/S0168-1656(00)00404-1] [PMID: 11245898]

[13] Moskowitz GJ, Merrick JM. Metabolism of poly-β-hydroxybutyrate. II. Enzymatic synthesis of D-(-) -β hydroxybutyryl coenzyme A by an enoyl hydrase from Rhodospirillum rubrum. Biochemistry 1969; 8(7): 2748-55.

[http://dx.doi.org/10.1021/bi00835a009] [PMID: 5808333]

[14] de Smet MJ, Eggink G, Witholt B, Kingma J, Wynberg H. Characterization of intracellular inclusions formed by *Pseudomonas oleovorans* during growth on octane. J Bacteriol 1983; 154(2): 870-8. [PMID: 6841319]

[15] Umeda F, Kitano Y, Murakami Y, Yagi K, Miura Y, Mizoguchi T. Cloning and sequence analysis of the poly (3-hydroxyalkanoic acid)-synthesis genes of *Pseudomonas acidophila.* Appl Biochem Biotechnol 1998; 70-72(1): 341-52. [http://dx.doi.org/10.1007/BF02920150] [PMID: 9627389]

[16] Miyamoto CM, Sun W, Meighen EA. The LuxR regulator protein controls synthesis of polyhydroxybutyrate in *Vibrio harveyi.* Biochim Biophys Acta 1998; 1384(2): 356-64. [http://dx.doi.org/10.1016/S0167-4838(98)00028-4] [PMID: 9659397]

[17] Steinbüchel A, Wieczorec R, Krüger N. PHA biosynthesis, its regulation of C1-utilizing microorganisms for polyester production. In: Lidstrom ME, Tabita FR, Eds. Microbial Growth on C1 Compounds. Dordrecht: Springer Netherlands 1996; pp. 237-44. [http://dx.doi.org/10.1007/978-94-009-0213-8_32]

[18] Zinn M. Dual (C,N) Nutrient limited growth of Pseudomonas oleovorans. PhD Thesis Switzerland: ETH Zürich 1998.

[19] Doi Y, Kawaguchi Y, Koyama N, *et al.* Synthesis and degradation of polyhydroxyalkanoates in *Alcaligenes eutrophus.* FEMS Microbiol Lett 1992; 103(2-4): 103-8. [http://dx.doi.org/10.1111/j.1574-6968.1992.tb05827.x]

[20] Kadouri D, Burdman S, Jurkevitch E, Okon Y. Identification and isolation of genes involved in poly(beta-hydroxybutyrate) biosynthesis in *Azospirillum brasilense* and characterization of a *phbC* mutant. Appl Environ Microbiol 2002; 68(6): 2943-9. [http://dx.doi.org/10.1128/AEM.68.6.2943-2949.2002] [PMID: 12039753]

[21] Ayub ND, Pettinari MJ, Ruiz JA, López NI. A polyhydroxybutyrate-producing Pseudomonas sp. isolated from Antarctic environments with high stress resistance. Curr Microbiol 2004; 49(3): 170-4. [http://dx.doi.org/10.1007/s00284-004-4254-2] [PMID: 15386099]

[22] Ayub ND, Pettinari MJ, Méndez BS, López NI. The polyhydroxyalkanoate genes of a stress resistant Antarctic *Pseudomonas* are situated within a genomic island. Plasmid 2007; 58(3): 240-8. [http://dx.doi.org/10.1016/j.plasmid.2007.05.003] [PMID: 17629557]

[23] Ayub ND, Tribelli PM, López NI. Polyhydroxyalkanoates are essential for maintenance of redox state in the Antarctic bacterium Pseudomonas sp. 14-3 during low temperature adaptation. Extremophiles 2009; 13(1): 59-66. [http://dx.doi.org/10.1007/s00792-008-0197-z] [PMID: 18931822]

[24] Obruca S, Marova I, Svoboda Z, Mikulikova R. Use of controlled exogenous stress for improvement of poly(3-hydroxybutyrate) production in *Wautersiaeutropha*: a preliminary study. Folia Microbiol (Praha) 2010; 55(1): 17-22. [http://dx.doi.org/10.1007/s12223-010-0003-z] [PMID: 20336499]

[25] Obruca S, Marova I, Stankova M, Mravcova L, Svoboda Z. Effect of ethanol and hydrogen peroxide on poly(3-hydroxybutyrate) biosynthetic pathway in *Cupriavidus necator* H16. World J Microbiol

Biotechnol 2010; 26(7): 1261-7.
[http://dx.doi.org/10.1007/s11274-009-0296-8] [PMID: 24026931]

[26] Vijayasankaran N, Carlson R, Srienc F. Synthesis of poly[(R)-3-hydroxybutyric acid) in the cytoplasm of *Pichia pastoris* under oxygen limitation. Biomacromolecules 2005; 6(2): 604-11.
[http://dx.doi.org/10.1021/bm0494141] [PMID: 15762620]

[27] Kamnev AA, Tugarova AV, Antonyuk LP, Tarantilis PA, Polissiou MG, Gardiner PH. Effects of heavy metals on plant-associated *rhizobacteria*: comparison of endophytic and non-endophytic strains of *Azospirillum brasilense*. J Trace Elem Med Biol 2005; 19(1): 91-5.
[http://dx.doi.org/10.1016/j.jtemb.2005.03.002] [PMID: 16240678]

[28] Passanha P, Kedia G, Dinsdale RM, Guwy AJ, Esteves SR. The use of NaCl addition for the improvement of polyhydroxyalkanoate production by Cupriavidus necator. Bioresour Technol 2014; 163: 287-94.
[http://dx.doi.org/10.1016/j.biortech.2014.04.068] [PMID: 24835740]

[29] Ibrahim MH, Steinbüchel A. Poly(3-hydroxybutyrate) production from glycerol by Zobellella denitrificans MW1 via high-cell-density fed-batch fermentation and simplified solvent extraction. Appl Environ Microbiol 2009; 75(19): 6222-31.
[http://dx.doi.org/10.1128/AEM.01162-09] [PMID: 19666728]

[30] Khare E, Chopra J, Arora NK. Screening for mcl-PHA-producing fluorescent pseudomonads and comparison of mcl-PHA production under iso-osmotic conditions induced by PEG and NaCl. Curr Microbiol 2014; 68(4): 457-62.
[http://dx.doi.org/10.1007/s00284-013-0497-0] [PMID: 24306149]

[31] Quillaguamán J, Guzmán H, Van-Thuoc D, Hatti-Kaul R. Synthesis and production of polyhydroxyalkanoates by halophiles: current potential and future prospects. Appl Microbiol Biotechnol 2010; 85(6): 1687-96.
[http://dx.doi.org/10.1007/s00253-009-2397-6] [PMID: 20024541]

[32] Yue H, Ling C, Yang T, *et al.* A seawater-based open and continuous process for polyhydroxyalkanoates production by recombinant *Halomonas campaniensis* LS21 grown in mixed substrates. Biotechnol Biofuels 2014; 7: 108.
[http://dx.doi.org/10.1186/1754-6834-7-108]

[33] Natarajan K, Kishore L, Babu CR. Sodium-chloride stress results in increased poly-bet--hydroxybutyrate production in *Rhizobium* DDSS-69. Microbios 1995; 82(331): 95-107.

[34] Pal A, Paul AK. Accumulation of polyhydroxyalkanoates by *rhizobacteria* under neath nickel-hyper accumulators from serpentine ecosystem. J Polymers Environ 2012; 20(1): 10-6.
[http://dx.doi.org/10.1007/s10924-011-0355-8]

[35] Chien CC, Wang LJ, Lin WR. Polyhydroxybutyrate accumulation by a cadmium-resistant strain of *Cupriavidus taiwanensis*. J Taiwan Inst Chem Eng 2014; 45(4): 1164-9.
[http://dx.doi.org/10.1016/j.jtice.2014.02.004]

[36] Follonier S, Henes B, Panke S, Zinn M. Putting cells under pressure: a simple and efficient way to enhance the productivity of medium-chain-length polyhydroxyalkanoate in processes with Pseudomonas putida KT2440. Biotechnol Bioeng 2012; 109(2): 451-61.
[http://dx.doi.org/10.1002/bit.23312] [PMID: 21858788]

[37] Kahar P, Tsuge T, Taguchi K, Doi Y. High yield production of polyhydroxyalkanoates from soybean oil by *Ralstonia eutropha* and its recombinant strain. Polym Degrad Stabil 2004; 83(1): 79-86. [http://dx.doi.org/10.1016/S0141-3910(03)00227-1]

[38] Obruca S, Marova I, Snajdar O, Mravcova L, Svoboda Z. Production of poly(3-hydroxybutyrate-c--3-hydroxyvalerate) by *Cupriavidus necator* from waste rapeseed oil using propanol as a precursor of 3-hydroxyvalerate. Biotechnol Lett 2010; 32(12): 1925-32. [http://dx.doi.org/10.1007/s10529-010-0376-8] [PMID: 20814716]

[39] Ciesielski S, Mozejko J, Pisutpaisal N. Plant oils as promising substrates for polyhydroxyalkanoates production. J Clean Prod 2015; 106: 408-21. [http://dx.doi.org/10.1016/j.jclepro.2014.09.040]

[40] Sharma YC, Singh B, Upadhyay SN. Advancements in development and characterization of biodiesel: a review. Fuel 2008; 87(12): 2355-73. [http://dx.doi.org/10.1016/j.fuel.2008.01.014]

[41] Obruca S, Petrik S, Benesova P, Svoboda Z, Eremka L, Marova I. Utilization of oil extracted from spent coffee grounds for sustainable production of polyhydroxyalkanoates. Appl Microbiol Biotechnol 2014; 98(13): 5883-90. [http://dx.doi.org/10.1007/s00253-014-5653-3] [PMID: 24652066]

[42] Meher LC, Dharmagadda VS, Naik SN. Optimization of alkali-catalyzed transesterification of *Pongamia pinnata* oil for production of biodiesel. Bioresour Technol 2006; 97(12): 1392-7. [http://dx.doi.org/10.1016/j.biortech.2005.07.003] [PMID: 16359862]

[43] Mozejko J, Przybyłek G, Ciesielski S. Waste rapeseed oil as a substrate for mcl-PHAs production. Eur J Lipid Sci Technol 2011; 113(12): 1550-7. [http://dx.doi.org/10.1002/ejlt.201100148]

[44] Sun Z, Ramsay JA, Guay M, Ramsay BA. Fermentation process development for the production of medium-chain-length poly-3-hyroxyalkanoates. Appl Microbiol Biotechnol 2007; 75(3): 475-85. [http://dx.doi.org/10.1007/s00253-007-0857-4] [PMID: 17453197]

[45] Reinecke F, Steinbüchel A. *Ralstonia eutropha* strain H16 as model organism for PHA metabolism and for biotechnological production of technically interesting biopolymers. J Mol Microbiol Biotechnol 2009; 16(1-2): 91-108. [http://dx.doi.org/10.1159/000142897] [PMID: 18957865]

[46] Ewering C, Brämer CO, Bruland N, Bethke A, Steinbüchel A. Occurrence and expression of tricarboxylate synthases in *Ralstonia eutropha.* Appl Microbiol Biotechnol 2006; 71(1): 80-9. [http://dx.doi.org/10.1007/s00253-005-0099-2] [PMID: 16133321]

[47] Riedel SL, Lu J, Stahl U, Brigham CJ. Lipid and fatty acid metabolism in *Ralstonia eutropha*: relevance for the biotechnological production of value-added products. Appl Microbiol Biotechnol 2014; 98(4): 1469-83. [http://dx.doi.org/10.1007/s00253-013-5430-8] [PMID: 24343766]

[48] Lu J, Brigham CJ, Rha C, Sinskey AJ. Characterization of an extracellular lipase and its chaperone from Ralstonia eutropha H16. Appl Microbiol Biotechnol 2013; 97(6): 2443-54. [http://dx.doi.org/10.1007/s00253-012-4115-z] [PMID: 22588499]

[49] Riedel SL, Bader J, Brigham CJ, *et al.* Production of poly(3-hydroxybutyrate-co-3-hydroxyhexanoate) by *Ralstonia eutropha* in high cell density palm oil fermentations. Biotechnol Bioeng 2012; 109(1): 74-83.
[http://dx.doi.org/10.1002/bit.23283] [PMID: 21809332]

[50] Cruz MV, Paiva A, Lisboa P, *et al.* Production of polyhydroxyalkanoates from spent coffee grounds oil obtained by supercritical fluid extraction technology. Bioresour Technol 2014; 157: 360-3.
[http://dx.doi.org/10.1016/j.biortech.2014.02.013] [PMID: 24594316]

[51] Kimura H, Takahashi T, Hiraka H, Iwana M, Takeishi M. Effective biosynthesis of poly(3-hydroxybutyrate) from plant oils by *Chromobacterium* sp. Polym J 1999; 31(2): 210-2.
[http://dx.doi.org/10.1295/polymj.31.210]

[52] Alias Z, Tan IK. Isolation of palm oil-utilising, polyhydroxyalkanoate (PHA)-producing bacteria by an enrichment technique. Bioresour Technol 2005; 96(11): 1229-34.
[http://dx.doi.org/10.1016/j.biortech.2004.10.012] [PMID: 15734309]

[53] Chee JY, Lau NS, Samian MR, Tsuge T, Sudesh K. Expression of *Aeromonas caviae* polyhydroxyalkanoate synthase gene in *Burkholderia* sp. USM (JCM15050) enables the biosynthesis of SCL-MCL PHA from palm oil products. J Appl Microbiol 2012; 112(1): 45-54.
[http://dx.doi.org/10.1111/j.1365-2672.2011.05189.x] [PMID: 22054430]

[54] Allen AD, Anderson WA, Ayorinde FO, Eribo BE. Biosynthesis and characterization of copolymer poly(3HB-co-3HV) from saponified *Jatropha curcas* oil by *Pseudomonas oleovorans.* J Ind Microbiol Biotechnol 2010; 37(8): 849-56.
[http://dx.doi.org/10.1007/s10295-010-0732-7] [PMID: 20467780]

[55] Walsh M, O'Connor K, Babu R, Woods T, Kenny S. Plant oils and products of their hydrolysis as substrates for polyhydroxyalkanoate synthesis. Chem Biochem Eng Q 2015; 29(2): 123-33.
[http://dx.doi.org/10.15255/CABEQ.2014.2252]

[56] Marsudi S, Unno H, Hori K. Palm oil utilization for the simultaneous production of polyhydroxyalkanoates and rhamnolipids by *Pseudomonas aeruginosa.* Appl Microbiol Biotechnol 2008; 78(6): 955-61.
[http://dx.doi.org/10.1007/s00253-008-1388-3] [PMID: 18299827]

[57] Chan PL, Yu V, Wai L, Yu HF. Production of medium-chain-length polyhydroxyalkanoates by *Pseudomonas aeruginosa* with fatty acids and alternative carbon sources. Appl Biochem Biotechnol 2006; 129-132: 933-41.
[http://dx.doi.org/10.1385/ABAB:132:1:933] [PMID: 16915701]

[58] Yun HS, Kim DY, Chung CW, Kim HW, Yang YK, Rhee YH. Characterization of a tacky poly (3-hydroxyalkanoate) produced by *Pseudomonas chlororaphis* HS21 from palm kernel oil. J Microbiol Biotechnol 2003; 13(1): 64-9.

[59] Thakor N, Trivedi U, Patel KC. Biosynthesis of medium chain length poly(3-hydroxyalkanoates) (mcl-PHAs) by *Comamonas testosteroni* during cultivation on vegetable oils. Bioresour Technol 2005; 96(17): 1843-50.
[http://dx.doi.org/10.1016/j.biortech.2005.01.030] [PMID: 16084364]

[60] Ashby RD, Solaiman DK, Foglia TA, Liu CK. Glucose/lipid mixed substrates as a means of

controlling the properties of medium chain length poly(hydroxyalkanoates). Biomacromolecules 2001; 2(1): 211-6.
[http://dx.doi.org/10.1021/bm000098+] [PMID: 11749175]

[61] Solaiman DK, Ashby RD, Foglia TA. Medium-chain-length poly(beta-hydroxyalkanoate) synthesis from triacylglycerols by *Pseudomonas saccharophila.* Curr Microbiol 1999; 38(3): 151-4.
[http://dx.doi.org/10.1007/PL00006779] [PMID: 9922465]

[62] Follonier S, Goyder MS, Silvestri AC, *et al.* Fruit pomace and waste frying oil as sustainable resources for the bioproduction of medium-chain-length polyhydroxyalkanoates. Int J Biol Macromol 2014; 71: 42-52.
[http://dx.doi.org/10.1016/j.ijbiomac.2014.05.061] [PMID: 24882726]

[63] Możejko J, Ciesielski S. Saponified waste palm oil as an attractive renewable resource for mcl-polyhydroxyalkanoate synthesis. J Biosci Bioeng 2013; 116(4): 485-92.
[http://dx.doi.org/10.1016/j.jbiosc.2013.04.014] [PMID: 23706994]

[64] Chandel AK, Singh OV. Weedy lignocellulosic feedstock and microbial metabolic engineering: advancing the generation of 'Biofuel'. Appl Microbiol Biotechnol 2011; 89(5): 1289-303.
[http://dx.doi.org/10.1007/s00253-010-3057-6] [PMID: 21181146]

[65] Wyman CE, Decker SR, Himmel ME, Brady JW, Skopec CE, Viikar L. Hydrolysis of cellulose and hemicellulose. In: Dumitriu S, Ed. Polysaccharides: Structural Diversity and Functional Versatility. 2nd ed. New York: CRC Press 2004; pp. 995-1034.
[http://dx.doi.org/10.1201/9781420030822.ch43]

[66] Bonnin E, Garnier C, Ralet M-C. Pectin-modifying enzymes and pectin-derived materials: applications and impacts. Appl Microbiol Biotechnol 2014; 98(2): 519-32.
[http://dx.doi.org/10.1007/s00253-013-5388-6] [PMID: 24270894]

[67] Gibson LJ. The hierarchical structure and mechanics of plant materials. J R Soc Interface 2012; 9(76): 2749-66.
[http://dx.doi.org/10.1098/rsif.2012.0341] [PMID: 22874093]

[68] Peng F, Peng P, Xu F, Sun R-C. Fractional purification and bioconversion of hemicelluloses. Biotechnol Adv 2012; 30(4): 879-903.
[http://dx.doi.org/10.1016/j.biotechadv.2012.01.018] [PMID: 22306329]

[69] Peters D. Raw Materials. In: Ulber R, Sell D, Eds. White Biotechnology, Advances in Biochemical Engineering/Biotechnology. Berlin: Springer 2007; pp. 1-30.

[70] Kumar R, Singh S, Singh OV. Bioconversion of lignocellulosic biomass: biochemical and molecular perspectives. J Ind Microbiol Biotechnol 2008; 35(5): 377-91.
[http://dx.doi.org/10.1007/s10295-008-0327-8] [PMID: 18338189]

[71] Kumar P, Barrett DM, Delwiche MJ, Stroeve P. Methods for pretreatment of lignocellulosic biomass for efficient hydrolysis and biofuel production. Ind Eng Chem Res 2009; 48(8): 3713-29.
[http://dx.doi.org/10.1021/ie801542g]

[72] Mizamoto K. Renewable biological systems for alternative sustainable energy production. FAO 1997.

[73] Lenihan P, Orozco A, O'Neill E, Ahmad MN, Rooney DW, Walker GM. Dilute acid hydrolysis of lignocellulosic biomass. Chem Eng J 2010; 156(2): 395-403.

[http://dx.doi.org/10.1016/j.cej.2009.10.061]

[74] Verardi A, De Bari I, Ricca E, Calabro V. Hydrolysis of lignolcellulosic biomass:current status of processes and technologies and future perspectives. 2012.

[75] Chandel AK, da Silva SS, Singh OV. Detoxification of lignocellulosic hydrolysates for improved bioethanol production. In: Aurelio M, Bernardes S, Eds. Biofuel Production-Recent Developments and Prospects, InTechOpen. 2011; pp. 225-46.

[76] Jönsson LJ, Alriksson B, Nilvebrant N-O. Bioconversion of lignocellulose: inhibitors and detoxification. Biotechnol Biofuels 2013; 6(1): 16. [http://dx.doi.org/10.1186/1754-6834-6-16] [PMID: 23356676]

[77] Kim D, Hahn J-S. Roles of the Yap1 transcription factor and antioxidants in *Saccharomyces cerevisiae*'s tolerance to furfural and 5-hydroxymethylfurfural, which function as thiol-reactive electrophiles generating oxidative stress. Appl Environ Microbiol 2013; 79(16): 5069-77. [http://dx.doi.org/10.1128/AEM.00643-13] [PMID: 23793623]

[78] Parawira W, Tekere M. Biotechnological strategies to overcome inhibitors in lignocellulose hydrolysates for ethanol production: review. Crit Rev Biotechnol 2011; 31(1): 20-31. [review]. [http://dx.doi.org/10.3109/07388551003757816] [PMID: 20513164]

[79] Alriksson B. Ethanol from lignocellulose: alkali detoxification of dilute-acid spruce hydrolysates 2006.

[80] Chandel AK, da Silva SS, Singh OV. Detoxification of lignocellulose hydrolysates: biochemical and metabolic engineering toward white biotechnology. BioEnergy Res 2013; 6(1): 388-401. [http://dx.doi.org/10.1007/s12155-012-9241-z]

[81] Wilson JJ, Deschatelets L, Nishikawa NK. Comparative fermentability of enzymatic and acid hydrolysates of steam-pretreated aspenwood hemicellulose by *Pichia stipitis* CBS 5776. Appl Microbiol Biotechnol 1989; 31(5): 592-6. [http://dx.doi.org/10.1007/BF00270801]

[82] Mussatto SI, Roberto IC. Alternatives for detoxification of diluted-acid lignocellulosic hydrolyzates for use in fermentative processes: a review. Bioresour Technol 2004; 93(1): 1-10. [http://dx.doi.org/10.1016/j.biortech.2003.10.005] [PMID: 14987714]

[83] Grzenia DL, Schell DJ, Ranil Wickramsinghe S. Detoxification of biomass hydrolysates by reactive membrane extraction. J Membr Sci 2010; 348(1-2): 6-12. [http://dx.doi.org/10.1016/j.memsci.2009.10.035]

[84] Gírio FM, Carvalheiro F, Duarte LC, Bogel-Łukasik R. Deconstruction of the hemicellulose fraction from lignocellulosic materials into simple sugars. In: Silva SS, Chandel AK, Eds. D-Xylitol. Berlin, Heidelberg: Springer Berlin Heidelberg 2012; pp. 3-37. [http://dx.doi.org/10.1007/978-3-642-31887-0_1]

[85] Persson P, Andersson J, Gorton L, Larsson S, Nilvebrant N-O, Jönsson LJ. Effect of different forms of alkali treatment on specific fermentation inhibitors and on the fermentability of lignocellulose hydrolysates for production of fuel ethanol. J Agric Food Chem 2002; 50(19): 5318-25. [http://dx.doi.org/10.1021/jf025565o] [PMID: 12207468]

[86] Prakasham RS, Rao RS, Hobbs PJ. Current trends in biotechnological production of xylitol and future prospects. Curr Trends Biotechnol Pharm 2009; 3(1): 8-36.

[87] Nilvebrant N-O, Reimann A, Larsson S, Jönsson LJ. Detoxification of lignocellulose hydrolysates with ion-exchange resins. Appl Biochem Biotechnol 2001; 91-93(1-9): 35-50. [http://dx.doi.org/10.1007/978-1-4612-0217-2_3]

[88] Liu ZL, Slininger PJ, Dien BS, Berhow MA, Kurtzman CP, Gorsich SW. Adaptive response of yeasts to furfural and 5-hydroxymethylfurfural and new chemical evidence for HMF conversion to 2,5-bi--hydroxymethylfuran. J Ind Microbiol Biotechnol 2004; 31(8): 345-52. [http://dx.doi.org/10.1007/s10295-004-0148-3] [PMID: 15338422]

[89] Heer D, Sauer U. Identification of furfural as a key toxin in lignocellulosic hydrolysates and evolution of a tolerant yeast strain. Microb Biotechnol 2008; 1(6): 497-506. [http://dx.doi.org/10.1111/j.1751-7915.2008.00050.x] [PMID: 21261870]

[90] Cesário MT, Raposo RS, de Almeida MC, van Keulen F, Ferreira BS, da Fonseca MM. Enhanced bioproduction of poly-3-hydroxybutyrate from wheat straw lignocellulosic hydrolysates. N Biotechnol 2014; 31(1): 104-13. [http://dx.doi.org/10.1016/j.nbt.2013.10.004] [PMID: 24157713]

[91] Lopes MS, Gosset G, Rocha RC, Gomez JG, Ferreira da Silva L. PHB biosynthesis in catabolite repression mutant of *Burkholderia sacchari.* Curr Microbiol 2011; 63(4): 319-26. [http://dx.doi.org/10.1007/s00284-011-9981-6] [PMID: 21761218]

[92] Keenan TM, Nakas JP, Tanenbaum SW. Polyhydroxyalkanoate copolymers from forest biomass. J Ind Microbiol Biotechnol 2006; 33(7): 616-26. [http://dx.doi.org/10.1007/s10295-006-0131-2] [PMID: 16761168]

[93] Pan W, Perrotta JA, Stipanovic AJ, Nomura CT, Nakas JP. Production of polyhydroxyalkanoates by *Burkholderia cepacia* ATCC 17759 using a detoxified sugar maple hemicellulosic hydrolysate. J Ind Microbiol Biotechnol 2012; 39(3): 459-69. [http://dx.doi.org/10.1007/s10295-011-1040-6] [PMID: 21953365]

[94] Pan W, Nomura CT, Nakas JP. Estimation of inhibitory effects of hemicellulosic wood hydrolysate inhibitors on PHA production by *Burkholderia cepacia* ATCC 17759 using response surface methodology. Bioresour Technol 2012; 125: 275-82. [http://dx.doi.org/10.1016/j.biortech.2012.08.107] [PMID: 23037882]

[95] Bowers T, Vaidya A, Smith DA, Lloyd-Jones G. Softwood hydrolysate as a carbon source for polyhydroxyalkanoate production. J Chem Technol 2014; 89(7): 1030-7.

[96] Silva JA, Tobella LM, Becerra J, Godoy F, Martínez MA. Biosynthesis of poly-β-hydroxyalkanoate by *Brevundimonas vesicularis* LMG P-23615 and *Sphingopyxis macrogoltabida* LMG 17324 using acid-hydrolyzed sawdust as carbon source. J Biosci Bioeng 2007; 103(6): 542-6. [http://dx.doi.org/10.1263/jbb.103.542] [PMID: 17630126]

[97] Salamanca-Cardona L, Ashe CS, Stipanovic AJ, Nomura CT. Enhanced production of polyhydroxyalkanoates (PHAs) from beechwood xylan by recombinant *Escherichia coli.* Appl Microbiol Biotechnol 2014; 98(2): 831-42. [http://dx.doi.org/10.1007/s00253-013-5398-4] [PMID: 24287934]

[98] Silva LF, Taciro MK, Michelin Ramos ME, Carter JM, Pradella JG, Gomez JG. Poly--hydroxybutyrate (P3HB) production by bacteria from xylose, glucose and sugarcane bagasse

hydrolysate. J Ind Microbiol Biotechnol 2004; 31(6): 245-54. [http://dx.doi.org/10.1007/s10295-004-0136-7] [PMID: 15221664]

[99] Yu J, Stahl H. Microbial utilization and biopolyester synthesis of bagasse hydrolysates. Bioresour Technol 2008; 99(17): 8042-8. [http://dx.doi.org/10.1016/j.biortech.2008.03.071] [PMID: 18474421]

[100] Gowda V, Shivakumar S. Agrowaste-based Polyhydroxyalkanoate (PHA) production using hydrolytic potential of *Bacillus thuringiensis* IAM 12077. Braz Arch Biol Technol 2014; 57(1): 55-61. [http://dx.doi.org/10.1590/S1516-89132014000100009]

[101] Alva Munoz LE, Riley MR. Utilization of cellulosic waste from tequila bagasse and production of polyhydroxyalkanoate (PHA) bioplastics by *Saccharophagus degradans*. Biotechnol Bioeng 2008; 100(5): 882-8. [http://dx.doi.org/10.1002/bit.21854] [PMID: 18383127]

[102] Obruca S, Benesova P, Petrik S, Oborna J, Prikryl R, Marova I. Production of polyhydroxyalkanoates using hydrolysate of spent coffee grounds. Process Biochem 2014; 49(9): 1409-14. [http://dx.doi.org/10.1016/j.procbio.2014.05.013]

[103] Prabu SC, Murugesan AG. Effective utilization and management of coir industrial waste for the production of poly-b- hydroxybutyrate (PHB) using the Bacterium *Azotobacter Beijerinickii*. Int J Environ Res 2010; 4(3): 519-24.

[104] Sindhu R, Silviya N, Binod P, Pandey A. Pentose-rich hydrolysate from acid pretreated rice straw as a carbon source for the production of poly-3-hydroxybutyrate. Biochem Eng J 2013; 78: 67-72. [http://dx.doi.org/10.1016/j.bej.2012.12.015]

[105] Zhang Y, Sun W, Wang H, Geng A. Polyhydroxybutyrate production from oil palm empty fruit bunch using *Bacillus megaterium* R11. Bioresour Technol 2013; 147: 307-14. [http://dx.doi.org/10.1016/j.biortech.2013.08.029] [PMID: 24001560]

[106] Kim S, Dale BE. Global potential bioethanol production from wasted crops and crop residues. Biomass Bioenergy 2004; 26(4): 361-75. [http://dx.doi.org/10.1016/j.biombioe.2003.08.002]

[107] Van-Thuoc D, Quillaguamán J, Mamo G, Mattiasson B. Utilization of agricultural residues for poly(3-hydroxybutyrate) production by *Halomonas boliviensis* LC1. J Appl Microbiol 2008; 104(2): 420-8. [PMID: 17887984]

[108] Cesário MT, Raposo RS, M D de Almeida MC, *et al.* Production of poly(3-hydroxybutyrate-c--4-hydroxybutyrate) by *Burkholderia sacchari* using wheat straw hydrolysates and gamma-butyrolactone. Int J Biol Macromol 2014; 71: 59-67. [http://dx.doi.org/10.1016/j.ijbiomac.2014.04.054] [PMID: 24811901]

[109] Saratale GD, Oh M-K. Characterization of poly-3-hydroxybutyrate (PHB) produced from *Ralstonia eutropha* using an alkali-pretreated biomass feedstock. Int J Biol Macromol 2015; 80: 627-35. [http://dx.doi.org/10.1016/j.ijbiomac.2015.07.034] [PMID: 26206741]

[110] Sawant SS, Salunke BK, Kim BS. Degradation of corn stover by fungal cellulase cocktail for production of polyhydroxyalkanoates by moderate halophile *Paracoccus sp.* LL1. Bioresour Technol 2015; 194: 247-55.

[http://dx.doi.org/10.1016/j.biortech.2015.07.019] [PMID: 26207871]

[111] Fischer G, Prieler S, van Velthuizen H, Berndes G, Faaij A, Londo M, *et al.* Biofuel production potentials in Europe: sustainable use of cultivated land and pastures, Part II. Biomass Bioenergy 2010; 34(2): 173-87.
[http://dx.doi.org/10.1016/j.biombioe.2009.07.009]

[112] Davis R, Kataria R, Cerrone F, *et al.* Conversion of grass biomass into fermentable sugars and its utilization for medium chain length polyhydroxyalkanoate (mcl-PHA) production by *Pseudomonas* strains. Bioresour Technol 2013; 150: 202-9.
[http://dx.doi.org/10.1016/j.biortech.2013.10.001] [PMID: 24177152]

[113] Koller M, Bona R, Hermann C, *et al.* Biotechnological production of poly(3-hydroxybutyrate) with *Wautersia eutropha* by application of green grass juice and silage juice as additional complex substrates. Biocatalysis Biotransform 2005; 23(5): 329-37.
[http://dx.doi.org/10.1080/10242420500292252]

[114] Radhika D, Murugesan AG. Bioproduction, statistical optimization and characterization of microbial plastic (poly 3-hydroxy butyrate) employing various hydrolysates of water hyacinth (*Eichhornia crassipes*) as sole carbon source. Bioresour Technol 2012; 121: 83-92.
[http://dx.doi.org/10.1016/j.biortech.2012.06.107] [PMID: 22858470]

[115] Zhang S, Norrlöw O, Wawrzynczyk J, Dey ES. Poly(3-hydroxybutyrate) biosynthesis in the biofilm of *Alcaligenes eutrophus*, using glucose enzymatically released from pulp fiber sludge. Appl Environ Microbiol 2004; 70(11): 6776-82.
[http://dx.doi.org/10.1128/AEM.70.11.6776-6782.2004] [PMID: 15528544]

[116] Nduko JM, Suzuki W, Matsumoto K, *et al.* Polyhydroxyalkanoates production from cellulose hydrolysate in *Escherichia coli* LS5218 with superior resistance to 5-hydroxymethylfurfural. J Biosci Bioeng 2012; 113(1): 70-2.
[http://dx.doi.org/10.1016/j.jbiosc.2011.08.021] [PMID: 21993429]

[117] Olbrich H. The Molasses. Berlin 1963.

[118] Solaiman DK, Ashby RD, Foglia TA, Marmer WN. Conversion of agricultural feedstock and coproducts into poly(hydroxyalkanoates). Appl Microbiol Biotechnol 2006; 71(6): 783-9.
[http://dx.doi.org/10.1007/s00253-006-0451-1] [PMID: 16708192]

[119] The Origins Of Molasses [homepage on the Internet]. Bremen: Hansa Melasse Handelsgesellschaft mbH [cited 11th September 2015]. Available from: http://www.melasse.de/index.php?id=originsofmo-lasses

[120] Gomaa E. Production of polyhydroxyalkanoates (PHAs) by *Bacillus subtilis* and *Escherichia coli* grown on cane molasses fortified with ethanol. Braz Arch Biol Technol 2014; 57(1): 145-54.

[121] El-Sheekh M, El-Abd M, El-Diwany A, Ismail A, Omar T. Poly-3-hydroxybutyrate (PHB) production by *Bacillus flexus* ME-77 using some industrial wastes. Rendiconti Lincei 2015; 26(2): 109-19.
[http://dx.doi.org/10.1007/s12210-014-0368-z]

[122] Page WJ, Manchak J, Rudy B. Formation of poly(hydroxybutyrate-co-hydroxyvalerate) by *Azotobacter vinelandii* UWD. Appl Environ Microbiol 1992; 58(9): 2866-73.
[PMID: 1444399]

[123] Kulpreecha S, Boonruangthavorn A, Meksiriporn B, Thongchul N. Inexpensive fed-batch cultivation for high poly(3-hydroxybutyrate) production by a new isolate of *Bacillus megaterium.* J Biosci Bioeng 2009; 107(3): 240-5.
[http://dx.doi.org/10.1016/j.jbiosc.2008.10.006] [PMID: 19269585]

[124] Albuquerque MG, Concas S, Bengtsson S, Reis MA. Mixed culture polyhydroxyalkanoates production from sugar molasses: the use of a 2-stage CSTR system for culture selection. Bioresour Technol 2010; 101(18): 7123-33.
[http://dx.doi.org/10.1016/j.biortech.2010.04.019] [PMID: 20434907]

[125] Omar S, Rayes A, Eqaab A, Voß I, Steinbüchel A. Optimization of cell growth and poly(3-hydroxybutyrate) accumulation on date syrup by a *Bacillus megaterium* strain. Biotechnol Lett 2001; 23(14): 1119-23.
[http://dx.doi.org/10.1023/A:1010559800535]

[126] Gouda MK, Swellam AE, Omar SH. Production of PHB by a *Bacillus megaterium* strain using sugarcane molasses and corn steep liquor as sole carbon and nitrogen sources. Microbiol Res 2001; 156(3): 201-7.
[http://dx.doi.org/10.1078/0944-5013-00104] [PMID: 11716209]

[127] Solaiman DK, Ashby RD, Hotchkiss AT Jr, Foglia TA. Biosynthesis of medium-chain-length poly(hydroxyalkanoates) from soy molasses. Biotechnol Lett 2006; 28(3): 157-62.
[http://dx.doi.org/10.1007/s10529-005-5329-2] [PMID: 16489492]

[128] Full TD, Jung DO, Madigan MT. Production of poly-β-hydroxyalkanoates from soy molasses oligosaccharides by new, rapidly growing *Bacillus* species. Lett Appl Microbiol 2006; 43(4): 377-84.
[http://dx.doi.org/10.1111/j.1472-765X.2006.01981.x] [PMID: 16965367]

[129] Law K, Leung Y, Lawford H, Chua H, Lo W, Yu P. Production of polyhydroxybutyrate by *Bacillus* species isolated from municipal activated sludge. Twenty-Second Symposium on Biotechnology for Fuels and Chemicals. Totowa, NJ: Humana Press 2001.
[http://dx.doi.org/10.1007/978-1-4612-0217-2_44]

[130] Liu F, Li W, Ridgway D, Gu T. Production of poly-β-hydroxybutyrate on molasses by recombinant *Escherichia coli.* Biotechnol Lett 1998; 4(20): 345-8.
[http://dx.doi.org/10.1023/A:1005367011378]

[131] Zhang H, Obias V, Gonyer K, Dennis D. Production of polyhydroxyalkanoates in sucrose-utilizing recombinant *Escherichia coli* and *Klebsiella* strains. Appl Environ Microbiol 1994; 60(4): 1198-205.
[PMID: 8017916]

[132] Mawson AJ. Bioconversions for whey utilization and waste abatement. Bioresour Technol 1994; 47(3): 195-203.
[http://dx.doi.org/10.1016/0960-8524(94)90180-5]

[133] Siso MI. The biotechnological utilization of cheese whey: a review. Bioresour Technol 1996; 57(1): 1-11.
[http://dx.doi.org/10.1016/0960-8524(96)00036-3]

[134] Kilara A, Vaqghela MN. Whey proteins. In: Yada RY, Ed. Proteins in food processing. Woodhead Publishing 2005; pp. 73-99.

[135] Panesar PS, Kennedy JF, Gandhi DN, Bunko K. Bioutilization of whey for lactic acid production. Food Chem 2007; 105(1): 1-14. [http://dx.doi.org/10.1016/j.foodchem.2007.03.035]

[136] Marwaha SS, Kennedy JF. Whey pollution problem and potential utilization. Int J Food Sci Technol 1988; 23(4): 323-36. [http://dx.doi.org/10.1111/j.1365-2621.1988.tb00586.x]

[137] Parris N, Anema G, Singh G, Creamer LK. Aggregation of whey proteins in heated sweet whey. J Agric Food Chem 1993; 41(3): 160-4. [http://dx.doi.org/10.1021/jf00027a021]

[138] Obruca S, Marova I, Melusova S, Mravcova L. Production of polyhydroxyalkanoates from cheese whey by employing *Bacillus megaterium* CCM 2037. Ann Microbiol 2011; 61(4): 947-53. [http://dx.doi.org/10.1007/s13213-011-0218-5]

[139] Evans MT, Gordon JF. Whey proteins. In: Grant RA, Ed. Applied Proteins. London: Applied Science Publisher 1980; pp. 31-67.

[140] Vasala A, Panula J, Neubauer P. Efficient lactic acid production from high salt containing dairy by-products by Lactobacillus salivarius ssp. salicinius with pre-treatment by proteolytic microorganisms. J Biotechnol 2005; 117(4): 421-31. [http://dx.doi.org/10.1016/j.jbiotec.2005.02.010] [PMID: 15876466]

[141] Margot A, Flaschel E, Renken A. Continuous monitoring of enzymatic whey protein hydrolysis. Correlation of base consumption with soluble nitrogen content. Process Biochem 1994; 29(4): 257-62. [http://dx.doi.org/10.1016/0032-9592(94)80066-9]

[142] Senthuran A, Senthuran V, Mattiasson B, Kaul R. Lactic acid fermentation in a recycle batch reactor using immobilized Lactobacillus casei. Biotechnol Bioeng 1997; 55(6): 841-53. [http://dx.doi.org/10.1002/(SICI)1097-0290(19970920)55:6<841::AID-BIT3>3.0.CO;2-G] [PMID: 18636595]

[143] Koller M, Bona R, Chiellini E, *et al.* Polyhydroxyalkanoate production from whey by Pseudomonas hydrogenovora. Bioresour Technol 2008; 99(11): 4854-63. [http://dx.doi.org/10.1016/j.biortech.2007.09.049] [PMID: 18053709]

[144] Marangoni C, Furigo A, Aragao GM. Production of poly(3-hydroxybutyrate-co-3-hydroxyvalerate) by *Ralstonia eutropha* in whey and inverted sugar with propionic acid feeding. Process Biochem 2002; 38(2): 137-41. [http://dx.doi.org/10.1016/S0032-9592(01)00313-2]

[145] Kosseva MR, Panesar PS, Kaur G, Kennedy JF. Use of immobilised biocatalysts in the processing of cheese whey. Int J Biol Macromol 2009; 45(5): 437-47. [http://dx.doi.org/10.1016/j.ijbiomac.2009.09.005] [PMID: 19766668]

[146] Amrane A. Analysis of the kinetics of growth and lactic acid production for *Lactobacillus helveticus* growing on supplemented whey permeate. J Chem Technol Biotechnol 2005; 80(3): 345-52. [http://dx.doi.org/10.1002/jctb.1211]

[147] Marova I, Certik M, Breierova E. Production of enriched biomass by carotenogenic yeasts-application of whole-cell yeast biomass to production of pigments and other lipid compounds. In: Matovic D, Ed.

Biomass - detection, production and usage. Rijeka: InTech 2011; pp. 345-84. [http://dx.doi.org/10.5772/19235]

[148] Amrane A. Effect of inorganic phosphate on lactate production by *Lactobacillus helveticus*. J Gen Appl Microbiol 2000; 49(3): 21-7. [PMID: 12682863]

[149] Liu C, Liu Y, Liao W, Wen Z, Chen S. Application of statistically-based experimental designs for the optimization of nisin production from whey. Biotechnol Lett 2003; 25(11): 877-82. [http://dx.doi.org/10.1023/A:1024009027255] [PMID: 12889798]

[150] Nikel PI, Pettinari MJ, Méndez BS, Galvagno MA. Statistical optimization of a culture medium for biomass and poly(3-hydroxybutyrate) production by a recombinant Escherichia coli strain using agroindustrial byproducts. Int Microbiol 2005; 8(4): 243-50. [PMID: 16562376]

[151] Khanafari A, Sepahei AA. Alginate biopolymer production *by Azotobacter chroococum* from whey degradation. Int J Environ Sci Technol 2007; 4(4): 427-32. [http://dx.doi.org/10.1007/BF03325977]

[152] Yellore V, Desai A. Production of poly-3-hydroxybutyrate from lactose and whey by *Methylobacterium* sp. ZP24. Lett Appl Microbiol 1998; 26(6): 391-4. [http://dx.doi.org/10.1046/j.1472-765X.1998.00362.x] [PMID: 9717306]

[153] Nath A, Dixit M, Bandiya A, Chavda S, Desai AJ. Enhanced PHB production and scale up studies using cheese whey in fed batch culture of *Methylobacterium* sp. ZP24. Bioresour Technol 2008; 99(13): 5749-55. [http://dx.doi.org/10.1016/j.biortech.2007.10.017] [PMID: 18032031]

[154] Pandian SR, Deepak V, Kalishwaralal K, Rameshkumar N, Jeyaraj M, Gurunathan S. Optimization and fed-batch production of PHB utilizing dairy waste and sea water as nutrient sources by *Bacillus megaterium* SRKP-3. Bioresour Technol 2010; 101(2): 705-11. [http://dx.doi.org/10.1016/j.biortech.2009.08.040] [PMID: 19744854]

[155] Dhanasekar R, Viruthagiri T, Sabarathinam PL. Biosynthesis of poly(3-hydroxybutyrate) from cheese whey using *Azotobacter vinelandii.* Ind J Chem Technol 2001; 8(1): 68-71.

[156] Pantazaki AA, Papaneophytou CP, Pritsa AG, Liakopoulus-Kyriakides M, Kyriakidis DA. Production of polyhydroxyalkanoates from whey by *Thermus thermophilus* HB8. Process Biochem 2009; 44(8): 847-53. [http://dx.doi.org/10.1016/j.procbio.2009.04.002]

[157] Povolo S, Romanelli MG, Basaglia M, *et al.* Polyhydroxyalkanoate biosynthesis by *Hydrogenophaga pseudoflava* DSM1034 from structurally unrelated carbon sources. N Biotechnol 2013; 30(6): 629-34. [http://dx.doi.org/10.1016/j.nbt.2012.11.019] [PMID: 23201074]

[158] Koller M, Hesse P, Salerno A, Reiterer A, Braunegg G. A viable antibiotic strategy against microbial contamination in biotechnological production of polyhydroxyalkanoates from surplus whey. Biomass Bioenergy 2011; 35(1): 748-53. [http://dx.doi.org/10.1016/j.biombioe.2010.10.008]

[159] Koller M, Hesse P, Bona R, Kutschera C, Atlić A, Braunegg G. Biosynthesis of High Quality

Polyhydroxyalkanoate Co- and Terpolyesters for Potential Medical Application by the Archaeon *Haloferax mediterranei.* Macromol Symp 2007; 253(1): 33-9.
[http://dx.doi.org/10.1002/masy.200750704]

[160] Koller M. Recycling of waste streams of the biotechnological poly(hydroxyalkanoate) production by *Haloferax mediterranei* on whey. Int J Polym Sci 2015; 1-8.
[http://dx.doi.org/10.1155/2015/370164]

[161] Obruca S, Benesova P, Oborna J, Marova I. Application of protease-hydrolyzed whey as a complex nitrogen source to increase poly(3-hydroxybutyrate) production from oils by *Cupriavidus necator.* Biotechnol Lett 2014; 36(4): 775-81.
[http://dx.doi.org/10.1007/s10529-013-1407-z] [PMID: 24243232]

[162] Povolo S, Toffano P, Basaglia M, Casella S. Polyhydroxyalkanoates production by engineered *Cupriavidus necator* from waste material containing lactose. Bioresour Technol 2010; 101(20): 7902-7.
[http://dx.doi.org/10.1016/j.biortech.2010.05.029] [PMID: 20537531]

[163] Ahn WS, Park SJ, Lee SY. Production of Poly(3-hydroxybutyrate) by fed-batch culture of recombinant Escherichia coli with a highly concentrated whey solution. Appl Environ Microbiol 2000; 66(8): 3624-7.
[http://dx.doi.org/10.1128/AEM.66.8.3624-3627.2000] [PMID: 10919830]

[164] Ahn WS, Park SJ, Lee SY. Production of poly(3-hydroxybutyrate) from whey by cell recycle fed-batch culture of recombinant *Escherichia coli.* Biotechnol Lett 2001; 23(8): 235-40.
[http://dx.doi.org/10.1023/A:1005633418161]

[165] Kim BS. Production of poly(3-hydroxybutyrate) from inexpensive substrates. Enzyme Microb Technol 2000; 27(10): 774-7.
[http://dx.doi.org/10.1016/S0141-0229(00)00299-4] [PMID: 11118585]

[166] Nikel PI, de Almeida A, Melillo EC, Galvagno MA, Pettinari MJ. New recombinant Escherichia coli strain tailored for the production of poly(3-hydroxybutyrate) from agroindustrial by-products. Appl Environ Microbiol 2006; 72(6): 3949-54.
[http://dx.doi.org/10.1128/AEM.00044-06] [PMID: 16751501]

[167] Lee SY, Choi J. Effect of fermentation performance on the economics of poly (3-hydroxybutyrate) production by *Alcaligenes latus.* Polym Degrad Stabil 1999; 59(1): 387-93.

[168] Chanprateep S. Current trends in biodegradable polyhydroxyalkanoates. J Biosci Bioeng 2010; 110(6): 621-32.
[http://dx.doi.org/10.1016/j.jbiosc.2010.07.014] [PMID: 20719562]

[169] Meher L. VIidyasagar D, Naik S. Technical aspects of biodiesel production by transesterification-a review. Renew Sustain Energy Rev 2006; 10(3): 248-68.
[http://dx.doi.org/10.1016/j.rser.2004.09.002]

[170] Rywińska A, Juszczyk P, Wojtatowicz M, *et al.* Glycerol as a promising substrate for *Yarrowia lipolytica* biotechnological applications. Biomass Bioenerg 2013; 48: 148-66.

[171] Schievano A, D'Imporzano G, Adani F. Substituting energy crops with organic wastes and agro-industrial residues for biogas production. J Environ Manage 2009; 90(8): 2537-41.

[http://dx.doi.org/10.1016/j.jenvman.2009.01.013] [PMID: 19254824]

[172] Tan HW, Abdul Aziz AR, Aroua MK. Glycerol production and its applications as a raw material: A review. Renew Sustain Ener Rev 2013; 27: 118-27.
[http://dx.doi.org/10.1016/j.rser.2013.06.035]

[173] Moita R, Freches A, Lemos PC. Crude glycerol as feedstock for polyhydroxyalkanoates production by mixed microbial cultures. Water Res 2014; 58(58): 9-20.
[http://dx.doi.org/10.1016/j.watres.2014.03.066] [PMID: 24731872]

[174] Thompson JC, He BB. Characterization of crude glycerol from biodiesel production from multiple feedstocks. Appl Eng Agric 2006; (22): 261-5.
[http://dx.doi.org/10.13031/2013.20272]

[175] Sdrula N. A study using classical or membrane separation in the biodiesel process. Desalination 2010; 250(3): 1070-2.
[http://dx.doi.org/10.1016/j.desal.2009.09.110]

[176] Amaral PF, Ferreira TF, Fontes GC, Coelho MA. Glycerol valorization: New biotechnological routes. Food Bioprod Process 2009; 87(3): 179-86.
[http://dx.doi.org/10.1016/j.fbp.2009.03.008]

[177] Cavalheiro JM, de Almeida MC, Grandfils C, da Fonseca MM. Poly(3-hydroxybutyrate) production by *Cupriavidus necator* using waste glycerol. Process Biochem 2009; 44(5): 509-15.
[http://dx.doi.org/10.1016/j.procbio.2009.01.008]

[178] Kachrimanidou V, Kopsahelis N, Papanikolaou S, *et al.* Sunflower-based biorefinery: poly(3-hydroxybutyrate) and poly(3-hydroxybutyrate-co-3-hydroxyvalerate) production from crude glycerol, sunflower meal and levulinic acid. Bioresour Technol 2014; 172: 121-30.
[http://dx.doi.org/10.1016/j.biortech.2014.08.044] [PMID: 25255188]

[179] García IL, López JA, Dorado MP, *et al.* Evaluation of by-products from the biodiesel industry as fermentation feedstock for poly(3-hydroxybutyrate-co-3-hydroxyvalerate) production by *Cupriavidus necator*. Bioresour Technol 2013; 130: 16-22.
[http://dx.doi.org/10.1016/j.biortech.2012.11.088] [PMID: 23280181]

[180] Zhu C, Nomura CT, Perrotta JA, Stipanovic AJ, Nakas JP. Production and characterization of poly--hydroxybutyrate from biodiesel-glycerol by *Burkholderia cepacia* ATCC 17759. Biotechnol Prog 2010; 26(2): 424-30.
[PMID: 19953601]

[181] Hermann-Krauss C, Koller M, Muhr A, Fasl H, Stelzer F, Braunegg G. Archaeal production of polyhydroxyalkanoate (PHA) co- and terpolyesters from biodiesel industry-derived by-products. Archaea 2013; 2013: 129268.
[http://dx.doi.org/10.1155/2013/129268] [PMID: 24453697]

[182] Castro JS. Bioconversion of commercial and waste glycerol into value-added polyhydroxyalkanoates by bacterial strains. J Microbiol 2014; 6(6)

[183] Palmeri R, Pappalardo F, Fragala M, Tomasello M, Damigella A, Catara AF. Polyhydroxyalkanoates (PHAs) production through conversion of glycerol by selected strains of *Pseudomonas Mediterranea* and *Pseudomonas Corrugata.* Chem Eng Trans 2012; 27: 121-6.

[184] Javers J, Karunanithy C. Polyhydroxyalkanoate production by *Pseudomonas putida* KT217 on a condensed corn solubles based medium fed with glycerol water or sunflower soapstock. Adv Microbiol 2012; 2(3): 241-51.
[http://dx.doi.org/10.4236/aim.2012.23029]

[185] Pappalardo F, Fragalà M, Mineo PG, *et al.* Production of filmable medium-chain-length polyhydroxyalkanoates produced from glycerol by *Pseudomonas mediterranea.* Int J Biol Macromol 2014; 65: 89-96.
[http://dx.doi.org/10.1016/j.ijbiomac.2014.01.014] [PMID: 24418340]

[186] Ashby RD, Foglia TA. Poly(hydroxyalkanoate) biosynthesis from triglyceride substrates. Appl Microbiol Biotechnol 1998; 49(4): 431-7.
[http://dx.doi.org/10.1007/s002530051194]

[187] Kenny ST, Runic JN, Kaminsky W, Woods T, Babu RP, O'Connor KE. Development of a bioprocess to convert PET derived terephthalic acid and biodiesel derived glycerol to medium chain length polyhydroxyalkanoate. Appl Microbiol Biotechnol 2012; 95(3): 623-33.
[http://dx.doi.org/10.1007/s00253-012-4058-4] [PMID: 22581066]

[188] Barnali B, Basu RK, Ash SN. Kinetic studies on acid catalysed hydrolysis of starch. J Sci Ind Res 2008; 67: 295-8.

[189] Doué G, Bédikou M, Koua G, Mégnanou R-M, Niamké S. Enzymatic and acid conversion of new starches from improved orphan crops: prospects for renewable materials uses in food and non-food industries. Springerplus 2014; 3(1): 498-504.
[http://dx.doi.org/10.1186/2193-1801-3-498] [PMID: 25279291]

[190] Lu DR. Starch-based completely biodegradable polymer materials. Express Polym Lett 2009; 3(6): 366-75.
[http://dx.doi.org/10.3144/expresspolymlett.2009.46]

[191] Rainbault M. General and microbiological aspects of solid substrate fermentation. Electron J Biotechnol 1998; 1(3)

[192] Alcázar-Alay SC, Meireles MA. Physicochemical properties, modifications and applications of starches from different botanical sources. Food Sci Technol (Campinas) 2015; 35(2): 215-36.
[http://dx.doi.org/10.1590/1678-457X.6749]

[193] Smith AM. The biosynthesis of starch granules. Biomacromolecules 2001; 2(2): 335-41.
[http://dx.doi.org/10.1021/bm000133c] [PMID: 11749190]

[194] Hii SL, Tan JS, Ling TC, Ariff AB. Pullulanase: role in starch hydrolysis and potential industrial applications. Enzyme Res 2012; 1-14.
[http://dx.doi.org/10.1155/2012/921362]

[195] Buléon A, Colonna P, Planchot V, Ball S. Starch granules: structure and biosynthesis. Int J Biol Macromol 1998; 23(2): 85-112.
[http://dx.doi.org/10.1016/S0141-8130(98)00040-3] [PMID: 9730163]

[196] Zeeman SC, Kossmann J, Smith AM. Starch: its metabolism, evolution, and biotechnological modification in plants. Annu Rev Plant Biol 2010; 61(1): 209-34.
[http://dx.doi.org/10.1146/annurev-arplant-042809-112301] [PMID: 20192737]

[197] Pervez S, Aman A, Iqbal S, Siddiqui NN, Ul Qader SA. Saccharification and liquefaction of cassava starch: an alternative source for the production of bioethanol using amylolytic enzymes by double fermentation process. BMC Biotechnol 2014; 14(14): 49.
[http://dx.doi.org/10.1186/1472-6750-14-49] [PMID: 24885587]

[198] Krueger CL, Radetski CM, Bendia AG, *et al.* Bioconversion of cassava starch by-product into *Bacillus* and related bacteria polyhydroxyalkanoates. Electron J Biotechnol 2012; 15(3)

[199] Poomipuk N, Reungsang A, Plangklang P. Poly-β-hydroxyalkanoates production from cassava starch hydrolysate by *Cupriavidus sp.* KKU38. Int J Biol Macromol 2014; 65: 51-64.
[http://dx.doi.org/10.1016/j.ijbiomac.2014.01.002] [PMID: 24412153]

[200] Blennow A, Wischmann B, Houborg K, *et al.* Structure function relationships of transgenic starches with engineered phosphate substitution and starch branching. Int J Biol Macromol 2005; 36(3): 159-68.
[http://dx.doi.org/10.1016/j.ijbiomac.2005.05.006] [PMID: 16024070]

[201] Shewry PR, Hawkesford MJ, Piironen V, *et al.* Natural variation in grain composition of wheat and related cereals. J Agric Food Chem 2013; 61(35): 8295-303.
[http://dx.doi.org/10.1021/jf3054092] [PMID: 23414336]

[202] Patni N, Pillai SG, Dwivedi AH. Wheat as a promising substitute of corn for bioethanol production. Proc Eng 2013; 51: 355-62.
[http://dx.doi.org/10.1016/j.proeng.2013.01.049]

[203] Ranum P, Peña-Rosas JP, Garcia-Casal MN. Global maize production, utilization, and consumption. Ann N Y Acad Sci 2014; 1312(1): 105-12.
[http://dx.doi.org/10.1111/nyas.12396] [PMID: 24650320]

[204] Ptaszek A, Berski W, Ptaszek P, Witczak T, Repelewicz U, Grzesik M. Viscoelastic properties of waxy maize starch and selected non-starch hydrocolloids gels. Carbohydr Polym 2009; 76(4): 567-77.
[http://dx.doi.org/10.1016/j.carbpol.2008.11.023]

[205] Lubiewski Z, Le Thanh J, Lewandowicz G. Increasing the efficiency of the enzymatic hydrolysis of potato starch in a membrane reactor. Biotechnol Bioprocess Eng 2010; 15(6): 917-22.
[http://dx.doi.org/10.1007/s12257-009-0094-z]

[206] Satyanarayana T, Noorwez SM, Kumar S, Rao JL, Ezhilvannan M, Kaur P. Development of an ideal starch saccharification process using amylolytic enzymes from thermophiles. Biochem Soc Trans 2004; 32(Pt 2): 276-8.
[http://dx.doi.org/10.1042/bst0320276] [PMID: 15046588]

[207] Lorenz K, Johnson JA. Starch hydrolysis under high temperatures and pressures. Cereal Chem 1972; 49: 616-28.

[208] Legin E, Copinet A, Duchiron F. A Single step high temperature hydrolysis of wheat starch. Starch 1998; 50(2-3): 84-9.
[http://dx.doi.org/10.1002/(SICI)1521-379X(199803)50:2/3<84::AID-STAR84>3.0.CO;2-4]

[209] Sarbatly R. The simultineous enzymatic hydrolysis of Tapioca starch for instant formation of glucose. J Appl Sci 2007; 7(15): 2057-62.
[http://dx.doi.org/10.3923/jas.2007.2057.2062]

[210] Wang P, Singh V, Xue H, Johnston D, Rausch K, Tumbleson ME. Comparison of raw starch hydrolyzing enzyme with conventional liquefaction and saccharification enzymes in dry-grind corn processing. Cereal Chem 2007; 84(1): 10-4.
[http://dx.doi.org/10.1094/CCHEM-84-1-0010]

[211] Li Y, Liu B, Song J, Jiang C, Yang Q. Utilization of potato starch processing wastes to produce animal feed with high lysine content. J Microbiol Biotechnol 2015; 25(2): 178-84.
[http://dx.doi.org/10.4014/jmb.1404.04035] [PMID: 25189407]

[212] Haas R, Jin B, Zepf FT. Production of Poly(3-hydroxybutyrate) from waste potato starch. Biosci Biotechnol Biochem 2008; 72(1): 253-6.
[http://dx.doi.org/10.1271/bbb.70503] [PMID: 18175895]

[213] Koutinas AA, Xu Y, Wang R, Webb C. Polyhydroxybutyrate production from a novel feedstock derived from a wheat-based biorefinery. Enzyme Microb Technol 2007; 40(5): 1035-44.
[http://dx.doi.org/10.1016/j.enzmictec.2006.08.002]

[214] Huang TY, Duan K-J, Huang S-Y, Chen CW. Production of polyhydroxyalkanoates from inexpensive extruded rice bran and starch by *Haloferax mediterranei.* J Ind Microbiol Biotechnol 2006; 33(8): 701-6.
[http://dx.doi.org/10.1007/s10295-006-0098-z] [PMID: 16491353]

[215] Chaleomrum N, Chookietwattana K, Dararat S. Production of PHA from cassava starch wastewater in sequencing batch reactor treatment system. APCBEE Procedia 2014; 8: 167-72.
[http://dx.doi.org/10.1016/j.apcbee.2014.03.021]

[216] Quillaguamán J, Hashim S, Bento F, Mattiasson B, Hatti-Kaul R. Poly(β-hydroxybutyrate) production by a moderate halophile, *Halomonas boliviensis* LC1 using starch hydrolysate as substrate. J Appl Microbiol 2005; 99(1): 151-7.
[http://dx.doi.org/10.1111/j.1365-2672.2005.02589.x] [PMID: 15960675]

[217] Song Y, Matsumoto K, Tanaka T, Kondo A, Taguchi S. Single-step production of polyhydroxybutyrate from starch by using α-amylase cell-surface displaying system of Corynebacterium glutamicum. J Biosci Bioeng 2013; 115(1): 12-4.
[http://dx.doi.org/10.1016/j.jbiosc.2012.08.004] [PMID: 22959444]

[218] Sheu D-S, Chen W-M, Yang J-Y, Chang R-C. Thermophilic bacterium *Caldimonas taiwanensis* produces poly(3-hydroxybutyrate-co-3-hydroxyvalerate) from starch and valerate as carbon sources. Enzyme Microb Technol 2009; 44(5): 289-94.
[http://dx.doi.org/10.1016/j.enzmictec.2009.01.004]

[219] Halami PM. Production of polyhydroxyalkanoate from starch by the native isolate *Bacillus cereus* CFR06. World J Microbiol Biotechnol 2008; 24(6): 805-12.
[http://dx.doi.org/10.1007/s11274-007-9543-z]

[220] Ward PG, Goff M, Donner M, Kaminsky W, O'Connor KE. A two step chemo-biotechnological conversion of polystyrene to a biodegradable thermoplastic. Environ Sci Technol 2006; 40(7): 2433-7.
[http://dx.doi.org/10.1021/es0517668] [PMID: 16649270]

[221] Goff M, Ward PG, O'Connor KE. Improvement of the conversion of polystyrene to polyhydroxyalkanoate through the manipulation of the microbial aspect of the process: a nitrogen

feeding strategy for bacterial cells in a stirred tank reactor. J Biotechnol 2007; 132(3): 283-6. [http://dx.doi.org/10.1016/j.jbiotec.2007.03.016] [PMID: 17559958]

[222] Tan G-Y, Chen C-L, Ge L, Li L, Tan SN, Wang J-Y. Bioconversion of styrene to poly(hydroxyalkanoate) (PHA) by the new bacterial strain *Pseudomonas putida* NBUS12. Microbes Environ 2015; 30(1): 76-85. [http://dx.doi.org/10.1264/jsme2.ME14138] [PMID: 25740622]

[223] Kenny ST, Runic JN, Kaminsky W, *et al.* Up-cycling of PET (polyethylene terephthalate) to the biodegradable plastic PHA (polyhydroxyalkanoate). Environ Sci Technol 2008; 42(20): 7696-701. [http://dx.doi.org/10.1021/es801010e] [PMID: 18983095]

[224] Guzik MW, Kenny ST, Duane GF, *et al.* Conversion of post consumer polyethylene to the biodegradable polymer polyhydroxyalkanoate. Appl Microbiol Biotechnol 2014; 98(9): 4223-32. [http://dx.doi.org/10.1007/s00253-013-5489-2] [PMID: 24413975]

[225] Bengtsson S, Hallquist J, Werker A, Welander T. Acidogenic fermentation of industrial wastewaters: Effects of chemostat retention time and pH on volatile fatty acids production. Biochem Eng J 2008; 40(3): 492-9. [http://dx.doi.org/10.1016/j.bej.2008.02.004]

[226] Venkata Mohan S, Reddy MV, Subhash GV, Sarma PN. Fermentative effluents from hydrogen producing bioreactor as substrate for poly(β-OH) butyrate production with simultaneous treatment: an integrated approach. Bioresour Technol 2010; 101(23): 9382-6. [http://dx.doi.org/10.1016/j.biortech.2010.06.109] [PMID: 20667721]

[227] Bengtsson S, Werker A, Christensson M, Welander T. Production of polyhydroxyalkanoates by activated sludge treating a paper mill wastewater. Bioresour Technol 2008; 99(3): 509-16. [http://dx.doi.org/10.1016/j.biortech.2007.01.020] [PMID: 17360180]

[228] Bengtsson S, Pisco AR, Reis MA, Lemos PC. Production of polyhydroxyalkanoates from fermented sugar cane molasses by a mixed culture enriched in glycogen accumulating organisms. J Biotechnol 2010; 145(3): 253-63. [http://dx.doi.org/10.1016/j.jbiotec.2009.11.016] [PMID: 19958801]

[229] Kek YK, Chang CW, Amirul AA, Sudesh K. Heterologous expression of *Cupriavidus* sp. USMAA2-4 PHA synthase gene in PHB^{-4} mutant for the production of poly(3-hydroxybutyrate) and its copolymers. World J Microbiol Biotechnol 2010; 26(9): 1595-603. [http://dx.doi.org/10.1007/s11274-010-0335-5]

[230] Taniguchi I, Kagotani K, Kimura Y. Microbial production of poly(hydroxyalkanoates)s from waste edible oils. Green Chem 2003; 5: 545-8. [http://dx.doi.org/10.1039/B304800B]

[231] Fukui T, Doi Y. Efficient production of polyhydroxyalkanoates from plant oils by *Alcaligenes eutrophus* and its recombinant strain. Appl Microbiol Biotechnol 1998; 49(3): 333-6. [http://dx.doi.org/10.1007/s002530051178] [PMID: 9581296]

[232] Wong YM, Brigham CJ, Rha C, Sinskey AJ, Sudesh K. Biosynthesis and characterization of polyhydroxyalkanoate containing high 3-hydroxyhexanoate monomer fraction from crude palm kernel oil by recombinant *Cupriavidus necator*. Bioresour Technol 2012; 121: 320-7. [http://dx.doi.org/10.1016/j.biortech.2012.07.015] [PMID: 22858502]

[233] Batcha AF, Prasad DM, Khan MR, Abdullah H. Biosynthesis of poly(3-hydroxybutyrate) (PHB) by *Cupriavidus necator* H16 from jatropha oil as carbon source. Bioprocess Biosyst Eng 2014; 37(5): 943-51.
[http://dx.doi.org/10.1007/s00449-013-1066-4] [PMID: 24078185]

[234] Verlinden RA, Hill DJ, Kenward MA, Williams CD, Piotrowska-Seget Z, Radecka IK. Production of polyhydroxyalkanoates from waste frying oil by *Cupriavidus necator*. AMB Express 2011; 1(1): 11.
[http://dx.doi.org/10.1186/2191-0855-1-11] [PMID: 21906352]

[235] Obruca S, Snajdar O, Svoboda Z, Marova I. Application of random mutagenesis to enhance the production of polyhydroxyalkanoates by *Cupriavidus necator* H16 on waste frying oil. World J Microbiol Biotechnol 2013; 29(12): 2417-28.
[http://dx.doi.org/10.1007/s11274-013-1410-5] [PMID: 23801326]

[236] Morais C, Freitas F, Cruz MV, Paiva A, Dionísio M, Reis MA. Conversion of fat-containing waste from the margarine manufacturing process into bacterial polyhydroxyalkanoates. Int J Biol Macromol 2014; 71: 68-73.
[http://dx.doi.org/10.1016/j.ijbiomac.2014.04.044] [PMID: 24794198]

[237] Kamilah H, Sudesh K, Yang TA. Characteristics of used palm olein and its bioconversion into polyhydroxybutyrate by *Cupriavidus necator* H16. Malaysian J Microbiol 2014; 10(2): 139-48.

[238] Ramsay JA, Hassan MC, Ramsay BA. Hemicellulose as a potential substrate for production of poly(β-hydroxyalkanoates). Can J Microbiol 1995; 41(13): 262-6.
[http://dx.doi.org/10.1139/m95-195]

[239] Lopes MS, Rocha RC, Zanotto SP, Gomez JG, Silva LF. Screening of bacteria to produce polyhydroxyalkanoates from xylose. World J Microbiol Biotechnol 2009; 25(10): 1751-6.
[http://dx.doi.org/10.1007/s11274-009-0072-9]

[240] Thirumala M. Production of polyhydroxybutyrate using cane molasses as a sole carbon substrate by *Bacillus* sp. 112A. IUP J Biotechnol 2012; 4(1): 25-33.

[241] Oliveira FC, Freire DM, Castilho LR. Production of poly(3-hydroxybutyrate) by solid-state fermentation with *Ralstonia eutropha*. Biotechnol Lett 2004; 26(24): 1851-5.
[http://dx.doi.org/10.1007/s10529-004-5315-0] [PMID: 15672227]

[242] Jelen P. Whey processing. In: Roginski H, Fuquay JW, Fox PF, Eds. Encyclopedia of dairy sciences. London: Academic Press 2003; pp. 2739-51.

[243] Aksu Z, Eren AT. Carotenoids production by the yeast *Rhodotorula mucilaginosa*: use of agricultural wastes as a carbon source. Process Biochem 2005; 40(9): 2985-91.
[http://dx.doi.org/10.1016/j.procbio.2005.01.011]

CHAPTER 2

PHA Biopolyester Production from Surplus Whey: Microbiological and Engineering Aspects

Martin Koller[1,2,*], **Lukas Marsalek**[3] and **Gerhart Braunegg**[2]

[1] *University of Graz, Office of Research Management and Service, c/o Institute of Chemistry, NAWI Graz, Heinrichstrasse 28/III, 8010 Graz, Austria*

[2] *ARENA Arbeitsgemeinschaft für Ressourcenschonende & Nachhaltige Technologien, Inffeldgasse 21b, 8010 Graz, Austria*

[3] *Department of Biotechnology, Institute of Applied Microbiology, University of Natural Resources and Life Sciences, Gregor-Mendel-Strasse 33, 1180 Wien, Austria*

Abstract: As the lactose-rich aqueous fraction that remains during the cheese production process after separating casein proteins and lipids from whole milk, whey accrues as an industrial by-product at vast quantity at dairies and cheese production manufactories.

The volumes of generated surplus whey almost equal those of the whole milk processed for cheese production, thus by far surmounting the needs of its restricted classical markets such as animal fodder, pharmaceutical matrices, or additives in food production. As a consequence, enormous quantities of whey are contemporarily disposed in aquatic environments, provoking prevailing ecological concern due to its considerable biochemical oxygen demand, and high expenses for the disposal. Hence, sustainable alternatives for a value-added application of surplus whey are needed; contemporarily, biotechnologists in various global regions are questing for such alternatives! Based on the catalytic action of selected microbial production strains, the spare carbon source contained in whey can be converted into value-added bio-products and at the same time, solve the industrial disposal problem.

[*] **Corresponding author Martin Koller:** Office of Research Management and Service, c/o Institute of Chemistry, University of Graz, NAWI Graz, Heinrichstrasse 28/III, 8010 Graz, Austria, and ARENA Arbeitsgemeinschaft für Ressourcenschonende & Nachhaltige Technologien, Inffeldgasse 21b, 8010 Graz, Austria; Tel: +43 316 380 5463; Email: martin.koller@uni-graz.at

Martin Koller (Ed.)

Polyhydroxyalkanoates (PHA), a group of biodegradable microbial polyesters with auspicious plastic-like material characteristics, are among the most promising target products accessible from whey´s bio-conversion. Using different production scales, production strains and production modes, excess whey can undergo upgrading to a carbon feedstock for generation of thermoplastic poly(3-hydroxybutyrate) (PHB) homopolyester or, by switching to advanced feeding strategies and application of powerful novel production strains, to tailor-made PHA co- and terpolyesters with customer-oriented material properties. The article covers microbiological, engineering, and economic aspects of whey based PHA production by meso- and extremophile wild-type or genetically engineered prokaryotic production strains. The route needed to make efforts in this field competitive with current PHA-production on prized feedstocks which in parallel interfere with nutrition for mankind, is presented. This competitiveness is addressed both in environmental and economic aspects.

Keywords: Agricultural waste, β-galactosidase, Carbon-rich waste, Cheese whey, Copolyester, Dairy industry, Dairy waste, Downstream processing, Green energy carriers, Homopolyester, Lactose, Nutraceuticals, Polyhydroxyalkanoates (PHA), Process design, Production strains, Substrates, Surplus materials, Terpolyester, Upstream processing, Whey.

THE NEED FOR VALUE-ADDED AND SUSTAINABLE UPGRADING OF WHEY: HOW TO TURN A SURPLUS PRODUCT INTO A USEFUL RAW MATERIAL

In cheese dairies or casein manufactories, whey accrues as the aqueous fraction that is generated as a side stream after the so called transformation, *i.e.*, the acidic or enzymatic coagulation of milk casein. This liquid fraction represents about 85-95% of the processed milk´s volume [1]; in addition, estimations report that cheese production generates approximately 9 L of whey per kg [2]. Reported amounts of the globally accruing quantities of whey are strongly fluctuating in dependence on the consulted source. Just to visualize the magnitude of generated whey in various countries, we present some selected estimations: While Peters reports an annual quantity of $1.15*10^8$ t [3], Audic estimates about $1.40*10^8$ t per year [4], which is almost twice as high as values reported by Prazers who reports $4.1 * 10^7$ annual tons [5]. Reliable data of OECD-FAO for 2008 are based on the assumption of even $1.60*10^8$ t per year, with an annual rise of 1-2% [1].

Whey accrues as a surplus material in various global regions [6] with highest quantities in North America and Europe. Due to the fact that cow milk is the globally predominant source for cheese production, it is clear that whey of bovine origin represents the lion´s share of the entire whey quantity. In the year 2008, about $4*10^7$ t of whey were generated in the USA, about $5*10^7$ t in the EU, and about $2.2*10^5$ t in Canada, which also constitutes a nation with considerable whey production (values for 1997) [7]. In addition of these countries located in the Northern hemisphere, also southern countries produce increasing volumes of whey; as a representative estimation, 270,000 tons annually are reported to accrue in India as a pure surplus stream without any use for further application [8].

On the one hand, the market for products containing whey include nutritionally relevant products *e.g.*, candies, whey powder used as dietary supplement for increasing the muscle size of body builders, drinks based on whey, food processing additives (*e.g.*, for manufacturing of ice cream or meat processing), chocolate substitute, infant food, cosmetics (*e.g.*, whey-containing shower gels, body lotions, or face crèmes) as well as products used for pharmaceutical purposes. However, due to its nature, these applications are restricted [8]. On the one hand, it has to be considered that whey *per se* has quite a lot of nutritive value based on its richness in proteins, carbohydrates, lipids, vitamins, and diverse essential minerals such as calcium (Table **1**); 100 g of consumed whey corresponds to an energy uptake exceeding 100 kJ. Further, a high number of people, about 75% of adults worldwide, suffer from hypolactasia, also known as lactose intolerance, due to insufficient β-galactosidase (lactase) activity [9]. This lacking lactase activity, also referred to as hypolactasia, can result in hypolactasia still during later life, which mainly occurs in countries of the Southern hemisphere and South European regions [10]. This frequent occurrence of deficient lactase activity impedes a broader application of whey for human nutrition.

Recent reports underline the need for new, innovative markets to valorise whey; at the end of July 2015, whey powder prices in the EU dropped to their lowest level since November 2010, only from March 2015 to July 2015, the prizes decreased by 28%, *inter alia* caused by a reduced export of whey powder to China [11]. This goes in parallel with the contemporary debate about the excessive amounts of milk which are on the European market; a valorisation of whey as the major by-product of the milk industry would provide a possibility to financially support milk producing companies, and generate stay-options for farmers in disadvantaged areas, thus generating socio-economic benefit.

Contemporarily, the predominant share of whey is disposed as waste, which often causes tremendous environmental pollution. Classical treatment strategies for whey are currently accomplished by various approaches: dumping into rivers, lakes, lagoons, or the sea, disposal in caves, application as porcine or bovine fodder, and dispersion over fields. Due to its high organic carbon content, whey is characterised by a considerable biochemical oxygen demand (BOD) exceeding 40,000 ppm, and a considerable chemical oxygen demand (COD) amounting to more than 50,000 ppm. This makes it clear that disposal of surplus whey leads to considerable efforts and costs [12, 13]. Lactose, whey´s most representative carbon compound, is, besides the proteinaceous retentate fraction, the component mainly responsible for these alarming BOD and COD values. The tremendous results of whey´s disposal in aquatic environments is manifested by the observed regression of aerobic marine species (both microbes and animals) caused by depletion of dissolved oxygen. Also the release of whey on fields, projected as a way to fertilize rural areas, is not as positive as it might look at a first glance; as a matter of fact, the disposal of whey on fields severely affects soil´s physical and chemical structure which results in reduced harvest yields of agricultural products [14]. The accruing whey quantities surmount those that are used as bovine or porcine feed applied in the most important agricultural applications. In addition, this application was recently prohibited in Italy in the case of production of porcine ham labelled as "*Denominazione Origine Protetta*" (DOP, protected origin) because feeding of pigs with whey is prohibited according to the related disciplinary regulation [15].

It has to be considered that the processing of a given volume of milk generates almost the same volume of whey which needs further treatment; each litre of generated whey contains up to 50 g of lactose. The problem resulting therefrom gets particularly prevailing because, as mentioned before, whey is frequently just discharged into the sea. For illustration one has to figure out that in the Po-region (Nothern Italy), the location of various big dairies, the quantities of whey to be disposed of per day amount to about 1,000 t, which corresponds to an impressive lactose quantity of 50 t! Hence, it is obvious that there exists an urgent need to develop new, sustainable strategies to transform whey safely and to generate value by this process.

In this context, a variety of such strategies are in status of development which aim at upgrading the surplus whey to a feedstock to produce valued products [16, 17]. In the case of biotechnological processes, the disaccharide lactose (full name: β-D-galactopyranosyl-(1→4)-D-glucose), whey′s major carbohydrate, can act as a carbon feedstock in various biomediated processes. According to the corresponding literature, the laboratory- to pilot scale production of the green energy carriers bioethanol [7, 18, 19], biogas [20], biohydrogen [21, 22], or biobutanol [23 - 25] is described. Further, the whey-based aerobic bioproduction of vinegar [26], bioactives like bacteriocins such as nisin, enterocins, or pediocin [27 - 30] and well-known fungal antibiotics like penicillin [31] was already investigated. In addition, whey-based production of diverse types of yeasts [32 - 34], surfactants like sophorolipids [35, 36], lactobionic acid [37, 38], single-cell protein [39], polysaccharides [40], the food additive lactic acid (E 270), for pharmaceutical matrices, as monomer for the production of polylactic acid (PLA) [12, 41], and, as the topic of this chapter, biopolymers with plastic-like properties like polyhydroxyalkanoates (PHAs) [42 - 50], was already assessed. In addition, contemporary efforts to use diverse microbial consortia for biotechnological conversion of surplus whey into the nutraceuticals lactic acid, lactobionic acid, lactosucrose, gluconic acid, or tagatose was comprehensively reviewed by Nath and colleagues. These authors summarize the application of free microbes in both discontinuous and continuous bioreactors, as well as immobilized cells in packed bed bioreactor for production of these nutraceuticals from whey [51]. Furthermore, inducing the production of recombinant proteins in high cell density cultivation [52], as well as the accumulation of lipids [53 - 55], can be accomplished by providing whey as feedstock; the same applies for production of insecticides like the *Bacillus thuringensis* toxin [56], the food colorant and vitamin riboflavin [57, 58], β-Galactosidase [59], or citric acid [60]. By chemical means, whey lactose can be transformed to products such as sweetening agents like lactitol (E 966) [61] or the laxative lactulose [62]. Table **1** provides a compilation of reported products which are accessible starting from the conversion of whey, either biotechnologically or by chemical means.

Table 1. Products produced starting from whey lactose as raw material.

Products	Production Technique	Microorganism(s)	Ref.
PHA	Biotechnology (conversion either directly or hydrolysis of substrate and biotransformation into PHA)	*Azotobacter* spp. Rec. *Escherichia coli, Hydrogenophaga psedoflava,* (convert lactose directly). Lactobacilli (conversion of whey to lactic acid), *Cupriavidus necator, etc.* (conversion of lactic acid to PHA). *Pseudomonas hydrogenovora, Haloferax mediterranei,* (conversion of hydrolysed lactose to PHA).	[42 - 50]
Poly(Lactic Acid) (PLA)	Biotechnology (anaerobic), followed by chemical polymerization towards PLA	Lactobacilli (from whey to lactic acid)	[12, 41, reviewed by 51]
Lactobionic Acid	Biotechnological conversion	*Pseudomonas* sp.	[37, 38, reviewed by 51]
Bioethanol ("Green Energy Carrier")	Biotechnology; anaerobic	*Klyveromyces lactis, Candida pseudotropicalis, Saccharomyces cerevisiae, Klyveromyces marxianus, etc.*	[7, 18, 19]
Biogas ("Green Energy Carrier")	Anaerobic biotechnological conversion	Methanogenic microbial consortia	[20]
Biohydrogen ("Green Energy Carrier")	Biotechnological conversion	Mesophilic mixed microbial cultures	[21, 22]
Biobutanol ("Green Energy Carrier")	Anaerobic biotechnological conversion	*Clostridium* sp.	[23 - 25]
Acetone	Anaerobic biotechnological conversion	*Clostridium* sp.	[24]
SCP (Single Cell Proteins)	Biotechnological	*Torulopsis bovina, Klyveromyces lactis, Klyceromyces fragilis, Candida intemedia*	[39]
Yeast	Biotechnological	*K. marxianus*	[32 - 34]

(Table 1) contd.....

Products	Production Technique	Microorganism(s)	Ref.
Vinegar	Biotechnology (2 steps)	*K. marxianus (anaerobic transformation) Acetobacter pasteurans (generation of acetic acid)*	[26]
Recombinant Proteins	Biotechnological conversion; whey as inducing factor for protein expression	Rec. *Escherichia coli*	[51]
Antibiotics (Bacteriocins and Antibiotics of Secondary Metabolism)	Biotechnology	*Lactococcus lactis* (production of bacteriocin); *Lactobacillus plantarum*, *Lactobacillus fermentum*, *Streptococcus lactis, Enterococcus faecium, etc.* Active against: *Listeria monocytogenes*, *Clostridium botulinum, etc.* *Penicillium chrysogenum* for penicillin production	[27 - 31]
Lipids	Biotechnological conversion	*Apiotrichum curvatum* (oleaginous yeast); Zygomycetes	[52 - 54]
Sophorolipids (Surfactants)	Biotechnology (2 steps)	*Cryptococcus curvatus*: production of single cell oil (SCO) in the first stage conversion of SCO to sophorolipids by *Candida bombicola* in the second stage)	[35, 36]
Insecticides (Protein Toxins)	Biotechnological conversion	*Bacillus thuringensis*	[56]
Exopolysaccharides (Thickening Agents)	Biotechnological conversion	*Enterobacter* A47	[40]
Riboflavin (Food Colorant, Vitamin B2)	Biotechnological conversion	*Lactobacillus* sp., *Ashbya gossypii*	[57, 58]
β-Galactosidase (Enzyme to Prepare Lactose-Free Dairy Products)	Biotechnological conversion	*Lactobacillus reuteri*	[59]
Citric Acid (Acidifier for *e.g.*, food technology)	Biotechnological conversion	*Aspergillus niger*	[60]
Lactitol (Artificial Sweetening Agents)	Reduction by chemistry or enzyme-hydrolysed hydrolysis	Enzymes from *Aspergillus* sp.or *Kluyveromyces* sp.	[61, reviewed by 51]

(Table 1) contd.....

Products	Production Technique	Microorganism(s)	Ref.
Lactulose (Laxative)	Chemical conversion	Chemical process	[62, reviewed by 51]
Tagatose (Sweetener and Versatile Neutraceutical)	Enzymatic or microbial conversion	Multi-step enzymatic route: Hydrolysis to glucose and galactose, isomerisation of galactose to tagatose *Lactobacillus diolivoras*, *Mycobacterium phleii*, *Geobacillus stearothermophilus*	Reviewed by [51]

WHAT CHARACTERIZES PHASS AS SO CALLED "GREEN PLASTICS"?

Talking about "Green Plastics": Clarifying some Inconsistencies

Nowadays, we witness that so called "*green plastic*" are definitely highly fashionable and well-accepted by the end consumer. As a matter of fact, the real characteristics of many of them are only partly or even not all according to generally recognized characterizations for the central features "*biobased*", "*biosynthesized*", "*compostable*", "*biodegradable*", and "*biocompatible*", which make them "*green*" [63]. These attributes are strictly defined by certificates and standard norms, as comprehensively summarized by Kržan *et al.* [64].

At first, it should be noted that even within the relevant scientific community, there exist diverging opinions about how to define "*biodegradability*" and "*compostability*" of materials, and, moreover, these two expressions are frequently mixed up. As a matter of fact, the respective norm EN-13432 which covers composting and biodegradation of plastic packaging clearly claims that, to label a material as "*biodegradable*", a minimum of 90% of its carbon share has to be metabolized within 180 days under standardized experimental conditions. The same norm allows the labelling of a solid material as "*compostable*", if a maximum of 10% of the solids remains in a sieve of 2 mm pore size after 180 days of composting under standardized conditions.

"*Biocompatibility*" of materials is assessed by using standardized methods for investigating its eco-toxicity; it does not negatively impact organisms or their surroundings where it is exposed, as postulated by the standardized norm ISO 10993. Together with biodegradability, biocompatibility is of utmost importance for the increasing field of tissue engineering and for polymer applications *in vivo* in the medical and surgical sector, *e.g.*, for implants, sutures, surgical pins and fibres, *etc.*

Concerning the raw material aspects, a polymer is considered "*biobased*", if the monomeric units are generated from renewable resources like whey. Such "*biobased*" materials, in contrast to petrochemical raw materials used for classic plastic production, stem from the metabolic action of living organisms and are embedded into nature´s closed carbon cycle. In the case of biopolymers, polymerization occurs either by chemical means, such as it is the case for PLA biosynthesis from lactic acid, or bio-mediated, see the *in vivo* hydroxyacyl-CoA polymerization by PHA synthase enzymes to generate PHAs, or during formation of starch by polymerization of glucose in pyrenoids. In the case of so called "*green plastics*", we have to consider that in the first given example (PLA), the conditions for classification as *"biosynthesized"* are not fulfilled. Because the monomer, lactic acid, can easily be produced from renewable resources such as whey by microbes (*e.g.*, lactobacilli), the definition for "*biobased*" is completely fulfilled; however, the polymerization of lactic acid is generally accomplished chemically by the use of often hazardous and priced catalysts, hence, the polymer is not "*biosynthesized*". Another typical example is the so called "*green*" or "*bio-based poly(ethylene)*" ("bio-PE") which is already commercialized at considerable quantities by the Brazilian company BRASKEM. In 2009, the annual plant capacity amounted to about 200,000 tons. The only "*green*" or "*bio*" step in the entire life cycle of these materials is the yeast-catalysed fermentative conversion of sucrose to ethanol; all subsequent steps, namely the reductive conversion of ethanol to ethylene, and the final polymerization of ethane to poly(ethylene), are chemical process steps which require the application of catalysts which contribute to the higher economic costs and negatively affect the environment. Moreover, the ultimate fate of used and disposed "bio-PE" items is completely the same as known for common petrol-based PE. Hence, also here, this so called "bioplastic" cannot be regarded as a "*green plastic*", although it is commercialized under this attribute. A similar example is provided by the so called "plant bottle packaging®"

recently introduced by the Coca Cola company to bottle their well-known CO_2-rich beverage. In 2015, the company presented the "*world's first poly(ethylene terephthalate) (PET) plastic bottle made entirely from plant materials*" at the Milan World Expo. As a matter of fact, at the moment, the only "*biobased*" content in these bottles is ethylene glycol which is obtained by the conversion of sucrose to "bio-ethanol", which later, after shipping the "bio-ethanol" from Brazil to India, is converted to ethylene glycol ("bio MEG"). This "bio MEG" is esterificated with terephtalic acid of non-biogenic origin in an esterification reactor, which, in a last step, is chemically poly-condensed to generate the polymer. Summing it up, the biogenic content of this "green bottle" amounts to modest 32.2 wt.-%, the production chain contains numerous non-biogenic steps, and, most importantly, the product is neither biodegradable, nor compostable, but a highly persistent, recyclable material like usual PET of petrochemical origin. Even if the terephtalic acid moiety was also replaced by "*bio-PTA*", which can already be obtained as a product on a small scale, the final product would not comply with the requirements needed to talk about a "*green plastic*".

Even apart from the five presented unambiguously defined categories which put the "green" character of the two case examples in a less positive light than claimed by the manufacturers, the classification of "bioplastics" as sustainable materials is often not a trivial task. In many cases, it is assumed that products based on renewable sources *per se* have an inherent environmental benefit over products stemming from fossil feedstocks. Also this claim still has to be validated by considering all ecological impacts which accrue starting from the production of a (plastic) material through its entire life span, encompassing all aspects of the raw material generation (fertilizers, water consumption, monocultures, eutrophication, *etc.*), polymer biosynthesis, polymer processing, until the delivery of the final product to the customer [65]. Only well-grounded and honest life cycle assessment (LCA) studies can provide for profound, complete, comparable and quantitative information on the ecological impact of these products. Only such LCA studies can reliably answer the question of any ecological superiority of "green plastics". Nevertheless, even such LCA studies related to PHA production do not always render a clear picture of their ecological performance in general and, in particular, in comparison with their fossil competitors. Surveying the still scarce LCA studies available in literature however gives reasons to assume that PHA can indeed surpass fossil competitors regarding their ecological performance, but only when their production is well optimized in regards to the life cycle and is particularly based on (agro) industrial by-products as feedstocks, and on the supply with energy (in particular electricity) stemming from clean sources [65 - 71].

Significance and Particularities of PHAs

Biologically, PHAs constitute a family of biodegradable microbial intracellular carbon- and energy storage materials; they are accumulated by a steadily growing number of prokaryotic genera starting from renewable carbon feedstocks [72 - 74]. Among all reported types of biopolymers with plastic-like material characteristics, PHAs are the only group with a closed, completely biological, synthesis and degradation pathway: Generation of the monomers from renewable materials occurs intracellularly in the cytosol by the action of enzymes present in the microbial cell. The same is valid for the polymerization of these monomers, and the conversion (remobilization) of the polymer by the cell as a carbon-and energy source. In addition, PHAs are well-known to be both biodegradable and compostable and, as evidenced by an increasing number of studies from the last years, biocompatible. Hence, all aspects to classify them as "green plastics" are fulfilled [63]!

Fig. (**1**) offers Scanning Transmission Electron Microscopy (STEM) pictures in different magnifications of cells of the eubacterial strain *Cupriavidus necator* grown on glucose; the cultivation was conducted in a continuous process consisting of a five-stage bioreactor cascade. PHA inclusions, recently also referred to as "carbonosomes" [75, 76], can be detected as white, refractive granules.

As previously mentioned, PHAs are produced starting from renewable resources which contain diverse convertible carbon compounds. In this context, purified monosaccharides, (waste) lipids, free fatty acids, methanol, agro-industrial waste like (ligno)cellulosics, starchy materials, beet- or cane sugar and molasses from their production, maltose, low-quality biodiesel, glycerol from the production of biodiesel, pentose sugars, methanol, methane, waste oils and other spent lipids, spent coffee ground, fruit pomaces, or, as the focus of this chapter, whey lactose from dairies, are available at ample amounts [72, 77 - 88]. A current topic of increasing interest is the photoautotrophic conversion of CO_2, preferably stemming from industrial gaseous emissions, to PHA biopolyesters by the biosynthetic action of metabolically specialized cyanobacteria [89, 90].

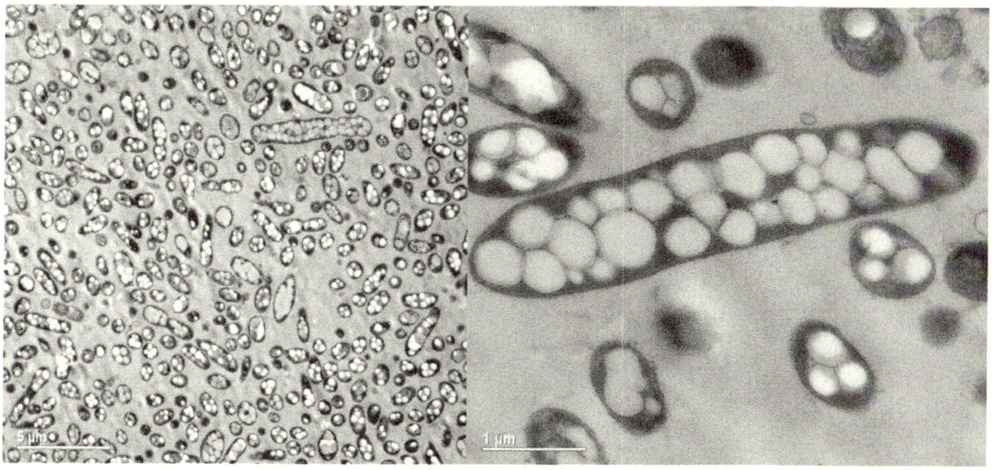

Fig. (1). STEM picture of *Cupriavidus necator* cells harbouring about 0.6 to 0.7 g/g PHB granules ("carbonosomes") in cell dry mass (CDM). Magnification: 1/20,000 (left) and 1/270,000 (right). Picture provided with courtesy of FELMI ZFE Graz, Dr. Elisabeth Ingolić.

PHAs play important physiological roles in living microbial cells. As the most important and best described function, they constitute high-molecular energy and carbon stocks which, under conditions of lacking exogenous carbon source, can be converted into a carbon and energy source [91 - 93]. In addition, PHAs contribute to the balancing of intracellular energy flow, maintenance of the cell´s osmotic functions, and the routing and distribution of the carbon flux to various metabolic pathways. During the last couple of years, a variety of additional important functions of PHAs in various ecosystems was elucidated, such as protection of cells against divers stress conditions such as unfavourable osmotic pressure, UV light, desiccation, high temperature, or oxidative stress, the involvement in special physiological processes like sporulation, cyst formation, germination, triggering nitrogen fixation in diazotrophic species (see review by

[74]), and the control of exopolysaccharide excretion [94].

As a general rule, PHA biosynthesis is boosted by an ample supply of the carbon source and, in parallel, with suboptimal supply of macro-components (nitrogen, phosphate or dissolved oxygen), or micro-components (magnesium, sulphate, and various metals) [73, 95, 96].

Composition and Production of PHAs

3-hydroxyalkanoates (3HAs) and, to a lower extend, 4-hydroxyalkanoates, *in vivo* polymerized to polyoxoxesters, are the predominant monomeric building blocks found in natural PHAs. Fig. (**2**) provides the general structure of PHA bio-polyesters.

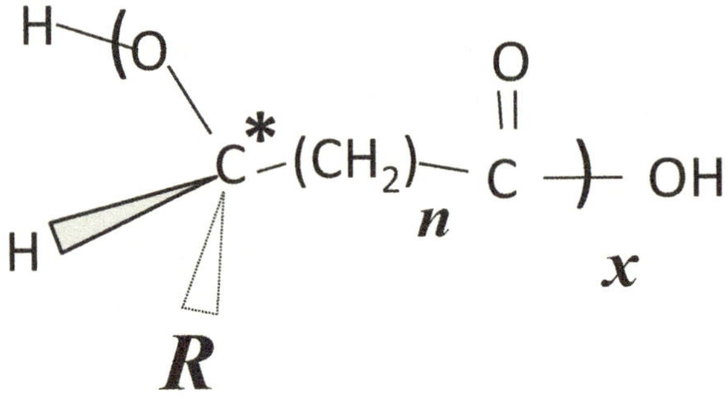

Fig. (2). Polyhydroxyalkanoates (PHAs): General structure. The asterisk (*) indicates the chiral center. ***R***: Side chain of building blocks, ***n***: methylene groups in the building block´s backbone, ***x***: degree of polymerization. ***R*** = $-CH_3$ and ***n*** = 1 represents 3-hydroxybutyrate (3HB), the monomer of PHB.

Polymer chains of microbial PHA contain about of 10^2 to 10^5 3HA monomers; this quantity is pre-defined during biosynthesis by different features like the microbial species, the carbon source, or the activity of the enzymatic system responsible for PHA synthesis during biosynthesis. The activity of the enzymes involved in the PHA metabolism is strongly influenced by process conditions such as temperature, pH-value, *etc.* Molar masses of natural PHAs range from approximately 60,000 to 1.000,000 Da with exceptions of low-molecular mass

PHA of only about 200 monomeric units [97], and the ultra-high molar mass PHA exceeding 3 MDa, as reported for *Methylocystis* sp. GB 25 (DSM 7674) [96]. In general, molar masses in the magnitude of 500,000 to 1,000,000 Da are needed for proper material processing; this is due to the fact that each processing step to manufacture *e.g.*, bars, foils, or other items occurs in a molten state of the polymer; this thermal stress reduces molar masses, thus minimizes the polyester quality [98].

3HAs found in natural PHAs are enantiomerically pure chiral compounds; in nature, they are exclusively *R*-configured. Poly([*R*]-3-hydroxybutyrate) (PHB), the well-known homopolyester of the building block (*R*)-3-hydroxybutyrate (3HB), is the most widely investigated and best characterized representative of the PHA family. PHB is described as a material with high degree of crystallinity (X_c), up to 70%, leading to high brittleness, low elasticity, and thus, limited processability. Especially the narrow temperature window between temperature of decomposition (approximately around 270°C) and melting temperature (approximately around 180°C) restricts the possible application for well-established techniques for polymer processing, such as melt extrusion, injection molding, and blow extrusion [88, 98]. These shortcomings can be overcome by incorporation of building blocks different from 3HB, *e.g.*, (*R*)-3-hydroxyvalerate (3HV) or the only important achiral building block 4-hydroxybutyrate (4HB), to interrupt the crystalline PHB matrix. This generates various PHA co- and terpolyesters which are characterized by enhanced material properties; especially parameters like the melting temperature, the glass transition point, or degree of crystallinity are lowered, whereas elongation at break typically increases. This allows for facilitated polymer processing towards marketable materials [99 - 101]. In contrast to the rather thermoplastic, short chain length PHAs (*scl*-PHA; building blocks consisting of 3 to 5 carbon atoms) described above, the so called medium chain length PHAs (*mcl*-PHAs; building blocks consisting of 6 or more carbon atoms) constitute rather elastomeric to resin- to latex-like materials, characterized by exceptionally low crystallinity, melting points (or, in many cases, melting ranges), and glass transition temperatures [102].

In dependence of the type of PHA (*scl*- or *mcl*-PHA), the precise material properties are determined by the exact composition at the monomeric level; this composition can be pre-defined during the bio-production by supplementing precursor substrates which are structurally related to the desired monomeric building blocks. For example, to provoke incorporation of 3HV building blocks into growing PHA chains, the microbial culture has to be supplied with 3HV-related precursors like propanoic acid or pentanoic acid, whereas production of 4HB requires precursors such as γ-butyrolactone, 4-hydroxybutyrate, 4-chlorobutyrate, or 1,4-butanediole [103, 104].

After the biotechnological production of PHA-rich microbial cells, suitable techniques have to be selected to recover these intracellular products from the surrounding biomass. This covers cell harvest (separation of PHA-rich biomass from the aqueous supernatant), product isolation (*via* chemical, mechanical or enzymatic cell disruption, solvent-mediated polymer extraction, or by combined processes) and, according to the purity needs, product refining [105 - 108].

PHAs: The Broad Range of their Applications

After recovery from residual cellular biomass, PHA can undergo processing towards vendible shapes, *e.g.*, to granulates or powders, with the same machineries which are used to process petrochemical plastics. Using these basic materials in a pure form or after combining them with compatible additives or additional miscible basic polymers, plastic-like items can be manufactured and commercialized as sustainable alternatives for various petrochemical plastics, *e.g.*, poly(ethylene) (PE), poly(propylene) (PP), poly(ethylene terephtalate) (PET), *etc.* [109 - 112]. Due to their versatility regarding composition and properties, PHAs nowadays increasingly attract the interest of various industries. As most widely described use, PHAs are potential candidates for production of packaging materials, especially where compostability and biodegradability of the packaging material is desired, such as enwrapping of food. Particularly in the case of perishable, oxygen sensitive goods, the extraordinary oxygen barrier of films made of PHA is highly desired [113]. Previously, also shampoo bottles (Wella, Germany) or engine oil containers (Castrol) made of PHAs have been commercialized. In addition, PHAs can be used to manufacture common all-day commodities such as razors, diapers, hygiene products, cups, dishes, cutleries, *etc.* (examples were produced by Metabolix, USA; BASF SE, Germany; ICI, UK, or

BIOPOL, Brazil), or can be used for paper coating [114]. For all these pruposes, the biopolyesters can be converted by conventional methods for plastic processing, *i.e.*, injection moulding, film blowing, *etc.*, resorting to the same machinery as for processing of petrochemical plastics.

Based on their biocompatibility, PHAs were widely explored for their potential application in medical, surgical, and pharmaceutical fields. Here, PHA and PHA-based blends and composites were used for tissue engineering, to manufacture bone implant materials and bone plates, for *in vivo* application as surgical pins, implants, screws, sutures, and meshes, production of artificial organs (heart valves, oesophagus, vein valves, *etc.*), meniscus regeneration devices, cages supporting spinal fusion, scaffolds for bone marrow, artificial blood vessels, nerve tissue engineering, and degradable matrices for controlled release of bioactive compounds (*e.g.*, antibiotics) [115 - 119]. Changing the composition of PHA on the monomeric level allows for manufacturing materials with tailor-made mechanical properties (*e.g.*, tensile strength and elasticity benchmarking the values of commonly used implant materials) and fine-tuned *in vivo* degradation rate [120]. As a matter fact, one can expect the fast increase of the number of medical applications for PHAs and its composites during the next few years. This is indicated by the approval of Poly-4-hydroxybutyrate (P4HB), 4HB´s considerably elastic homopolyester, and the respective Poly(3HB-*co*-4HB) copolyesters, as implant material by the company Tepha [121]. New fascinating aspects of PHA to be used as biocompatible implants to cure femoral fractures were only recently reported by Meischel and colleagues [120]. In experiments with growing rats, the response of the bone to PHA-ZrO_2-Herafill – composites, shaped into implant pins, was monitored by *in vivo* micro-focus computed tomography (CT), *ex vivo* CT, and surface investigations of explanted implant pins. Although the implants did not reveal considerable degradation after 36 weeks *in vivo*, the pins resulted in excellent osteogenesis around the implant. In addition, the determined parameters for quantifying elasticity, tensile strength and strain properties of the PHA-composite pins were close to that of the bone and, as an additional benefit, these materials did not cause any acidification as well-known from degradation of other clinically applied implant materials such as PLA or poly(glycolic acid). Hence, the results might show a new route to generate coating for rapidly degrading implants [120].

In addition, PHAs containing chemically reactive monomers (predominately unsaturated groups) suggest their applications as so-called "functional materials" which can be used for defined applications in technological niches, *e.g.*, as special adhesives, degradable latexes, or functional gels [122, 123]; further, such functional PHAs can undergo coupling reactions to carry bio-active compounds, as demonstrated by the preparation of *mcl*-PHA surfaces linked to zoosteric acid, as used by Hany and colleagues to prevent bio-fouling on those parts of ships and boats which are exposed to water [124].

Apart from the direct application of PHA as polymeric material, breakdown of the polyesters to the monomeric units creates a cocktail of chiral monomers which can be applied as synthons in organo-chemical synthesis of fine compounds, *e.g.*, pheromones, fragrance compounds, aromatics, vitamins, or antibiotics [111]. In this context, 3HB and oligomers of 3HB therapeutically significant effects, *e.g.*, they impact cell proliferation, avoid necrotic cell death, can be used as anti-glaucoma drugs, or against pathogenic viruses and bacteria [125]. In addition, it was shown that the linking of DP18L peptide to (*R*)-3-hydroxydecanoate (3HD), obtained from hydrolysis of the microbial *mcl*-PHA poly[(*R*)-3-hydroxydecanoate], enhances the anti-cancer activity of this peptide [126]. To compare this effect with the impact of linking DP18L to other 3HAs, recent studies carried out by Szwej and colleagues report the dependence of the biological structure and activity of the peptide on the chain length of the 3HA; 3-hydroxynonanoate and 3-hydroxydecanoate turned out to be most beneficial to enhance anti-cancer activity of the 3HA-peptide conjugates [127].

A new field of application for PHA which, at a first glance, might appear rather strange was identified in the transesterification of these polymers with methanol towards alkyl esters to generate new "green energy carriers". This conversion generates 3-hydroxyalkanoate methyl esters (3HAME) with a chemical composition similar to biodiesel originating from transesterification of lipids [128]. As a matter of fact, also 3HAMEs were investigated as combustion engine fuels, and revealed combustion heats similar to gasoline. Realistically, conversion of PHA to "green energy carriers" only seems to be reasonable from an economic point of view when using PHA fractions of modest material properties such as low molar mass or extraordinary high polydispersity; this can be the case for PHA blends produced by mixed cultures, *e.g.*, cultivated in waste water bodies. PHA accumulated by such cultures are often *mcl*-PHAs, predominately consinsting of (*R*)-3-hydroxyhexanoate (3HHx) and longer monomers. Generally, longer carbon chains of the 3HA results in higher combustion heat for the novel biofuel [112, 128, 129].

Factors Impacting the Economic Viability of PHA Production

Although ambitious efforts are contemporarily undertaken all over the world to make PHAs competitive with their well-established petrochemical counterparts, PHA production has still not met the benchmarks regarding economic competitiveness [73, 130, 131]. To a large extent, production cost is determined by the expenses of raw materials. Therefore, the use of inexpensive raw materials rich in carbon and abundantly available, such as surplus whey, is generally regarded a promising strategy to drastically reduce costs related to PHA production. Considering the tremendous quantities of whey it is clear that it increasingly attracts the attention of researchers and industry globally [72, 77 - 88]. Apart from the costs of the main carbon source, costs for precursor substrates needed to generate PHA copolyesters also considerably contribute to the overall production costs and therefore call for the exploration of inexpensive precursor compounds [73, 132, 133].

In parallel with the substrate expenses, PHA recovery and refining has to be optimized economically. PHAs as intracellular polyesters first need to be isolated from residual cell material which predominantly consists of proteins, phospholipids, triacylglycerides, nucleic acids and different polysaccharides. As the most widely applied technique, high quantities of often precarious organic solvents such as chloroform are used to extract PHA from cells; these extraction techniques are also connected to considerable energy requirements. Therefore, numerous efforts are contemporarily underway to develop inexpensive and sustainable recovery methods which reduce the use of large amounts of solvents and energy requirements, and thus effectively lead to a closed-loop material cycle within the recovery process. It is also reported that the intracellular mass fraction of PHA also significantly determines the costs for PHA recovery; generally, higher PHA shares in biomass (at least 60 wt.-%) positively impact the recovery process [106].

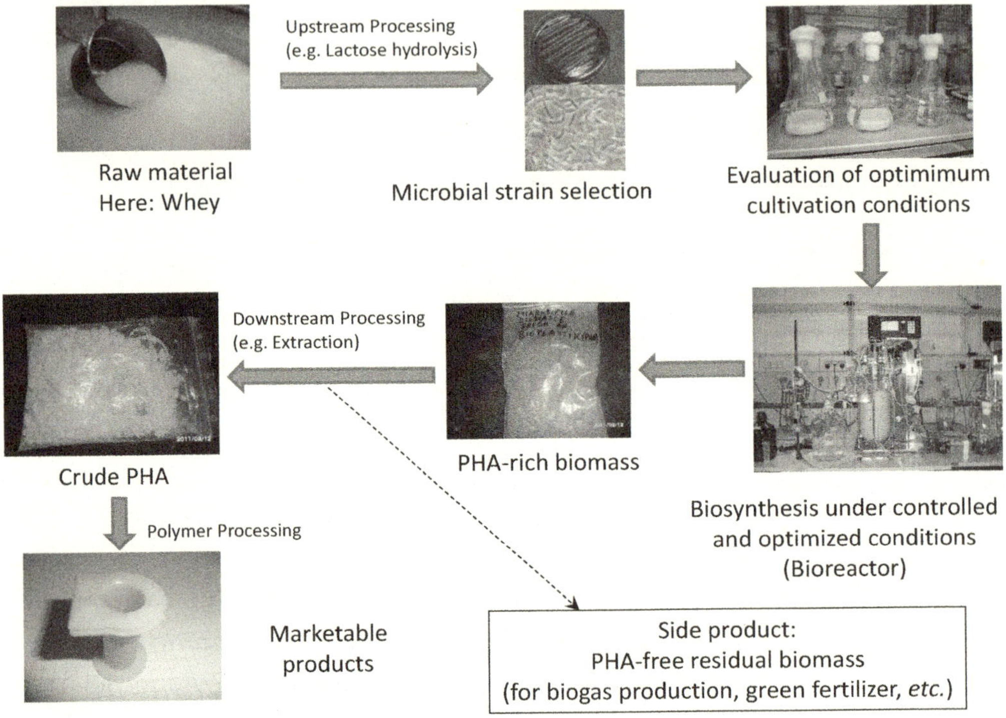

Fig. (3). General process scheme for PHA biopolymer production.

In addition to the selected raw materials and the recovery method, volumetric productivity of the PHA production process was identified as the next crucial cost factor. Currently, PHA production is managed in a rather simple process design in dis- or semi continuous production mode. This inherently reduces the overall efficiency of the process by the time demanding procedures such as pre-treatment of the bioreactor equipment, period for microbial adaptation, and post-treatment of the bioreactor (revamping time) [42, 134 - 138]. As a way out, continuous production is recognized as a viable technique to obtain high productivities and constant product quality. Numerous research activities were already conducted to investigate the potential of continuous PHA production processes in single-, two-, and multistage facilities [137, 139, 140], as reviewed by [141, 142]. As the maybe most remarkable development in the field of continuous PHA production, a cascade of five continuously stirred bioreactors was investigated for glucose-based chemostat PHA production by the bacterial strain *Cupriavidus necator*. For this process, high productivities for PHA, namely 1.85 g/(Lh), were reported during extended time periods and constant product quality regarding molar masses and molar mass distribution [137].

Fig. (3) provides a schematic of the general process chain for PHA production, indicating the economic hot-spots.

PHA Research: *Quo Vadis?*

At the moment, the PHA-related R&D activities focus on several key topics, as summarized in the subsequent sections:

Raw Materials

During the biotechnological production process, carbon feedstocks which are available at vast quantity are increasingly applied to provide the needed substrate for both microbial growth and PHA-biosynthesis. In addition to inexpensive carbon sources, the application of nitrogen-rich additives that decrease the phase for biomass production (microbial lag- or adaptation phase) is needed to enhance volumetric productivity. Cheap additives are available from agriculture, *e.g.*, rejects from green grass cultivation, and displayed promising results on laboratory scale [143]. In this context, a combination of utilization of the carbohydrate and the protein fraction of whole whey was recently reported by Obruca and colleagues [144]; these authors hydrolyzed whole whey by the enzyme alcalase, and tested the hydrolysate as nitrogen supplement for PHB production by *Cupriavidus necator* H16; waste frying oil was used as main carbon source. Addition of the hydrolyzate supported both growth and PHB synthesis; PHB yields obtained on shaking flask scale were more than 3.5-fold higher than in control cultivations without supplementation of the whey hydrolysate. This positive impact was confirmed by experiments under controlled conditions on a bioreactor scale. *C. necator* produced 28.1 g/L PHB corresponding to about 0.9 g PHB per g biomass, and excellent yield coefficient (0.94 g PHB per g oil); this was about 40% higher than results from the control cultivations. According to the authors, next experiments should be designed in such a way that the waste oils are replaced by a whey permeate fraction as a main carbon source, combined with a different production strain able to convert lactose in the whey permeate. This would on the one hand result in lower product yields (based on the fact that less PHA can be produced per g sugar than per g lipids), but, on the other hand, provides a convenient strategy for holistic conversion of the whole whey [144].

For a production of high-quality PHA-copolyesters, expensive precursor compounds are normally needed in order to provide for production of PHA-building blocks differing from 3HB. Such precursors can be generated from cheap, abundant raw materials like low-quality biodiesel [133] or even from products of wood liquefaction processes [132]. In the case of microbial specialists, structurally unrelated substrates available at low price, maybe even waste streams, can be directly applied as feedstocks PHA copolyesters production, without the need for parallel feeding with structurally related precursors. This is described *e.g.*, for the copolyester production by *Haloferax mediterranei* on whey or glycerol [45 - 47, 50, 88, 145], or *Hydrogenophaga pseudoflava* on whey-based materials [146] as sole carbon substrates.

New Powerful Production Strains and Genetic Engineering

Technological challenges of the production process itself generally involve high energy requirements for sterility precautions which can be overcome by applying robust bacterial strains; they have to remain genetically stable for extended time periods in continuous cultivation setups, and, in parallel, can minimize the hazard of microbial contamination. Extremophile microbes, *e.g.*, the halophile archaeon *H. mediterranei* present a solution by minimizing the normally necessary, extremely energy demanding sterility precaution during for PHA production [46, 145]. Nowadays, there exists an increasing trend to explore additional powerful PHA-producers by isolating strains from various environmental habitats [147, 148]. These microbes are expected to be highly productive under extreme cultivation conditions (salinity, pH-value, temperature, substrate concentration), or to produce PHA from inexpensive feedstocks available at zero expenses, like *e.g.*, certain cyanobacteria which produce PHA solely from atmospheric CO_2 [89, 90]. Further, Gram-positive strains are more and more explored as PHA-producers. Although such strains are scarce in comparison to the myriad of Gram-negative PHA-producers described in literature, they attract increasing attention due to the fact that PHA produced by these strains is free of endotoxins, a group of lipopolysaccharides predominately produced by Gram-negative microbes. Such endotoxins are well-known to cause inflammatory effects when applied *in vivo*, *e.g.*, as implants. For that reason, PHA from Gram-negative strains has to be excessively purified prior to being suitable for *in vivo* applications [115, 117, 149]. Therefore, the chapter at hand contains its own section dedicated to the accomplished efforts for whey-based PHA-production by Gram-positive *Bacilli*.

Research endevors in genetic engineering predominately aim at increased volumetric productivity and increased PHA molecular masses. This is possible by knock-outs of those genes encoding for PHA-degrading enzymes (depolymerases), or by deleting genes encoding for production of unwanted carbon-containing by-products such as extracellular polysaccharides or spores, as reviewed in [15].

In order to enhance PHA productivity by the phototrophic non-sulfur bacterium *Rhosospirillum rubrum*, genes encoding for twelve different PHA synthases were heterologously expressed in two generated *phaC* deletion mutants of *R. rubrum* in a study by Klask *et al.* Eight different PHA synthase genes of class I, and four of class IV were selected for heterologous expression. All recombinant *R. rubrum* strains showed significant PHA synthesis and accumulation, although PHA contents in the recombinant strains of the single mutant *R. rubrum* Δ*phaC2* were generally higher in comparison to those of the double mutant *R. rubrum* Δ*phaC1*Δ*phaC2*. It was possible to increase the accumulation of PHA by up to 25% due to heterologous expression of PHA synthase genes compared to the not genetically engineered wild type organism [150].

Enhancing oxygen uptake by microbes can be accomplished by inserting catalase or peroxidase encoding genes. This is a promising strategy to accelerate microbial growth and to yield higher final biomass density, thus enhancing the overall productivity [151].

Another way of reaching higher PHA productivities is to block metabolic pathways which hamper the utilization of inexpensive carbon sources. Recently, Volodina *et al.* described different approaches to genetically engineer the PHA producer *Ralstonia eutropha* H16 in order to make it capable of using carbon-rich industrial by-products and residual compounds like glycerol, plant oils and raw sugar-rich materials like molasses, starch and lignocellulosics [152].

Similar investigations were carried out for whey conversion; here, *Cupriavidus necator*, a not lactose utilizing PHA producer, was engineered genetically. This was accomplished by choosing one gene of the intracellular PHA depolymerases (*phaZ1*) as an insertion site for the *Escherichia coli* genes *lacZ, lacI,* and *lacO* in *C. necator* cells. The obtained recombinant strain was able to convert lactose from cheese whey for cellular growth and PHA biosynthesis. At the same time, a part of *C. necator*´s intracellular PHA degradation system was removed, thus reducing the intracellular PHA conversion which resulted in increased PHA productivity [44].

As demonstrated by Heinrich and colleagues, also the composition of PHBHV copolyesters can be influenced at the monomeric level by means of genetic engineering in order to produce biopolymers of tailored composition [153]. In *Rhodospirillum rubrum* S1, different genes encoding for the enzymes pyridine nucleotide transhydrogenases (*pntAB*, *udhA*) and *phaB*, encoding for acetoacetyl-CoA reductase, (EC 1.1.1.36) were overexpressed. The resulting recombinant strain, equipped with the *Escherichia coli* MG1655 gene for membrane-bound transhydrogenase PntAB plus the acetoacetyl-CoA reductase encoding *R. eutropha* H16 *phaB1* gene, produced PHBHV with 3HV fraction of up to 13 mol-% from fructose; this was considerably (13 times) higher in comparison to the wild-type organism. As an additional benefit, the engineered *R. rubrum* strain synthesized PHBHV with a 3HV molar fraction of 0.56 from CO_2 and CO from syngas. According to the authors, this increased 3HV formation was due to excess propionyl-CoA generated from threonine and other amino acids as the strain´s strategy to compensate an intracellular redox imbalance. Similar experiments were carried out by Zhang *et al.*, who genetically manipulated the methylcitric acid cycle and the methylmalonyl-CoA pathway of *C. necator*. The authors demonstrated that, in contrast to the wild type strain, the engineered strains were able to synthesize PHBHV from fructose as a sole carbon source without the need for a 3HV-related precursor compound [154].

Genetic modification can also be implemented to improve downstream processing. PHA producing strains can be engineered in such a way that they excrete of the enzyme nuclease at high quantity; after cracking the cells by mechanic techniques, the nuclease degrades nucleic acids excreted into the fermentation medium. This degradation of nucleic acids considerably lowers the viscosity of the broth, which further facilitates separating PHA granules from the aqueous phase, which can be done using centrifugation, floatation, or flocculation. This was successfully demonstrated by the insertion of *Staphylococcus aureus* nuclease gene into various PHA-producing bacteria such as the powerful PHA production strains *C. necator* or *Pseudomonas* sp. It turned out that *S. aureus* nuclease gene was excellently inserted into the *C. necator* chromosome and well expressed particularly in *Pseudomonas* sp. Expressed nuclease was directed to the cell´s periplasm and, after cell disintegration, released into the culture medium, without negatively affecting the strain´s stability or specific PHA productivity [155].

Mixed Culture Approaches

An auspicious opportunity to save costs on PHA production is the implementation of environmental principles for strain selection as frequently used in waste water treatment to acquire enriched microbial cultures of outstanding capacity for PHA production. Therefore, increasing efforts are contemporarily devoted to the application of mixed microbial consortia instead of monoseptic cultivation techniques [156, 157]. The use of mixed microbial cultures is increasingly attracting attention regarding their capability to accumulate significant amounts of PHA at rather low cost; generally, monoseptic cultures are *per se* more expensive than the application of mixed cultures. This is due to minor sterility demands, and lower requirements for equipment and control devices. Such mixed consortia can utilize a wide range of inexpensive substrates such as (agro)industrial and domestic waste streams [156, 158]. Enrichment of PHA-storing organisms in a mixed microbial culture can be achieved by applying operational conditions which exert a selective pressure on the microbial population. Such enrichment processes, using a selective pressure and alternating availability and lacking of exogenous substrates, are often termed "feast–and-famine approaches" or "dynamic feeding" [158]. Such external substrates often constitute volatile fatty acids (VFAs) or carbohydrates. This was demonstrated in the past by a non-sterile multi-stage process for selectively enriching PHA producing microbes from diverse natural samples by exposing them towards extended periods of carbon starvation [159]. This approach used the advantage of PHA-containing cells to survive under limitation of external carbon-source as criterion for selection. The first bioreactor of the cascade was supplied with a medium rich in carbon and poor in nitrogen; the system was run at considerably low dilution rates. Starting with the second bioreactor in the cascade, the cells were forced to handle limitation both of carbon and nitrogen; consequently those cells not accumulating PHA in the first bioreactor died off. As a result, 9 bacterial species able to produce PHA were obtained by the enrichment experiments from diverse samples [159].

As a further example, Fradinho *et al.* showed that PHA production from pure and mixed organic acids (acetate, propionate, butyrate, lactate, malate and citrate) can be achieved. The addition of VFAs acetate and butyrate resulted in PHB formation, whereas propionate induced the synthesis of 3HV building blocks. The presence of acetate boosted the uptake rates of both butyrate and propionate [160]. Hence, optimized PHA accumulation by bacterial enrichment cultures consists of three steps: First, a pre-treatment to convert readily biodegradable compounds of high (bio)chemical oxygen demand, *e.g.*, whey lactose, into feedstocks accepted for PHA production by many production strains, such as VFAs. As the next step, VFAs are used to run a feast/famine feeding regime. Finally, this enrichment culture is supplied with VFAs as a substrate and at the same time limited with at least one nutrient essential for growth, in order to generate microbial cell mass with maximized PHA mass fraction [158].

Johnson *et al.* worked on enrichment experiments to obtain mixed bacterial cultures with outstanding capacity for PHA accumulation as an alternative to genetically engineered strains which produce PHA from pure substrates. Process optimization efforts resulted in high PHA fractions in cell dry mass (CDM) and high productivity, although rather expensive substrates, equipment and high energy input were required. Best results reported from this study reached a maximum PHB fraction in CDM of 0.89 g/g after 7.6 h of continuous feeding with acetate [161]. In addition, the possibility to integrate PHA production with the process water treatment from a sugar production plant was demonstrated; here, the cultivation of PHA-producing consortia did not only reduce the amount of carbon from the waste water, but also resulted in nitrogen removal [162].

Surplus whey, wastewater from olive oil production, and a synthetic mixture of acetate and propionate were applied as substrates for the cultivation of mixed microbial consortia. The capability of these consortia to synthesize PHB and PHBHV on the mentioned substrates was assessed by Hilliou and colleagues. In addition, different recovery methods were applied in this study which yielded PHAs of different purity affecting the melt processability of these materials. On the one side, impurities led to unwanted thermal PHA degradation at lower temperature, but, on the other side, they also stimulated melting at low temperature by acting as thermo-stabilizers, improving viscosity of melt and enhancing shear thinning. Furthermore, it was shown that impurities did not influence crystallinity of aged polymers, but PHA crystallites increased in size. In addition, those PHA samples of higher protein content showed better thermal stability, in contrast to samples with higher inorganic fraction which displayed enhanced properties regarding melt viscoelasticity. It was suggested that the changes in diverse polymer properties by the presence of different organic and inorganic residues were, although sometimes in contrary to each other, as a whole beneficial for processing towards marketable polymeric items [163].

Downstream Processing

Sustainable recovery of PHA from microbial biomass requires environmentally safe and efficient solvents, in parallel with novel enzymatic or chemical methods for cell lysis, and enhanced strategies for mechanical cell disruption with low energy demands. As an important direction of impact, the application of novel, non-halogenated solvents for PHA extraction under optimized extraction conditions (temperature, pressure) is investigated to an increasing extent [164 - 166]. For efficient recovery of the biopolymer, an intracellular polymer mass fraction of at least 0.6 g/g, combined with PHA granules as large as possible is beneficial to optimize the recovery process; these aspects are preassigned *in statu nascendi* of PHA bio-production firstly by the nutrient supply and, secondly, by the strain adaptation to the cultivation conditions [122].

Only recently, Lopez-Abelairas and colleagues compared acid treatment (H_2SO_4) of PHB-rich *C. necator* H16 biomass with alkaline treatment (NaClO, NaOH, and with combining NaOH with chlorinated solvents) for polymer recovery. Purity, recovery yield, and polymer degradation obtained after each treatment were reported, and an assessment of the economic and environmental impacts of each treatment was provided. Lowest costs for PHA recovery were achieved with treatments with NaOH or with H_2SO_4 (1.02 €/kg or 1.11 €/kg, respectively). CO_2-emissions for the process were also estimated; they amount to only 18% of those based on NaClO application. Best purity (98%) and minimum PHA degradation resulted from acidic treatment without the need for organic solvents [167].

As a by-product of the PHA-recovery process, residual non-PHA cell mass (NPCM) needs a sustainable, value-adding strategy for further utilization. Here, research and development is predominately directed to the anaerobic NPCM transformation for methane-rich biogas production or, in addition, NPCM is exploited as a carbon- and nitrogen rich mixture which can be supplied in subsequent cultivation setups after treating NPCM with chemical or enzymatic hydrolysis. Alternatively, there exist agricultural applications for NCPM as a so called "green fertilizer"; this was demonstrated for a production of PHA homo- and copolyesters, commercialized as BIOPOL®, on a pilot scale at PHB/ISA in Brazil [136].

Improved Process Design

The growing commercial importance of PHA requires the development of new, more efficient production processes [138, 141, 142]. This can only be achieved by considering the kinetics of PHA accumulation in bioreactors, and, indeed, efforts in this area have been increasing during the last couple of years [168]. Such kinetic analysis and the resulting mathematical modelling of PHA production processes provides for a reduced quantity of laboratory experimental trials during development of the bioprocess, and as the case arises, shows the route towards optimized feeding regimes and process design. Despite the promising theoretical fundaments of continuous PHA production regarding productivity and product quality [169], semi- and pilot-scale PHA production is still exclusively accomplished in discontinuous mode, mainly by applying fed-batch feeding strategies. In future, continuous PHA production, based on the auspicious results already elaborated on a laboratory bioreactor scale, is expected to provide for enhanced PHA production processes regarding volumetric productivity and quality of the bioproduct when compared to discontinuous or semi-continuous processes used currently on industrial scale [138, 141, 142]. As mentioned before, a flexible multistage bioreactor cascade as developed by Atlić and associates [137] has the potential to act as a widely used tool for advanced process engineering. These authors designed this set-up as a five-stage bioreactor cascade. The idea for this set-up came up from theoretical considerations of Braunegg and colleagues, who concluded that the combination of a continuous stirred tank reactor and a tubular plug flow reactor, or, alternatively, a multi-stage bioreactor cascade, might constitute the ideal process engineering design for a production of secondary metabolites like PHA, allowing for a high-throughput production of highly uniform polymers. In addition to the promising performance of this cascade in long-term experimental runs [137], recently designed low structured [170] and metabolic [171] models for PHA biosynthesis using this equipment propose further improvements in the operation of this cascade system, leading even to more optimized productivity and product quality. In addition, such multistage continuous systems could also be used to conveniently produce blocky structured PHA (*b*-PHA), which constitute completely new biomaterials with various fascinating polymer characteristics, as recently demonstrated *e.g.*, in the case of P(3HB-*b*-3HHx) [172].

Although a rather old technology, solid state fermentation (SSF) has been rediscovered during the last couple of years as an alternative to submerged cultivation techniques in liquid media. The principle investment costs for SSF are usually lower than for equipment used for submerged fermentation, and, most of all, the cost of raw materials for SSF can be cheap since this technique is based on the use of polymeric agricultural residues such as straw as fermentation beds and substrates. Therefore, several authors investigated the potential of SSF also in respect to cost reduction of PHA production which only recently was comprehensively reviewed by Sindhu and colleagues [173]. Although several positive factors (investment, substrates) could suggest at a first glance that SSF might constitute a viable technique for cost-efficient PHAs production, a more holistic assessment of the process makes it clear that this method is still by far too immature for commercialization. The major bottlenecks that have to be overcome are related to culture maintenance, and insufficient productivity [173].

Follow-up Materials: Advanced Blends and Composites Based on PHA

Recently, a lot of R&D activities are devoted to design and preparation of polymeric composites and blends which are both biocompatible and biodegradable; this field has become one of the key research areas in the PHA field with a tremendously growing number of respective scientific publications and patent filings. The enhancement of the material properties is needed to allow the processing of the materials by injection moulding or film blowing. PHAs can undergo processing combined with diverse compatible materials like natural fibres, organic compounds, inorganics, or other compatible polymeric materials; this strategy results in designing and developing new blends and composites based on PHA with often astonishing material characteristics. Therefore, the utilizations of polymers like poly(ε-caprolactone) (PCL), poly(vinyl alcohol) (PVA), or others, encompassing also synthetic PHA analogues like atactic PHB (see [174]), inorganic filler materials (diverse clays, montmorillonite, sephiolites, $CaCO_3$), and organic filler materials from agriculture (straw, wood dust, *etc.*) were already tested [69, 175]. More recently, the blending of PHA with poly(lactic acid) (PLA) was optimized by improving the miscibility of these biopolymers through changing the ratio between PHA and PLA in the blend, or by mixing the blend with compatibilisers, peroxides, and transesterification catalysts. This way, the extruded films displayed high compatibility and extraordinarily high barrier properties as desired for *e.g.*, food packaging applications [176]. The high compatibility of PHBHV with other biopolymers like PLA was also demonstrated by melt spinning experiments carried out by Hufenus *et al.* [176]. These experiments were aimed to overcome the major problem of PHBHV, namely its rather low crystallization rate which constitutes a major bottleneck during the melt-processing of PHBHV fibres. Therefore, PHBHV and PLA were combined in a configuration consisting of a PHBHV core covered by a PLA sheath, and underwent a new spin pack concept. The generated PHBHV plus PLA fibers displayed considerable tensile strength (up to 0.34 GPa) and excellent E-moduli (up to 7.1 GPa); hence, reaching extremely high firmness as desired. Biocompatibility tests carried out additionally revealed high cytocompatibility of these novel materials [177].

Blends of PHB and poly(3-hydroxyoctanoate) (PHO) are currently investigated to produce bio-resorbable nerve guidance conduits, aiming at the generation of systems for repairing periphal nerves. By determining the optimum ratio of PHB and PHO (75/25), mechanical blend properties were enhanced corresponding to those of peripheral nerves [118].

Regarding inexpensive fillers from agriculture surplus streams, the application of lignocelluloses such as bagasse from sugar production, cereal flour, fruit residues, crop fibers, dust from sawing different types of wood, or cereal straw is comprehensively described [175]. Such PHAs blended with natural fibre composites often reveal enhanced mechanical properties beneficial for diverse purposes even though the final product density is considerably reduced. Recently it was successfully demonstrated that new formulations based on PHA copolyesters (PHBHV), blends with miscible biodegradable polymer (ECOFLEX®), hazelnut shell powder as lignocellulosic filler, biologically benign plasticisers (citric acid esters) and melt strength enhancers (Joncryl), chain extenders, nucleating agents (calcium carbonate), and antioxidants (ANOX 20® and ALKANOX 240®) constitute materials with excellent processability, thermo-mechanical properties, high elongation at break, and good biodegradation behaviour under mature compost conditions; these materials were suggested to be used for injection moulding and blow extrusion processes [178]. Due to the nature of the investigated filler materials from agriculture, also biodegradation of the final polymeric material are accelerated. Because those natural fibre materials often constitute real waste streams, their application often causes a considerable decrease of production cost of the finally obtained vendible product (reviewed by [69]).

Tannic acid was used as a natural polyphenolic thermal and processing stabilizer for PHB. The thermal stability of both native PHB and PHB combined with tannic acid was investigated by rheology and calorimetry. As an outcome, it was shown that pure PHB degrades considerably, whereas addition of tannic acid improves the thermal stability, thus widening the processing window of this biopolymer [179]. PHBHV was also combined by thermoforming with lignin, an abundant raw material with restricted market use, to develop novel advanced composites of compatible materials to be used for packaging purposes. Composite sheets obtained by thermoforming were produced and assessed regarding the influence of different shares of lignin on the resulting thermo-oxidative stability, the melting and crystallization behaviour, mechanical properties, viscoelasticity, and gas permeability (CO_2 and O_2) of the sheets. As a result, Kraft lignin turned out as well compatible with the bacterial polyester (PHBHV), and acts as an active additive, with a reinforcing effect. Also these materials are suggested to act as packaging for various easily perishable goods [180].

In addition to the above described natural fibre composites and composites with additional polymeric materials, PHA-based nanocomposites constitute a further emerging field. Such nanocomposites have the potential to tailor specific material characteristics, such as crystallization behaviour, thermal properties, gas permeability, and mechanical characteristics (toughness, *etc.*); it was demonstrated that such nanofillers can act as nucleating agents, thus providing higher thermal stability to the polymeric matrix [181]. To design nanocomposites, only reduced quantities of the filler material, usually organophilically modified clays, or, as a recently emerging development, cellulose nanofibers, can be applied to efficiently enhance material properties. In the case of PHBHV processed together with microcellulosic nanowhiskers, features like material toughness, Young's modulus, and tensile strength were considerably increased by these nanoadditives [182]. PHA can also be processed to functional nanoparticles carrying pharmaceutically active compounds, as demonstrated by Rodriguez-Contreras and colleagues, who prepared doxycycline-loaded micro-and nanospheres of PHB by different techniques, and demonstrated outstanding anti-bacterial activity of the generated particles [183].

THE ENTIRE PROCESS TO CONVERT THE RAW MATERIAL (MILK) TO PHA BIOPOLYMERS

General Aspects

Production of cheese and casein generates whey as the pre-dominant by-product. A simplified process scheme, starting from the feedstock milk and ending with biotechnological PHA production, is provided in Fig. (**4**). By enzymatic or acidic treatment in the so called "transformation" process, milk's casein fraction denatures and precipitates as "curd cheese". In this step, full fat whey remains as a liquid residue. Full fat whey further undergoes a skimming process in order to remove the major part of lipids. The remaining sweet skimmed whey is concentrated by removing about 80% of its water content, and gets pasteurized. Finally, the obtained "whey concentrate" undergoes separation by ultra-filtration technique, resulting in the carbohydrate-rich whey permeate fraction, and the protein-rich whey retentate fraction; the latter still contains substantial amounts of lactose [2, 5, 15, 184].

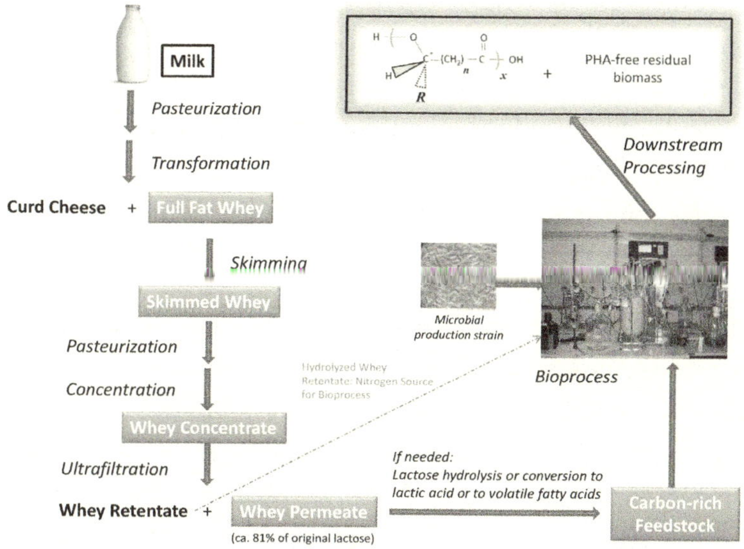

Fig. (4). From the feedstock milk to full fat, skimmed and concentrated whey to PHA biopolyesters. The hydrolysis of concentrated whey to whey permeate and whey retentate results in a carbon rich fraction (permeate) to be used as a feedstock, and a proteinaceous fraction (retentate) which can be used for generation of a complex nitrogen source for PHA-accumulating microbes.

Significance of Whey´s Protein Fraction

As the whey´s permeate fraction, containing around 80% of the carbohydrates present at the beginning in the raw material milk, constitutes a precious carbonaceous feedstock for production of valued biomaterials like PHA or green energy carriers. Besides, also the proteinaceous retentate fraction contains ingredients of interest for diverse applications [184]. New findings from the medical field report the general beneficial role of whey proteins on intestinal enteroendocrine cells to stimulate cell growth and to increase the production of hormones [184, 185]. α-Lactalbumin and β-lactoglobulin are the major proteins in whey retentate constituting precious resources for human and animal nutritional supplements. By means of biotechnology, these proteins can also be supplied, with or without prior hydrolysis to peptides and amino acids, as a nitrogen-rich supplement for optimized farming of microbes [144], analogous to other proteinaceous additives [143, 186], such additives can positively contribute to specific microbial growth rates and the synthesis of higher concentrations of catalytically active biomass. This is in contrast to the today's application of the minor proteins contained in whey which have high market value. Here, the enzymes lactoferrin (a transferrin with antibiotic and antiviral activity which plays a significant role in the innate immune system), lactoperoxidase (an effective anti-microbial compound), the immunologically active proteins IgG or osteopontin, as well as lactoferricin, (displays anti-microbial and anti-cancer activity), are of significance in the pharmaceutical and medicinal field and even for food conservation [185].

As a new development, plastic films made of PE or PET were coated with whey proteins in order to design recyclable bio materials characterized by minimum permeability for O_2 and moisture (reviewed by [113]). These materials are designated to be used for food packaging purposes [187]. Especially the high oxygen barrier (oxygen transfer rate of 1 cm^3/m^2d bar at standard conditions), and the water vapour transfer rate (2 $g/(m^2d)$) make these new materials auspicious for packaging of such food where shelf life extension is aspired. These layer-by-layer film materials can conveniently be recycled and display reduced environmental impact if compared to the original plastic material. Preliminary evaluation indicates that these materials comply with food safety regulations according to respective norms. At the moment, the new materials are tested and evaluated as packaging solution, *inter alia* for food [188], especially regarding the potential of commonly used materials like ethylene vinyl alcohol copolymers [189]. It was also demonstrated that the major drawback of the whey protein coating, namely the rather high brittleness, typical for denatured proteins, can be overcome by combining the whey protein with hydrolysed whey protein together with an eco-friendly plasticizer like glycerol [190]. Further studies also assed the biodegradability of the whey protein layer itself; this is of importance since natural polymers often become non-biodegradable by cross-linking or blending with not degradable additives. It turned out that the materials undergo fast biodegradation and composting even after attachment on the commercial plastic film, and do not induce any detrimental influence on quality of compost [191]. Properties of these films can be further enhanced in terms of reduced solid migration rate and reduced film swelling by applying whey protein films combined with Montmorrilonite particles on a nano-scale [192]. Recent developments report the enhancement of thermal, structural, morphological and mechanical properties. Elasticity and plasticity of such whey protein-montmorillonite nanocomposites are especially improved by additional incorporation of optimized quantities of citric acid [193, 194]. A combination of such advanced protein-based materials with PHA matrices stemming from the conversion of the whey permeate fraction instead of newly manufactured plastic surfaces might open the door for completely novel and sustainable materials dedicated to food packaging purposes.

Table 2 summarizes the contents present in different whey categories: Sweet and fermented (sour) whey, whey permeate and, in addition, proteinaceous whey retentate.

Table 2. Ingredients in milk, diverse categorizes of whey and whey follow up products (based on [81, 229]).

Compound [% (w/w)]	Milk	Sweet Whey	Dry Sweet Whey (Powder)	Fermented (Acid) Whey (Powder)	Dry Sweet Whey (Powder)	Whey Permeate	Whey Retentate
Lactose	ca. 4.8	4.7–4.9	73.5	4.5–4.9	65.6	23	14
Lactic Acid	-	traces	traces	0.5	7	-	-

(Table 2) contd.....

Compound [% (w/w)]	Milk	Sweet Whey	Dry Sweet Whey (Powder)	Fermented (Acid) Whey (Powder)	Dry Sweet Whey (Powder)	Whey Permeate	Whey Retentate
Proteins	3.4-3.5	0.75–1.1	12.9	0.45	12.3	0.75	13
Lipids	max. 4.2	0.15–0.2	1.1	traces	1.0	-	3-4
Inorganic Compounds* (minerals like *e.g.*, calcium, phosphate, *etc.*)	ca. 0.7	ca. 0.7	8	0.6-0.7	11	ca. 2.7	ca. 0.7

* expressed as ash content.

Case Examples for PHA Production Starting from Whey by Diverse Microbes

Since the last comprehensive review on whey-based PHA biosynthesis was published [15], several new developments in this field were reported worldwide, as detailed in the next sections.

Transforming the eco-pollutant whey into marketable bio-products like PHA brings a commercial benefit (a new raw material is available at zero expenses, no more disposal costs) and at the same time reduce prevailing environmental problems. PHA biosynthesis based on the catabolism of different sugars (glucose, fructose, *etc.*) is well described; this starts from the classical metabolic route of glycolysis or the 2-keto-3-desoxy-6-phosphogluconate (KDPG) pathway to yield a pyruvate molecule which undergoes decarboxylation to Acetyl-CoA as the central metabolite. The further condensation of the C2-unit Acetyl-CoA to the C4-unit Acetoacetyl-CoA under carbon-excess and concomitant limitation of other growth essential nutrients, and its final polycondensation to PHB by the catalytic action of the PHA polymerases is exhaustively described in literature [80, 91, 93, 195]. However, by working over the respective literature, it is noteworthy that the number of microbes described to directly metabolize lactose for biomass formation and PHA accumulation is quite manageable [47, 48, 196].

Between the years 2001 and 2004, the EU-FP5 funded the project WHEYPOL ("Dairy industry waste as a source for sustainable polymeric material production", [197]), which is an unprecedented R&D approach, investigated the holistic process chain for whey-based microbial PHA bio-production. This was accomplished by an interdisciplinary international consortium from academia and industry; this consortium shined a light on all relevant aspects of upgrading whey towards PHA, encompassing the monitoring of fluctuations in the raw material composition, aspects of microbiology and genetic engineering in order to identify the most powerful prokaryotic production strains, fermentation technology in laboratory- to pilot scale bioreactors, advanced downstream processing, detailed characterization of the produced polyesters, processing of PHA to prototype items, and holistic economic and ecological process appraisal [63, 197].

The following sections provide an overview of meso- and extremophile Gram-negative and Gram-positive wild-type strains which accumulate PHA on whey-based media, and summarize the collective efforts to enhance whey-based PHA production by techniques of genetic modification.

Gram-Negative Wild Type Organisms as Whey-Based PHA-Producers

Looking in the literature, it is obvious that the majority of strains described for whey-based PHA production belong to the group of Gram-negative eubacteria. In 1994, the first investigation of PHB biosynthesis from whey-derived lactose by the wild-type organism *Pseudomonas cepacia* (nowadays called *Burkholderia cepacia*) was published by Young and associates. On a shaking flask scale, these authors assessed the potential of this organism in terms of growth and PHB productivity on lactose, and compared these results to the strain´s performance on glucose and xylose. In the case of lactose, *P. cepacia* accumulated PHA to a concentration of almost 5 g/L, and interacellular mass fractions amounting to 0.56 g PHB per g cell material. The accumulated biopolyesters were recovered from biomass and subjected to molecular mass determination [196]. These investigations continued some years later by Wallner and colleagues, who cultivated this organism under conditions favouring PHB accumulation on 50/50 glucose/galactose mixtures, mimicking a solution obtained by lactose hydrolysis. In contrast to the pioneering investigations conducted by Young and associates, these experiments were performed at controlled cultivation parameters in laboratory reactors. The outcomes by far outperformed the data of Young *et al.* for shaking flask cultures on lactose: PHB productivity amounted to 0.18 g/(Lh), the PHB concentration at the end of the experiment to 4.0 g/L. Hence, applying hydrolysed lactose is clearly beneficial for this strain compared to the use of alternative substrates regarding biomass formation, and, therefore, the authors proposed enzymatic hydrolysis of whey lactose to develop a cost efficient whey-based PHA production process [198].

From an Indian pond, *Methylobacterium* sp. ZP24 was isolated by Yellore and Desai. The organism was cultivated using purified lactose in shaking flask cultivations and accumulated up to 3.1 g/L PHB and PHB contents in biomass up to 0.59 g/g. Later, pure lactose was replaced by whey as a carbon substrate, and the nitrogen supply was optimized in order to enhance PHB production. These first experiments on whey were somewhat lower than previously achieved from lactose-based set-ups; the experiments resulted in PHB concentrations of up to 2.6 g/L, and PHB fractions in biomass of up to 0.44 g/g [199]. Repetition of these whey-based experiments under controlled parameters in laboratory bioreactors were later carried out by Nath and colleagues. In simple batch-set ups, PHB concentrations at the end of the experiments were reported with 2.07 g/L, together with polymer contents in biomass of 0.67 g/g, and a polymer productivity amounting to 0.06 g/(Lh) [200].

The direct conversion of whey lactose without prior hydrolysis, towards PHA, based on a systematic exploration of various wild type microorganisms, was accomplished in shaking flask cultures carried out by Povolo and Casella [201]. These authors reported lactose-based PHA production by the culture collection bacteria *Hydrogenophaga pseudoflava* DSM 1034, *Paracoccus denitrificans* DSM 413, and *Sinorhizobium meliloti* 41. Among them, *S. meliloti* and *H. pseudoflava* were able to produce PHA directly from cheese whey permeate, and especially *H. pseudoflava* turned out to be a promising candidate for whey-based PHA production. In addition, the authors investigated their own microbial isolates which turned out to constitute *Sinorhizobium* sp., *Bacillus megaterium* and *Bacillus* sp., and examined their growth and PHA production on cheese whey permeate. Promising results were reported also for these strains [201].

If a strain´s natural activity of ß-galactosidase (lactase, E.C. 3.2.1.23), responsible for the hydrolysis of lactose to D-(+)-glucose and D-(+)-galactose as the respective monomers, is not sufficiently high to efficiently use lactose as a sole carbon source, lactose can undergo chemical or enzymatic hydrolysis to yield equimolar mixtures of the monomeric sugars. In comparison with lactose, the monomers are transformed by considerably more microbes, and typically display better conversion kinetics [48]. In addition, the solubility of glucose/galactose mixtures is considerably higher compared to the solubility of lactose which has its upper limit at about 200 g/L (this is the typical lactose concentration in whey permeate).

Another more complex approach starts from anaerobic lactose transformation to lactic acid. This is done by using lactobacilli which are able to generate lactic acid at outstanding yields; using homofermentative lactobacilli, at least 0.9 g lactic acid can be expected per g of converted lactose. Lactic acid can be stored as a stable intermediary product, and, in a subsequent aerobic cultivation process, can be used as a carbon substrate for cultivation of numerous powerful microbial production strains. Among these strains, also the best-known PHA producers are found, such as *C. necator*, *Azohydromonas lata,* or *Azotobacter vinelandii.* As an alternative, lactic acid is recovered, purified and chemically polymerized by ring-opening polymerization to PLA.

Last, but not least, a multi-step whey-based production process can also be applied to generate free VFAs from whey lactose by the action of mixed cultures. In a subsequent step, these VFAs undergo conversion towards PHA by well-known strains like *C. necator*. Domingos and colleagues presented this strategy by converting whey lactose to VFAs such as propionic acid, butyric acid, capronic acid, valeric acid and acetic acid using an immobilized acidogenic mixed culture under anoxic cultivation conditions. The authors concluded that the obtained mixture of VFAs, due to its contents in both even (butyric, acid, capronic) and odd (propionic, valeric) numbered VFAs, might constitute a precious substrate cocktail for a production of PHBHV copolyesters [202].

Based on experiments in laboratory bioreactors, Koller and colleagues compared and assessed the potential for whey-based PHA production of three Gram-negative wild type strains. In this work, the eubacterial strains *Pseudomonas hydrogenovora* (today *Burkholderia funghorum*) and *H. pseudoflava*, and the archaeal haloarchebacterium *Haloferax mediterranei* were investigated [46].

Among these candidates, *H. mediterranei* displayed the highest potential for whey-based PHA production on industrial scale. The most outstanding properties of this strain were its high robustness and stability of the culture; due to the high salinity needed for efficient cultivation of *H. mediterranei*, the risk of contamination by other microbes during the cultivation is negligible; thus, sterility precautions (sterilization of bioreactor equipment and cultivation media) are restricted to an absolute minimum. Only recently, *H. mediterranei* cultivations for PHA production were carried out without prior sterilization of the bioreactor; no microbial contamination was observed [88, 203]. Additionally, based on particularities of the PHA-biosynthesis pathway of this strain which generate high intracellular pools of the 3HV-precursor propionyl-CoA [204], *H. mediterranei* produces the copolyester P(3HB-*co*-8-10%-3HV) from 3HV-unrelated substrates like sugars or glycerol as sole carbon sources, thus saving the considerable costs for the 3HV-related precursor substrates (*e.g.*, propanoic acid or pentanoic acid) as. Only recently, Han and colleagues reported that blocky structured copolyesters of random distribution are produced by this organism [205]. The strain grew well on whey permeate after hydrolysis (50/50 mixture glucose/galactose as carbon source) with max. rate for specific growth of μ_{max} = 0.11 1/h, which is an outstanding value for halophilic archaea. PHBHV copolyester was synthetized at max. rate for specific production q_P = 0.08 g/(g h); the yield of whey sugar conversion to the biopolyester amounted to 0.33 g/g [46]. By additional optimization of process conditions in follow-up experiments, it was possible to increase the strain´s volumetric productivity for PHBHV on pre-treated (hydrolysed) whey permeate up to 0.09 g/(Lh) (q_P: 0.15 g/(g h)), and the obtained biomass concentration to 16.8 g/L with 73 g/g PHBHV in CDM [47]. Further experiment were carried out in order to improve the polyester quality; based on whey permeate hydrolysate as main carbon substrate, a high-quality terpolyester P(3HB-*co*-21.8%-3HV-*co*-5.1%-4HB) was accumulated by *H. mediterranei* by supplying precursor compounds for 3HV (pentanoic acid) and for 4HB (γ-butyrolactone). The accumulated polyester was recovered from biomass and subjected to a detailed thermal-analytic characterization and molecular mass determination. Based on the promising results obtained for polymer characterization (low crystallinity, low glass transition temperature, low melting point, high molar masses, low molar mass distribution), the authors assumed that the terpolyester could by significant for high-performance uses, *e.g.*, in the medicinal area [145]. The transformation of whey hexoses to 3-hydroxyvalerate (3HV) building blocks, the superb polymer characteristics (melting temperature, glass transition temperature, and polydispersity lower than for copolyesters produced by this strain), in synopsis with a facilitated, inexpensive recovery process based on cell-disintegration in hypotonic media make this strain an attractive candidate organisms for PHA biosynthesis based on whey and other inexpensive carbon feed stocks such as crude glycerol from biodiesel production [88], extruded starch [206], rice-based ethanol stillage residues [207], or rice bran [208]. Cost estimations for whey-based PHA production, combined with a solvent-free downstream processing, suggest a production price of € 2.82 per kg PHA [46]. In this context, it has to be emphasized that the high salinity of the cultivation medium (about 20 g/L NaCl) presupposes special requirements for the bioreactor material and the measuring sensors; steel of the highest quality or polymeric materials like PEEK should be applied [209]. In addition, for economic and sustainability reasons, it was of interest to investigate a possible recyclability of the highly saline side streams of this process. It was recently shown that spent fermentation broth of whey-based PHA-production using *H. mediterranei* can replace a significant fraction of new salt-rich cultivation medium in succeeding fermentation setups. Further, 29% of the costly growth additive yeast extract, acting as a source for nitrogen- and phosphate to make cultivations of the microbes more efficiently, can be substituted by using cell debris from the previous cultivation setup. Hence, strategies are developed for combining cost reduction in PHA production with reducing environmental hazards by waste stream recycling [203].

The last remaining issue that needs to be addressed for whey-based PHA production by *H. mediterranei* is the insufficient conversion of galactose in the presence of glucose. This was recently solved by Pais and colleagues who discovered the high importance of optimum supply of trace elements to enhance the enzymatic system responsible for galactose conversion [50].

The multifaceted aspects of whey-based PHA production by *H. mediterranei* provoked a question for a holistic life cycle assessment (LCA) study in order to asses both the environmental impact and the economic feasibility of this process. Therefore, during the process development based on data from a 300 L pilot fermentation set-up, an assessment of the "sustainable process index" (SPI) was accomplished for process optimization regarding environmental sustainability. It was clearly shown that the calculated ecological footprint was in a similar range like fossil plastics, if the above described process side streams are recycled. In addition, further optimization approaches to enhance the ecological feasibility of the process were suggested. Additionally, the developed process for biopolyester production was compared with a process for whey powder, today′s most frequently implemented use of surplus whey. Also in this case, despite its immature state of development, the novel PHA production process was superior according to the SPI calculations [71]. The complete decoding of the *H. mediterranei* genome in 2012 [210] gives reasons to expect a rapid development of genetic improvements of this organism in order to further enhance its capability to produce PHA and other valued bio-products. For example, genetic engineering can be applied to avoid the generation of unwanted by-products, such as extracellular polysaccharides (EPS). In *H. mediterranei* cultures, EPS formation occurs in parallel to PHA biosynthesis and causes a considerable loss of carbon source, and an unwanted high viscosity of the fermentation medium [94]. Therefore, EPS-negative mutants were already designed which show enhanced conversion yields of carbon source towards PHA [211].

According to the studies presented by Koller *et al.*, two other investigated strains also demonstrated promising performance for whey-based PHA production, but were inferior in direct comparison to *H. mediterranei*. The cells of *P. hydrogenovora* (*B. funghorum*) turned out to contain rather low PHA mass fractions in CDM, expressing low volumetric productivity and modest yields because of the shift of carbon flux towards other metabolites such as organic acids. In these experiments, the mass fraction of PHB in CDM did not exceed 0.12 g/g at the end of the process, the specific volumetric productivity q_p amounted to modest 2.9 mg/g h. After supplementation with pentanoic acid as a 3HV-precursor in follow-up experiments, the strain accumulated poly(3HB-*co*-21%-3HV) at a mass fraction of 0.12 g/g biomass at q_p of 2.0 mg/(gh); this

PHBHV copolyester displayed advanced polymer properties regarding crystallinity and melting temperature [46 - 48].

The third candidate investigated in this study, *H. pseudoflava*, accumulated PHA of enhanced quality, especially regarding the excellent molar masses and high polymer homogeneity. Although in this case, PHA production occurred from non-hydrolyzed lactose at higher q_P rates and higher yields than it was shown for *B. funghorum*, the strain was inferior to *H. mediterranei* regarding stability and strain robustness. *E.g.*, by adding pentanoic (valeric) acid as a 3HV-precursor, this strain accumulated poly(3HB-*co*-5%-HV) at mass fractions of 0.40 g/g in CDM at q_p of 12.5 mg/(gh). If no pentanoic acid was supplied, the homopolyester PHB was produced with contents of 0.30 g PHB per CDM; q_p amounted to 16.0 mg/(gh) [46, 48]. Some years later, additional experiments with this organism on sucrose, lactose and whey were carried out by Povolo and colleagues who interestingly noticed the production of PHBHV copolyesters with small amounts of 3HV by cultivation on lactose or sucrose. When whey was used as a carbon source, even the incorporation of 4HB was detected leading to the production of about 0.6 g/L of poly(3HB-*co*-3HV-*co*-4HB) (PHBHV4HB) terpolyesters. The same terpolyester production was also noticed using lactose or sucrose after additional supplementation of yeast extract. Based on these findings, the authors assumed that one or more unknown components, present in both yeast extract and in whey, might cause 4HB production and incorporation in the biopolyester. It will be highly intriguing to trace these 4HB-precursor compounds and the biochemical pathways responsible for the production and assembly of copolyesters containing 3HV building blocks [146].

A number of more recent publications report on PHA biosynthesis by novel microbial species; this suggests that the capability of new isolates from various environments for PHA production based on whey in is still underestimated, and provides a broad field for further investigation.

As an example, the thermophile eubacterium *Thermus thermophilus* HB8 (DSM 579) turned out to utilize whey-based lactose at optimum cultivation conditions of 70°C. The strain converts as well glucose/galactose mixtures for PHA biosynthesis under nitrogen limitation [212]. PHA was accumulated at polymer contents in cells up to 0.35 g/g. As the most remarkable finding, the produced PHA was a new heteropolymer as confirmed by GC-MS and ^{13}C-NMR, containing the *scl*-PHA building block 3HV (molar fraction of 0.38 mol/mol in total PHA) as well as the *mcl*-PHA building blocks 3-hydroxyheptanoate (3HHp; molar 3HHp fraction 0.10 mol/mol), 3-hydroxynanoate (3HN; fraction 0.17 mol/mol) and the unsaturated monomer 3-hydroxyundecanoate (3HU; fraction 0.35 mol/mol). Despite the rather low molar mass of the produced PHA, its unique composition invites for further investigation of this special polymer. Apart from the unusual polymer composition, the application of this thermophile organism provides additional advantages such as reduced risk for microbial contamination, a better homogenization of remaining whey lipids in the cultivation medium, and a higher stability of the culture due to a higher resistance of the strain´s enzymes to physical/chemical denaturation [212].

Moreover, *Azohydromonas lata* DSM 1123 was investigated for PHB production using pentanoic acid as a 3HV-related precursor to obtain poly(3HB-*co*-27% - HV). Cheese whey pre-treated by acidification and thermal treatment was used as a substrate. The authors reported that the application of whey hydrolysate for PHA production by this strain was beneficial compared to the application of pure sugars. Furthermore, they investigated the effect of age of the inoculum culture on the cultivation kinetics and determined the optimum temperature for respective cultivations at 30°C. The maximum PHA concentration obtained amounted to rather modest 1.66 g/L [213].

Domingos *et al.* reported cheese whey powder-based production of VFAs from in a packed-bed biofilm reactor (PBBR) containing ceramic support elements; this PBBR was operated continuously under anaerobic acidogenic conditions. The lactose present in the whey powder solution was completely metabolized to a mixture of VFAs with a yield of about 80%. Despite the rather high obtained VFA yield, a considerable decrease in propionic acid formation was observed during the process. Hence, process optimization is needed in this case in order to produce PHBHV copolyesters in subsequent aerobic cultivations [201]. Based on a mixed culture approach, a recent study conducted by Valentino and colleagues reported on the fate and effect of β-hexachlorocyclohexane (β-HCH) during diverse process steps of PHA production and recovery; cheese whey artificially contaminated with β-HCH was used as a feedstock for a production of VFAs, which later were subjected to aerobic conversion to PHA. The predominant part of β-HCH was adsorbed on solid particles of whey and removed after acidogenesis, when these residual solids were separated together with the active biomass from the liquid-phase. PHA purification steps were also beneficial for a more complete removal of β-HCH; a total of about 99.9% of the original β-HCH quantity was removed. In addition, it turned out that β-HCH had neither inhibiting effect on anaerobic acidogenesis nor on the subsequent aerobic PHA accumulation by mixed microbial cultures [214].

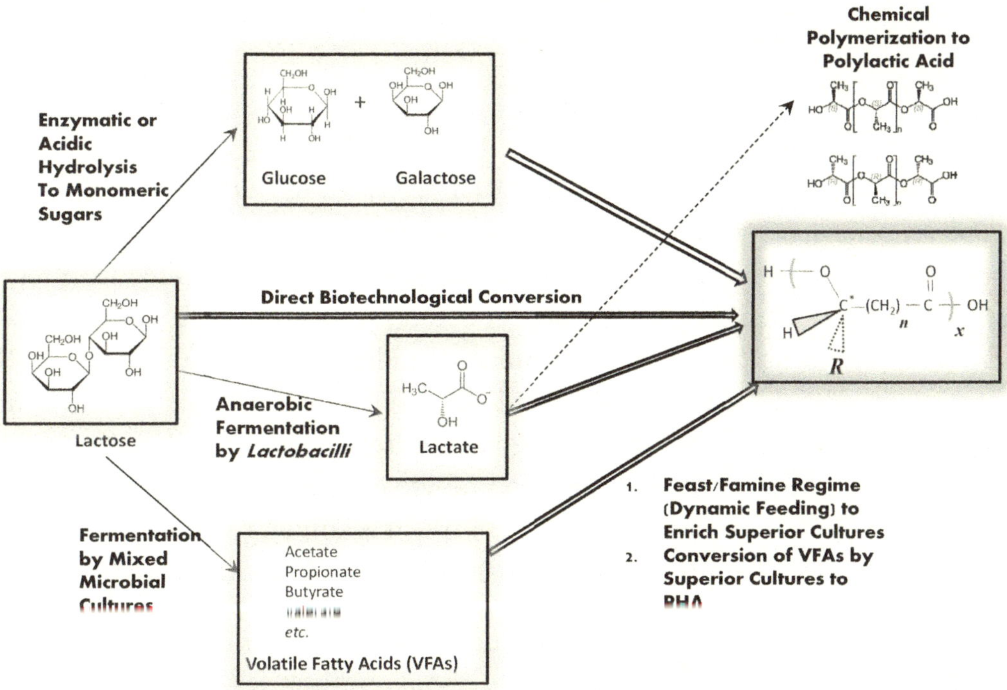

Fig. (5). Four different routes from whey lactose to PHA: conversion without prior hydrolysis, whey lactose hydrolysis to the monomeric sugars, conversion to lactic acid by lactobacilli, or conversion of lactose by mixed cultures to volatile fatty acids (VFAs). Alternatively, the intermediary product lactic acid can undergo chemical polymerization towards poly(lactic acid) (PLA).

All the examples detailed above show that the selection of the microbial production strain or a mixed microbial consortium determines the required pre-treatment of whey lactose, hence the need for hydrolysis, or even conversion to lactic acid or VFAs (Fig. 5).

Gram-Positive Wild Type Organisms as Whey-Based PHA-Producers

All Gram-positive PHA producers able to utilize whey are as of today members of the genus *Bacillus*.

Omar and colleagues investigated a range of carbon sources as substrates for *Bacillus megaterium*–mediated PHA production; this strain was isolated from sewage treatment plant sludge. As it turned out, a date syrup was found to be a suitable substrate for this strain whereas the outcomes using lactose as the carbon source in whey was less promising; on a shaking flask scale, a rather low polymer content of 26% PHB in CDM, and CDM concentration not exceeding1.8 g/L was reported [215].

A gram positive PHA-producer strain SRKP-3 with high genetic similarity to *Bacillus megaterium* was isolated from brackish water by Pandian *et al.* Medium optimization based on a multifactorial experimental design was applied in order to assess the strain´s potential for PHA production on surplus whey from dairies, sea water, and rice bran. From the optimized medium at pH-value 9, a PHB concentration of 6.37 g/L was achieved on a shaking flask scale. The same medium was utilized for fed-batch bioreactor studies under controlled conditions. Dairy waste was supplied several times to provide excess carbon source, resulting in a maximum PHB concentration of 11.32 g/L. Dissolved oxygen (DO) turned out as a major growth-limiting factor to positively affect PHB accumulation [216].

In 2011, Obruca and colleagues investigated the new isolate *Bacillus megaterium* CCM 2032 for PHB production on whey on a shaking flask scale. By optimization of the medium composition, the authors succeeded to improve the obtained PHB yields by about 50 times; best results for catalytically active biomass and PHB were reported to be 2.82 g or 1.05 g/L, in comparison to the original, not optimized medium based on whey. After further optimization including the supplementation of small quantities of ethanol as exogenous stress factor, PHB contents in biomass were increased by approximately 40% after addition of 1% ethanol to the cultivation medium at the start of stationary growth (CDM amounted to 2.87 g/L, PHB to 1.48 g/L) [49]. These experiments originated from the application of exogenous stress conditions on *C. necator*, as reported before by the same group of authors, who applied hydrogen peroxide and ethanol to boost PHA productivity in this organism [217].

Using different inexpensive carbon sources like whey, bagasse, starch, and rice bran, Nagamani and colleagues produced a PHBHV copolyester using the new Gram-positive isolate *Bacillus* OU40^T which had, based on genotypic investigations, a 60% similarity to *Bacillus cereus*. Also in this case, no 3HV-related precursors were needed to incorporate 3HV-building blocks into the polyesters. For whey as single C-source, the strain achieved a CDM concentration of 3 g/L in shaking flask experiments, containing about 0.6 g PHBHV per g biomass with a considerably high 3HV share of about 0.14 g 3HV per g PHA [218]. These experiments were later repeated by the same group of authors for another isolate, namely *Bacillus* OU73^T. As a carbon substrate, the supernatant of acidified whey was applied in shaking flask set-ups, and compared to the utilization of other inexpensive carbon sources like rice bran, starch, and bagasse. Also this isolate turned out to accumulate a 3HV-containing copolyester from all investigated inexpensive carbon sources (starch, rice bran, whey, bagasse). In the case of whey, a CDM concentration of about 3 g/L was obtained, containing 0.4 g Poly(3HB-*co*-17.9%-3HV) per g CDM [219].

Genetically Engineered Organisms as Whey-Based PHA-Producers

The first really promising reports on lactose-based PHA production involved genetically engineered organisms. Here, the direct lactose converter *E. coli* was "furnished" with the PHA synthesis genes from the natural PHA producer *C. necator* (in the original literature named *A. eutrophus*). These experiments are significant because of the extraordinarily high productivity achieved, thus providing for large-scale lactose-based PHA production [218, 219]. Lee and colleagues obtained a concentration of 5.2 g/L PHB by cultivating the recombinant *E. coli* strain GCSC 4401 (pSYL107), carrying *C. necator* PHA synthesis genes, on bovine whey powder solution on a shaking flask scale, and 4.5 g/L PHB by using rec. *E. coli* GCSC 6576 (pSYL107). These high concentrations were enabled by the applied strategy for genetic manipulation: only the PHA-synthesis genes were transferred into the host organism, not the genes encoding for the degradation of PHA [220].

The rec. *E. coli* strain GCSC 6576, containing the *C. necator* genes for PHA biosynthesis at high-copy-number plasmid on a plasmid plus the *ftsZ* gene from *E. coli*, was used for whey-based PHB production. The strain was cultivated in pH-stat fed-batch bioreactor using whey powder as the carbon feedstock. 109 g/L of CDM and 50 g/L of PHB were obtained after 47 h of cultivation. Switching to the whey solution with a lactose concentration of 210 g/L as a carbon substrate, 87 g/L of CDM and 69 g/L of PHB were obtained after 49 h by the same pH-stat fed-batch cultivation conditions; here the PHB fraction in CDM amounted to 0.80 g/g. These results suggest that high-throughput PHB production by cultivations of rec. *E. coli* in fed-batch mode on whey concentrate is a viable strategy [221].

Soon after these observations, the application of whey permeate as carbon substrate for rec. *E. coli* strains was also examined by other researchers [42, 43, 222]. As of today, these experiments yielded most auspicious outcomes for PHA production on whey by rec. *E. coli* CGSC 4401: after 36.5 h of cultivation, PHA concentration amounted to 168 g/L, which corresponds to PHA productivity of 4.6 g/(Lh) and 0.87 g PHA per g CDM, as reported by Ahn and colleagues [43]. These experiments were carried out in a cell-recycle setup under pH-stat conditions using an external module and a highly concentrated whey solution with a lactose content of 280 g/L as a substrate feed. When using non-concentrated whey permeate as a substrate (210 g/L lactose content), the obtained values were somewhat lower, but still excellent in comparison to previous reports on whey-based PHA-production reaching a PHA concentration of 100 g/L at the end of the process, which corresponds to a productivity of 2.6 g/(Lh), and 0.67 g PHA per g CDM as polymer mass fraction. This article demonstrates a strategy to tackle problems typically originating from repeated addition of diluted whey in fed-batch cultivation mode, which results in a quickly growing volume of cultivation medium inside the bioreactor facility [43].

The research group of Prof. Casella approached the topic of whey-based PHA production by genetically engineered strains the other way round: these researchers genetically designed an engineered *C. necator* strain which harboured the *E. coli lacZ*, *lacI* and *lacO* genes needed for metabolic transformation of lactose. By these recombinant organisms it was possible to directly produce PHA from lactose, and, in addition, from whey permeate with and without preceding hydrolysis. On lactose, 0.21 g/L PHB were obtained in shake flasks cultivations; these results were improved by using non-hydrolyzed whey permeate (1.4 g/L PHB) or hydrolysed whey permeate (2.1 g/L PHB). Additionally, as described above, insertion of the *lac* operon into the *phaZ* gene provides for reduction of the PHA depolymerization, thus improving final polymer yields [44].

In addition, Catalán and associates designed a modified *Herbaspirillum seropedicae* Z69 which carried the *E. coli* lactose-converting genes *lacZ*, *lacI* and *lacO* in the plasmid pHM3. These authors obtained 1.8 g/L PHB and 5 g/L CDM on shaking flask scale [223].

By using a transposition-mediated mechanism, *Escherichia coli* strain CML3-1 was engineered by Pais and colleagues by inserting the PHB synthesis genes from *C. necator* into the chromosome under the control of lactose-inducible promoter. It was demonstrated that the PHA synthesis by this engineered construct on cheese whey was rather modest because of a shift of the carbon flux towards concomitant synthesis of organic acids. Therefore, the method called "proton suicide" was used as a way to create an *E. coli* mutant strain which produced lower quantities of organic acids and converted a higher share of the whey-stemming carbon source towards PHA [224]. In these experiments, a total of 13 different *E. coli* mutants were developed and investigated in shaking flask cultivations, either on complex or minimal lactose-containing media. As most auspicious candidate for fed-batch cultivations in bioreactors on whey as single carbon substrate, the mutant strain P8-X8 was selected. Although cellular growth kinetics were considerably lower for the mutant strain in comparison to the wild-type, less organic acids (OA) were produced from whey lactose (0.04 CmolOA/CmolLac), and higher PHB productivity (18.88 g PHB/L) was reported. For the wild-type strain, these values amounted to 0.11 CmolOA/CmolLac and 7.8 g PHB/L. Hence, his engineering approach was efficient since it reduced organic acid production and at the same time improved the PHB yield on lactose (0.28 CmaolPHB/Cmollac *vs* 0.10 CmolPHB/Cmollac of the original strain) [225].

Mathematical Modelling of Whey-Based PHA Production

Considering the multifaceted intracellular regulation mechanisms involved in the PHA metabolism, and despite the complexity of the carbon source whey, mathematical models of PHA synthesis provides a potent instrument to predict diverse metabolic ongoings and to reduce the quantity of required laboratory experiments. Different types of mathematical models are contemporarily available to describe microbial PHA production [168]. These models range from discontinuous PHA biosynthesis [225] to continuous processes of pure microbial cultures in multistage bioreactor cascades [170, 171]. Different models are also available for dynamic feeding of mixed cultures to obtain consortia with superior PHA-accumulation capabilities [226], encompassing low as well as high structured modelling approaches [227]. A lot of progress was reached during the last couple of years in the areas of metabolic flux analysis and modelling for metabolic control (reviewed by [227]). Kinetic, dynamic, cybernetic, and structured metabolic models are not described yet for better understanding of whey-based PHA synthesis. Such advanced models can be understood as "structured kinetic models"; they are developed to simulate effects of concentrations of enzymes and diverse metabolites on microbial PHB synthesis rate [228]. They give us an idea about how the cell´s physiological state and intracellular enzyme and metabolite levels determine PHB synthesis kinetics. Sophisticated experimental tools, extensive computational methods and detailed information about metabolic fluxes are needed to develop such high structured models. Therefore, for practical purposes in laboratories and in industry, some model structure simplification is required. Simulations obtained by simplified models are not as accurate as it is the case for more detailed, high structured models. Nevertheless, such simplified modelling approach might constitute a viable tool *e.g.*, when metabolic pathways and/or the corresponding rate determining mechanisms are not sufficiently described, which hampers development of metabolic flux models [169].

Process modelling using whey or whey sugars is still very scarce [15]. It was shown that various types of models can be used to describe such processes. The selection of the most adequate model type depends on various factors such as the chemical composition of whey (native whey, whey permeate, partially or completely fermented whey, hydrolysed or intact whey) and on particularities of the applied production strain (growth-, partially, or non-growth associated PHA synthesis). Native whey is rich in lactose as carbon source, in casein as a complex nitrogen- and carbon source, contains a lot of phosphates, and, because of anaerobic conversion of lactose, it also contains lactic acid plus some additional compounds of the anaerobic metabolism. Based on the high nitrogen and phosphate content, whole whey is not really a suitable substrate for microbial PHA synthesis in simple-one-stage cultivation setups. This is especially the case for PHA producing microbes with insufficient or lacking lactase expression. This is in contrast to casein-free, hydrolysed whey permeate rich in glucose and galactose. However, independent from the type of whey used as carbon source, it will be necessary to apply such models which encompass multi-substrate kinetics for generation of models of satisfying predictive power [15].

For modelling of P(3HB-*co*-3HV-*co*-4HB) biosynthesis from whey and respective 3HV- and 4HB precursors by *H. mediterranei*, a formal kinetic model and a low structured model for whey-based PHB synthesis by *P. hydrogenovora* (today: *B. funghorum*) were developed, as summarized in the following paragraphs [223]:

H. mediterranei cultivations in batch and fed-batch mode were modelled. At that time, the metabolic pathways active in this archaeon were not elucidated yet, only some fundamental metabolic routes were described basically. Particularly, the strain's pathway(s) towards degradation of galactose were not known at all. This insufficient knowledge on the strain´s metabolism gave reason to develop a formal-kinetic model instead of choosing more sophisticated low structured or even high structured metabolic flux modelling approaches. As outcome, the simulated data for process variables obtained *in silico* fitted well, but not excellently to the data obtained by the bioreactor experiments. This is particularly valid for PHB concentration in late exponential growth phase, and for glucose concentration during the transient time period between exponential and stationary growth phase.

In addition, derived from experimental data and *in silico* results of the formal-kinetic model described above, a low structured metabolic model was developed for a fed-batch process with *B. funghorum*. PHB production from glucose/galactose mixtures was studied, whereas ammonia and hydrolysed whey casein acted as nitrogen sources. This strain is a prime example for organisms where PHB synthesis only starts after nitrogen limitation ("non-growth-associated PHA production"). Complete uncoupling of growth phase and the phase of PHB synthesis was demonstrated by the calculated data.

For both above cases, the authors emphasize that a better understanding of the complex intracellular situation during PHA biosynthesis from mixed carbon sources is enabled by the developed models, without the requirement to analyse the involved enzyme reactions in all details [225].

Table **3**, an extension of a previous tabular summary publisged in [15], provides a compilation of the data discussed in this chapter for PHA production based on lactose, whey and diverse whey-derived materials. Table **3** presents the production strains, the pre-treatment of whey to make it accessible to the microbes, production rates, CDM and PHA concentrations, production scale (bioreactors or shaking flask experiments), and, if provided by the individual authors, information on polymer characterization.

Table 3. PHA production from lactose, whey and whey-derived substrates: Overview of literature data. ***Nota bene*: this Table is an updated extention of a previous one published in [15].**

Microbial Production Strain	Type of Carbon Source	Composition of PHA Produced on Monomeric Level	PHA Vol. Productivity [g/(Lh)]	CDM at the End of Process [g/L]	PHA at the End of Process [g/L]	Characterization of PHA Properties Available?	Production Facility and - Scale	Ref.
Pseudomonas cepacia ATCC 17759	Pure lactose	PHB homopolyester	0.03	4.9	2.75	Molar mass characterization	Shaking flasks	[196]
Pseudomonas cepacia ATCC 17759	Mixtures glucose/galactose (equimolar)	PHB homopolyester	0.18	16.7	4.0	-	Shaking flasks	[196]
Sinorhizobium melioti 41	Non-hydrolyzed whey permeate	PHB homopolyester	0.00018	0.483	0.017	-	Shaking flasks	[201]
Hydrogenophaga pseudoflava DSM 1034	Non-hydrolyzed whey permeate	PHB homopolyester	0.00018	0.375	0.017	-	Shaking flasks	[201]
Hydrogenophaga pseudoflava DSM 1034	Non-hydrolyzed whey permeate + 3HV precursor pentanoic acid	P(3HB-*co*-4.6%-3HV) copolyester	0.05	6.7	2.7	Thermal analysis, Molar mass characterization	Bioreactor (10L) operated in fed-batch mode	[46]

(Table 3) contd.....

Microbial Production Strain	Type of Carbon Source	Composition of PHA Produced on Monomeric Level	PHA Vol. Productivity [g/(Lh)]	CDM at the End of Process [g/L]	PHA at the End of Process [g/L]	Characterization of PHA Properties Available?	Production Facility and - Scale	Ref.
Hydrogenophaga pseudoflava DSM 1034	Pure lactose, whole whey	P(3HB-*co*-3HV-*co*-4HB) terpolyester	0.008	2.03	0.60	GPC, DSC, TGA, GC and Molar mass characterization	Shaking flasks	[146]
Pseudomonas hydrogenovora DSM 1749	Whey permeate (enzymatically hydrolyzed)	PHB homopolyester	0.03	10.6	1.27	Thermal analysis, Molar mass characterization	Bioreactor (2L) operated in fed-batch mode	[46, 48]
Pseudomonas hydrogenovora DSM 1749	Whey permeate (enzymatically hydrolyzed) plus 3HV precursor pentanoic acid	P(3HB-*co*-21%-3HV) copolyester	0.05	11.7	1.44	Thermal analysis, Molar mass characterization	Bioreactor (2L) operated in fed-batch mode	[48]
Methylobacterium sp. ZP24	Pure lactose	PHB homopolyester	0.08	5.25	3.1	-	Shaking flasks	[199]
Methylobacterium sp. ZP24	Supernatant of whey	PHB homopolyester	0.017	3.8	0.8	-	Shaking flasks	[199]
Methylobacterium sp. ZP24	Whey supernatant plus nitrogen source	PHB homopolyester	0.054	5.9	2.6	-	Shaking flasks	[199]
Methylobacterium sp. ZP24	Whole whey	PHB homopolyester	0.023	5.1	1.1	-	Shaking flasks	[199]
Methylobacterium sp. ZP24	Whole whey plus nitrogen source	PHB homopolyester	0.008	7.1	0.4	-	Shaking flasks	[199]
Methylobacterium sp. ZP24	Processed whey from cheese manufacturing	PHB homopolyester	0.06	3.34	2.07	-	Jar bioreactor, semiautomatic, (2L), operated in batch mode	[200]
Methylobacterium sp. ZP24	Processed whey from cheese manufacturing	PHB homopolyester	0.09	5.53	3.54	-	Jar bioreactor, semiautomatic, (2L), operated in fed-batch mode	[200]
Methylobacterium sp. ZP24	Processed whey from cheese manufacturing	PHB homopolyester	0.056	n.r.	3.91	-	Jar bioreactor, semiautomatic, (30L), operated in fed-batch mode	[200]
Azohydromonas lata DSM 1123	Whole whey	P(3HB-*co*-12.7%-3HV) copolyester	1.66	9.2	1.66	-	Shaking flasks	[213]
Thermus thermophiles HB8	Supernatant of acidified whey	*scl/mcl*-heteropolyester	0.012	1.92	0.29	^{13}C and ^{1}H NMR, GC-MS, Molar mass characterization	Shaking flasks	[212]

(Table 3) contd.....

Microbial Production Strain	Type of Carbon Source	Composition of PHA Produced on Monomeric Level	PHA Vol. Productivity [g/(Lh)]	CDM at the End of Process [g/L]	PHA at the End of Process [g/L]	Characterization of PHA Properties Available?	Production Facility and - Scale	Ref.
Cupriavidus necator H16	Hydrolyzed whole whey to support PHA production from waste frying oil	PHB homopolyester	0.96	31.0	28.1	Molar mass characterization	Bioreactor (2L) operated in fed-batch mode	[144]
Archaea								
Haloferax mediterranei DSM 1411	Whey permeate (enzymatically hydrolyzed)	P(3HB-*co*-8--3HV) copolyester	0.05	11.0	5.5	Thermal analysis, Molar mass characterization	Bioreactor (42L) operated in fed-batch mode	[46]
Haloferax mediterranei DSM 1411	Whey permeate (enzymatically hydrolyzed)	P(3HB-*co*-6--3HV)	0.09	16.8	12.2	Thermal analysis, Molar mass characterization	Bioreactor (42L) operated in fed-batch mode	[47]
Haloferax mediterranei DSM 1411	Whey permeate (enzymatically hydrolyzed)plus 3HV precursor pentanoic acid + 4HB precursor γ-butyrolactone	P-(3HB-*co*-21.8%-3HV-*co*-5.1%-4HB) terpolyester	0.14	16.8	14.7	Thermal analysis, Molar mass characterization	Bioreactor (10L) operated in fed-batch mode	[48]
Haloferax mediterranei DSM 1411	Whey permeate (enzymatically hydrolyzed)	P(3HB-*co*-8--3HV) copolyester				-	Bioreactor (10L) operated in fed-batch mode	[203]
Haloferax mediterranei DSM 1411	Whey permeate (acidically hydrolyzed)	P(3HB-*co*-1.5%-3HV) copolyester	0.17	7.54	4.00	Molar mass characterization	Bioreactor (2L) operated in fed-batch mode	[50]
Bacillus megaterium	Pure lactose	PHB homopolyester	0.01	1.8	0.47	-	Shaking flasks	[215]
Bacillus megaterium SRKP-3	Dairy waste	PHB homopolyester	0.31	n.r.	11.32	-	Bioreactor (3L) operated in fed-batch mode	[216]
Bacillus megaterium CCM2037	Whole whey	PHB homopolyester		2.87	1.46	-	Shaking flasks	[49]
Bacillus OU40T	Whole whey (no further information provided)	P(3HB-*co*-14%-3HV) copolyester	n.r.	3.0	1.8	FTIR	Shaking flasks	[218]
Bacillus OU73T	Supernatant of acidified whey	P(3HB-*co*-18%-3HV) copolyester	0.025	3.0	1.2	FTIR	Shaking flasks	[219]
rec. *E. coli GCSC 4401 (pSYL107)*	Solution of whey powder of bovine origin	PHB homopolyester	n.r.	6.4	5.2	-	Shaking flasks	[220]

(Table 3) contd.....

Microbial Production Strain	Type of Carbon Source	Composition of PHA Produced on Monomeric Level	PHA Vol. Productivity [g/(Lh)]	CDM at the End of Process [g/L]	PHA at the End of Process [g/L]	Characterization of PHA Properties Available?	Production Facility and - Scale	Ref.
rec. *E. coli GCSC 6576 (pSYL107)*	Solution of whey powder of bovine origin	PHB homopolyester	n.r.	5.7	4.5	-	Shaking flasks	[220]
rec. *Cupriavidus necator* mRePT	Pure lactose	PHB homopolyester	0.04	8.1	0.21	-	Shaking flasks scale	[44]
rec. *Cupriavidus necator* mRePT	Whey permeate (hydrolyzed)	PHB homopolyester	0.05	8.0	2.4	-	Shaking flask	[44]
rec. *Cupriavidus necator* mRePT	Whole whey permeate	PHB homopolyester	0.03	6,5	1.4	-	Shaking flasks	[42]
rec. *Escherichia coli* CGSC 4401 harbouring pJC4	Solution of processed whey powder of bovine origin; controlled DOC	PHB homopolyester	1.15	83.1	46.8	-	Jar bioreactor (6.6L); operated in fed-batch mode	[42]
rec. *Escherichia coli* CGSC 4401 harbouring pJC4	Solution of processed whey powder of bovine origin; higher concentration of whey feed	PHB homopolyester	1.42	102.9	59.6	-	Jar bioreactor (6.6L); operated in fed-batch mode	[42]
rec. *Escherichia coli* CGSC 4401 harbouring pJC4	Solution of processed whey powder of bovine origin; controlled DOC	PHB homopolyester	2.57	119.5	96.2	-	Jar bioreactor (6.6L); operated in fed-batch mode	[42]
rec. *Escherichia coli* CGSC 4401 harbouring pJC4	Solution of processed whey powder of bovine origin	PHB homopolyester	4.6	194	168	-	Jar bioreactor (6.6L); operated in fed-batch mode; equipped with cell recycle membrane module system	[43]
rec. *Escherichia coli* strain CML3-1 harbouring *C. necator* synthesis genes	Powder from cheese whey	PHB homopolyester	0.33	48.85	18.88	-	Laboratory bioreactor	[224]
Herbaspirillum seropedicae Z69 (pHM3)	Pure lactose	PHB homopolyester	0.09	5	1.8	-	Shaking flasks	[223]

CONCLUDING REMARKS

The review presents strategies how surplus materials from industry, such as whey as an example for other carbon-rich (agro)industrial by-products which do not interfere with human nutrition purposes, can take the role of raw materials in bio-mediated production processes of various vendible goods. Implementation of such waste streams like dairy whey should be understood as only one, but a definitely crucial step to finally make PHA production processes market-fit. In addition to applying inexpensive feedstocks like whey, fundamental improvements in the process design are needed such as switching from time-demanding and low productive discontinuous production strategies to advanced single- or multistage continuous production modes. These approaches constitute another pivotal step to reach higher productivity and allow manufacturing bioplastics of tailored composition at constant and reproducible quality. Further, the quest for new powerful production strains with excellent growth and PHA-accumulation kinetics and physiological particularities which allow them to outperform established strains still has to go on. In this respect, numerous promising candidates, especially among extremophiles, have been found and suggested for advanced whey-based PHA production processes. By mimicking waste water treatment strategies, mixed microbial cultures can be used to convert the BOC in whey-polluted water bodies to organic acids which subsequently act as a carbon source for PHA accumulation by enrichment cultures of superior PHA storage capacity. Here, the development of advanced dynamic feeding regimes is needed. In addition to the feedstock expenses and the process design, the applied downstreaming technique to recover PHA from microbial biomass is a cost-decisive factor in the PHA production lines, which, despite all the efforts published worldwide, is still not optimized for large-scale PHA production.

Independent from the selected production strains and the developed production modes, whey-based PHA production on a relevant scale has to be indispensably combined with already available process lines in existing dairy plants, where whey directly accumulates as a surplus stream. This should enable minimizing manufacturing costs. In addition, it is of utmost importance to close material cycles during the production process, especially regarding environmentally hazardous material streams. Successful efforts in this direction were demonstrated in the case of the highly saline side stream that accumulates during the whey-based PHA production process using highly halophile microorganisms.

In any case, in order to develop economically effective PHA production processes, the close collaboration of different scientific disciplines will indispensably be required. The know-how of chemical engineers, enzymologists, genetic engineers microbiologists, polymer scientists, and experts for development of mathematical models and LCA and cleaner production studies has to be combined. This way, synergism can be created which is needed to finally close the obvious gaps between the myriad of promising data obtained on a laboratory scale and their implementation on an industrial scale. By combining all our knowledge and collective efforts to improve every individual process step, we strongly believe that substantial progress towards a sustainable, cost-efficient technology can definitely be accomplished. Regarding the manifold positive results that have been elaborated globally, one can conclude that the research community is on the right track towards combining ecological benefit of auspicious biopolyesters with economic feasibility; this will finally attract interest for their broader industrial application and market breakthrough. Considering the available LCA studies which assess the ecological impact of whey-based PHA production, together with economic calculation, one can claim that the efforts in this direction are definitely on a promising route. Finally, it will be needed to integrate also the scientific fields of psychology and societal sciences in order to generate a broader acceptance in the public for goods of higher sustainability, even if today such materials do not for 100% meet the benchmark of petrol-based plastics in terms of material performance and price.

CONFLICT OF INTEREST

The authors confirm that they have no conflict of interest to declare for this publication.

ACKNOWLEDGEMENTS

The authors acknowledge the financial support by the European Commission, granting the FP5 and FP7 projects WHEYPOL (GRD2-2000-30385) and ANIMPOL (Contract No: 245084). In addition, the authors gratefully acknowledge the providing of the electron microscopic picture of *C. necator* cells by Dr. Elisabeth Ingolić, FELMI-ZFE Graz.

REFERENCES

[1] Guimarães PM, Teixeira JA, Domingues L. Fermentation of lactose to bio-ethanol by yeasts as part of integrated solutions for the valorisation of cheese whey. Biotechnol Adv 2010; 28(3): 375-84. [http://dx.doi.org/10.1016/j.biotechadv.2010.02.002] [PMID: 20153415]

[2] Kosikowski FV. Whey utilisation and whey products. J Dairy Sci 1979; 62: 1149-60. [http://dx.doi.org/10.3168/jds.S0022-0302(79)83389-5]

[3] Peters D. Carbohydrates for fermentation. Biotechnol J 2006; 1(7-8): 806-14. [http://dx.doi.org/10.1002/biot.200600041] [PMID: 16937414]

[4] Audic JL, Chaufer B, Daufin G. Non-food applications of milk components and dairy co-products: A review. Lait 2003; 83(6): 417-38. [http://dx.doi.org/10.1051/lait:2003027]

[5] Prazeres AR, Carvalho F, Rivas J. Cheese whey management: a review. J Environ Manage 2012; 110: 48-68. [http://dx.doi.org/10.1016/j.jenvman.2012.05.018] [PMID: 22721610]

[6] Illanes A. Whey upgrading by enzyme biocatalysis. Electron J Biotechnol 2011; 14(6): 15. [http://dx.doi.org/10.2225/vol14-issue6-fulltext-11]

[7] Ghaley AE, El-Taweel AA. Kinetic modelling of continuous production of ethanol from cheese whey. Biomass Bioenergy 1997; 12(6): 461-72. [http://dx.doi.org/10.1016/S0961-9534(97)00012-3]

[8] Aneja RP, Mathur BN, Chandan RC, Banerjee AK. Technology of Indian milk products: Handbook on process technology modernization for professionals, entrepreneurs and scientists. Dairy India Yearbook 2002.

[9] Heyman MB. Lactose intolerance in infants, children, and adolescents. Pediatrics 2006; 118(3): 1279-86. [http://dx.doi.org/10.1542/peds.2006-1721] [PMID: 16951027]

[10] Bulhões AC, Goldani HA, Oliveira FS, Matte US, Mazzuca RB, Silveira TR. Correlation between lactose absorption and the C/T-13910 and G/A-22018 mutations of the lactase-phlorizin hydrolase (LCT) gene in adult-type hypolactasia. Braz J Med Biol Res 2007; 40(11): 1441-6. [http://dx.doi.org/10.1590/S0100-879X2007001100004] [PMID: 17934640]

[11] Mintec. Available from: https://www.mintecglobal.com/fr/dairy-eggs/eu-prices-whey-4-months/news/ 2007.

[12] Kim HO, Wee YJ, Kim JN, Yun JS, Ryu HW. Production of lactic acid from cheese whey by batch and repeated batch cultures of *Lactobacillus* sp. RKY2. Appl Biochem Biotechnol 2006; 129-132: 694-704. [http://dx.doi.org/10.1385/ABAB:131:1:694] [PMID: 16915680]

[13] Viñas M, Borzacconi L, Martínez J. Anaerobic treatment of yeast manufacturing wastewater in UASB reactors. Environ Technol 1994; 15: 79-85. [http://dx.doi.org/10.1080/09593339409385406]

[14] Gonzales Siso MI. The biotechnological utilization of cheese whey: A review. Bioresour Technol

1996; 57: 1-11.
[http://dx.doi.org/10.1016/0960-8524(96)00036-3]

[15] Koller M, Muhr A, Reiterer A, *et al.* Whey lactose as a raw material for microbial production of biodegradable polyesters. In: Saleh HE, Ed. Polyester Intech. Open Access Publisher 2012.
[http://dx.doi.org/10.5772/48737]

[16] Mollea C, Bosco F, Marmo L. Valorisation of cheese whey, a by-product from the dairy industry. In: Muzzalupo I, Ed. Food Industry Intech. Open Access Publisher 2013.
[http://dx.doi.org/10.5772/53159]

[17] Pescuma M, de Valdez GF, Mozzi F. Whey-derived valuable products obtained by microbial fermentation. Appl Microbiol Biotechnol 2015; 99(15): 6183-96.
[http://dx.doi.org/10.1007/s00253-015-6766-z] [PMID: 26124070]

[18] Zafar S, Owais M. Ethanol production from crude whey by *Kluyveromyces marxianus*. Biochem Eng J 2006; 27(3): 295-8.
[http://dx.doi.org/10.1016/j.bej.2005.05.009]

[19] Jin Y. Co-fermentation of wheat and whey permeate for ethanol production and chemical characteristics of the resulting dried distillers' grains with solubles. PhD dissertation University of Alberta 2014.

[20] Comino E, Riggio VA, Rosso M. Biogas production by anaerobic co-digestion of cattle slurry and cheese whey. Bioresour Technol 2012; 114: 46-53.
[http://dx.doi.org/10.1016/j.biortech.2012.02.090] [PMID: 22444637]

[21] Carrillo-Reyes J, Celis LB, Alatriste-Mondragón F, Razo-Flores E. Different start-up strategies to enhance biohydrogen production from cheese whey in UASB reactors. Int J Hydrogen Energy 2012; 37(7): 5591-601.
[http://dx.doi.org/10.1016/j.ijhydene.2012.01.004]

[22] Venetsaneas N, Antonopoulou G, Stamatelatou K, Kornaros M, Lyberatos G. Using cheese whey for hydrogen and methane generation in a two-stage continuous process with alternative pH controlling approaches. Bioresour Technol 2009; 100(15): 3713-7.
[http://dx.doi.org/10.1016/j.biortech.2009.01.025] [PMID: 19231170]

[23] Maddox IS. Production of n-butanol from whey filtrate using *Clostridium acetobutylicum* NCIB 2951. Biotechnol Lett 1980; 2(11): 493-8.
[http://dx.doi.org/10.1007/BF00129545]

[24] Welsh FW, Veliky IA. Production of acetone-butanol from acid whey. Biotechnol Lett 1984; 6(1): 61-4.
[http://dx.doi.org/10.1007/BF00128231]

[25] Raganati F, Olivieri G, Procentese A, Russo ME, Salatino P, Marzocchella A. Butanol production by bioconversion of cheese whey in a continuous packed bed reactor. Bioresour Technol 2013; 138: 259-65.
[http://dx.doi.org/10.1016/j.biortech.2013.03.180] [PMID: 23619138]

[26] Parrondo J, Herrero M, García LA, Díaz M. A Note—Production of Vinegar from Whey. J Inst Brew 2003; 109(4): 356-8.

[http://dx.doi.org/10.1002/j.2050-0416.2003.tb00610.x]

[27] Flôres SH, Alegre RM. Nisin production from *Lactococcus lactis* A.T.C.C. 7962 using supplemented whey permeate. Biotechnol Appl Biochem 2001; 34(Pt 2): 103-7.
[http://dx.doi.org/10.1042/BA20010030] [PMID: 11592916]

[28] Kumar M, Jain AK, Ghosh M, Ganguli A. Industrial whey utilization as a medium supplement for biphasic growth and bacteriocin production by probiotic *Lactobacillus casei* LA-1. Probiotics Antimicrob Proteins 2012; 4(3): 198-207.
[http://dx.doi.org/10.1007/s12602-012-9104-0] [PMID: 26782046]

[29] Schirru S, Favaro L, Mangia NP, *et al.* Comparison of bacteriocins production from *Enterococcus faecium* strains in cheese whey and optimized commercial MRS medium. Ann Microbiol 2014; 64(1): 321-31.
[http://dx.doi.org/10.1007/s13213-013-0667-0]

[30] Adegbehingbe KT, Bello M. Antibacterial activities of fermented whey on some selected enteropathogenic bacteria. Int J Curr Microbiol App Sci 2014; 3(9): 152-61.

[31] Jónás Á, Fekete E, Flipphi M, *et al.* Extra- and intracellular lactose catabolism in *Penicillium chrysogenum*: phylogenetic and expression analysis of the putative permease and hydrolase genes. J Antibiot 2014; 67(7): 489-97.
[http://dx.doi.org/10.1038/ja.2014.26] [PMID: 24690910]

[32] De Palma Revillion JP, Brandelli A, Záchia Ayub MA. Production of yeast extract from whey using *Kluyveromyces marxianus*. Braz Arch Biol Technol 2003; 46(1): 121-7.
[http://dx.doi.org/10.1590/S1516-89132003000100017]

[33] Siso MI, Ramil E, Cerdan ME, Freire Picos MA. Respirofermentative metabolism in *Kluyveromyces lactis*: ethanol production and the Crabtree effect. Enzyme Microb Technol 1996; 18: 585-91.
[http://dx.doi.org/10.1016/0141-0229(95)00151-4]

[34] Daniel Ferrari M, Bianco R, Froche C, Loperena ML. Baker's yeast production from molasses/cheese whey mixtures. Biotechnol Lett 2001; 23: 1-4.
[http://dx.doi.org/10.1023/A:1026778503871]

[35] Daniel HJ, Reuss M, Syldatk C. Production of sophorolipids in high concentration from deproteinized whey and rapeseed oil in a two stage fed batch process using *Candida bombicola* ATCC 22214 and *Cryptococcus curvatus* ATCC 20509. Biotechnol Lett 1998; 20(12): 1153-6.
[http://dx.doi.org/10.1023/A:1005332605003]

[36] Daniel HJ, Otto RT, Binder M, Reuss M, Syldatk C. Production of sophorolipids from whey: development of a two-stage process with *Cryptococcus curvatus* ATCC 20509 and *Candida bombicola* ATCC 22214 using deproteinized whey concentrates as substrates. Appl Microbiol Biotechnol 1999; 51(1): 40-5.
[http://dx.doi.org/10.1007/s002530051360] [PMID: 10077820]

[37] Alonso S, Rendueles M, Díaz M. Efficient lactobionic acid production from whey by *Pseudomonas taetrolens* under pH-shift conditions. Bioresour Technol 2011; 102(20): 9730-6.
[http://dx.doi.org/10.1016/j.biortech.2011.07.089] [PMID: 21862326]

[38] Miyamoto Y, Ooi T, Kinoshita S. Production of lactobionic acid from whey by *Pseudomonas* sp.

LS13-1. Biotechnol Lett 2000; 22(5): 427-30.
[http://dx.doi.org/10.1023/A:1005617903152]

[39] Schultz N, Chang L, Hauck A, Reuss M, Syldatk C. Microbial production of single-cell protein from deproteinized whey concentrates. Appl Microbiol Biotechnol 2006; 69(5): 515-20.
[http://dx.doi.org/10.1007/s00253-005-0012-z] [PMID: 16133331]

[40] Antunes S, Alves VD, Grandfils C, Freitas F, Reis M. Production of extracellular polysaccharide by *Enterobacter* A47 using cheese whey as the sole carbon source. 1st EPNOE Junior Scientists Meeting. The Netherlands, Wageningen. 2015.

[41] Prasad S, Srikanth K, Limaye AM, Sivaprakasam S. Homo-fermentative production of D-lactic acid by *Lactobacillus* sp. employing casein whey permeate as a raw feed-stock. Biotechnol Lett 2014; 36(6): 1303-7.
[http://dx.doi.org/10.1007/s10529-014-1482-9] [PMID: 24563313]

[42] Ahn WS, Park SJ, Lee SY. Production of Poly(3-hydroxybutyrate) by fed-batch culture of recombinant *Escherichia coli* with a highly concentrated whey solution. Appl Environ Microbiol 2000; 66(8): 3624-7.
[http://dx.doi.org/10.1128/AEM.66.8.3624-3627.2000] [PMID: 10919830]

[43] Ahn WS, Park SJ, Lee SY. Production of poly(3-hydroxybutyrate) from whey by cell recycle fed-batch culture of recombinant *Escherichia coli.* Biotechnol Lett 2001; 23: 235-40.
[http://dx.doi.org/10.1023/A:1005633418161]

[44] Povolo S, Toffano P, Basaglia M, Casella S. Polyhydroxyalkanoates production by engineered *Cupriavidus necator* from waste material containing lactose. Bioresour Technol 2010; 101(20): 7902-7.
[http://dx.doi.org/10.1016/j.biortech.2010.05.029] [PMID: 20537531]

[45] Koller M, Bona R, Braunegg G, *et al.* Production of polyhydroxyalkanoates from agricultural waste and surplus materials. Biomacromolecules 2005; 6(2): 561-5.
[http://dx.doi.org/10.1021/bm049478b] [PMID: 15762613]

[46] Koller M, Hesse PJ, Bona R, *et al.* Various archae- and eubacterial strains as potential polyhydroxyalkanoate producers from whey lactose. Macromol Biosci 2007; 7: 218-2266.
[http://dx.doi.org/10.1002/mabi.200600211] [PMID: 17295410]

[47] Koller M, Atlic A, Gonzalez-Garcia Y, Kutschera C, Braunegg G. Polyhydroxyalkanoate (PHA) biosynthesis from whey lactose. Macromol Symp 2008; 27: 287-92.

[48] Koller M, Bona R, Chiellini E, *et al.* Polyhydroxyalkanoate production from whey by *Pseudomonas hydrogenovora.* Bioresour Technol 2008; 99(11): 4854-63.
[http://dx.doi.org/10.1016/j.biortech.2007.09.049] [PMID: 18053709]

[49] Obruca S, Marova I, Melusova S, Mravcova M. Production of polyhydroxyalkanoates from cheese whey employing *Bacillus megaterium* CCM 2037. Ann Microbiol 2011; 61: 947-53.
[http://dx.doi.org/10.1007/s13213-011-0218-5]

[50] Pais J, Serafim LS, Freitas F, Reis MA. Conversion of cheese whey into poly(3-hydroxybutyrate-c--3-hydroxyvalerate) by *Haloferax mediterranei.* N Biotechnol 2016; 33(1): 224-30.
[http://dx.doi.org/10.1016/j.nbt.2015.06.001] [PMID: 26134839]

[51] Nath A, Verasztó B, Basak S, Koris A, Kovács Z, Vatai G. Synthesis of lactose-derived nutraceuticals from dairy waste whey—a review. Food Bioproc Technol 2016; 9(1): 16-48. [http://dx.doi.org/10.1007/s11947-015-1572-2]

[52] Viitanen MI, Vasala A, Neubauer P, Alatossava T. Cheese whey-induced high-cell-density production of recombinant proteins in *Escherichia coli.* Microb Cell Fact 2003; 2(1): 2. [http://dx.doi.org/10.1186/1475-2859-2-2] [PMID: 12740045]

[53] Ykema A, Verbree EC, Kater MM, Smit H. Optimization of lipid production in the oleaginous yeast *Apiotrichum curvatum* in whey permeate. App Microbiol Biot 1998; 29(2-3): 211-8.

[54] Vamvakaki AN, Kandarakis I, Kaminarides S, Komaitis M, Papanikolaou S. Cheese whey as a renewable substrate for microbial lipid and biomass production by Zygomycetes. Eng Life Sci 2010; 10(4): 348-60. [http://dx.doi.org/10.1002/elsc.201000063]

[55] Taskin M, Saghafian A, Aydogan MN, Arslan NP. Microbial lipid production by cold-adapted oleaginous yeast *Yarrowia lipolytica* B9 in non-sterile whey medium. Biofuels Bioprod Biorefin 2015; 9: 595-605. [http://dx.doi.org/10.1002/bbb.1560]

[56] Purnawati R, Sunarti TC, Syamsu K, Rahayuningsih M. Characterization of novel *Bacillus thuringiensis* isolated from Attacus atlas and its growth kinetics in the cultivation media of tofu whey for bioinsecticide production. J Biol Agric Healthc 2015; 4(16): 33-9.

[57] Ertrk E, Erkmen O, Oner MD. Effects of various supplements on riboflavin production by *Ashbya gossypii* in whey. J Environ Eng Sci 1998; 22(2): 371-6.

[58] Guru V, Viswanathan K. Riboflavin production in milk whey using probiotic bacteria-*Lactobacillus acidophilus* and *Lactococcus lactis.* Ind J Fund Appl Life Sci 2013; 3(4): 169-76.

[59] Li M, Zhu H, Qin H, Zhang Y, Yang H. Optimization of a whey containing medium for β-Galactosidase production by *Lactobacillus reuteri.* In: Zhang T-C, Nakajima M, Eds. Advances in Applied Biotechnology. Berlin, Heidelberg: Springer 2015; pp. 599-608. [http://dx.doi.org/10.1007/978-3-662-46318-5_62]

[60] El-Holi MA, Al-Delaimy S. Citric acid production from whey with sugars and additives by *Aspergillus niger.* Afr J Biotechnol 2004; 2(10): 356-9.

[61] Gänzle MG, Haase G, Jelen P. Lactose: crystallization, hydrolysis and value-added derivatives. Int Dairy J 2008; 18: 685-94. [http://dx.doi.org/10.1016/j.idairyj.2008.03.003]

[62] Nahla TK, Musa TN. Chemical isomerization of whey lactose to lactulose by using batch reaction. Pak J Nutr 2015; 14(5): 255-8. [http://dx.doi.org/10.3923/pjn.2015.255.258]

[63] Koller M, Salerno A, Muhr A, Reiterer A, Braunegg G. Polyhydroxyalkanoates: Biodegradable polymers and plastics from renewable resources. Mater Tehnol 2013; 47(1): 5-12.

[64] Kržan A, Hemjinda S, Miertus S, Corti A, Chiellini E. Standardization and certification in the area of environmentally degradable plastics. Polym Degrad Stabil 2006; 91(12): 2819-33. [http://dx.doi.org/10.1016/j.polymdegradstab.2006.04.034]

[65] Narodoslawsky M, Shazad K, Kollmann R, Schnitzer H. LCA of PHA production-identifying the ecological potential of bio-plastic. Chem Biochem Eng Q 2015; 29(2): 299-305. [http://dx.doi.org/10.15255/CABEQ.2014.2262]

[66] Shahzad K, Kettl KH, Titz M, *et al.* Comparison of ecological footprint for biobased PHA production from animal residues utilizing different energy resources. Clean Technol Envir 2013; 15(3): 525-36. [http://dx.doi.org/10.1007/s10098-013-0608-4]

[67] Akiyama M, Tsuge T, Doi Y. Environmental life cycle comparison of polyhydroxyalkanoates produced from renewable carbon resources by bacterial fermentation. Polym Degrad Stabil 2003; 80: 183-94. [http://dx.doi.org/10.1016/S0141-3910(02)00400-7]

[68] Harding KG, Dennis JS, von Blottnitz H, Harrison ST. Environmental analysis of plastic production processes: comparing petroleum-based polypropylene and polyethylene with biologically-based poly-β-hydroxybutyric acid using life cycle analysis. J Biotechnol 2007; 130(1): 57-66. [http://dx.doi.org/10.1016/j.jbiotec.2007.02.012] [PMID: 17400318]

[69] Pietrini M, Roes L, Patel MK, Chiellini E. Comparative life cycle studies on poly(3-hydroxybutyrate--based composites as potential replacement for conventional petrochemical plastics. Biomacromolecules 2007; 8(7): 2210-8. [http://dx.doi.org/10.1021/bm0700892] [PMID: 17583946]

[70] Titz M, Kettl KH, Shahzad K, Koller M, Schnitzer H, Narodoslawsky M. Process optimization for efficient biomediated PHA production from animal-based waste streams. Clean Technol Envir 2012; 14: 495-503. [http://dx.doi.org/10.1007/s10098-012-0464-7]

[71] Koller M, Sandholzer D, Salerno A, Braunegg G, Narodoslawsky M. Biopolymer from industrial residues: Life cycle assessment of poly (hydroxyalkanoates) from whey. Resour Conserv Recycling 2013; 73: 64-71. [http://dx.doi.org/10.1016/j.resconrec.2013.01.017]

[72] Koller M, Atlić A, Dias M, Reiterer A, Braunegg G. Microbial PHA Production from Waste Raw Materials. In: Chen G-Q, Ed. Plastics from Bacteria Microbiology Monographs. Berlin: Springer Heidelberg 2010; pp. 83-119. [http://dx.doi.org/10.1007/978-3-642-03287-5_5]

[73] Koller M, Salerno A, Miranda de Sousa Dias M, Reiterer A, Braunegg G. Modern biotechnological polymer synthesis: a review. Food Technol Biotechnol 2010; 48(3): 255-69.

[74] Koller M, Gasser I, Schmid F, Berg G. Linking ecology with economy: insights into polyhydroxyalkanoate-producing microorganisms. Eng Life Sci 2011; 11(3): 222-37. [http://dx.doi.org/10.1002/elsc.201000190]

[75] Jendrossek D. Polyhydroxyalkanoate granules are complex subcellular organelles (carbonosomes). J Bacteriol 2009; 191(10): 3195-202. [http://dx.doi.org/10.1128/JB.01723-08] [PMID: 19270094]

[76] Jendrossek D, Pfeiffer D. New insights in the formation of polyhydroxyalkanoate granules (carbonosomes) and novel functions of poly(3-hydroxybutyrate). Environ Microbiol 2014; 16(8):

2357-73.
[http://dx.doi.org/10.1111/1462-2920.12356] [PMID: 24329995]

[77] Follonier S, Goyder MS, Silvestri AC, *et al.* Fruit pomace and waste frying oil as sustainable resources for the bioproduction of medium-chain-length polyhydroxyalkanoates. Int J Biol Macromol 2014; 71: 42-52.
[http://dx.doi.org/10.1016/j.ijbiomac.2014.05.061] [PMID: 24882726]

[78] Follonier S, Riesen R, Zinn M. Pilot-scale production of functionalized mcl-PHA from grape pomace supplemented with fatty acids. Chem Biochem Eng Q 2015; 29(2): 113-21.
[http://dx.doi.org/10.15255/CABEQ.2014.2251]

[79] Le Meur S, Zinn M, Egli T, Thöny-Meyer L, Ren Q. Production of medium-chain-length polyhydroxyalkanoates by sequential feeding of xylose and octanoic acid in engineered *Pseudomonas putida* KT2440. BMC Biotechnol 2012; 12(1): 53.
[http://dx.doi.org/10.1186/1472-6750-12-53] [PMID: 22913372]

[80] Lee SY. Bacterial polyhydroxyalkanoates. Biotechnol Bioeng 1996; 49(1): 1-14.
[http://dx.doi.org/10.1002/(SICI)1097-0290(19960105)49:1<1::AID-BIT1>3.3.CO;2-1] [PMID: 18623547]

[81] Braunegg G, Koller M, Varila P, *et al.* Production of plastics from waste derived from agrofood industry. In: Graziani M, Fornasiero P, Eds. Renewable Resources and Renewable Energy, A Global Challenge. 2nd ed. Florida, USA: CRC Press 2007; pp. 119-35.

[82] Obruca S, Benesova P, Marsalek L, Marova I. Use of lignocellulosic materials for PHA production. Chem Biochem Eng Q 2015; 29(2): 135-44.
[http://dx.doi.org/10.15255/CABEQ.2014.2253]

[83] Obruca S, Benesova P, Kucera D, Petrik S, Marova I. Biotechnological conversion of spent coffee grounds into polyhydroxyalkanoates and carotenoids. N Biotechnol 2015; 32(6): 569-74.
[http://dx.doi.org/10.1016/j.nbt.2015.02.008] [PMID: 25721970]

[84] Khosravi-Darani K, Mokhtari ZB, Amai T, Tanaka K. Microbial production of poly(hydroxybutyrate) from C1 carbon sources. Appl Microbiol Biotechnol 2013; 97(4): 1407-24.
[http://dx.doi.org/10.1007/s00253-012-4649-0] [PMID: 23306640]

[85] Riedel SL, Jahns S, Koenig S, *et al.* Polyhydroxyalkanoates production with *Ralstonia eutropha* from low quality waste animal fats. J Biotechnol 2015; 214: 119-27.
[http://dx.doi.org/10.1016/j.jbiotec.2015.09.002] [PMID: 26428087]

[86] Muhr A, Rechberger EM, Salerno A, *et al.* Biodegradable latexes from animal-derived waste: Biosynthesis and characterization of *mcl*-PHA accumulated by *Ps. citronellolis.* React Funct Polym 2013; 73(10): 1391-8.
[http://dx.doi.org/10.1016/j.reactfunctpolym.2012.12.009]

[87] Muhr A, Rechberger EM, Salerno A, *et al.* Novel description of *mcl*-PHA biosynthesis by *Pseudomonas chlororaphis* from animal-derived waste. J Biotechnol 2013; 165(1): 45-51.
[http://dx.doi.org/10.1016/j.jbiotec.2013.02.003] [PMID: 23467001]

[88] Hermann-Krauss C, Koller M, Muhr A, Fasl H, Stelzer F, Braunegg G. Archaeal production of polyhydroxyalkanoate (PHA) co-and terpolyesters from biodiesel industry-derived by-products.

Archaea 2013; 1-10.

[89] Drosg B, Fritz I, Gattermayr F, Silvestrini L. Photo-autotrophic production of poly (hydroxyalkanoates) in *cyanobacteria*. Chem Biochem Eng Q 2015; 29(2): 145-56. [http://dx.doi.org/10.15255/CABEQ.2014.2254]

[90] Koller M, Marsalek M. Cyanobacterial polyhydroxyalkanoate production: *Status Quo* and *Quo Vadis*? Curr Biotechnol 2015; 4(4): 464-80. [http://dx.doi.org/10.2174/2211550104666150917010849]

[91] Steinbüchel A, Hein S. Biochemical and molecular basis of microbial synthesis of polyhydroxyalkanoates in microorganisms. Adv Biochem Eng Biotechnol 2001; 71: 81-123. [http://dx.doi.org/10.1007/3-540-40021-4_3] [PMID: 11217418]

[92] Steinbuchel A. Recent advances in the knowledge of the metabolism of bacterial polyhydroxyalkanoic acids and potential impacts on the production of biodegradable thermoplastics. Acta Biotechnol 1991; 11: 419-27. [http://dx.doi.org/10.1002/abio.370110504]

[93] Steinbüchel A, Valentin HE. Diversity of bacterial polyhydroxyalkanoic acids. FEMS Microbiol Lett 1995; 128: 219-28. [http://dx.doi.org/10.1016/0378-1097(95)00125-O]

[94] Koller M, Chiellini E, Braunegg G. Study on the production and re-use of poly(3-hydroxybutyrate-*co*-3-hydroxyvalerate) and extracellular polysaccharide by the archaeon *Haloferax mediterranei* strain DSM 1411. Chem Biochem Eng Q 2015; 29(2): 87-98. [http://dx.doi.org/10.15255/CABEQ.2014.2058]

[95] Kim YB, Lenz RW. Polyesters from microorganisms. Adv Biochem Eng Biotechnol 2001; 71: 51-79. [http://dx.doi.org/10.1007/3-540-40021-4_2] [PMID: 11217417]

[96] Helm J, Wendlandt KD, Jechorek M, Stottmeister U. Potassium deficiency results in accumulation of ultra-high molecular weight poly-beta-hydroxybutyrate in a methane-utilizing mixed culture. J Appl Microbiol 2008; 105(4): 1054-61. [http://dx.doi.org/10.1111/j.1365-2672.2008.03831.x] [PMID: 18422550]

[97] Reusch RN, Shabalin O, Crumbaugh A, Wagner R, Schröder O, Wurm R. Posttranslational modification of *E. coli* histone-like protein H-NS and bovine histones by short-chain poly-(R)-3-hydroxybutyrate (cPHB). FEBS Lett 2002; 527(1-3): 319-22. [http://dx.doi.org/10.1016/S0014-5793(02)03236-2] [PMID: 12220682]

[98] Koller M, Hesse P, Kutschera C, *et al.* Sustainable embedding of the bioplastic poly-(3-hydroxybutyrate) into the sugarcane industry: Principles of a future-oriented technology in Brazil. In: Eyerer P, Ed. Polymers-Opportunities and Risks II. Berlin, Heidelberg: Springer 2010; pp. 81-96.

[99] Laycock B, Halley P, Pratt S, Werker A, Lant P. The chemomechanical properties of microbial polyhydroxyalkanoates. Prog Polym Sci 2014; 39(2): 397-442. [http://dx.doi.org/10.1016/j.progpolymsci.2013.06.008]

[100] Ye HM, Wang Z, Wang HH, Chen GQ, Xu J. Different thermal behaviors of microbial polyesters poly (3-hydroxybutyrate-co-3-hydroxyvalerate-co-3-hydroxyhexanoate) and poly (3-hydroxybutyrate-c--3-hydroxyhexanoate). Polymer (Guildf) 2010; 51(25): 6037-46.

[http://dx.doi.org/10.1016/j.polymer.2010.10.030]

[101] Aziz NA, Sipaut CS, Abdullah AA. Improvement of the production of poly(3-hydroxybutyrate-c--3-hydroxyvalerate-co-4-hydroxybutyrate) terpolyester by manipulating the culture condition. J Chem Technol Biotechnol 2012; 87(11): 1607-14.
[http://dx.doi.org/10.1002/jctb.3817]

[102] Zinn M. Biosynthesis of medium-chain-length poly(R)-3-hydroxyalkanoates. In: Chen G-Q, Ed. Plastics from Bacteria Microbiology Monographs. Berlin, Heidelberg: Springer 2010; pp. 213-36.
[http://dx.doi.org/10.1007/978-3-642-03287-5_9]

[103] Doi Y, Kunioka M, Nakamura Y, Soga K. Nuclear magnetic resonance studies on unusual bacterial copolyesters of 3-hydroxybutyrate and 4-hydroxybutyrate. Macromolecules 1988; 21(9): 2722-7.
[http://dx.doi.org/10.1021/ma00187a012]

[104] Kunioka M, Kawaguchi Y, Doi Y. Production of biodegradable copolyesters of 3-hydroxybutyrate and 4-hydroxybutyrate by *Alcaligenes eutrophus.* App Microbiol Biot 1989; 30(6): 569-73.
[http://dx.doi.org/10.1007/BF00255361]

[105] Kunasundari B, Sudesh K. Isolation and recovery of microbial polyhydroxyalkanoates. Express Polym Lett 2011; 5(7): 620-34.
[http://dx.doi.org/10.3144/expresspolymlett.2011.60]

[106] Koller M, Niebelschütz H, Braunegg G. Strategies for recovery and purification of poly[(*R*)-3-hydroxyalkanoates] (PHA) biopolyesters from surrounding biomass. Eng Life Sci 2013; 13(6): 549-62.
[http://dx.doi.org/10.1002/elsc.201300021]

[107] Madkour MH, Heinrich D, Alghamdi MA, Shabbaj II, Steinbüchel A. PHA recovery from biomass. Biomacromolecules 2013; 14(9): 2963-72.
[http://dx.doi.org/10.1021/bm4010244] [PMID: 23875914]

[108] Jacquel N, Lo CW, Wei YH, Wu HS, Wang SS. Isolation and purification of bacterial poly (3-hydroxyalkanoates). Biochem Eng J 2008; 39(1): 15-27.
[http://dx.doi.org/10.1016/j.bej.2007.11.029]

[109] Reddy CS, Ghai R, Rashmi KV, Kalia VC. Polyhydroxyalkanoates: an overview. Bioresour Technol 2003; 87(2): 137-46.
[http://dx.doi.org/10.1016/S0960-8524(02)00212-2] [PMID: 12765352]

[110] Khanna S, Srivastava AK. Recent advances in microbial polyhydroxyalkanoates. Process Biochem 2005; 40: 607-19.
[http://dx.doi.org/10.1016/j.procbio.2004.01.053]

[111] Ren Q, Grubelnik A, Hoerler M, *et al.* Bacterial poly(hydroxyalkanoates) as a source of chiral hydroxyalkanoic acids. Biomacromolecules 2005; 6(4): 2290-8.
[http://dx.doi.org/10.1021/bm050187s] [PMID: 16004474]

[112] Chen GQ. A microbial polyhydroxyalkanoates (PHA) based bio- and materials industry. Chem Soc Rev 2009; 38(8): 2434-46.
[http://dx.doi.org/10.1039/b812677c] [PMID: 19623359]

[113] Koller M. Poly(hydroxyalkanoates) for food packaging: Application and attempts towards implementation. App Food Biotechnol 2014; 1(1): 3-15.

[114] Lauzier CA, Monasterios CJ, Saracovan I, Marchessault RH, Ramsay BA. Film formation and paper coating with poly ([beta]-hydroxyalkanoate), a biodegradable latex. Tappi J 1993; 76(5)

[115] Zinn M, Witholt B, Egli T. Occurrence, synthesis and medical application of bacterial polyhydroxyalkanoate. Adv Drug Deliv Rev 2001; 53(1): 5-21. [http://dx.doi.org/10.1016/S0169-409X(01)00218-6] [PMID: 11733115]

[116] Chen GQ, Wu Q. The application of polyhydroxyalkanoates as tissue engineering materials. Biomaterials 2005; 26(33): 6565-78. [http://dx.doi.org/10.1016/j.biomaterials.2005.04.036] [PMID: 15946738]

[117] Luef KP, Stelzer F, Wiesbrock F. Poly (hydroxyalkanoates) in medical applications. Chem Biochem Eng Q 2015; 29(2): 287-97. [http://dx.doi.org/10.15255/CABEQ.2014.2261]

[118] Lizarraga-Valderrama LR, Nigmatullin R, Taylor C, *et al.* Nerve tissue engineering using blends of poly (3□hydroxyalkanoates) for peripheral nerve regeneration. Eng Life Sci 2015; 15(6): 612-21. [http://dx.doi.org/10.1002/elsc.201400151]

[119] Chen GQ, Wu Q. The application of polyhydroxyalkanoates as tissue engineering materials. Biomaterials 2005; 26(33): 6565-78. [http://dx.doi.org/10.1016/j.biomaterials.2005.04.036] [PMID: 15946738]

[120] Meischel M, Eichler J, Martinelli E, *et al.* Adhesive strength of bone-implant interfaces and in-vivo degradation of PHB composites for load-bearing applications. J Mech Behav Biomed Mater 2016; 53: 104-18. [http://dx.doi.org/10.1016/j.jmbbm.2015.08.004] [PMID: 26318571]

[121] Martin DP, Williams SF. Medical applications of poly-4-hydroxybutyrate: a strong flexible absorbable biomaterial. Biochem Eng J 2003; 16(2): 97-105. [http://dx.doi.org/10.1016/S1369-703X(03)00040-8]

[122] Chen G-Q. Plastics completely synthesized by bacteria: Polyhydroxyalkanoates. In: Chen G-Q, Ed. Plastics from Bacteria Microbiology Monographs. Berlin, Heidelberg: Springer 2010; pp. 17-37. [http://dx.doi.org/10.1007/978-3-642-03287-5_2]

[123] Chen G-Q. Industrial Production of PHA. In: Chen G-Q, Ed. Plastics from Bacteria. Berlin, Heidelberg: Microbiology Monographs Springer 2010; pp. 121-32. [http://dx.doi.org/10.1007/978-3-642-03287-5_6]

[124] Hany R, Böhlen C, Geiger T, Schmid M, Zinn M. Toward non-toxic antifouling: synthesis of hydroxy-, cinnamic acid-, sulfate-, and zosteric acid-labeled poly[3-hydroxyalkanoates]. Biomacromolecules 2004; 5(4): 1452-6. [3-hydroxyalkanoates]. [http://dx.doi.org/10.1021/bm049962e] [PMID: 15244464]

[125] Ruth K, Grubelnik A, Hartmann R, Egli T, Zinn M, Ren Q. Efficient production of (R)-3-hydroxycarboxylic acids by biotechnological conversion of polyhydroxyalkanoates and their purification. Biomacromolecules 2007; 8(1): 279-86. [http://dx.doi.org/10.1021/bm060585a] [PMID: 17206818]

[126] O'Connor S, Szwej E, Nikodinovic-Runic J, *et al.* The anti-cancer activity of a cationic anti-microbial peptide derived from monomers of polyhydroxyalkanoate. Biomaterials 2013; 34(11): 2710-8.

[http://dx.doi.org/10.1016/j.biomaterials.2012.12.032] [PMID: 23343631]

[127] Szwej E, Devocelle M, Kenny S, *et al.* The chain length of biologically produced (R)-3-hydroxyalkanoic acid affects biological activity and structure of anti-cancer peptides. J Biotechnol 2015; 204: 7-12.
[http://dx.doi.org/10.1016/j.jbiotec.2015.02.036] [PMID: 25820126]

[128] Zhang X, Luo R, Wang Z, Deng Y, Chen G-Q. Application of (R)-3-hydroxyalkanoate methyl esters derived from microbial polyhydroxyalkanoates as novel biofuels. Biomacromolecules 2009; 10(4): 707-11.
[http://dx.doi.org/10.1021/bm801424e] [PMID: 19249855]

[129] Gao X, Chen JC, Wu Q, Chen G-Q. Polyhydroxyalkanoates as a source of chemicals, polymers, and biofuels. Curr Opin Biotechnol 2011; 22(6): 768-74.
[http://dx.doi.org/10.1016/j.copbio.2011.06.005] [PMID: 21705209]

[130] Choi J, Lee SY. Factors affecting the economics of polyhydroxyalkanoate production by bacterial fermentation. Appl Microbiol Biotechnol 1999; 51: 13-21.
[http://dx.doi.org/10.1007/s002530051357]

[131] Sudesh K, Iwata T. Sustainability of biobased and biodegradable plastics. Clean 2008; 36(5-6): 433-42.

[132] Koller M, Miranda Dias M, Rodríguez-Contreras A, *et al.* Liquefied wood as inexpensive precursor-feedstock for bio-mediated incorporation of (*R*)-3-hydroxyvalerate into polyhydroxyalkanoates. Materials (Basel) 2015; 8(9): 6543-57.
[http://dx.doi.org/10.3390/ma8095321]

[133] Koller M, Salerno A, Strohmeier K, *et al.* Novel precursors for production of 3-hydroxyvalerate-containing poly [(*R*)-hydroxyalkanoate]s. Biocatalysis Biotransform 2014; 32(3): 161-7.
[http://dx.doi.org/10.3109/10242422.2014.913580]

[134] Kim BS, Lee SC, Lee SY, Chang HN, Chang YK, Woo SI. Production of poly(3-hydroxybutyric acid) by fed-batch culture of *Alcaligenes eutrophus* with glucose concentration control. Biotechnol Bioeng 1994; 43(9): 892-8.
[http://dx.doi.org/10.1002/bit.260430908] [PMID: 18615882]

[135] Ryu HW, Hahn SK, Chang YK, Chang HN. Production of poly(3-hydroxybutyrate) by high cell density fed-batch culture of *Alcaligenes eutrophus* with phospate limitation. Biotechnol Bioeng 1997; 55(1): 28-32.
[http://dx.doi.org/10.1002/(SICI)1097-0290(19970705)55:1<28::AID-BIT4>3.0.CO;2-Z] [PMID: 18636441]

[136] Nonato RV, Mantelatto PE, Rossell CE. Integrated production of biodegradable plastic, sugar and ethanol. Appl Microbiol Biotechnol 2001; 57(1-2): 1-5.
[http://dx.doi.org/10.1007/s002530100732] [PMID: 11693904]

[137] Atlić A, Koller M, Scherzer D, *et al.* Continuous production of poly([*R*]-3-hydroxybutyrate) by *Cupriavidus necator* in a multistage bioreactor cascade. Appl Microbiol Biotechnol 2011; 91(2): 295-304.
[http://dx.doi.org/10.1007/s00253-011-3260-0] [PMID: 21503760]

[138] Kaur G, Roy I. Strategies for large-scale production of polyhydroxyalkanoates. Chem Biochem Eng Q 2015; 29(2): 157-72.
[http://dx.doi.org/10.15255/CABEQ.2014.2255]

[139] Zinn M, Weilenmann HU, Hany R, Schmid M, Egli T. Tailored synthesis of poly([R]-3-hydroxybutyrate-co-3-hydroxyvalerate) (PHB/HV) in *Ralstonia eutropha* DSM 428. Acta Biotechnol 2003; 23: 309-16.
[http://dx.doi.org/10.1002/abio.200390039]

[140] Sun Z, Ramsay JA, Guay M, Ramsay BA. Fermentation process development for the production of medium-chain-length poly-3-hyroxyalkanoates. Appl Microbiol Biotechnol 2007; 75(3): 475-85.
[http://dx.doi.org/10.1007/s00253-007-0857-4] [PMID: 17453197]

[141] Koller M, Muhr A. Continuous production mode as a viable process-engineering tool for efficient poly(hydroxyalkanoate) (PHA) bio-production. Chem Biochem Eng Q 2014; 28(1): 65-77.

[142] Koller M, Braunegg G. Potential and prospects of continuous polyhydroxyalkanoate (PHA) production. Bioengineering 2015; 2(2): 94-121.
[http://dx.doi.org/10.3390/bioengineering2020094]

[143] Koller M, Bona R, Hermann C, *et al.* Biotechnological production of poly(3-hydroxybutyrate) with *Wautersia eutropha* by application of green grass juice and silage juice as additional complex substrates. Biocatalysis Biotransform 2005; 23(5): 329-37.
[http://dx.doi.org/10.1080/10242420500292252]

[144] Obruca S, Benesova P, Oborna J, Marova I. Application of protease-hydrolyzed whey as a complex nitrogen source to increase poly(3-hydroxybutyrate) production from oils by *Cupriavidus necator.* Biotechnol Lett 2014; 36(4): 775-81.
[http://dx.doi.org/10.1007/s10529-013-1407-z] [PMID: 24243232]

[145] Koller M, Hesse PJ, Bona R, Kutschera C, Atlic A, Braunegg G. Biosynthesis of high quality polyhydroxyalkanoate co- and terpolyesters for potential medical application by the archaeon *Haloferax mediterranei.* Macromol Symp 2007; 253: 33-9.
[http://dx.doi.org/10.1002/masy.200750704]

[146] Povolo S, Romanelli MG, Basaglia M, *et al.* Polyhydroxyalkanoate biosynthesis by *Hydrogenophaga pseudoflava* DSM1034 from structurally unrelated carbon sources. N Biotechnol 2013; 30(6): 629-34.
[http://dx.doi.org/10.1016/j.nbt.2012.11.019] [PMID: 23201074]

[147] González-García Y, Nungaray J, Córdova J, *et al.* Biosynthesis and characterization of polyhydroxyalkanoates in the polysaccharide-degrading marine bacterium *Saccharophagus degradans* ATCC 43961. J Ind Microbiol Biotechnol 2008; 35(6): 629-33.
[http://dx.doi.org/10.1007/s10295-007-0299-0] [PMID: 18193466]

[148] Rodríguez-Contreras A, Koller M, Braunegg G, Marqués-Calvo MS. Poly[(*R*)-3-hydroxybutyrate] production under different salinity conditions by a novel *Bacillus megaterium* strain. N Biotechnol 2016; 33(1): 73-7.
[http://dx.doi.org/10.1016/j.nbt.2015.08.006] [PMID: 26344348]

[149] Valappil SP, Boccaccini AR, Bucke C, Roy I. Polyhydroxyalkanoates in Gram-positive bacteria: insights from the genera Bacillus and Streptomyces. A Van Leeuw 2007; 91(1): 1-17.

[150] Klask C, Raberg M, Heinrich D, Steinbüchel A. Heterologous expression of various PHA synthase genes in *Rhodospirillum rubrum.* Chem Biochem Eng Q 2015; 29(2): 75-85. [http://dx.doi.org/10.15255/CABEQ.2014.2249]

[151] Ouyang SP, Liu Q, Fang L, Chen GQ. Construction of *pha*-operon-defined knockout mutants of *Pseudomonas putida* KT2442 and their applications in poly(hydroxyalkanoate) production. Macromol Biosci 2007; 7(2): 227-33. [http://dx.doi.org/10.1002/mabi.200600187] [PMID: 17295412]

[152] Volodina E, Raberg M, Steinbüchel A. Engineering the heterotrophic carbon sources utilization range of *Ralstonia eutropha* H16 for applications in biotechnology. Crit Rev Biotechnol 2015; 2: 1-14. [http://dx.doi.org/10.3109/07388551.2015.1079698] [PMID: 26329669]

[153] Heinrich D, Raberg M, Steinbüchel A. Synthesis of poly(3-hydroxybutyrate-co-3-hydroxyvalerate) from unrelated carbon sources in engineered *Rhodospirillum rubrum.* FEMS Microbiol Lett 2015; 362(8): fnv038. [http://dx.doi.org/10.1093/femsle/fnv038] [PMID: 25761750]

[154] Zhang L, Shi ZY, Wu Q, Chen GQ. Microbial production of 4-hydroxybutyrate, poly-4-hydroxybutyrate, and poly(3-hydroxybutyrate-co-4-hydroxybutyrate) by recombinant microorganisms. Appl Microbiol Biotechnol 2009; 84(5): 909-16. [http://dx.doi.org/10.1007/s00253-009-2023-7] [PMID: 19434404]

[155] Boynton ZL, Koon JJ, Brennan EM, *et al.* Reduction of cell lysate viscosity during processing of poly(3-hydroxyalkanoates) by chromosomal integration of the staphylococcal nuclease gene in *Pseudomonas putida.* Appl Environ Microbiol 1999; 65(4): 1524-9. [PMID: 10103246]

[156] Queirós D, Lemos PC, Rossetti S, Serafim LS. Unveiling PHA-storing populations using molecular methods. App Microbiol Biot 2015; pp. 1-14.

[157] Valentino F, Karabegovic L, Majone M, Morgan-Sagastume F, Werker A. Polyhydroxyalkanoate (PHA) storage within a mixed-culture biomass with simultaneous growth as a function of accumulation substrate nitrogen and phosphorus levels. Water Res 2015; 77: 49-63. [http://dx.doi.org/10.1016/j.watres.2015.03.016] [PMID: 25846983]

[158] Queirós D, Rossetti S, Serafim LS. PHA production by mixed cultures: a way to valorize wastes from pulp industry. Bioresour Technol 2014; 157: 197-205. [http://dx.doi.org/10.1016/j.biortech.2014.01.099] [PMID: 24556373]

[159] Renner G, Schellauf F, Braunegg G. Selective enrichment of bacteria accumulating polyhydroxyalkanoates in multistage continuous culture. Food Technol Biotechnol 1998; 36: 203-8.

[160] Fradinho JC, Oehmen A, Reis MA. Photosynthetic mixed culture polyhydroxyalkanoate (PHA) production from individual and mixed volatile fatty acids (VFAs): substrate preferences and co-substrate uptake. J Biotechnol 2014; 185: 19-27. [http://dx.doi.org/10.1016/j.jbiotec.2014.05.035] [PMID: 24915131]

[161] Johnson K, Jiang Y, Kleerebezem R, Muyzer G, van Loosdrecht MC. Enrichment of a mixed bacterial culture with a high polyhydroxyalkanoate storage capacity. Biomacromolecules 2009; 10(4): 670-6. [http://dx.doi.org/10.1021/bm8013796] [PMID: 19193058]

[162] Anterrieu S, Quadri L, Geurkink B, *et al.* Integration of biopolymer production with process water treatment at a sugar factory. N Biotechnol 2014; 31(4): 308-23.
[http://dx.doi.org/10.1016/j.nbt.2013.11.008] [PMID: 24361532]

[163] Hilliou L, Machado D, Oliveira CS, *et al.* Impact of fermentation residues on the thermal, structural, and rheological properties of polyhydroxy (butyrate-co-valerate) produced from cheese whey and olive oil mill wastewater. J Appl Polym Sci 2015; 133(2)

[164] Koller M, Bona R, Chiellini E, Braunegg G. Extraction of short-chain-length poly-[(*R*)-hydroxyalkanoates] (*scl*-PHA) by the "anti-solvent" acetone under elevated temperature and pressure. Biotechnol Lett 2013; 35(7): 1023-8.
[http://dx.doi.org/10.1007/s10529-013-1185-7] [PMID: 23525946]

[165] Samorì C, Basaglia M, Casella S, *et al.* Dimethyl carbonate and switchable anionic surfactants: two effective tools for the extraction of polyhydroxyalkanoates from microbial biomass. Green Chem 2015; 17(2): 1047-56.
[http://dx.doi.org/10.1039/C4GC01821D]

[166] Samorì C, Abbondanzi F, Galletti P, *et al.* Extraction of polyhydroxyalkanoates from mixed microbial cultures: Impact on polymer quality and recovery. Bioresour Technol 2015; 189: 195-202.
[http://dx.doi.org/10.1016/j.biortech.2015.03.062] [PMID: 25889806]

[167] López-Abelairas M, García-Torreiro M, Lú-Chau T, Lema JM, Steinbüchel A. Comparison of several methods for the separation of poly (3-hydroxybutyrate) from *Cupriavidus necator* H16 cultures. Biochem Eng J 2015; 93: 250-9.
[http://dx.doi.org/10.1016/j.bej.2014.10.018]

[168] Novak M, Koller M, Braunegg M, Horvat P. Mathematical modelling as a tool for optimized PHA production. Chem Biochem Eng Q 2015; 29(2): 183-220.
[http://dx.doi.org/10.15255/CABEQ.2014.2101]

[169] Braunegg G, Lefebvre G, Renner G, Zeiser A, Haage G, Loidl-Lanthaler K. Kinetics as a tool for polyhydroxyalkanoate production optimization. Can J Microbiol 1995; 41(13): 239-48.
[http://dx.doi.org/10.1139/m95-192]

[170] Horvat P, Vrana Špoljarić I, Lopar M, Atlić A, Koller M, Braunegg G. Mathematical modelling and process optimization of a continuous 5-stage bioreactor cascade for production of poly [-(*R*)-3-hydroxybutyrate] by *Cupriavidus necator.* Bioprocess Biosyst Eng 2013; 36(9): 1235-50.
[http://dx.doi.org/10.1007/s00449-012-0852-8] [PMID: 23135491]

[171] Lopar M, Špoljarić IV, Atlić A, Koller M, Braunegg G, Horvat P. Five-step continuous production of PHB analyzed by elementary flux, modes, yield space analysis and high structured metabolic model. Biochem Eng J 2013; 79: 57-70.
[http://dx.doi.org/10.1016/j.bej.2013.07.003]

[172] Chen GQ, Hajnal I, Wu H, Lv L, Ye J. Engineering biosynthesis mechanisms for diversifying polyhydroxyalkanoates. Trends Biotechnol 2015; 33(10): 565-74.
[http://dx.doi.org/10.1016/j.tibtech.2015.07.007] [PMID: 26409776]

[173] Sindhu R, Pandey A, Binod P. Solid-state fermentation for the production of poly (hydroxyalkanoates). Chem Biochem Eng Q 2015; 29(2): 173-81.

[http://dx.doi.org/10.15255/CABEQ.2014.2256]

[174] Rydz J, Wolna-Stypka K, Adamus G, Janeczek H, *et al.* Forensic engineering of advanced polymeric materials. Part 1–Degradation studies of polylactide blends with atactic poly [(R, S)-–hydroxybutyrate] in paraffin. Chem Biochem Eng Q 2015; 29(2): 247-59. [http://dx.doi.org/10.15255/CABEQ.2014.2258]

[175] Chiellini E, Cinelli P, Chiellini F, Imam SH. Environmentally degradable bio-based polymeric blends and composites. Macromol Biosci 2004; 4(3): 218-31. [http://dx.doi.org/10.1002/mabi.200300126] [PMID: 15468211]

[176] Jost V, Kopitzky R. Blending of polyhydroxybutyrate-co-valerate with polylactic acid for packaging applications–Reflections on miscibility and effects on the mechanical and barrier properties. Chem Biochem Eng Q 2015; 29(2): 221-46. [http://dx.doi.org/10.15255/CABEQ.2014.2257]

[177] Hufenus R, Reifler FA, Maniura-Weber K, Spierings A, Zinn M. Biodegradable bicomponent fibers from renewable sources: Melt-spinning of poly (lactic acid) and poly. Macromol Mater Eng 2012; 297(1): 75-84. [(3-hydroxybutyrate)-co-(3-hydroxyvalerate)]. [http://dx.doi.org/10.1002/mame.201100063]

[178] Pérez Amaro L, Chen H, Barghini A, Corti A, Chiellini E. High performance compostable biocomposites based on bacterial polyesters suitable for injection molding and blow extrusion. Chem Biochem Eng Q 2015; 29(2): 261-74. [http://dx.doi.org/10.15255/CABEQ.2014.2259]

[179] Auriemma M, Piscitelli A, Pasquino R, Cerruti P, Malinconico M, Grizzuti N. Blending poly (3-hydroxybutyrate) with tannic acid: Influence of a polyphenolic natural additive on the rheological and thermal behavior. Eur Polym J 2015; 63: 123-31. [http://dx.doi.org/10.1016/j.eurpolymj.2014.12.021]

[180] Kovalcik A, Machovsky M, Kozakova Z, Koller M. Designing packaging materials with viscoelastic and gas barrier properties by optimized processing of poly (3-hydroxybutyrate-*co*-3-hydroxyvalerate) with lignin. React Funct Polym 2015; 94: 25-34. [http://dx.doi.org/10.1016/j.reactfunctpolym.2015.07.001]

[181] Martínez-Sanz M, Villano M, Oliveira C, *et al.* Characterization of polyhydroxyalkanoates synthesized from microbial mixed cultures and of their nanobiocomposites with bacterial cellulose nanowhiskers. N Biotechnol 2014; 31(4): 364-76. [http://dx.doi.org/10.1016/j.nbt.2013.06.003] [PMID: 23827196]

[182] Ten E, Turtle J, Bahr D, Jiang L, Wolcott M. Thermal and mechanical properties of poly (3-hydroxybutyrate-co-3-hydroxyvalerate)/cellulose nanowhiskers composites. Polymer (Guildf) 2010; 51(12): 2652-60. [http://dx.doi.org/10.1016/j.polymer.2010.04.007]

[183] Rodríguez-Contreras A, Canal C, Calafell-Monfort M, Ginebra MP, Julio-Moran G, Marqués-Calvo MS. Methods for the preparation of doxycycline-loaded phb micro-and nano-spheres. Eur Polym J 2013; 49(11): 3501-11. [http://dx.doi.org/10.1016/j.eurpolymj.2013.08.010]

[184] de Wit JN. Marschall Rhône-Poulenc Award Lecture. Nutritional and functional characteristics of

whey proteins in food products. J Dairy Sci 1998; 81(3): 597-608.
[http://dx.doi.org/10.3168/jds.S0022-0302(98)75613-9] [PMID: 9565865]

[185] Gillespie AL, Calderwood D, Hobson L, Green BD. Whey proteins have beneficial effects on intestinal enteroendocrine cells stimulating cell growth and increasing the production and secretion of incretin hormones. Food Chem 2015; 189: 120-8.
[http://dx.doi.org/10.1016/j.foodchem.2015.02.022] [PMID: 26190610]

[186] Page WJ, Cornish A. Growth of *Azotobacter vinelandii* UWD in fish peptone medium and simplified extraction of poly-β-hydroxybutyrate. Appl Environ Microbiol 1993; 59(12): 4236-44.
[PMID: 16349121]

[187] Whey layer: Barrier biopolymers for sustainable packaging. Available from: http://www.wheylayer.eu/project.html 2015.

[188] Schmid M, Dallmann K, Bugnicourt E, *et al.* Properties of whey-protein-coated films and laminates as novel recyclable food packaging materials with excellent barrier properties. Int J Polym Sci 2012.
[http://dx.doi.org/10.1155/2012/562381]

[189] Bugnicourt E, Schmid M, Nerney O, *et al.* Processing and validation of whey-protein-coated films and laminates at semi-industrial scale as novel recyclable food packaging materials with excellent barrier properties. Adv Mater Sci Eng 2013.
[http://dx.doi.org/10.1155/2013/496207]

[190] Schmid M, Hinz LV, Wild F, Noller K. Effects of hydrolysed whey proteins on the techno-functional characteristics of whey protein-based films. Materials (Basel) 2013; 6(3): 927-40.
[http://dx.doi.org/10.3390/ma6030927]

[191] Cinelli P, Schmid M, Bugnicourt E, *et al.* Whey protein layer applied on biodegradable packaging film to improve barrier properties while maintaining biodegradability. Polym Degrad Stabil 2014; 108: 151-7.
[http://dx.doi.org/10.1016/j.polymdegradstab.2014.07.007]

[192] Wakai M, Almenar E. Effect of the presence of montmorillonite on the solubility of whey protein isolate films in food model systems with different compositions and pH. Food Hydrocoll 2015; 43: 612-1.
[http://dx.doi.org/10.1016/j.foodhyd.2014.07.022]

[193] Azevedo VM, Silva EK, Pereira CF, da Costa JM, Borges SV. Whey protein isolate biodegradable films: Influence of the citric acid and montmorillonite clay nanoparticles on the physical properties. Food Hydrocoll 2015; 43: 252-8.
[http://dx.doi.org/10.1016/j.foodhyd.2014.05.027]

[194] Azevedo VM, Dias MV, Borges SV, *et al.* Development of whey protein isolate bio-nanocomposites: Effect of montmorillonite and citric acid on structural, thermal, morphological and mechanical properties. Food Hydrocoll 2015; 48: 179-88.
[http://dx.doi.org/10.1016/j.foodhyd.2015.02.014]

[195] Braunegg G, Lefebvre G, Genser KF. Polyhydroxyalkanoates, biopolyesters from renewable resources: physiological and engineering aspects. J Biotechnol 1998; 65(2,3): 127-61.

[196] Young FK, Kastner JR, May SW, May W. Microbial production of poly-β-hydroxybutyric acid from

d-xylose and lactose by *Pseudomonas cepacia.* Appl Environ Microbiol 1994; 60(11): 4195-8. [PMID: 16349449]

[197] Community research and development service. Available from: http://cordis.europa.eu/project/rcn/60146_en.html 2015.

[198] Wallner E, Haage G, Bona R, Schellauf F, Brauengg G. The production of poly-3-hydroxybutyrate-co-3-hydroxyvalerate with *Pseudomona cepacia* ATCC 17759 on various carbon sources. In: Chiellini E, Ed. Biorelated polymers: Sustainable polymer science and technology New York. Springer, US 2001; pp. 139-45. [http://dx.doi.org/10.1007/978-1-4757-3374-7_12]

[199] Yellore V, Desai A. Production of poly-3-hydroxybutyrate from lactose and whey by *Methylobacterium* sp. ZP24. Lett Appl Microbiol 1998; 26(6): 391-4. [http://dx.doi.org/10.1046/j.1472-765X.1998.00362.x] [PMID: 9717306]

[200] Nath A, Dixit M, Bandiya A, Chavda S, Desai AJ. Enhanced PHB production and scale up studies using cheese whey in fed batch culture of *Methylobacterium* sp. ZP24. Bioresour Technol 2008; 99(13): 5749-55. [http://dx.doi.org/10.1016/j.biortech.2007.10.017] [PMID: 18032031]

[201] Povolo S, Casella S. Bacterial production of PHA from lactose and cheese whey permeate. Macromol Symp 2003; 197: 1-9. [http://dx.doi.org/10.1002/masy.200350701]

[202] Domingos JM, Martinez GA, Scoma A, Bertin L, Reis MA, Fava F. Production of volatile fatty acids from cheese whey with immobilized cells. Environ Eng Manag J 2013; 12 (Suppl. 11): 101-4.

[203] Koller M. Recycling of waste streams of the biotechnological poly (hydroxyalkanoate) production by *Haloferax mediterranei* on whey. Int J Polym Sci 2015.

[204] Han J, Hou J, Zhang F, *et al.* Multiple propionyl coenzyme A-supplying pathways for production of the bioplastic poly (3-hydroxybutyrate-co-3-hydroxyvalerate) in *Haloferax mediterranei.* Appl Environ Microb 2013; 79(9): 2922-31.

[205] Han J, Wu LP, Hou J, Zhao D, Xiang H. Biosynthesis, characterization, and hemostasis potential of tailor-made poly(3-hydroxybutyrate-co-3-hydroxyvalerate) produced by *Haloferax mediterranei.* Biomacromolecules 2015; 16(2): 578-88. [http://dx.doi.org/10.1021/bm5016267] [PMID: 25559462]

[206] Chen CW, Don TM, Yen HF. Enzymatic extruded starch as a carbon source for the production of poly (3-hydroxybutyrate-co-3-hydroxyvalerate) by *Haloferax mediterranei.* Process Biochem 2006; 41(11): 2289-96. [http://dx.doi.org/10.1016/j.procbio.2006.05.026]

[207] Bhattacharyya A, Saha J, Haldar S, Bhowmic A, Mukhopadhyay UK, Mukherjee J. Production of poly-3-(hydroxybutyrate-co-hydroxyvalerate) by *Haloferax mediterranei* using rice-based ethanol stillage with simultaneous recovery and re-use of medium salts. Extremophiles 2014; 18(2): 463-70. [http://dx.doi.org/10.1007/s00792-013-0622-9] [PMID: 24442255]

[208] Huang TY, Duan KJ, Huang SY, Chen CW. Production of polyhydroxyalkanoates from inexpensive extruded rice bran and starch by *Haloferax mediterranei.* J Ind Microbiol Biotechnol 2006; 33(8):

701-6.
[http://dx.doi.org/10.1007/s10295-006-0098-z] [PMID: 16491353]

[209] Hezayen FF, Rehm BH, Eberhardt R, Steinbüchel A. Polymer production by two newly isolated extremely halophilic archaea: application of a novel corrosion-resistant bioreactor. Appl Microbiol Biotechnol 2000; 54(3): 319-25.
[http://dx.doi.org/10.1007/s002530000394] [PMID: 11030566]

[210] Han J, Zhang F, Hou J, *et al.* Complete genome sequence of the metabolically versatile halophilic archaeon *Haloferax mediterranei*, a poly(3-hydroxybutyrate-co-3-hydroxyvalerate) producer. J Bacteriol 2012; 194(16): 4463-4.
[http://dx.doi.org/10.1128/JB.00880-12] [PMID: 22843593]

[211] Zhao D, Cai L, Wu J, *et al.* Improving polyhydroxyalkanoate production by knocking out the genes involved in exopolysaccharide biosynthesis in *Haloferax mediterranei.* Appl Microbiol Biotechnol 2013; 97(7): 3027-36.
[http://dx.doi.org/10.1007/s00253-012-4415-3] [PMID: 23015099]

[212] Pantazaki AA, Papaneophytou CP, Pritsa AG, Liakopoulou-Kyriakides M, Kyriakidis DA. Production of polyhydroxyalkanoates from whey by *Thermus thermophilus* HB8. Process Biochem 2009; 44: 847-53.
[http://dx.doi.org/10.1016/j.procbio.2009.04.002]

[213] Baei MS, Najafpour GD, Lasemi Z, *et al.* Optimization PHAs production from dairy industry wastewater (cheese whey) by *Azohydromonas lata* DSMZ 1123. Iran J Energy Environm 2010; 1(2): 132-6.

[214] Valentino F, Riccardi C, Campanari S, Pomata D, Majone M. Fate of β-hexachlorocyclohexane in the mixed microbial cultures (MMCs) three-stage polyhydroxyalkanoates (PHA) production process from cheese whey. Bioresour Technol 2015; 192: 304-11.
[http://dx.doi.org/10.1016/j.biortech.2015.05.083] [PMID: 26048084]

[215] Omar S, Rayes A, Eqaab A, Viss I, Steinbuchel A. Optimization of cell growth and poly(3-hydroxybutyrate) accumulation on date syrup by a *Bacillus megaterium* strain. Biotechnol Lett 2001; 23: 1119-23.
[http://dx.doi.org/10.1023/A:1010559800535]

[216] Pandian SR, Deepak V, Kalishwaralal K, Rameshkumar N, Jeyaraj M, Gurunathan S. Optimization and fed-batch production of PHB utilizing dairy waste and sea water as nutrient sources by *Bacillus megaterium* SRKP-3. Bioresour Technol 2010; 101(2): 705-11.
[http://dx.doi.org/10.1016/j.biortech.2009.08.040] [PMID: 19744854]

[217] Obruca S, Marova I, Svoboda Z, Mikulikova R. Use of controlled exogenous stress for improvement of poly(3-hydroxybutyrate) production in *Cupriavidus necator.* Folia Microbiol (Praha) 2010; 55(1): 17-22.
[http://dx.doi.org/10.1007/s12223-010-0003-z] [PMID: 20336499]

[218] Nagamani P, Mahmood SK. Production of poly(3-hydroxybutyrate-co-3hydroxyvalerate) by a novel *Bacillus* OU40T from inexpensive carbon sources. Int J Pharma Bio Sci 2013; 4(1): 182-93.

[219] Nagamani P, Chaitanya K, Mahmood SK. Production and characterization of polymer from *Bacillus* OU73T from in-expensive carbon sources. Int J Curr Microbiol App Sci 2015; 4(7): 833-40.

[220] Lee SY, Middelberg AP, Lee YK. Poly(3-hydroxybutyrate) production from whey using recombinant *Escherichia coli.* Biotechnol Lett 1997; 19(10): 1033-5.
[http://dx.doi.org/10.1023/A:1018411820580]

[221] Wong HH, Lee SY. Poly-(3-hydroxybutyrate) production from whey by high-density cultivation of recombinant *Escherichia coli.* Appl Microbiol Biotechnol 1998; 50(1): 30-3.
[http://dx.doi.org/10.1007/s002530051252] [PMID: 9720197]

[222] Kim BS. Production of poly(3-hydroxybutyrate) from inexpensive substrates. Enzyme Microb Technol 2000; 27(10): 774-7.
[http://dx.doi.org/10.1016/S0141-0229(00)00299-4] [PMID: 11118585]

[223] Catalán AI, Ferreira F, Gill PR, Batista S. Production of polyhydroxyalkanoates by *Herbaspirillum seropedicae* grown with different sole carbon sources and on lactose when engineered to express the *lacZlacY* genes. Enzyme Microb Technol 2007; 40: 1352-7.
[http://dx.doi.org/10.1016/j.enzmictec.2006.10.008]

[224] Pais J, Farinha I, Freitas F, *et al.* Improvement on the yield of polyhydroxyalkanotes production from cheese whey by a recombinant *Escherichia coli* strain using the proton suicide methodology. Enzyme Microb Technol 2014; 55: 151-8.
[http://dx.doi.org/10.1016/j.enzmictec.2013.11.004] [PMID: 24411458]

[225] Koller M, Horvat P, Hesse P, *et al.* Assessment of formal and low structured kinetic modeling of polyhydroxyalkanoate synthesis from complex substrates. Bioprocess Biosyst Eng 2006; 29(5-6): 367-77.
[http://dx.doi.org/10.1007/s00449-006-0084-x] [PMID: 17031672]

[226] Tamis J, Marang L, Jiang Y, van Loosdrecht MC, Kleerebezem R. Modeling PHA-producing microbial enrichment cultures--towards a generalized model with predictive power. N Biotechnol 2014; 31(4): 324-34.
[http://dx.doi.org/10.1016/j.nbt.2013.11.007] [PMID: 24333144]

[227] Horvat P, Koller M, Braunegg G. Recent advances in elementary flux modes and yield space analysis as useful tools in metabolic network studies. World J Microbiol Biotechnol 2015; 31(9): 1315-28.
[http://dx.doi.org/10.1007/s11274-015-1887-1] [PMID: 26066363]

[228] Franz A, Song H, Ramkrishna D, Kienle A. Experimental and theorethical analysiys of poly (β-hydroxybutyrate) formation and consumptions in *Ralstonia eutropha.* Biochem Eng J 2011; 55: 49-58.
[http://dx.doi.org/10.1016/j.bej.2011.03.006]

[229] Sawyer MW. Controlling the mineral content of sweet whey powder in an industrial setting. PhD dissertation San Luis Obispo: California Polytechnic State University 2010.

CHAPTER 3

Biopolyesters and Related Valuable Products in Phototrophic Microbes

Péter B. Kós[1,2], **András Fülöp**[3] and **Gábor Rákhely**[1,4,*]

[1] *Department of Biotechnology, University of Szeged, Közép fasor 52, H-6726 Szeged, Hungary*

[2] *Institute of Plant Biology, Biological Research Centre, Hungarian Academy of Sciences, Közép fasor 52, H-6726 Szeged, Hungary*

[3] *Bay Zoltán Nonprofit Ltd. for Applied Research, Division for Biotechnology, Közép fasor 52, H-6726 Szeged, Hungary*

[4] *Institute of Biophysics, Biological Research Centre, Hungarian Academy of Sciences, Közép fasor 52, H-6726 Szeged, Hungary*

Abstract: Poly(3-hydroxyalkanoates) are widely used environmentally sound bioplastic produced by numerous microbes including anoxygenic and oxygenic photosynthetic bacteria. Similarly to heterotroph microbes, the photosynthetic microbes can accumulate these polyester-type storage materials under nutrient limitation such as nitrogen, phosphorous or sulfur starvation. A number of purple non-sulfur, sulfur photosynthetic and cyanobacteria can produce PHA of various compositions. The amount, type, length and monomer composition of the PHA strongly depend on the nutrients and the growth conditions, and the light intensity also seriously affects the yield. While anoxygenic photosynthetic microbes need external inorganic or organic electron source, for photosynthetic CO_2 fixation, the oxygenic microbes, cyanobacteria can gain electron from water splitting. In principle, anoxygenic photochemo-litoautotrophic and cyanobacteria can produce PHA from CO_2 as sole carbon source, but addition of organic substrates significantly improves the yield. PHA can also be decomposed by the cells under certain conditions and the PHA metabolism has connections to other metabolic routes including glycogen metabolism. There are several approaches for improving the production rate and yield and – for this purpose –

[*] **Corresponding author Gábor Rákhely:** Department of Biotechnology, University of Szeged, and Institute of Biophysics, Biological Research Center, Hungarian Academy of Sciences, Szeged, Hungary; Tel: +36 62 546940; Fax: +36-62 544352; E-mail: rakhely.gabor@brc.mta.hu.

Martin Koller (Ed.)

a number of artificially genetically modified microorganisms, metabolic routes were constructed.

Keywords: Bacteria, Biocompatible bioplastics, CO_2 fixation, Cyanobacteria, Microbial storage materials, Photosynthetic microbes, Polyhydroxyalkanoates (PHA), Purple photosynthetic bacteria.

INTRODUCTION

In unbalanced growth conditions with very frequent substrate fluctuations a widespread strategy is to store surplus substrates. In microorganisms, the accumulation of certain compounds offers selective benefits when the growth conditions are not optimal. For example, zero-valence sulfur, glycogen, polyhydroxyalkanoates (PHA) and polyphosphate can be accumulated depending on the growth conditions [1]. PHA represents carbon and energy storages in more than 300 Gram-positive, Gram-negative bacterial and archaeal species [2].

These polymers are synthesized by the microbes during impaired cultivation conditions, when the cell proliferation is hampered by a nutrient limitation (generally nitrogen, phosphate, or oxygen).

More than 150 hydroxyalkanoic acid monomers are involved in PHAs compositions (Fig. **1**) [3]. PHAs can be grouped into two classes according to the number of carbon atoms in the monomers: short chain length PHAs (*scl*-PHA), containing 3 to 5 carbon atoms in the monomer and medium chain length PHAs (*mcl*-PHA), synthesized from monomers composed of 6 to 15 carbon atoms.

PHA synthesis has been an important field of biotechnology for over three decades by now [4]. Since the feasibility of the technology depends on the economic benefits, all factors influencing the profitability are to be considered. The cost of the raw materials may total up to more than half of the production costs of PHA [5], therefore bacteria growing on cheap substrates present an evident route in searching for viable technologies.

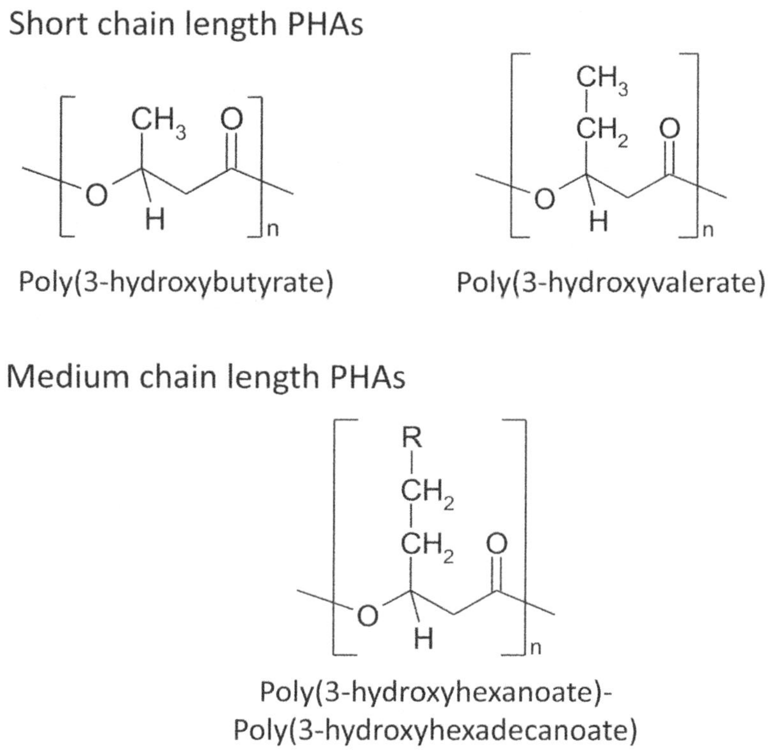

Fig. (1). Chemical structure of the short and medium chain PHAs. The R alkyl group is built up from 1-11 carbons.

Although, most of the studies focus on the PHA production by non-photosynthetic bacteria (for recent reviews see [6 - 9]), both anoxygenic and oxygenic photosynthetic bacteria can also produce and decompose PHA. While in heterotrophic fermentative bacteria, the cells produce PHA from various cheap substrates, such as: sugars, starch, (hemi)cellulose, whey, oil and glycerol, photochemolitoautotrophic microbes and cyanobacteria can utilize CO_2 for PHA production, although the yield can be substantially improved by supplementing the media with organic compounds including volatile organic acids. Nevertheless, CO_2 would be the cheapest and environmentally best substrate for production of polyester bioplastics. Therefore, in the following section, the current status of the research on production of PHA by photosynthetic microbes is overviewed.

PHA Production and Metabolism in Anoxygenic Phototrophic Microbes

A physiological survey of 41 strains of proteobacteria, many of which are photosynthetic, indicated that sulfur and non-sulfur photosynthetic bacteria are able to synthesize C4 and C5 PHA copolyesters from a variety of carbon sources. The main genera with this capacity are *Rhodobacter*, *Rhodopseudomonas* and *Rhodospirillum* from non-sulfur and *Allochromatium*, *Thiocapsa* and *Ectothiorhospira* from sulfur purple bacteria. From the tested species, only *Rubrivivax gelatinosus* 2150 was able to accumulate 1.3% 3-hydroxyheptanoate if the cells were propagated on a heptanoate-containing medium [8].

PHA Production by Anaerobic Photosynthetic Bacteria

It has been known for several decades that *Rhodobacter capsulatus* can grow photoheterotrophically on organic acids like acetate, propionate, or butyrate [10 - 12]. Weaver and his colleagues have tested 33 *R. capsulatus* strains on various substrates and in different growth conditions. The cells' growth rate on fatty acids (C2-C12) decreases with increased fatty acid chain length. In the wild type *R. capsulatus* B10, acetone, caproate and hexanoate were the best substrates for PHA accumulation, resulting in PHA content in the cell dry mass (CDM): 42% [100% poly(3-hydroxybutyrate), (PHB)], 90% (100% PHB) and 40% [consisting of 3-hydroxybutyrate (3HB) and 3-hydroxyvalerate (3HV) in the ratio 60:40 plus trace quantities of 3-hydroxycaproate (a.k.a 3-hydroxyhexanoate, 3HHx)], respectively. The ability of this bacterium to accumulate PHA, even in the absence of *phaAB* genes (see below), clearly indicates that acetoacetyl-CoA and β-hydroxybutyr--CoA can be produced through alternative pathways.

PHB accumulation in *Rhodopseudomonas palustris* cells during photoheterotrophic growth is independent of nitrogen source (N_2 or NH_4Cl), but strongly depends on nitrogen starvation. The reducing power saved as PHA lasts until the nitrogenases are fully expressed and then it is quickly degraded [8].

Rhodospirillum rubrum is capable of aerobic, anaerobic and autotrophic growth. There was linear correlation between growth rate and PHA accumulation in the strain. On CO, CO_2 and acetate it accumulated 30% PHA in CDM, with a composition of 99.4% 3HB and 0.6% 3HV [13]. Growth on different n-alkanoic acids (C2-C10) as the sole carbon sources revealed that only the C2-C7 compounds can be used by *R. rubrum* for PHA synthesis. *R. rubrum* flexibly incorporates not only C4 (3HB), but also C5 and C6 monomers (3HV, 3HHx) into its storage polymer, depending on carbon source length and concentration [14].

Addition of external carbon sources (acetate, lactate, and propionate) to the medium of anaerobic purple bacteria cultivated under illumination resulted in 2-5 times more PHA in the cells as compared to those grown in the dark [15]. The same diel pattern was also found in *Rhodobacter sphaeroides*, which could produce more PHA - during anoxic cultivation applying high intensity illumination - than under oxygenic dark growth conditions [16, 17].

Testing the influence of ammonium, sulfur and phosphorus (0.01-10mM) on PHA production, it was found that phosphorus had the highest and sulfate the least effect on intracellular polymer content. The infeasibility of acetate accumulation due to the lack of a glyoxylate shunt in *R. spaeroides* and *R. rubrum* [18] results in excess carbon source accumulation in the form of polyester. In *R. spaeroides* RV strain, pyruvate might also be used for PHB production, despite the fact that it is primarily considered to take part in glycogen accumulation [19].

Thiocapsa roseopersicina is a photochemoautotrophic purple sulfur bacterium (PSB) with high hydrogen metabolic versatility, containing four hydrogenases [20, 21]. The primary electron sources are reduced sulfur compounds and simple organic substrates, such as acetate, succinate, pyruvate and glucose, which can also be utilized as carbon, energy and electron sources. This bacterium, grown on acetate-containing medium, can accumulate up to 30% of the CDM of PHA, which can be used for H_2 production [22]. The accumulated polyester was identified as PHA copolymer. The cells propagated on 5 g/L acetate under nitrogen limitation resulted in a composition of 69.8 mol-% of 3HB and 30.2 mol-% of 3HV. Increasing the acetate concentration to 10 g/L under nitrogen-fixing conditions caused the molar ratio to change dramatically: 94.7 mol-% 3HB and 5.3 mol-% 3HV. Surprisingly, the lack of a glycogen synthesis gene leads to a substantial decrease in amount of accumulated PHAs: 5 g/L acetate, nitrogen-fixing condition just 2 wt.-%; 10 g/L acetate, and photochemolotiautotrophic nitrogen fixing conditions resulted in merely 11.2 wt.-% total PHA content in biomass. In contrast with our expectations, the lack of glycogen synthesis gene, *glgA*, in *T. roseopersicina* led to a reduced ability to accumulate PHA. When 10 g/L acetate was supplemented with a limited amount of glutamate (0.2 g/L) as an "energy saving" (16 ATP per mol fixed N_2) nitrogen source, the total PHA content increased by only 3% as compared to the values obtained for nitrogen fixing condition. The 3HB:3HV molar ratio did not change significantly (Fülöp *et al.* unpublished data).

PHA Production by Aerobic Anoxygenic Phototrophic Bacteria

The aerobic anoxygenic phototrophic bacteria (AAPB) are primarily heterotrophic organisms, but can also utilize light energy for growth. They are closely associated with phytoplankton and growing much faster (up to three times larger cell size) as compared to other heterotrophic bacteria. They are major contributors to the dissolved organic carbon (DOC) in the marine environment [23, 24].

Electron-transparent particles, assumed to be PHB, were observed in several *Sandaracinobacter*, *Roseococcus* and *Erythromicrobium* strains.

Erytromicrobium sibiricum (formerly *S. sibiricus*) RB16-17 cells, propagated on urea, nitrate or nitrogen-fixing conditions, accumulated 40 to 50 v/v% PHB of the cell volume [25].

In the case of *Dinoroseobacter* sp. JL1447, acetic acids, fatty acids, butyric acids and sugars are proper substrates for polyester accumulation. At high C/N ratios (>3) substrates that led directly to increased acetyl-CoA and yielded the most PHA production (72% and 68% respectively) [26].

Obviously, light is another key factor in polyester accumulation in phototrophic bacteria. Examining PHA content change in photosynthetic microbial mats (green algae and PSB dominated layers too) revealed a polyester content decrease during the day followed by an increase during the dark period. In the pink layer, dominated by PSB, PHA content doubled overnight.

In *Dinoroseobacter sp.* JL1447, light metabolism and respiration share the same electron transfer system. Under illumination, the NADH-dependent respiration is detained creating high NADH/NAD ratios [25, 27], thus favoring PHA accumulation.

Biosynthetic Routes

The diversity of PHAs (Fig. **1**) is maintained *via* 14 known pathways (Fig. **2**) [28]. These can be divided into natural and engineered pathways. The first 4 pathways belong to naturally occurring PHA synthesis routes.

Pathway I includes the metabolic steps leading to acetyl-CoA, acetoacetyl-CoA and 3-hydroxybutyryl-CoA. The latter is the substrate in the polymerization reaction leading to the formation of PHB.

In pathway II, fatty acid substrates are converted to R-3-hydroxyacyl-CoA monomer *via* β-oxidation pathway which are further used primarily for mclPHA synthesis.

Pathway III uses acetyl-CoA to malonyl-CoA to 3-ketoacyl-ACP, also resulting in R-3-hydroxyacyl-CoA monomers.

Pathway IV starts from butyric acid to S-3-hydroxybutyryl-CoA and finally acetyl-CoA.

Pathways V to XIV are engineered pathways resulting in different PHAs (Fig. **2**).

The lack of precise control of PHA composition results in PHA monomer variations.

There are several genes involved in PHB biosynthesis, which might be localized in distinct gene clusters. One cluster is *phaARPB,* which encodes for β-ketothiolase, PHB synthesis regulatory protein, PHB granule associated phasin and NADPH-dependent acetoacetyl-CoA reductase, respectively. Additionally, *phaCE* genes encode for a two component PHB synthase.

Multiple alignments of the translated proteins in the case of *phaARPB* genes show relative identities values between 44-91% (PhaA), 76-80% (PhaR), 23-71% (*A. vinosum* PhaP) and 41-87% (PhaB), respectively. For PhaCE proteins, the relative identities values were 51-89% (PhaC) and 24-72% (PhaE), respectively.

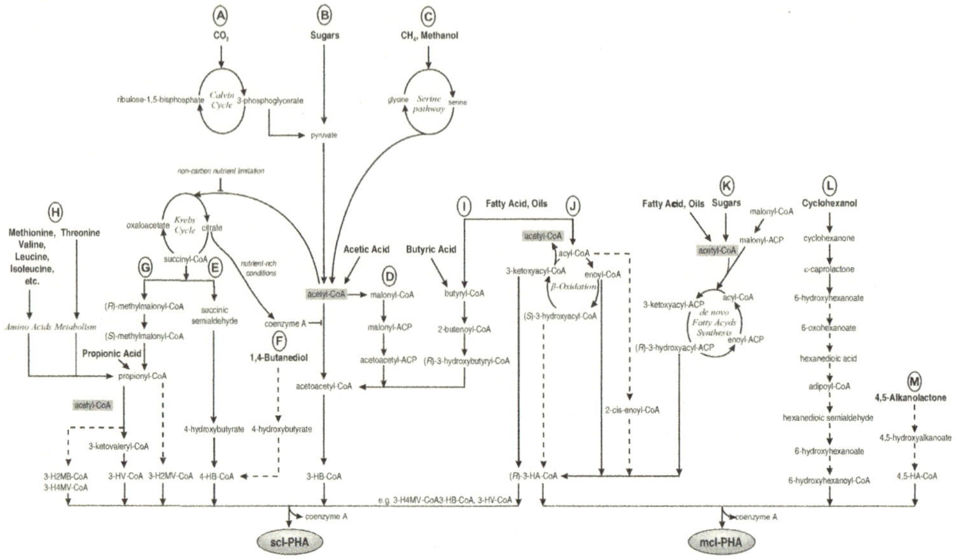

Fig. (2). PHA biosynthetic routes. Dotted lines illustrate hypothetical pathways (Adapted from [187]).[1] (3H2MB-CoA, 3-hydroxy-2-methylbutyryl-CoA; 3H4MV-CoA, 3-hydroxy-4-methylvaleryl-CoA; 3HV-CoA, 3-hydroxyvaleryl-CoA; 3-H2MV-CoA, 3-hydroxy-2-methylvaleryl-CoA; 4HB-CoA, 4-hydroxybutyryl-CoA; 3HB-CoA, 3-hydroxybutyryl-CoA; (*R*)-3-HA-CoA, (*R*)-3-hydroxyacyl-CoA; 4,5-HA-CoA, 4,5-hydroxyacyl-CoA) (adapted from [187]).

Biosynthetic pathway I Fig. (**2**) is one of the most common and studied pathways with a huge literature, especially in *Cupriavidus necator* (previously named as *Ralstonia eutropha* or *Alcaligenes eutrophus*) [29 - 33].

The metabolism of sugars and small organic compounds (acetate, butyrate *etc.*) begins with acetyl-CoA. First, two acetyl-CoA molecules are condensed into acetoacetyl-CoA by β-ketoacyl-CoA thiolase or β-ketothiolase enzyme (*phbA* gene product). Next, the acetoacetyl-CoA is stereospecifically reduced to (*R*)-3-hydroxybutyryl-CoA by an NADPH-dependent acetoacetyl-CoA dehydrogenase (or reductase), which is coded by the *phbB* gene. Finally, the polymerization of (*R*)-3-hydroxybutyryl-CoA enantiomers into poly-3-hydroxybutyrate by PHB polymerase/synthase (PhbC) takes place.

Pathway II was described in *Rhodopseudomonas rubrum* [34, 35]. In *R.rubrum*, β-oxidation of fatty acids results in the monomers of PHA. In the first reaction, similarly to the pathway in *C. necator*, acetoacetyl-CoA is formed, which is reduced into L-(+)-3-hydroxybutyryl-CoA by a NADH-dependent reductase. Finally, D-(−)-3-hydroxybutyryl-CoA is produced from L-(+)-3-hydroxybuty-yl-CoA by the assistance of two enol-CoA hydratase enzymes.

Polyester Synthases

Two component class III type PHA synthases are present in several phototrophic purple bacteria (and cyanobacteria), including *Allochromatium vinosum* [36, 37], *Thiocystis violacea* [38], *Thiocapsa pfennigii* [39], *Synecocystis* sp PCC 6803 [40], *Thiocapsa roseopersicina* [22], and *Ectothiorhodospira shaposnikovii* [41]. These PHA synthases prefer the monomers of the short chain length PHAs (*scl*-PHA) [33, 42]. These enzymes are characteristic of *γ-Proteobacteria,* but are widespread in cyanobacteria and halophilic archea, as well [42, 43].

Class III type PHA synthases are comprised of two hetero-subunits: PhaC and PhaE [22]. PhaC subunit (M_w 40 kDa) has a sequence similarity of 21-28% to class I and class II (M_w 40 kDa) enzymes, however, and PhaE subunit does not resemble polyester synthases. In all microorganisms, the two genes coding for the components of class III PHA synthases *(phaC* and *phaE)*, co-occur in their genomes and likely constitute a single operon [2].

Rhodospirillum rubrum has three homologous polymerase synthase genes (*phaC1*, *phaC2*, *phaC3*). On acetate and hexanoate, PhaC2 is the major PHA producer incorporating C5 and C6 monomers. PhaC1 has an inhibitory effect on PHA synthesis [34, 35].

As a pioneering work in 1992, Steinbuchel's lab produced findings, that all PhaC proteins harbour a lipase box [45] G-X-[S/C]-X-G and, according to the numbering in *Allochromatium vinosum*, the essential serine-87 from the lipases is replaced by Cys-149 [46]. Considering this high similarity to *Burkholderia cepacia* lipase, a high resolution structural model was constructed between residues 131-175 which encompass the α/β-hydrolase domain including the lipase box (Figs. **3** and **4**) [47].

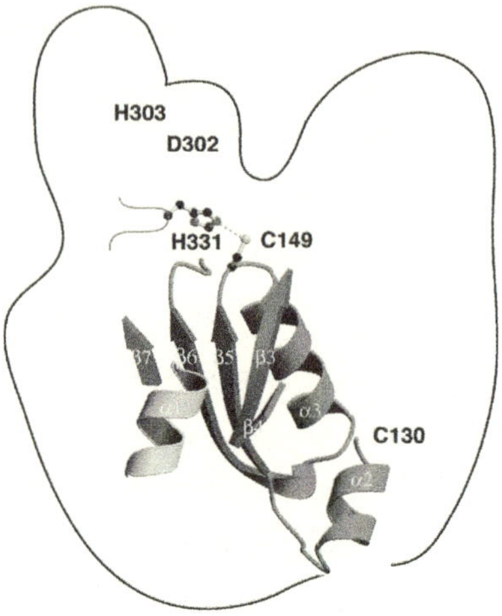

Fig. (3). Scheme of the structural model for Class III PHA synthases. The purple colour indicates region of highest similarity to bacterial lipases. H331 and C149 (in ball-and-stick) are proposed as catalytic residues. The remaining part of the model is outlined, and the estimated positions of C130, D302, and H303 are shown. [Reprinted with permission from [47] Copyright (2000) American Chemical Society].

Mutational analysis of PHA synthases of PhaC from *A. vinosum* undoubtedly demonstrated that the Cys-149, His-331 and Asp-302 residues participated in covalent catalysis. Substitutions of these amino acids almost abolished enzymatic activity [47]. Jia and coworkers created PHA synthase PhaC mutants in *A. vinosum* in order to determine the accuracy of the previously proposed C130 catalytic residue [46], as well as its location within the modelled structure. They also aimed to elucidate the role of other conserved amino residues, H331, H303, D 302 and C149, taking into account the similarities to α/β-hydrolases.

Measuring the CoA release activity of various mutants clarified that C130 is not important for the catalysis. H331 is found to be essential, together with previously reported C149 [46]. D302, as a general base catalyst, is likely required for proposed polyoxoester formation which deprotonates the hydroxyl group of 3-hydroxybutyryl-CoA. Replacement of the highly conserved tryptophan-425 by alanine, apparently diminished the enzymatic activity. Tryptophan-425 seemed to promote the dimerization of the subunits *via* hydrophobic protein-protein interactions [48].

The corresponding residues of *Thiocapsa rosepersicina* PhaC protein (Cys-148, His-330 and Asp-301) can also be found, most probably involved in the same functions.

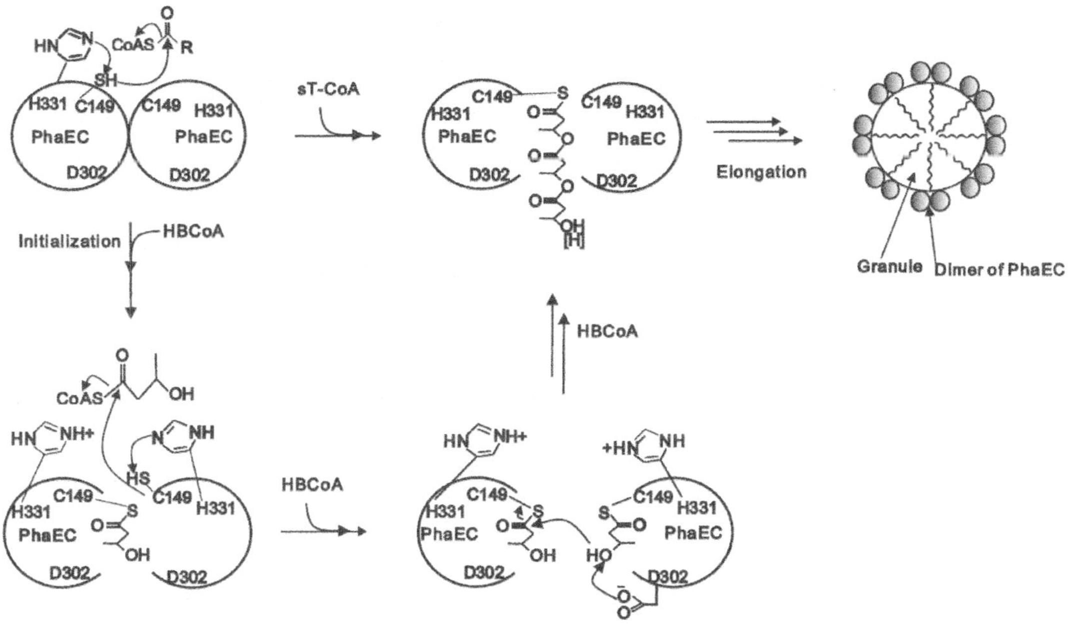

Fig. (4). Molecular model of the reaction catalyzed by the PHA synthase. The active site is located at the interface of the subunits providing reactive thiol group (C149). H331 has a role in activation of the C149. D302 converts the 3-hydroxyl group of HBCoA or a bound HB for nucleophilic attack. sT-CoA is suggested to turn from the initiation process. The synthase is likely dimers of trimers. For easy visualization, only the heterodimeric active form of PhaEC is displayed. [Reprinted with permission from [47]. Copyright (2000) American Chemical Society].

Experiments with the cysteine149-alaninc mutant of *A. vinosum* and the saturated trimer of 3HB-CoA, in which the hydroxyl group was labelled with tritium ([^{3}H]-sT-CoA), provided the [^{14}C]-3HB-CoA, the first proof for the importance of Cys-149 in covalent catalysis [49]. In addition, it was suggested that aspartate-302 plays a role as the 3-hydroxyl group activator of 3HB-CoA, thus contributing to chain elongation [50].

The two-component synthase of *Thiocapsa pfennigii*, with high identity to class III PhaC (up to 85%) [2], is an exception of this classification. The hybrid class III synthase generated by interchanging PhaC and PhaE subunits from *A. vinosum* and *T. pfennigii* revealed that the PhaC subunit governed the specificity of the substrate usage [39]. The absence of PhaE subunit almost completely abolished the PhaC activity [46].

Detailed comparisons of the 88 synthases from different microorganisms unravelled that these enzymes show high variations in identities ranging from 8-96% [3]. Eight amino acids, identified in all synthases, might be crucial in enzyme activities. The highly hydrophobic and conserved C-terminal domain from the class I and II synthases might be involved in binding to the hydrophobic polyester core [29]. Unlike, class III and IV synthases, the hydrophobic C terminus existing in PhaE and PhaR subunits, which also might be attributed to polyester core binding.

The polyester particle size and gel filtration chromatography analyses indicated that the active synthase has two-unit-composition *in vitro* and the granule associated form has a decameric composition in *A. vinosum* [45]. Considering the data of Jia *et al.* [47] and the two-subunit-composition of these synthases, the observed dodecamer might constitute a hexameric protein complex.

Factors Controlling the Formation of PHA

In *Rhodobacter capsulatus,* the nitrogen regulatory cascade is not involved in PHA production [12]. The regulation of PHA synthesis might occur posttranslationally. This conclusion was also suggested for *C. necator*, where control is proposed to occur by mechanisms related to allosteric effects and substrate availability for the biosynthetic enzymes [48 - 53].

Rhodobacter sphaeroides has a primarily $NADPH_2$-dependent acetyl-CoA reductase, in contrast to *Rhodospirillum rubrum* where acetoacetyl-CoA is mostly $NADH_2$-dependent [54].

Previous investigation of PHB accumulation by photosynthetic bacteria [8, 54 - 57] and cyanobacteria [55, 57, 58] have reported PHB accumulating activity under nitrogen-limited conditions, which is similar to heterotrophic bacteria.

In the non-sulfur photosynthetic bacterium, *Rhodospirillum rubrum*, the amount of accumulated PHA is inversely associated with phosphorus, ammonium and sulfate concentrations (0.01-10 mM). Phosphorus limitation resulted in the highest PHA accumulation (15% of the CDM), followed by nitrogen (13% of the CDM) and then sulfur (8% of the CDM) [14].

Brandl and his co-workers [16] also investigated the influence of several factors, such as acetate and light intensity under ammonium limiting conditions on cell growth and PHA storage. For *R. sphaeroides*, acetate concentrations between 90 to 120 mM were determined to be optimal for cell growth and polymer formation. Cell growth was affected by decreasing light intensities, in contrast to polymer content which remained constant at a broad range of light intensities (≈ 8-700 $\mu E/(m^2 s)$). A decrease in accumulated polyester level was only observed at very low light intensities.

Liebergesell *et al.* [8] has analyzed the PHA content of nitrogen-starved cultures in 30 non-sulfur and sulfur photosynthetic bacteria. PHB accumulation by *Rhodobacter sphaeroides* strain RV could be observed even in the presence of nitrogen. The highest PHB content in RV strain reached about 40% of cell dry weight. Upon nitrogen starvation in *Rhodopseudomonas palustris*, the consumption of 35±6% of the electrons accounted for over 30% of the CDM of accumulated as PHB.

The most intensive PHB accumulation occurred in the first 50 hours of nitrogen starvation, following a similar trend to that of acetate consumption and H_2 and CO_2 production [59].

In *Rp. palustris* and other purple non-sulfur bacteria starved for phosphate [60] or sulfur [61], PHB synthesis was greatly favoured over H_2 production.

PHA accumulation, under pH-stat and pH-uncontrolled growth conditions, was measured in *R. sphaeroides*. Uncontrolled conditions resulted in the highest PHB accumulation. The optimal growth rate, at pH-value 7.5, and the optimal PHB accumulating conditions, at pH-value 8.0 and 8.5, could be easily distinguished [62].

Nitrogen limited conditions favour the production of cyclopropane fatty acids, supposed stress response mediators in acid shock in *E. coli* [63]. They are also proposed to function as a transcription activator for a potential cyclopropane fatty acid synthase under singlet oxygen induced oxidative stress in *Rhodobacter sphaeroides* [64].

In Vivo Formation of Polyester Particles, Self-Assembly and Structure

The need for a biodegradable polyester resulted in a great deal of research related to PHA metabolism. One of the most important outcomes is that the number of polyester granule associated proteins (PGAPs) goes far beyond the need of PHA synthesis. The bacterial frontrunner in PHA research is *C. necator* (previously *R. eutropha*), which is also used in several biotechnological applications [65]. The native granule (nPHA) isolated from *C. necator* contained 16 specifically PHB granule-associated proteins. The strictly PGAPs were composed of intracellular PHB depolymerase (PhaZA1), six phasin proteins (PhaP2 to PhaP7) and a PhaC2 (second PHB synthase-like protein). Several proteins showed multiple locations: the phasin PhaP1, the PHB synthase PhaC1, the regulator PhaR and the PhaM protein, recently discovered, [30, 36, 64 - 66]. There are four recently described PGAPs with unknown function, the PhaZE1, the PhaZE2, a putative phospholipase and a hypothetical protein [67]. These findings strongly suggest that the PHA granules are just simple inclusions bodies rather than subcellular carbonosome organelles [30].

Immunoelectron microscopic studies revealed that *A. vinosum* (PhaC and PhaE subunits) and *C. necator* synthases can be localized on the granule surface under PHA storage condition [45, 68].

In most bacteria, the major protein components within the granule fraction are all small proteins (like PhaP) and the synthases represent just a minor protein fraction [30, 69, 70]. In *Rhodococcus ruber*, the PHB granules are covered by a 14 kDa protein (PhaP) layer [69].

Throughout the whole PHA synthesis process, the covalently bound PHA synthase incorporates (*R*)-3-hydroxyacyl-CoA monomers into the growing polyester chains until metabolic and spatial constraints terminate the process [49, 71]. Two models have been proposed for formation of PHA granules: the micelle model and the budding model [3].

The micelle model is supported by *in vitro* formation of PHA granules using purified *C. necator* PHA synthase and substrate in the absence of membranes. The resulting artificial, spherical granules, with 3 μm diameter, are six times larger than their native counterparts. It has also been disclosed that the molecular sizes of the polymer depended on initial PHA synthase concentration. The (*R*)-3-hydroxybutyryl coenzyme A without the adenosine 3',5'-bisphosphate moiety was not recognized by the PHA synthase, as it is an essential structural element for catalysis [68].

In the budding model, soluble PHA synthases, localized close to the inner membrane, catalyse the polymerization reaction in the intramembrane space, where the nascent polyester granules are surrounded by an inner membrane budded from the phospholipid monolayer [3]. Examination of early stage polyester granule localization supports for budding model, despite the fact that experimental evidence for the presence of *in vivo* lipid layer is still missing. Micrographs of Nile Red stained cells and/or phasin-EGFP fusion in *Caryophanon latum*, *C. necator* and *R. rubrum* cells showed granule localization near cell poles [72, 73]. The near polar localization of newly synthetized polyester granules was also supported by fluorescence microscopy using a PHA synthase-GFP fusion [74].

Catabolism of Polyester in Purple Bacteria

The survival of bacteria in environments without external carbon sources depends on internal carbon sources. In these bacteria, there is a cyclic metabolic pathway where the cells synthesize 3HB-CoA from acetyl-CoA through acetoacetyl-CoA and then PHB; the catabolism of PHB takes place *via* 3HB with the assistance of nPHB depolymerase. 3HB is dehydrogenated by 3HB-dehydrogenase to acetoacetate. The CoA transfer to acetoacetate is performed by acetoacetate:succinyl-CoA transferase. In the last step, ketothiolase cleaves acetoacetyl-CoA to acetyl-CoA [75 - 77]. The simultaneous synthesis and degradation of PHB Fig. (**5**) was first reported in 1990 [78]. The 55% CDM polyester content (100 mol-% 3HB) of *C. necator* cells grown in the presence of pentoic acid as the sole carbon source, under nitrogen-depleted conditions, is converted into PHBHV (51 mol-% 3HB and 49 mol-% 3HV). In 96 hours, the

3HV content increased dramatically from 0 to 49 mol-%. The transfer of the cells with 50% PHBHV/CDM random copolymer (44 mol-% 3HB and 56 mol-% 3HV) content into nitrogen free butyric acid media led to a copolymer with 81 mol-% 3HB and 19 mol-% 3HV units composition.

The turnover was confirmed by a ^{14}C-radiolabelled glucose experiment where the labelled ^{14}C atom was incorporated at a higher rate during PHB degradation than PHB accumulation [79].

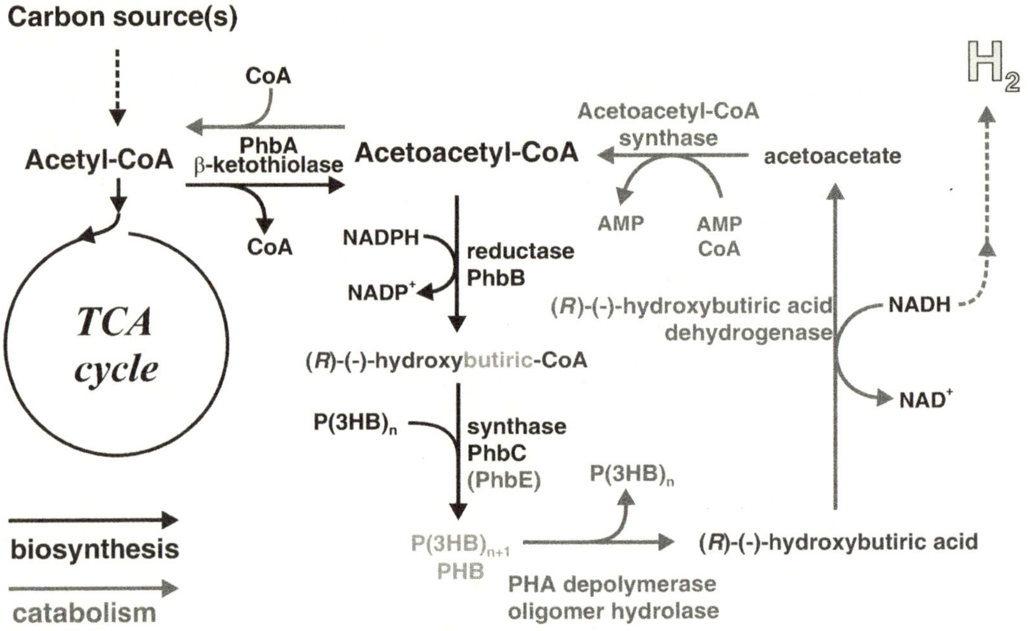

Fig. (5). Polyhydroxybutyrate metabolism in purple (sulfur) photosynthetic bacteria.

As it was shown, the decomposition of PHA finally produces acetyl-CoA in a concomitant formation of reducing power, electrons (Fig. **5**). This reducing power can be incorporated into the biomass *via* NAD^+/NADH or the excess of electrons must be removed. One of the ultimate solutions for the cells is to produce H_2 *via* either nitrogenase or hydrogenase. During nitrogen limitation (in the absence of NH_4^+), the cellular nitrogenases are upregulated and consume huge amount of reducing power and ATP. In the absence of any nitrogen compound in the medium, the microbe will fix molecular N_2 and produce H_2. It was shown for a purple sulfur photosyntetic bacterium that PHA degradation can drive the nitrogenase-linked H_2 production in dark [22, 60]. Therefore, biodegradation of this type of bioplastic might be utilized for biofuel production (Fig. **5**).

PHA Depolymerases

Depolymerases are classified in four main groups according to the physical states of the polyester granules. Native PHA granules - composed of polymers, covered by a surface layer of protein and phospholipids - as an endogenous carbon reservoir, can be degraded by intracellular PHA depolymerases. Further classification into n*scl*-PHA and n*mcl*-PHA depolymerases is based on the carbon atom numbers of the monomers which the enzymes can recognize.

The partially crystalline, so-called denatured, granules, lack a surface layer and are decomposed by extracellular PHA depolymerases, also classified into d*scl*-PHA and d*mcl*-PHA [77, 80, 81].

There are at least two exceptions of this classification, *e.g.* an extracellular n*scl*-PHA depolymerase of *Paucimonas lemoignei* and a periplasmic enzyme from *Rhodospirillum rubrum* [82 - 84].

All PHA depolymerases (except for few intracellular n*scl*-PHA depolymerases), have a catalytic center composed of a "serine – histidine – aspartic acid" triad. The catalytic serine is embedded in the so-called 'lipase box' (GxSxG), similar to almost all serine hydrolases [85]. Additionally, it has a second conserved histidine, near the oxyanion hole, as it was also described in lipases [80, 81]. The oxyanion hole is a pocket-like structure, stabilizing the hydrolysis created transition state negative charge on an oxyanion. Stabilizing the transition state *via* a hydrogen bridge between a histidine and a negatively charged oxyanion lowers the activation energy necessary for the reaction [81, 86]. Depending on the lipase box location (N- or C-terminal) relative to the sequence localization of the oxyanion hole, two categories of catalytic regions can be distinguished: lipase box type 1 and lipase box type 2, respectively. The factors promoting the depolymerases to degrade PHAs with high specificity are still not clear. Despite the fact that among PHA depolymerases and other known α/β-hydrolases, such as lipases and esterases belonging to the similar fold family and having highly conserved residues in their active site, the overall sequence similarities are low and their substrate specificity substantially varies. Based on the huge PHA Depolymerase Engineering Database, containing sequences of more than 587 depolymerases (DED, http://www.ded.uni-stuttgart.de;), the PHA depolymerases make up 8 superfamilies according to their function and structure [80]: Intracellular n*scl*-PHA depolymerases (no lipase box), intracellular n*scl*-PHA depolymerases (lipase box), intracellular nPHAMCL depolymerases, periplasmatic PHA depolymerases, extracellular d*scl*-PHA depolymerases (catalytic domain type 1), extracellular d*scl*-PHA depolymerases (catalytic domain type 2), extracellular n*scl*-PHA depolymerases, extracellular d*mcl*-PHA depolymerases.

The best characterized d*scl*-PHA depolymerases have complex domain structure composed of:

- 22-58 amino acid-long signal peptide, for secretion of the polypeptide through the cytoplasmic membrane, which is removed at the end of transport;
- Large, catalytic, N-terminal domain with strictly conserved catalytic amino acids - serine, aspartate and histidine. The catalytic serine makes up part of the lipase box;
- Linker domain connection the catalytic and C-terminal domain of *scl*-PHA depolymerases;
- C-terminal substrate binding domain (SBD). The C-terminal 40-60 amino acids fulfill the dPHB binding function.

The PHA depolymerases which are truncated at C-terminal domain (59 amino acids) lack the dPHB degrading activity [86].

Two varieties of SBD can be distinguished. Several conserved amino acids, His-49, Arg-44 and Cys-2, are present in both SBDs. In type 1, a second conserved cysteine occurs at position 57 or 59 [81].

Cyanobacterial Production of Polyhydroxyalkanoates

Cyanobacteria are photosynthetic bacteria that can utilize CO_2 as the sole carbon source. Although they are unicellular organisms, the cells of several orders form filaments, colonies or biofilms. Cyanobacteria can be found in almost all conceivable niches around the globe, from oceans to deserts. This wide variety of inhabited habitats goes hand in hand with a remarkable variability of their metabolism. What is of pivotal importance for the sake of the current review is that several genera are capable of fixing inorganic nitrogen. Hence, cultivation of these bacteria may need rather cheap media, making them excellent choice of biotechnological applications such as synthesis of materials of high value [137]. Another important feature is that several genera can grow under mixotrophic condition and capable of utilizing various carbon sources, including glucose and acetate, as well as maltose, fructose and ethanol [93] that may fundamentally increase the productivity.

The main carbon source is only one of the important factors influencing the economics of PHA production; and the productivity, the possible ways of recovery as well as the purity and possible downstream applications are also to be considered. The main PHA variety that cyanobacteria produce is PHB homopolymer, that is in pure form highly crystalline and brittle, therefore special formulation and functionalization is also to be considered for diverse applications [87, 88]. Here, we will introduce the cyanobacterial strains capable of synthesizing PHA and the efforts toward improving their productivity.

Cyanobacterial Strains Capable of Synthesizing PHA

Similarly to heterotroph bacteria, cyanobacteria can accumulate PHA, as a storage material, in the presence of surplus carbon source relative to other essential nutrients such as nitrogen or phosphorous. The first report demonstrating cyanobacterial PHA production was published a half century ago [89] in *Chlorogloea fritschii*, a nitrogen fixing filamentous strain. The PHB polymer forms spherical and sometimes elongated granules with the diameter ranging from 100 to 800 nm when the cells are cultivated in a special medium with soil extract [90]. Since this early observation several strains of cyanobacteria have been demonstrated to be capable of synthesizing PHA by during oxygenic photosynthesis [91]. These are mostly diazotrophic strains, such as unicellular *Gloeothece* [58] and *Gloeocapsa* [92] as well as filamentous *Nostoc* [93], *Trichodesmium thiebautii* [94], *Aulosira* [95] and *Anabaena* [96] genera.

Nevertheless several strains that do not fix nitrogen has also been found to have this feature like, *e.g.* filmentous *Arthrospira* [97], *Oscillatoria* [98] and *Spirulina* genera [99 - 102] as well as unicellular *Synechocystis* [103, 104], *Synechococcus* [105, 106], *Thermosynechococcus* [107] and *Phormidium* [108] strains. Capability of fixing atmospheric molecular nitrogen is an important trait since limitation in combined nitrogen is known to induce carbon reservation. It is easily achieved when nitrogen source is simply omitted from the culture medium of cyanobacteria capable for nitrogen fixation. For non-diazotrophic bacteria, however, two stage culturing is required [105] which makes the production economically less feasible.

The ability of accumulating PHA is nevertheless strain-specific. While *Synechococcus* sp. MA19, a thermophilic strain is capable of PHA production, *Synechococcus* sp. PCC7002, a marine bacterium and *Synechococcus* sp. PCC7942, a fresh water strain, are not.

Screening for PHA Synthesizing Capability

The capability is not rare among cyanobacteria. In an elaborated survey several methods, including Southern blot, Western blot, sequencing of *phaC*-specific PCR amplicon and cloning of PHA synthase genes were applied for proving the existence of a type-III PHA synthase, hence the capability of synthesizing PHB, in eleven strains that included *Synechococcus* sp. strains MA19 and PCC 6715, *Chlorogloeopsis fritschii* PCC 6912, *Anabaena cylindrica* SAG 1403-2, *Cyanothece sp.* strains PCC 7424, PCC 8303 and PCC 8801, and *Gloeocapsa* sp. strain PCC 7428 [44]. A more simple traditional method for screening for PHA production is the maintenance of the isolates and then staining the intracellular PHB granules using oxazine dyes such as Nile Red that shows fluorescence in the granules. Three out of ten strains [109] and in a repeated survey eleven of fifteen isolates [108] of marine and fresh water origin showed PHB production. This method nevertheless depends on the amount of the produced (stored) polymer, thus may necessitate the maintenance of the cultures in nutrient limited or otherwise specialized culture media [110] prior to staining. For decreasing this burden and increasing the throughput a simple PCR-based method was recently elaborated. Using the known DNA sequences of cyanobacterial *phaC*, one of the four key enzymes in PHB synthesis, the conserved positions allowed the design of a primer pair that is specific to this gene in all so far tested strains. The positive result of the PCR amplification correlates well with the Nile Red staining, making the screening for PHB synthesizing cyanobacterial isolates much faster [111].

Production of PHB from CO_2 in Cyanobacteria

Generally, the level PHA production of cyanobacteria under optimal growth conditions is rather low as compared to heterotrophic bacteria. The highest PHB content under photoautotrophic growth has been about 8.5% in CDM reported for *Nostoc muscorum* [93, 112]. Nevertheless, the exact data are hard to compare as early works rarely described all details of the culture conditions. Even the actual growth phase also has a significant impact on the yield: the PHB content of *N. muscorum* was about 150% and 200% in the exponential and logarithmic phase, respectively, of that in the lag phase [93], in agreement with the earlier observation on the similar behavior of *Oscillatoria limosa* and *Gloeothece* sp [58], showing that the newly inoculated bacteria use up the carbon reserve in the lag phase to reorganize their metabolism and gain full photoautotrophic growth potential.

In order to maximize the polymer synthesis and establish heteropolymer production with preferential monomer ratio, several parameters of the culture conditions need to be considered and optimized (pH-value, phosphate and nitrate concentration, acetate, glucose and PHA monomers, illumination, temperature, *etc.*) [7]. Alternatively, the gene content and the organization and regulation of the metabolic pathways leading to PHA production can be altered.

Efforts toward Improved PHA Production

Optimization of Culture Conditions

As mentioned previously, PHA is produced and used as carbon reserve when the carbons source is abundant and there are growth limiting conditions. This phenomenon is widely investigated [55, 91, 93, 103, 113 - 118] and used by several research groups for significantly increased PHA production using such limiting conditions. The most widely investigated nutrient deprivation is the nitrogen deficiency. Several cyanobacterial strains are capable of storing nitrogen in the cell by producing cyanophycin, a non-ribosomally synthesized amino acid polymer composed of aspartic acid and arginine, when cells have sufficient nitrogen supply. Upon nitrogen deprivation this nitrogen reserve is used up followed by decomposition of some pigments resulting in chlorosis, before the overall metabolism gets accommodated to the limited nitrogen supply. This process, which has been investigated on molecular level recently [97, 113, 119 - 121], may take several days, depending on the strain and the conditions. Accordingly, two and three weeks of nitrogen deficiency resulted in 1.9 and 3.5 fold increase, respectively, in the PHB content of *N. muscorum* as compared to the nitrate supplemented control [93]. Nitrogen deprivation was more effective for carbon reserve production, including up to 13.5% PHB in *Synechocystis* than phosphorus, sulfur, iron or calcium deprivation. Upon nitrogen deprivation their PHB content of *Spirulina* and *Synechococcus* was increased to 10% and 14% of the dry weight, respectively. When these cultures were spiked with nitrate this carbon reserve was utilized. 80% of the PHB content degraded in 8 days in *Spirulina* cells and somewhat slower, 57% was used up in *Synechococcus* [122, 123]. In the dark the mobilization of the PHB is half as fast as in light [102] in spite that in the latter case photosynthetic carbon fixation provides extra carbon. It points out the preferential photosynthetic metabolism of *Spirulina* over heterotrophic growth using its carbon reserve.

Since PHB production needs energy and carbon, and during photoautotrophic growth both come from photosynthesis, one would suspect that maximal PHB production needs continuous high light intensity. This idea is corroborated by the observation that irrespective of the initial cell density of the culture the final biomass and cellular PHB concentration reached similar values in *A. fertilissima* [124]. Self-shading in the dense cultures was claimed to be the reason of equilibration of the production. Contrarily, in *Synechocysis* PCC6803 only the biomass production was promoted by illumination but the PHB synthesis was not [104]. It was found that dark periods are essential for PHB accumulation in both *N. muscorum* [93] and *A. fertilissima* [124]. It seems possible that first glycogen is formed as carbon reserve and it is later converted to PHB in the dark. Nevertheless, in the dark *O. limosa* metabolized its glycogen content in the dark, and conversion of glycogen to PHA was not observed. From the results the authors concluded that PHA (which is PHV in this case) may function exclusively as carbon storage whereas glycogen is used for carbon and energy reserve [98]. It was also shown that in *Synechococcus* sp. MA19 grown in dark, nitrogen-depleted conditions, the PHB content raised to 27% because cell mass was reduced owing to glycogen catabolism [105]. Microarray analysis of *Synechocystis* PCC 6803 also pointed out that glycogen catabolism were induced by nitrogen depletion [121]. In the light of these findings it is plausible to suspect that the relative increase of PHA in the cell dry weight may be partly attributed to the consumption of glycogen during weeks of darkness. Considering that in *Synechocystis* PCC 6803 the glycogen reserve under appropriately optimized condition may count more than 40% of CDM [114], the contribution of the loss of the total glycogen reserve to the relative PHB production may be significant.

Sulfate is also an essential nutrient for cyanobacteria and its limitation also arrests cell division and growth of the culture. In *Synechococcus* PCC6803, sulfur deprivation resulted in 3.5 fold increase of the produced polymer [57, 58]. Still, the increase in PHB content of *Aulosira fertilissima* was only obvious after four days and it ceased completely by the 12th day [124]. This filamentous diazotroph have heterocysts, specialized nitrogen fixing cells devoid of photosynthesis, with strictly anaerobic milieu inside them. It is hypothesized that the transient increase in the production was attributed to inactivation of the nitrogenase enzyme in the heterocysts, hence the sulfate deprivation caused enhanced nitrogen limitation.

Phosphate limitation is also known to increase PHB accumulation. This effect is nevertheless not abrupt either, since similar to cyanophycin granules, cyanobacteria tend to store phosphorous in polyphosphate bodies in the cytosol under ideal growth conditions. Freshwater cyanobacteria are frequently cultured in BG-11 medium [125] that contains much higher concentration of phosphate than necessary. Cells grown in such rich media are capable to store so high amount of phosphate that is sufficient for several cell divisions [126]. *Spirulina maxima* produces low amount of PHB, which is greatly stimulated by phosphorus deficient conditions after the intracellular phosphate reserves got depleted [57, 127]. A related strain, *Spirulina platensis* as well as *Synechococcus* sp. MA19 [128] and *Nostoc muscorum* [129] also produce PHB under phosphate limitation.

When it was combined with glucose and acetate supplementation in *N. muscorum*, the cellular PHB content could be increased to 21.5% in 8 days [130]. Similarly, phosphate deficient conditions and concomitant supplementation of acetate also have stimulatory effect on PHA accumulation of PHA up to 29% in *Synechocystis* sp. PCC 6803 grown in glucose containing medium [103].

The PHA production usually does not have specific pH-value or temperature requirement. *N. muscorum* that prefers roughly neutral pH-value for growth had highest production at pH-value 8.5 and increasing the pH-value decreased the produced amount [93]. Similarly, pH-value 8.5 was optimal for the highest amount of PHBHV production, with increasing 3HV monomer ratio toward more acidic region [131]. *Spirulina platensis* isolates with growth preference of pH 10 value produced more PHA at pH-value 9 than pH-value 11 and pH-value 12, and much reduced amount in neutral and acidic pH, while the bacterium adjusted the pH-value of the medium toward pH-value 10 [102]. Mesophilic strains usually produce the highest concentration between 20°C and 30°C while the thermophilic *Synechococcus* sp. MA19 prefers 50°C [129].

Although cyanobacteria are photoautotrophic organisms, most strains can take up and utilize organic carbon compounds. The imported organic material such as monosaccharides can be used for mixotrophic or heterotrophic growth [132, 133]. Moreover, it is noted that when precursors of PHA synthesis (such as acetate for 3HB and valerate for 3HV formation) are supplied to the culture medium, these compounds are readily taken up and incorporated to the produced PHA pool [96, 134]. In this case, the polymer is not produced photoautrophically from the sole CO_2 carbon source, but this method is also widely considered as improvement of the cyanobacterial production. This can have profound impact on the polymer synthesis as not only the produced amount can be greatly improved, but also formation of heteropolymers in lieu of PHB, and the composition and ratio of the incorporated monomers into the so-formed heteropolymer can be significantly influenced at will. A combination of this monomer supply with mixotrophy can further increase the production, such as addition of acetate and some selected monosaccharides could boost the PHB production to 18-35% of CDM of *N. muscorum* [93].

It is noteworthy that *O. limosa* is an exceptional PHA-producer that accumulates solely poly(3-hydroxyvalerate) (PHV); in this case the polymer production is not increased upon acetate supplementation [135].

The complexity of the contribution of each of several factors affecting the production needs systematic multifactor process optimization, which may lead to either improvement in the production yield or the reduction of the production costs by decreasing the necessary input material [110]. Such optimization resulted in an increase to as high as 78% CDM in *N. muscorum* [131], the highest such value so far published. Moreover, the hydroxyvalerate ratio in the polymer PHBHV could be increased from nil to 27% that results in less crystalline hence less brittle and more elastic and heat tolerant material [117]. Similar thorough optimization resulted in up to 50% HV content in the polymer that may total up to 72% in *A. fertilissima*. These high yields and high hydroxyvalerate contents are already comparable to the best heterotrophic PHA producers, emphasizing that this strain and generally cyanobacteria can be promising candidates for PHA production. Nevertheless for this outstanding performance strictly maintained conditions and the supplementation with pure chemicals were necessary. Therefore the cheap production from solely CO_2 as carbon source still needs further research.

Genetic Modification for Higher Yield

Beyond the above summarized optimization of the culture conditions, the possible genetic modification of the metabolism of the bacterium may provide an extra tool for improve the PHA production. Using the synthetic biology approach cyanobacteria has been engineered to produce several materials [136, 137]. Accordingly, the establishment and improvement of PHA production in cyanobacteria by genetic engineering is a plausible approach.

The key enzymes for the most common PHB production pathway [52, 53] are β-ketothiolase, acetoacetyl-CoA reductase and a two component P3HB synthase (refer to route B on Fig. **2**). For strains devoid of these genes, their introduction from heterologous source establishes the synthesis of PHB monomer. For heteropolymer production alternative pathways can be introduced and the metabolic fluxes can be adjusted for optimizing the quantity and composition of the polymer.

As the first attempt, PHB production could be successfully established in *Synechococcus* sp. PCC7942 by introducing the genes of PHB-synthesizing enzymes from *Alcaligenes eutrophus*. The acetate supplementation combined with nitrogen starvation enhanced the PHB content by more than 25%. The properties of the PHB product from this mutant cyanobacterium were similar to those from an *E. coli* transformant with the same genes [115].

Synechococcus sp. PCC7002 also lacks the genes involved in the biosynthetic pathway for poly-3-hydroxybutyrate. When these genes from *Chlorogloeopsis fritschii* PCC 9212 were introduced into *Synechococcus* sp. PCC7002, the mutant was able to accumulate 2.77% of total CDM as PHB. When the *phaABEC* operon coding for the acetyl-CoA actyltransferase from *C. fritschii* PCC 9212 was expressed from a strong promoter in this bacterium, the PHB synthesis was not induced by nitrogen limitation [138], which may indicate that in other cyanobacteria the nitrate limitation increases the PHB production *via* upregulating the *pha* operon, which cannot be further increased using this operon. Alternatively, since PHB biosynthetic pathway is present *Synechococcus* sp. MA19 but missing from the wild-type *Synechococcus* sp. PCC7002, this could be due to differences in the regulatory mechanisms in these two strains.

More importantly, the repressor of the carbon concentrating mechanism [139] was inactivated in this strain, which resulted in 43% higher P3HB production relative to the original mutant. Transcriptomic analysis of the resulted strain revealed overexpression of the bicarbonate concentrating genes but also the *nifJ* gene that catalyzes the oxidative decarboxylation of pyruvate to acetyl-CoA suggesting that the key component in the improved polymer synthesis is the increase in acetyl-CoA. Additionally, by introducing 4-hydroxybutyrate dehydrogenase and 4-hydroxybutyryl-CoA transferase from *Porphyromonas gingivalis* that convert succinic semialdehyde to 4-hydroxybutyryl-CoA (4HB-CoA) (route E in Fig. **2**), it was possible to achieve production of poly(3-hydroxybutyrate-*co*-4-hydroxybutyrate) (PHB4HB) copolymer from CO_2 without precursor supply [137]. This polymer has lower melting temperature (150°C), lower crystallinity

(45%) and very high extension to brake value (444%) compared to PHB (177°C, 60% and 5%, respectively), which can be beneficial in most applications.

In another very successful attempt, the *pha* genes from *Wautersia eutropha* were cloned along with the *recA* gene from *E. coli* under the regulation of the strong promoter of RuBisCo to a *recA* null mutant *Synechococcus* sp. PCC7002 stain. It allowed not only up to 52% CDM accumulation of PHB under nitrate limitation, but the resulting mutant is maintained in an antibiotics-free manner, which is a rather important trait from environmental point of view [140].

In strains naturally capable of producing PHA, the endogenous pathways and the regulation fluctuates. Horvat *et al.* also considered these metabolic aspects as these may present existing tools as well as limitations for the production [141]. Just like in the above example, for producing copolymers with better physical parameters than those of pure PHB homopolymer, alternative metabolic pathways can be introduced that lead to production of different monomers, which together result in copolymer formation.

For the purpose of such genetic manipulations, *Synechocystis* PCC6803 is well-suited, since its genome is sequenced, its genes are fully annotated, countless mutants have been created to investigate its metabolic paths, several transcriptome profiling datasets are available from microchip and RNA-seq techniques and generally speaking this is arguably the most thoroughly investigated cyanobacterial strain [142]. This bacterium possesses the genes for PHB production [39, 143]. Under photoautotrophic conditions, the PHB production is only about 3%, but by concomitantly applying all tricks of nutrient deprivation together with acetate and sugar supplementation it can be greatly boosted to up to 38% CDM [144]. In another study, where glycogen and lipid production were also considered, the total productivity of these products could be increased to 61.5% with 13.5% PHB content using only nitrogen deprivation. This could not be further increased using 0.4% glucose in photoheterotrophic condition while the lipid and glycogen content was boosted [114].

In the first attempts to increase the production by genetic modification, PHB biosynthetic genes were introduced from *C. necator (previously R. eutropha or A.*

eutrophus). Although, the PHA synthase activity was higher in the mutant than in the wild type, this extra activity could only slightly increase the PHA accumulation, showing that PHA synthase activity was not the bottleneck of the process [145]. Contrarily to some extent, Hondo and coworkers could achieve about 7% PHB accumulation in *Synechocystis* PCC 6803 transformed with a plasmid carrying a hybrid *phaECAB* operon comprised of *phaEC* genes from *Synechocystis* sp. PCC 6803 and *phaAB* genes from and *A. eutrophus* [146].

Instead of relying exclusively on the β-ketothiolase-catalyzed acetoacetyl-CoA production (similarly to as described above), an extremely active acetoacetyl-CoA synthase from *Streptomyces sp.* CL190 was introduced in the PHB biosynthesis pathway [147]. The product of this gene catalyzes the irreversible condensation of acetyl-CoA and malonyl-CoA yielding acetoacetyl-CoA, which supposedly have pivotal importance for improved formation of PHA. The CO_2 evolution during the condensation reaction also helps to shift the equilibrium of the reaction toward the production of acetoacetyl-CoA [142]. Additionally, an extremely efficient PHA synthase from *Chromobacterium* sp. USM2 (PHACCs) was also introduced along with the above gene to promote the photosynthetic production of PHB in *Synechocystis* sp. The mutant obtained was compared to the control strain by means of several approaches including the assessment of PHB production level and transcriptomic profiling *via* RNA-seq analysis and quantitative RT-PCR [148]. Interestingly, the expression levels of dozens of genes changed significantly, which shows a central position of acetyl-CoA not only in the metabolism of the organism but also in the regulation of several cellular processes. Nevertheless, the transcript levels of the native genes in the PHB pathway did not correlate with the increased PHB production. Indeed, the expression of *phaA* and *phaB* genes were two-fold repressed. It suggests that the concentration of these enzymes is not the bottleneck in achieving high PHA accumulation, which could reach 14% photoautotrophically and 41% with 0.4% acetate supplementation. The authors also suggest that the total flux of carbon is the potential driving force for the anabolism of PHA [143]. These findings are in agreement with the above mentioned works with the *C. necator* genes [145] and the *Synechococcus* sp. PCC7002 engineering [138].

The effect of nitrogen limitation on gene expression has been recently thoroughly investigated pointing out the gradual rearrangement of the gene expression profile and the corresponding control of transcription for two major metabolic activities of nitrogen-starved and light-limited growth [119, 120]. The most suitable strain for this purpose is arguably *Synechocystis* PCC6803 due to the wealth of detailed information on its genetics as mentioned above. Nevertheless it is important to note that there may be significant differences in the connection of nitrogen depletion and PHA production related metabolism of different genera. The most obvious difference stems from the ability to fix nitrogen. With this respect the unicellular diazotroph strains that fix nitrogen in the dark phase of the day differ greatly from the filamentous strains where the heterocysts enable nitrogen fixation in the light and provide continuous combined nitrogen for the vegetative cells.

It is known that the sugar catabolic pathways and nitrogen limitation are interconnected in a rather complex way that among others includes the SigE sigma factor of *Synechocystis* PCC6803 [118], and the change in gene expression, translation and protein activation is also time-dependent [149, 150], in concert with the kinetics of nitrogen reserve utilization mentioned above. Testing the potential in the gene regulation, the SigE sigma factor was overexpressed in a glucose-tolerant *Synechocystis* PCC6803. As a consequence, the gene expression profile of the bacterium changed in a rather complex way, which eventually resulted in more than double as high PHB accumulation under nitrogen limitation in the mutants than in the wild type. It is also noteworthy that the results also suggested that the PhaABCE proteins are post-transcriptionally regulated in *Synechocystis* 6803 [150].

Another innovative approach is the production of mixture of monomers in the culture medium that can be polymerized in a later step. These monomers are produced by sugar-based reactions and chemically in petroleum-based chemical industry, hence the photoautotrophic production would also be beneficial reactions. It was shown that upon inactivation of the PHB polymerase coding enzymes the monomers are readily secreted to the culture medium. By introducing metabolic paths for alternative monomers, the mixture is produced in the broth [152].

Recovery of PHA from Cyanobacteria

The recovery process of cellular products may have significant impact on the quality of the product, the overall economics of the production [153] and the sustainability, which places this step of the production into the focus of interest [154]. For this reason, it is important to have efficient and economically feasible methods for separation and recovery of PHB with no deterioration of the product quality and minimal environmental impact. Moreover, the optimal methods are strain specific due to the differences in cell wall composition as well as intracellular amount and composition of PHA granules. Accordingly, several recovery methods have been tested on cyanobacteria and among them for each strain the most suitable, economical and efficient recovery method should be considered with the overall economics of PHA production in mind [5].

The recovery methods can be based upon the disruption of the cells and separating the cell debris from the liquid containing the PHB granules. The hard cell wall of the Gram-negative cyanobacteria can be broken using physical force in bead mills. While this process is widely used in lab scale for obtaining the intracellular content of cyanobacteria, the scale up to industrial scale has not been fully elaborated. The optimization usually includes the systematic investigation of several variables, such as the size and density of the beads, the time of operation, the speed and geometry of the mill as well as the load of cells and beads. Contrary of the complexity, pilot scale investigations show promising results towards energy consumption optimized process [155].

As an alternative of physical disruption of the cells, the recovery methods are mostly based upon extraction of PHB by organic solvents like trichloromethane or chloroform, solvents that also help separating the PHB content from other cell constituents by more or less selective extraction. This approach was first developed for *C. necator* (called *Ralstonia eutropha* or *Alcaligenes eutrophus* before reclassification) [156], the bacterium that is capable of producing PHB up to 80% of its dry weight. The dried cells were first treated with 30% hypochlorite, and then equal volume of chloroform was added to the aqueous phase. This resulted in 91% recovery and 97% purity of the extracted material and about 20% decrease in the molecular mass, from the original 1.27×10^6 to 1.02×10^6 [151]. The modified version used a 50% sodium hypochlorite/chloroform dispersion solution resulting in similar results for that bacterium [157]. In contrast, 4% hypochlorite was sufficient for liberating PHB from *Spirulina* LEB18. When this concentration was increased to 10% and combined with trichloromethane extraction, it resulted in about 27 fold decrease in the molecular mass from 4.36×10^6 to 1.59×10^5 [158]. This extraction method often yields PHB of green color due to co-extracted chlorophylls and possibly other pigments. This feature can be beneficial, depending on the downstream applications, as the non-PHB impurities that count around 30% has beneficial impact on the physical characteristics of the product and the biological effects of these compounds can be useful in clinical applications [159 - 161]. Nevertheless, if the pigment content of the product is to be avoided and pure PHB separation is necessary, a methanol extraction prior to drying the cells can remove all colors. The PHA that is then extracted by hot chloroform, precipitated with diethyl ether, and finally washed with acetone that results in highly pure preparation.

A recent method for extraction of the PHA from biomass is accomplished at high temperature and pressure using "anti-solvent" acetone extraction method, which also produces highly pure polymer [151] by simply cooling the solution that leads to the precipitation of the product. Nevertheless, this method has not been tested with cyanobacterial samples displaying lower polymer and relatively high pigment content.

Another innovative separation process uses immiscible ionic liquids (ILs) for extraction instead of organic solvents that need to be evaporated for recovery of the extracts, a step of high energy demand and environmentally hazard. These materials can be regarded as molten salts of melting point less than 100°C and unique properties, such as high ionic conductivity and high thermostability. Extraction of useful materials from cyanobacteria was presented using methods that include heating [162, 163], which is energetically unfavorable. For decreasing the energy consumption several ILs were investigated with respect to their hydrogen-bonding receipt and donating abilities (α and β values, respectively) and the effect of these parameters on the solubility of cyanobacterial cells and PHB granules. Among the best candidates 1-ethyl-3-methylimidazolium methylphosphonate ([C2mim][MeO(H)PO_2]) was found to solubilize cyanobacterial components, but not PHB. It enables the recovery of the PHB granules by filtration and allows the recycling of the solvent [164]. The method has proven effective by recovery of 98% of the PHB content; nevertheless it is still in lab scale and needs verification with higher volumes.

Practical Applications of Cyanobacterial PHA

A rather innovative medical application of plastics is their use in regenerative medicine. Some tissues need an extracellular matrix scaffold for regenerate. These matrices provide anchorage to the cells and help in the development of the tissues in the right orientation and extent and stimulate the production of the right structures. In the modern tissue engineering, biocompatible and biodegradable plastics are preferentially used in order to decrease the risk of tissue rejection as compared to traditional scaffold materials and for saving the patients from the needless operation for removing these objects after healing. PHB from cyanobacteria is suitable for creating textile fibers and it has been proposed that such materials from *Nostoc* sp. be used in medical applications such as surgical meshes [112]. One the most advanced method of the production of such bioplastics fibers is the technique called electrospinning [165 - 167], yielding materials with high porosity and pore connectivity which, together with the low fiber diameter, result in high surface area to volume ratio that is crucial in cell adhesion. This formulation provides further benefits by enabling incorporation of drugs [168] into the material during production. These drugs may have various functions such substrates or growth factors stimulating cell proliferation [169, 170], analgesics [171] as well as antibacterial [172, 173] or antifungal [174] effects. Since the active materials are administered locally, low doses suffice as compared to conventional techniques and the drugs are released gradually in longer timescale, as another advantage.

Due to the above benefits, this technique is getting increasing attention worldwide using various composite materials, mainly based on poly-lactic acid and related polymers [168, 175, 176], although several other materials are also thoroughly tested [177-179]. According to recent investigations PHB and its copolymers are equally excellent choices for such applications [180], with the polymers from cyanobacteria and especially from *Spirulina* being exceptionally well-suited [181,

182]. The capability of *Arthrospira* sp. (formerly belonging to the *Spirulina* genus) to produce PHB were discovered as early as in 1992 [55] and this phenomenon has been investigated ever since in these important strains. Recently, the properties of the electrospun nanofibers were compared to commercial counterparts and significant differences were found, mostly favoring the cyanobacterial products. The measurements included the establishment of physical parameters such as elasticity, filament diameter, conductivity, degradation temperature range, crystallinity, among others as the functions of the composition [181]. The PHB extracted from *Spirulina* was compared to commercial PHB and PHBHV containing 5% and 12% 3HV. For added beneficial biological effects in tissue growth, the polymers were biofunctionalized using the residual biomass of the cyanobacterium that was dried after the removal of PHB [183], or spiked with sodium chloride for altered conductivity.

An important observation was that the addition of biomass allowed the application of as low as 7% solution of the cyanobacterial polymer as compared to the generally applied 20-22% concentration. This resulted in the preparation of fibers of significantly reduced diameter, which may greatly reduce production costs as well as fundamentally influence the features of the membranes or other final products made of the fibers. The nanofibers made of commercial PHB was assessed decades ago [184]; in comparison, the pure cyanobacterial PHB was highly elastic, having over twice as high elasticity, tensile strength and braking elongation than those of the commercial copolymer counterparts [181]. Nevertheless, the tensile strength was over an order of magnitude lower than what could be achieved using additives and melt-spun production technology [185]. Moreover, when *Spirulina* biomass was added in comparable amount (5%) as biofunctionalizer to the 7% PHB containing solution prior to electrospinning procedure, both the elasticity and the tensile strength decreased greatly in response. On the other hand, embedding purified phycocyanin into electrospun PHB increased the thermal stability of this pigment, emphasizing that all traits of such bioplastic products need to be taken into consideration with respect to the biological effects, the physical stresses and production prices, when the right production and formulation is chosen. Since the cyanobacterial PHA synthase is thylakoid membrane associated [128, 186], endogenous metabolic systems seem promising with due culture conditions, supply of monomers or precursors when pigment content of the polymer is considered beneficial; whereas recombinant strains harbouring appropriately tuned metabolic pathways, possibly using genes from heterotrophic bacteria, may otherwise be preferential.

CONCLUDING REMARKS

Polyhydroxyalkanoates are biodegradable polyesters which can be produced *via* microbial fermentation processes. These materials are widely used as bioplastics since they are renewable, biocompatible and environmentally friendly. However, the cost of bulk production of PHA might be a substantial economic factor limiting their applications. Industrially, dark fermentation processes are developed for production of PHA from various sugars (sugar, starch, hemicellulose) but oil and glycerol might be also good substrates for this purpose. In contrast, photosynthetic microbes can fix CO_2 by using light energy and produce polyhydroxyalkanoates. Photochemolitoautotrophic strains and cyanobacteria can grow using CO_2 as sole carbon source and many bacteria can synthesize various storage materials including biopolyesters, such as PHA under nutrient (nitrogen, phosphorous, sulfur) limitation. It should be noted that the yield can be substantially improved by adding organic substrates to the medium. Nevertheless, for climatic reasons, CO_2 capture is of primary importance and technologies using CO_2 for producing valuable and environmentally sound materials have attracted much attention. In the sections above, it was clearly demonstrated that photosynthetic bacteria can synthesize PHA both homo- and heteropolymers and might be serious candidates in industrial application. The composition and the yield might be varied by the growth conditions and light is also an important factor influencing the yield. Although, the efficacy of the process is significantly lower as compared to the technologies applying fermentative microbes, it can be further improved by genetically modified strains and/or bioprocess optimization. Alternatively, waste utilization combined with CO_2 fixation might also lead to a reasonable approach.

CONFLICT OF INTEREST

The authors confirm that they have no conflict of interest to declare for this publication.

ACKNOWLEDGEMENTS

Declared none.

NOTES

[1]Enzyme numbers according to [187]: 1: Glyceraldehyde-3-phosphate dehydrogenase; 2: Pyruvate dehydrogenase complex; 3: 3-Ketothiolase (PhaA); 4: NADPH-dependent acetoacetyl-CoA reductase (PhaB); 5: PHA synthase (PhaC); 6: Acetyl-CoA carboxylase (ACC); 7: Malonyl-CoA:ACP transacylase (FabD); 8: 3-Ketoacyl carrier protein synthase (FabH); 9: NADPH-dependent 3-Ketoacyl reductase (FabG); 10: Succinic semialdehyde dehydrogenase (SucD); 11: 4-Hydroxybutyrate dehydrogenase (4HbD); 12: 4-Hydroxybutyrate-CoA:CoA transferase (OrfZ); 13: Alcohol dehydrogenase (putative); 14: Hydroxyacyl-CoA synthase (putative); 15: Methylmalonyl-CoA mutase (Sbm); 16: Methylmalonyl-CoA racemase; 17: Methylmalonyl-CoA decarboxylase (YgfG); 18: Ketothiolase (putative); 19: 3-Ketothiolase (BktB); 20: Ketothiolase (putative); 21: NADPH-dependent acetoacetyl-CoA reductase; 22: Acyl-CoA synthetase (FadD); 23: Acyl-CoA oxidase(putative); 24: Enoyl-CoA hydratase (putative); 25: (R)-Enoy--CoA hydratase (PhaJ); 26: Epimerase; 27: 3-Ketoacyl-CoA thiolase (FadA); 28: 3-Hydroxyacyl-ACP:CoA transacylase (PhaG); 29: Cyclohexanol dehydrogenase (ChnA); 30: Cyclohexanone monooxygenases (ChnB); 31: Caprolactone hydrolase (ChnC); 32: 6-hydroxyhexanoate dehydrogenase (ChnD); 33: 6-Oxohexanoate dehydrogenase (ChnE); 34: Semialdehyde dehydrogenase (putative); 35: 6-hydroxyhexanoate dehydrogenase (putative); 36: Hydroxyacyl-CoA synthase (putative); 37: Lactonase (putative).

REFERENCES

[1] Anderson AJ, Dawes EA. Occurrence, metabolism, metabolic role, and industrial uses of bacterial polyhydroxyalkanoates. Microbiol Rev 1990; 54(4): 450-72.
[PMID: 2087222]

[2] Rehm BH. Polyester synthases: natural catalysts for plastics. Biochem J 2003; 376(Pt 1): 15-33.
[http://dx.doi.org/10.1042/bj20031254] [PMID: 12954080]

[3] Rehm BH. Biogenesis of microbial polyhydroxyalkanoate granules: a platform technology for the production of tailor-made bioparticles. Curr Issues Mol Biol 2007; 9(1): 41-62.

[PMID: 17263145]

[4] Byrom D. Polymer synthesis by microorganisms: technology and economics. Trends Biotechnol 1987; 5(9): 246-50.
[http://dx.doi.org/10.1016/0167-7799(87)90100-4]

[5] Choi J, Lee SY. Factors affecting the economics of polyhydroxyalkanoate production by bacterial fermentation. Appl Microbiol Biotechnol 1999; 51(1): 13-21.
[http://dx.doi.org/10.1007/s002530051357]

[6] Bugnicourt E. Polyhydroxyalkanoate (PHA): Review of synthesis, characteristics, processing and potential applications in packaging. Express Polym Lett 2014; 8(11): 791-808.
[http://dx.doi.org/10.3144/expresspolymlett.2014.82]

[7] Singh Saharan B, Grewal A, Kumar P. Biotechnological Production of Polyhydroxyalkanoates: A Review on Trends and Latest Developments. Chinese J Biol 2014; 1-18.
[http://dx.doi.org/10.1155/2014/802984]

[8] Liebergesell M, Hustede E, Timm A, Steinbüchel A, Fuller RC, Lenz RW, *et al.* Formation of poly(3-hydroxyalkanoates) by phototrophic and chemolithotrophic bacteria. Arch Microbiol 1991; 155(5): 415-21.
[http://dx.doi.org/10.1007/BF00244955]

[9] Madison LL, Huisman GW. Metabolic engineering of poly(3-hydroxyalkanoates): from DNA to plastic. Microbiol Mol Biol Rev 1999; 63(1): 21-53.
[PMID: 10066830]

[10] Eidels L, Preiss J. Carbohydrate metabolism in Rhiodopseudomonas capsulata: Enzyme titers, glucose metabolism, and polyglucose polymer synthesis. Arch Biochem Biophys 1970; 140(1): 75-89.

[11] Weaver PF, Wall JD, Gest H. Characterization of Rhodopseudomonas capsulata. Arch Microbiol 1975; 105(3): 207-16.
[http://dx.doi.org/10.1007/BF00447139] [PMID: 1103769]

[12] Kranz RG, Gabbert KK, Locke TA, Madigan MT. Polyhydroxyalkanoate production in Rhodobacter capsulatus: genes, mutants, expression, and physiology. Appl Environ Microbiol 1997; 63(8): 3003-9.
[PMID: 9251189]

[13] Jin H, Nikolau BJ. Evaluating PHA productivity of bioengineered Rhodosprillum rubrum. PLoS One 2014; 9(5): e96621.
[http://dx.doi.org/10.1371/journal.pone.0096621] [PMID: 24840941]

[14] Brandl H, Knee EJ Jr, Fuller RC, Gross RA, Lenz RW. Ability of the phototrophic bacterium Rhodospirillum rubrum to produce various poly (beta-hydroxyalkanoates): potential sources for biodegradable polyesters. Int J Biol Macromol 1989; 11(1): 49-55.
[http://dx.doi.org/10.1016/0141-8130(89)90040-8] [PMID: 2518731]

[15] Rothermich MM, Guerrero R, Lenz RW, Goodwin S. Characterization, seasonal occurrence, and diel fluctuation of poly(hydroxyalkanoate) in photosynthetic microbial mats. Appl Environ Microbiol 2000; 66(10): 4279-91.
[http://dx.doi.org/10.1128/AEM.66.10.4279-4291.2000] [PMID: 11010871]

[16] Brandl H. Gross R a., Lenz RW, Lloyd R, Fuller RC. The accumulation of poly(3-hydroxyalkanoates)

in *Rhodobacter sphaeroides*. Arch Microbiol 1991; 155(4): 337-40.
[http://dx.doi.org/10.1007/BF00243452]

[17] Lorrungruang C, Martthong J, Sasaki K, Noparatnaraporn N. Selection of photosynthetic bacterium *Rhodobacter sphaeroides* 14F for polyhydroxyalkanoate production with two-stage aerobic dark cultivation. J Biosci Bioeng 2006; 102(2): 128-31.
[http://dx.doi.org/10.1263/jbb.102.128] [PMID: 17027875]

[18] Willison JC. Biochemical genetics revisited: the use of mutants to study carbon and nitrogen metabolism in the photosynthetic bacteria. FEMS Microbiol Rev 1993; 10(1-2): 1-38.
[http://dx.doi.org/10.1111/j.1574-6968.1993.tb05862.x] [PMID: 8431308]

[19] Khatipov E, Miyake M, Miyake J, Asada Y. Accumulation of poly-β-hydroxybutyrate by *Rhodobacter sphaeroides* on various carbon and nitrogen substrates. FEMS Microbiol Lett 1998; 162(1): 39-45.

[20] Kovács KL, Kovács AT, Maróti G, *et al.* The hydrogenases of *Thiocapsa roseopersicina*. Biochem Soc Trans 2005; 33(Pt 1): 61-3.
[http://dx.doi.org/10.1042/BST0330061] [PMID: 15667265]

[21] Maróti J, Farkas A, Nagy IK, *et al.* A second soluble Hox-type NiFe enzyme completes the hydrogenase set in *Thiocapsa roseopersicina* BBS. Appl Environ Microbiol 2010; 76(15): 5113-23.
[http://dx.doi.org/10.1128/AEM.00351-10] [PMID: 20543059]

[22] Fülöp A, Béres R, Tengölics R, Rákhely G, Kovács KL. Relationship between PHA and hydrogen metabolism in the purple sulfur phototrophic bacterium *Thiocapsa roseopersicina* BBS. Int J Hydrogen Energy 2012; 37(6): 4915-24.
[http://dx.doi.org/10.1016/j.ijhydene.2011.12.019]

[23] Jiao N, Zhang Y, Zeng Y, *et al.* Distinct distribution pattern of abundance and diversity of aerobic anoxygenic phototrophic bacteria in the global ocean. Environ Microbiol 2007; 9(12): 3091-9.
[http://dx.doi.org/10.1111/j.1462-2920.2007.01419.x] [PMID: 17991036]

[24] Liu R, Zhang Y, Jiao N. Diel variations in frequency of dividing cells and abundance of aerobic anoxygenic phototrophic bacteria in a coral reef system of the South China Sea. Aquat Microb Ecol 2010; 58(3): 303-10.
[http://dx.doi.org/10.3354/ame01371]

[25] Yurkov VV, Beatty JT. Aerobic anoxygenic phototrophic bacteria. Microbiol Mol Biol Rev 1998; 62(3): 695-724.
[PMID: 9729607]

[26] Xiao N, Jiao N. Formation of polyhydroxyalkanoate in aerobic anoxygenic phototrophic bacteria and its relationship to carbon source and light availability. Appl Environ Microbiol 2011; 77(21): 7445-50.
[http://dx.doi.org/10.1128/AEM.05955-11] [PMID: 21908634]

[27] Beatty JT. On the natural selection and evolution of the aerobic phototrophic bacteria. Photosynth Res 2002; 73(1-3): 109-14.
[http://dx.doi.org/10.1023/A:1020493518379] [PMID: 16245110]

[28] Meng DC, Shen R, Yao H, Chen JC, Wu Q, Chen GQ. Engineering the diversity of polyesters. Curr Opin Biotechnol 2014; 29(1): 24-33.
[http://dx.doi.org/10.1016/j.copbio.2014.02.013] [PMID: 24632193]

[29] Rehm BH, Antonio R V, Spiekermann P, Amara A, Steinbüchel A. Steinbuchel A. Molecular characterization of the poly(3-hydroxybutyrate) (PHB) synthase from *Ralstonia eutropha*: *In vitro* evolution, site-specific mutagenesis and development of a PHB synthase protein model. Biochim Biophys Acta - Protein Struct Mol Enzymol 2002; 1594(1): 178-90.

[30] Jendrossek D. Polyhydroxyalkanoate granules are complex subcellular organelles (carbonosomes). J Bacteriol 2009; 191(10): 3195-202. [http://dx.doi.org/10.1128/JB.01723-08] [PMID: 19270094]

[31] Pötter M, Madkour MH, Mayer F, Steinbüchel A. Regulation of phasin expression and polyhydroxyalkanoate (PHA) granule formation in Ralstonia eutropha H16. Microbiology 2002; 148(Pt 8): 2413-26. [http://dx.doi.org/10.1099/00221287-148-8-2413] [PMID: 12177335]

[32] Sudesh K, Abe H, Doi Y. Synthesis, structure and properties of polyhydroxyalkanoates: Biological polyesters. Prog Polym Sci 2000; 25(10): 1503-55. [http://dx.doi.org/10.1016/S0079-6700(00)00035-6]

[33] Budde CF, Mahan AE, Lu J, Rha C, Sinskey AJ. Roles of multiple acetoacetyl coenzyme A reductases in polyhydroxybutyrate biosynthesis in Ralstonia eutropha H16. J Bacteriol 2010; 192(20): 5319-28. [http://dx.doi.org/10.1128/JB.00207-10] [PMID: 20729355]

[34] Clemente T, Shah D, Tran M, *et al.* Sequence of PHA synthase gene from two strains of Rhodospirillum rubrum and *in vivo* substrate specificity of four PHA synthases across two heterologous expression systems. Appl Microbiol Biotechnol 2000; 53(4): 420-9. [http://dx.doi.org/10.1007/s002530051636] [PMID: 10803898]

[35] Jin H, Nikolau BJ. Role of genetic redundancy in polyhydroxyalkanoate (PHA) polymerases in PHA biosynthesis in Rhodospirillum rubrum. J Bacteriol 2012; 194(20): 5522-9. [http://dx.doi.org/10.1128/JB.01111-12] [PMID: 22865850]

[36] Liebergesell M, Schmidt B, Steinbüchel A. Isolation and identification of granule-associated proteins relevant for poly(3-hydroxyalkanoic acid) biosynthesis in Chromatium vinosum D. FEMS Microbiol Lett 1992; 78(2-3): 227-32. [http://dx.doi.org/10.1111/j.1574-6968.1992.tb05572.x] [PMID: 1490603]

[37] Liebergesell M, Steinbüchel A. Cloning and nucleotide sequences of genes relevant for biosynthesis of poly(3-hydroxybutyric acid) in Chromatium vinosum strain D. Eur J Biochem 1992; 209(1): 135-50. [http://dx.doi.org/10.1111/j.1432-1033.1992.tb17270.x] [PMID: 1396692]

[38] Liebergesell M, Steinbüchel A. Cloning and molecular analysis of the poly(3-hydroxybutyric acid) biosynthetic genes of Thiocystis violacea. Appl Microbiol Biotechnol 1993; 38(4): 493-501. [http://dx.doi.org/10.1007/BF00242944] [PMID: 7763384]

[39] Liebergesell M, Rahalkar S, Steinbüchel A. Analysis of the Thiocapsa pfennigii polyhydroxyalkanoate synthase: subcloning, molecular characterization and generation of hybrid synthases with the corresponding Chromatium vinosum enzyme. Appl Microbiol Biotechnol 2000; 54(2): 186-94. [http://dx.doi.org/10.1007/s002530000375] [PMID: 10968631]

[40] Hein S, Tran H, Steinbüchel A. *Synechocystis* sp. PCC6803 possesses a two-component polyhydroxyalkanoic acid synthase similar to that of anoxygenic purple sulfur bacteria. Arch

Microbiol 1998; 170(3): 162-70.
[http://dx.doi.org/10.1007/s002030050629] [PMID: 9683655]

[41] Zhang S, Kolvek S, Goodwin S, Lenz RW. Poly(hydroxyalkanoic acid) biosynthesis in Ectothiorhodospira shaposhnikovii: Characterization and reactivity of a type III PHA synthase. Biomacromolecules 2004; 5(1): 40-8.
[http://dx.doi.org/10.1021/bm034171i] [PMID: 14715006]

[42] Yuan W, Jia Y, Tian J, *et al.* Class I and III polyhydroxyalkanoate synthases from Ralstonia eutropha and Allochromatium vinosum: characterization and substrate specificity studies. Arch Biochem Biophys 2001; 394(1): 87-98.
[http://dx.doi.org/10.1006/abbi.2001.2522] [PMID: 11566031]

[43] Han J, Hou J, Liu H, *et al.* Wide distribution among halophilic archaea of a novel polyhydroxyalkanoate synthase subtype with homology to bacterial type III synthases. Appl Environ Microbiol 2010; 76(23): 7811-9.
[http://dx.doi.org/10.1128/AEM.01117-10] [PMID: 20889776]

[44] Hai T, Hein S, Steinbüchel A. Multiple evidence for widespread and general occurrence of type-III PHA synthases in cyanobacteria and molecular characterization of the PHA synthases from two thermophilic cyanobacteria: Chlorogloeopsis fritschii PCC 6912 and Synechococcus sp. strain MA19. Microbiology 2001; 147(Pt 11): 3047-60.
[http://dx.doi.org/10.1099/00221287-147-11-3047] [PMID: 11700355]

[45] Liebergesell M, Sonomoto K, Madkour M, Mayer F, Steinbüchel A. Purification and characterization of the poly(hydroxyalkanoic acid) synthase from Chromatium vinosum and localization of the enzyme at the surface of poly(hydroxyalkanoic acid) granules. Eur J Biochem 1994; 226(1): 71-80.
[http://dx.doi.org/10.1111/j.1432-1033.1994.tb20027.x] [PMID: 7957260]

[46] Müh U, Sinskey AJ, Kirby DP, Lane WS, Stubbe J. PHA synthase from *chromatium vinosum*: cysteine 149 is involved in covalent catalysis. Biochemistry 1999; 38(2): 826-37.
[http://dx.doi.org/10.1021/bi9818319] [PMID: 9888824]

[47] Jia Y, Kappock TJ, Frick T, Sinskey AJ, Stubbe J. Lipases provide a new mechanistic model for polyhydroxybutyrate (PHB) synthases: characterization of the functional residues in *Chromatium vinosum* PHB synthase. Biochemistry 2000; 39(14): 3927-36.
[http://dx.doi.org/10.1021/bi9928086] [PMID: 10747780]

[48] Gerngross TU, Snell KD, Peoples OP, *et al.* Overexpression and purification of the soluble polyhydroxyalkanoate synthase from *Alcaligenes eutrophus*: evidence for a required posttranslational modification for catalytic activity. Biochemistry 1994; 33(31): 9311-20.
[http://dx.doi.org/10.1021/bi00197a035] [PMID: 8049232]

[49] Tian J, Sinskey AJ, Stubbe J. Class III polyhydroxybutyrate synthase: involvement in chain termination and reinitiation. Biochemistry 2005; 44(23): 8369-77.
[http://dx.doi.org/10.1021/bi050331u] [PMID: 15938626]

[50] Tian J, Sinskey AJ, Stubbe J. Detection of intermediates from the polymerization reaction catalyzed by a D302A mutant of class III polyhydroxyalkanoate (PHA) synthase. Biochemistry 2005; 44(5): 1495-503.
[http://dx.doi.org/10.1021/bi047734z] [PMID: 15683234]

[51] Kessler B, Witholt B. Factors involved in the regulatory network of polyhydroxyalkanoate metabolism. J Biotechnol 2001; 86(2): 97-104.
[http://dx.doi.org/10.1016/S0168-1656(00)00404-1] [PMID: 11245898]

[52] Lee SY, Choi J, Wong HH. Recent advances in polyhydroxyalkanoate production by bacterial fermentation: mini-review. Int J Biol Macromol 1999; 25(1-3): 31-6.
[http://dx.doi.org/10.1016/S0141-8130(99)00012-4] [PMID: 10416647]

[53] Lee SY. Plastic bacteria? Progress and prospects for polyhydroxyalkanoate production in bacteria. Trends Biotechnol 1996; 14(11): 431-8.
[http://dx.doi.org/10.1016/0167-7799(96)10061-5]

[54] Hustede E, Steinbüchel A, Schlegel HG. Relationship between the photoproduction of hydrogen and the accumulation of PHB in non-sulfur purple bacteria. Appl Microbiol Biotechnol 1993; 39(1): 87-93.
[http://dx.doi.org/10.1007/BF00166854]

[55] De Philippis R, Sili C, Vincenzini M. Glycogen and poly-β-hydroxybutyrate synthesis in *Spirulina maxima.* J Gen Microbiol 1992; 138(8): 1623-8.
[http://dx.doi.org/10.1099/00221287-138-8-1623]

[56] Dierstein R, Drews G. Nitrogen-limited continuous culture of *Rhodopseudomonas capsulata* growing photosynthetically or heterotrophically under low oxygen tensions. Arch Microbiol 1974; 99(2): 117-28.
[http://dx.doi.org/10.1007/BF00696228] [PMID: 4604821]

[57] De Philippis R, Ena A, Guastiini M, Sili C, Vincenzini M. Factors affecting poly-β-hydroxybutyrate accumulation in cyanobacteria and in purple non-sulfur bacteria. FEMS Microbiol Lett 1992; 103(2-4): 187-94.

[58] Stal LJ. Poly(hydroxyalkanoate) in cyanobacteria: an overview. FEMS Microbiol Lett 1992; 103(2-4): 169-80.
[http://dx.doi.org/10.1111/j.1574-6968.1992.tb05835.x]

[59] McKinlay JB, Oda Y, Rühl M, Posto AL, Sauer U, Harwood CS. Non-growing *Rhodopseudomonas palustris* increases the hydrogen gas yield from acetate by shifting from the glyoxylate shunt to the tricarboxylic acid cycle. J Biol Chem 2014; 289(4): 1960-70.
[http://dx.doi.org/10.1074/jbc.M113.527515] [PMID: 24302724]

[60] Vincenzini M, Marchini A, Ena A, De Philippis R. H2 and poly-β-hydroxybutyrate, two alternative chemicals from purple non sulfur bacteria. Biotechnol Lett 1997; 19(8): 759-62.
[http://dx.doi.org/10.1023/A:1018336209252]

[61] Melnicki MR, Eroglu E, Melis A. Changes in hydrogen production and polymer accumulation upon sulfur-deprivation in purple photosynthetic bacteria. Int J Hydrogen Energy 2009; 34(15): 6157-70.
[http://dx.doi.org/10.1016/j.ijhydene.2009.05.115]

[62] Suzuki T, Tsygankov A, Miyake J, Tokiwa Y, Asada Y. Accumulation of poly-(hydroxybutyrate) by a non-sulfur photosynthetic bacterium, *Rhodobacter sphaeroides* RV at different pH. Biotechnol Lett 1995; 17(4): 395-400.
[http://dx.doi.org/10.1007/BF00130796]

[63] Cronan JE Jr. Phospholipid modifications in bacteria. Curr Opin Microbiol 2002; 5(2): 202-5.

[http://dx.doi.org/10.1016/S1369-5274(02)00297-7] [PMID: 11934618]

[64] Nam T-W, Ziegelhoffer EC, Lemke RA, Donohue TJ. Proteins needed to activate a transcriptional response to the reactive oxygen species singlet oxygen. MBio 2013; 4(1): e00541-12. [http://dx.doi.org/10.1128/mBio.00541-12] [PMID: 23300250]

[65] Reinecke F, Steinbüchel A. *Ralstonia eutropha* strain H16 as model organism for PHA metabolism and for biotechnological production of technically interesting biopolymers. J Mol Microbiol Biotechnol 2009; 16(1-2): 91-108. [http://dx.doi.org/10.1159/000142897] [PMID: 18957865]

[66] Pötter M, Müller H, Reinecke F, *et al.* The complex structure of polyhydroxybutyrate (PHB) granules: four orthologous and paralogous phasins occur in *Ralstonia eutropha.* Microbiology 2004; 150(Pt 7): 2301-11. [http://dx.doi.org/10.1099/mic.0.26970-0] [PMID: 15256572]

[67] Pfeiffer D, Wahl A, Jendrossek D. Identification of a multifunctional protein, PhaM, that determines number, surface to volume ratio, subcellular localization and distribution to daughter cells of poly(3-hydroxybutyrate), PHB, granules in *Ralstonia eutropha* H16. Mol Microbiol 2011; 82(4): 936-51. [http://dx.doi.org/10.1111/j.1365-2958.2011.07869.x] [PMID: 22023320]

[68] Gerngross TU, Reilly P, Stubbe J, Sinskey AJ, Peoples OP. Immunocytochemical analysis of poly---hydroxybutyrate (PHB) synthase in *Alcaligenes eutrophus* H16: localization of the synthase enzyme at the surface of PHB granules. J Bacteriol 1993; 175(16): 5289-93. [PMID: 8349571]

[69] Pieper-Fürst U, Madkour MH, Mayer F, Steinbüchel A. Purification and characterization of a 14-kilodalton protein that is bound to the surface of polyhydroxyalkanoic acid granules in *Rhodococcus ruber.* J Bacteriol 1994; 176(14): 4328-37. [PMID: 8021220]

[70] Wieczorek R, Pries A, Steinbüchel A, Mayer F. Analysis of a 24-kilodalton protein associated with the polyhydroxyalkanoic acid granules in *Alcaligenes eutrophus.* J Bacteriol 1995; 177(9): 2425-35. [PMID: 7730274]

[71] Grage K, Jahns AC, Parlane N, *et al.* Bacterial polyhydroxyalkanoate granules: biogenesis, structure, and potential use as nano-/micro-beads in biotechnological and biomedical applications. Biomacromolecules 2009; 10(4): 660-9. [http://dx.doi.org/10.1021/bm801394s] [PMID: 19275166]

[72] Jendrossek D, Selchow O, Hoppert M. Poly(3-hydroxybutyrate) granules at the early stages of formation are localized close to the cytoplasmic membrane in *Caryophanon latum.* Appl Environ Microbiol 2007; 73(2): 586-93. [http://dx.doi.org/10.1128/AEM.01839-06] [PMID: 17085698]

[73] Jendrossek D. Fluorescence microscopical investigation of poly(3-hydroxybutyrate) granule formation in bacteria. Biomacromolecules 2005; 6(2): 598-603. [http://dx.doi.org/10.1021/bm049441r] [PMID: 15762619]

[74] Peters V, Rehm BH. *In vivo* monitoring of PHA granule formation using GFP-labeled PHA synthases. FEMS Microbiol Lett 2005; 248(1): 93-100. [http://dx.doi.org/10.1016/j.femsle.2005.05.027] [PMID: 15963662]

[75] Oeding V, Schlegel HG. Beta-ketothiolase from *Hydrogenomonas eutropha* H16 and its significance in the regulation of poly-beta-hydroxybutyrate metabolism. Biochem J 1973; 134(1): 239-48. [http://dx.doi.org/10.1042/bj1340239] [PMID: 4198758]

[76] Senior PJ, Dawes EA. The regulation of poly-beta-hydroxybutyrate metabolism in Azotobacter beijerinckii. Biochem J 1973; 134(1): 225-38. [http://dx.doi.org/10.1042/bj1340225] [PMID: 4723225]

[77] Jendrossek D, Handrick R. Microbial degradation of polyhydroxyalkanoates. Annu Rev Microbiol 2002; 56: 403-32. [http://dx.doi.org/10.1146/annurev.micro.56.012302.160838] [PMID: 12213937]

[78] Doi Y, Segawa A, Kawaguchi Y, Kunioka M. Cyclic nature of poly(3-hydroxyalkanoate) metabolism in Alcaligenes eutrophus. FEMS Microbiol Lett 1990; 55(1-2): 165-9. [http://dx.doi.org/10.1111/j.1574-6968.1990.tb13856.x] [PMID: 2328906]

[79] Taidi B, Mansfield DA, Anderson AJ. Turnover of poly(3-hydroxybutyrate) (PHB) and its influence on the molecular mass of the polymer accumulated by Alcaligenes eutrophus during batch culture. FEMS Microbiol Lett 1995; 129: 201-5.

[80] Knoll M, Hamm TM, Wagner F, Martinez V, Pleiss J. The PHA Depolymerase Engineering Database: A systematic analysis tool for the diverse family of polyhydroxyalkanoate (PHA) depolymerases. BMC Bioinformatics 2009; 10: 89. [http://dx.doi.org/10.1186/1471-2105-10-89] [PMID: 19296857]

[81] Tokiwa Y, Calabia BP. Degradation of microbial polyesters. Biotechnol Lett 2004; 26(15): 1181-9. [http://dx.doi.org/10.1023/B:BILE.0000036599.15302.e5] [PMID: 15289671]

[82] Handrick R, Reinhardt S, Kimmig P, Jendrossek D. The "intracellular" poly(3-hydroxybutyrate) (PHB) depolymerase of *Rhodospirillum rubrum* is a periplasm-located protein with specificity for native PHB and with structural similarity to extracellular PHB depolymerases. J Bacteriol 2004; 186(21): 7243-53. [http://dx.doi.org/10.1128/JB.186.21.7243-7253.2004] [PMID: 15489436]

[83] Handrick R, Reinhardt S, Schultheiss D, *et al.* Unraveling the function of the *Rhodospirillum rubrum* activator of polyhydroxybutyrate (PHB) degradation: the activator is a PHB-granule-bound protein (phasin). J Bacteriol 2004; 186(8): 2466-75. [http://dx.doi.org/10.1128/JB.186.8.2466-2475.2004] [PMID: 15060050]

[84] Handrick R, Technow U, Reichart T, Reinhardt S, Sander T, Jendrossek D. The activator of the *Rhodospirillum rubrum* PHB depolymerase is a polypeptide that is extremely resistant to high temperature (121 ° C) and other physical or chemical stresses. FEMS Microbiol Lett 2004; 230(2): 265-74. [http://dx.doi.org/10.1016/S0378-1097(03)00919-4] [PMID: 14757249]

[85] Jaeger KE, Ransac S, Dijkstra BW, Colson C, van Heuvel M, Misset O. Bacterial lipases. FEMS Microbiol Rev 1994; 15(1): 29-63. [http://dx.doi.org/10.1111/j.1574-6976.1994.tb00121.x] [PMID: 7946464]

[86] Briese BH, Jendrossek D. Biological basis of enzyme-catalyzed polyester degradation: 59 C-terminal amino acids of poly(3-hydroxybutyrate) (PHB) depolymerase a from *Pseudomonas lemoignei* are

sufficient for PHB binding. Macromol Symp 1998; 130(1): 205-16.
[http://dx.doi.org/10.1002/masy.19981300119]

[87] Drosg B, Fritz I, Gattermayr F, Silvestrini L. Photo-autotrophic Production of Poly (hydroxyalkanoates) in Cyanobacteria. Chem Biochem Eng Q 2015; 29(2): 145-56.
[http://dx.doi.org/10.15255/CABEQ.2014.2254]

[88] Koller M, Marsalek L. Cyanobacterial Polyhydroxyalkanoate production: *Status Quo* and *Quo vadis?* Curr Biotechnol 2015; 4(4): 464-80.
[http://dx.doi.org/10.2174/2211550104666150917010849]

[89] Carr NG. The occurrence of poly-beta-hydroxybutyrate in the blue-green alga, *Chlorogloea fritschii.* Biochim Biophys Acta 1966; 120(2): 308-10.
[http://dx.doi.org/10.1016/0926-6585(66)90353-0] [PMID: 5962514]

[90] Jensen TE, Sicko LM. Fine structure of poly-beta-hydroxybutyric acid granules in a blue-green alga, *Chlorogloea fritschii.* J Bacteriol 1971; 106(2): 683-6.
[PMID: 4102335]

[91] Samantaray S, Bhati R, Mallick N. Cyanobacterial polyhydroxyalkanoates: an alternative source for plastics. Cyanobacteria: an economic perspective. Available from: http://medcontent.metapress.com/index/A65RM03P4874243N.pdf 2014; (2): 227-44. 2014;
[http://dx.doi.org/10.1002/9781118402238.ch14]

[92] Rippka R, Neilson A, Kunisawa R, Cohen-Bazire G. Nitrogen fixation by unicellular blue-green algae. Arch Mikrobiol 1971; 76(4): 341-8.
[http://dx.doi.org/10.1007/BF00408530] [PMID: 4996394]

[93] Sharma L, Mallick N. Accumulation of poly-beta-hydroxybutyrate in Nostoc muscorum: regulation by pH, light-dark cycles, N and P status and carbon sources. Bioresour Technol 2005; 96(11): 1304-10.
[http://dx.doi.org/10.1016/j.biortech.2004.10.009] [PMID: 15734319]

[94] Siddiqui PJ, Bergman B, Björkman PO, Carpenter EJ. Ultrastructural and chemical assessment of poly-beta-hydroxybutyric acid in the marine cyanobacterium *Trichodesmium thiebautii.* FEMS Microbiol Lett 1992; 73(1-2): 143-8.
[PMID: 1325936]

[95] Samantaray S, Nayak JK, Mallick N. Wastewater utilization for poly-β-hydroxybutyrate production by the cyanobacterium *Aulosira fertilissima* in a recirculatory aquaculture system. Appl Environ Microbiol 2011; 77(24): 8735-43.
[http://dx.doi.org/10.1128/AEM.05275-11] [PMID: 21984242]

[96] Lama L, Nicolaus B, Calandrelli V, Manca MC, Romano I, Gambacorta A. Effect of growth conditions on endo- and exopolymer biosynthesis in *Anabaena cylindrica* 10 C. Phytochemistry 1996; 42(3): 655-9.
[http://dx.doi.org/10.1016/0031-9422(95)00985-X]

[97] Deschoenmaeker F, Facchini R, Leroy B, Badri H, Zhang C-C, Wattiez R. Proteomic and cellular views of *Arthrospira* sp. PCC 8005 adaptation to nitrogen depletion. Microbiology 2014; 160(Pt 6): 1224-36.
[http://dx.doi.org/10.1099/mic.0.074641-0] [PMID: 24648480]

[98] Stal LJ, Heyer H, Jacobs G. Occurrence and role of poly-hydroxy-alkanoate in the cyanobacterium *Oscillatoria limosa*. Novel Biodegradable Microbial Polymers. Springer 1990; pp. 435-8. [http://dx.doi.org/10.1007/978-94-009-2129-0_38]

[99] Vincenzini M, Sili C, de Philippis R, Ena A, Materassi R. Occurrence of poly-beta-hydroxybutyrate in *Spirulina* species. J Bacteriol 1990; 172(5): 2791-2. [PMID: 2332408]

[100] Shrivastav A, Mishra SK, Mishra S. Polyhydroxyalkanoate (PHA) synthesis by *Spirulina subsalsa* from Gujarat coast of India. Int J Biol Macromol 2010; 46(2): 255-60. [http://dx.doi.org/10.1016/j.ijbiomac.2010.01.001] [PMID: 20060853]

[101] Campbell J III, Stevens SE Jr, Balkwill DL. Accumulation of poly-beta-hydroxybutyrate in *Spirulina platensis*. J Bacteriol 1982; 149(1): 361-3. [PMID: 6798024]

[102] Jau M-H, Yew SP, Toh PS, *et al.* Biosynthesis and mobilization of poly(3-hydroxybutyrate) [P(3HB)] by *Spirulina platensis*. Int J Biol Macromol 2005; 36(3): 144-51. [http://dx.doi.org/10.1016/j.ijbiomac.2005.05.002] [PMID: 16005060]

[103] Panda B, Jain P, Sharma L, Mallick N. Optimization of cultural and nutritional conditions for accumulation of poly-β-hydroxybutyrate in *Synechocystis* sp. PCC 6803. Bioresour Technol 2006; 97(11): 1296-301. [http://dx.doi.org/10.1016/j.biortech.2005.05.013] [PMID: 16046119]

[104] Wu GF, Wu QY, Shen ZY. Accumulation of poly-beta-hydroxybutyrate in cyanobacterium *Synechocystis* sp. PCC6803. Bioresour Technol 2001; 76(2): 85-90. [http://dx.doi.org/10.1016/S0960-8524(00)00099-7] [PMID: 11131804]

[105] Miyake M, Erata M, Asada Y. A thermophilic cyanobacterium, *Synechococcus* sp. MA19, capable of accumulating poly-β-hydroxybutyrate. J Ferment Bioeng 1996; 82(5): 512-4. [http://dx.doi.org/10.1016/S0922-338X(97)86995-4]

[106] Akiyama H, Okuhata H, Onizuka T, *et al.* Antibiotics-free stable polyhydroxyalkanoate (PHA) production from carbon dioxide by recombinant cyanobacteria. Bioresour Technol 2011; 102(23): 11039-42. [http://dx.doi.org/10.1016/j.biortech.2011.09.058] [PMID: 21983412]

[107] Eberly JO, Ely RL. Photosynthetic accumulation of carbon storage compounds under CO□ enrichment by the thermophilic cyanobacterium Thermosynechococcus elongatus. J Ind Microbiol Biotechnol 2012; 39(6): 843-50. [http://dx.doi.org/10.1007/s10295-012-1092-2] [PMID: 22383176]

[108] Gopi K, Balaji S, Muthuvelan B. Isolation, purification and screening of biodegradable polymer PHB producing cyanobacteria from marine and fresh water resources. Iran J Energy & Environ 2014; 5(1): 94-100.

[109] Balaji S, Gopi K, Lavanya B, Muthuvelan B. Isolation and optimization of poly β hydroxybutyrate producing cyanobacterial strains. Int J Appl Biol Pharm Technol 2012; 3(1): 137-45.

[110] Sharma L, Kumar Singh A, Panda B, Mallick N, Singh AK, Panda B, *et al.* Process optimization for poly-β-hydroxybutyrate production in a nitrogen fixing cyanobacterium, *Nostoc muscorum* using

response surface methodology. Bioresour Technol 2007; 98(5): 987-93.
[http://dx.doi.org/10.1016/j.biortech.2006.04.016] [PMID: 16765593]

[111] Lane CE, Benton MG. Detection of the enzymatically-active polyhydroxyalkanoate synthase subunit gene, *phaC*, in cyanobacteria *via* colony PCR. Mol Cell Probes 2015; 29(6): 454-60.
[http://dx.doi.org/10.1016/j.mcp.2015.07.001] [PMID: 26162861]

[112] Banerjee A, Sharma S. Study of biodegradable polyesters from algal sources for use in future textile fiber applications. Proc Annual Internat Conf & Exhibit. 35-44.

[113] Hauf W, Schlebusch M, Hüge J, Kopka J, Hagemann M, Forchhammer K. Metabolic Changes in *Synechocystis* PCC6803 upon Nitrogen-Starvation: Excess NADPH Sustains Polyhydroxybutyrate Accumulation. Metabolites 2013; 3(1): 101-18.
[http://dx.doi.org/10.3390/metabo3010101] [PMID: 24957892]

[114] Monshupanee T, Incharoensakdi A. Enhanced accumulation of glycogen, lipids and polyhydroxybutyrate under optimal nutrients and light intensities in the cyanobacterium *Synechocystis* sp. PCC 6803. J Appl Microbiol 2014; 116(4): 830-8.
[http://dx.doi.org/10.1111/jam.12409] [PMID: 24299499]

[115] Takahashi H, Miyake M, Tokiwa Y, Asada Y. Improved accumulation of poly-3-hydroxybutyrate by a recombinant cyanobacterium. Biotechnol Lett 1998; 20(2): 183-6.
[http://dx.doi.org/10.1023/A:1005392827791]

[116] Sharma L, Panda B, Singh AK, Mallick N. Studies on poly-beta-hydroxybutyrate synthase activity of *Nostoc muscorum*. J Gen Appl Microbiol 2006; 52(4): 209-14.
[http://dx.doi.org/10.2323/jgam.52.209] [PMID: 17116969]

[117] Bhati R, Mallick N. Poly (3-hydroxybutyrate- co -3-hydroxyvalerate) copolymer production by the diazotrophic cyanobacterium *Nostoc muscorum* Agardh: Process optimization and polymer characterization. Algal Res 2015; 7: 78-85.
[http://dx.doi.org/10.1016/j.algal.2014.12.003]

[118] Balaji S, Gopi K, Muthuvelan B. A review on production of poly β hydroxybutyrates from cyanobacteria for the production of bio plastics. Algal Res 2013; 2(3): 278-85.
[http://dx.doi.org/10.1016/j.algal.2013.03.002]

[119] Aguirre von Wobeser E, Ibelings BW, Bok J, Krasikov V, Huisman J, Matthijs HC. Concerted changes in gene expression and cell physiology of the cyanobacterium *Synechocystis* sp. strain PCC 6803 during transitions between nitrogen and light-limited growth. Plant Physiol 2011; 155(3): 1445-57.
[http://dx.doi.org/10.1104/pp.110.165837] [PMID: 21205618]

[120] Krasikov V, Aguirre von Wobeser E, Dekker HL, Huisman J, Matthijs HC. Time-series resolution of gradual nitrogen starvation and its impact on photosynthesis in the cyanobacterium *Synechocystis* PCC 6803. Physiol Plant 2012; 145(3): 426-39.
[http://dx.doi.org/10.1111/j.1399-3054.2012.01585.x] [PMID: 22289076]

[121] Osanai T, Imamura S, Asayama M, *et al.* Nitrogen induction of sugar catabolic gene expression in *Synechocystis* sp. PCC 6803. DNA Res 2006; 13(5): 185-95.
[http://dx.doi.org/10.1093/dnares/dsl010] [PMID: 17046957]

[122] Toh PS, Jau M-H, Yew S-P, Abed RM, Sudesh K. Comparison of Polyhydroxyalkanoates biosynthesis, mobilization and the effects on cellular morphology in *Spirulina platensis* and *Synechocystis* sp. UNIWG. J Biosci 2008; 19(2): 21-38.

[123] Yew SP, Jau M-H, Yong KH, Abed RM, Sudesh K. Morphological Studies of *Synechocystis* sp. UNIWG under Polyhydroxyalkanoate accumulating conditions. Malaysian J Microbiol 2005; 1(1): 48-52.

[124] Samantaray S, Mallick N. Impact of various stress conditions on Poly-β-hydroxybutyrate (PHB) accumulation in *Aulosira fertilissima* CCC 444. Curr Biotechnol 2015; 04(3): 366-72. [http://dx.doi.org/10.2174/2211550104666150806000642]

[125] Allen MM. Simple conditions for growth of unicellular blue-green algae in plates. J Phycol 1968; 4(1): 1-4. [http://dx.doi.org/10.1111/j.1529-8817.1968.tb04667.x] [PMID: 27067764]

[126] Oliver RL, Hamilton DP, Brookes JD, Ganf GG. Physiology, blooms and prediction of planktonic cyanobacteria. In: Whitton BA, Ed. Ecology of Cyanobacteria II. SE: Springer Netherlands 2012; Vol. 6: pp. 155-94. [http://dx.doi.org/10.1007/978-94-007-3855-3_6]

[127] de Philippis R, Sili C, Vincenzini M. Glycogen and poly-β-hydroxybutyrate synthesis in *Spirulina maxima.* J Gen Microbiol 1992; 13: 1623-8. [http://dx.doi.org/10.1099/00221287-138-8-1623]

[128] Nishioka M, Nakai K, Miyake M, Asada Y, Taya M. Production of poly-β-hydroxybutyrate by thermophilic cyanobacterium, *Synechococcus* sp. MA19, under phosphate-limited conditions. Biotechnol Lett 2001; 3: 1095-9. [http://dx.doi.org/10.1023/A:1010551614648]

[129] Panda B, Sharma L, Mallick N. Poly-beta-hydroxybutyrate accumulation in *Nostoc muscorum* and *Spirulina platensis* under phosphate limitation. J Plant Physiol 2005; 162(12): 1376-9. [http://dx.doi.org/10.1016/j.jplph.2005.05.002] [PMID: 16425456]

[130] Haase SM, Huchzermeyer B, Rath T. PHB accumulation in *Nostoc muscorum* under different carbon stress situations. J Appl Phycol 2012; 24(2): 157-62. [http://dx.doi.org/10.1007/s10811-011-9663-6]

[131] Samantaray S, Mallick N. Role of cultural variables in tailoring poly(3- hydroxybutyrate-co-3-hydroxyvalerate) copolymer synthesis in the diazotrophic cyanobacterium *Aulosira fertilissima* CCC 444. J Appl Phycol 2015; 27(1): 197-203. [http://dx.doi.org/10.1007/s10811-014-0299-1]

[132] Kurian D, Jansèn T, Mäenpää P. Proteomic analysis of heterotrophy in *Synechocystis* sp. PCC 6803. Proteomics 2006; 6(5): 1483-94. [http://dx.doi.org/10.1002/pmic.200500413] [PMID: 16421936]

[133] Anderson SL, McIntosh L. Light-activated heterotrophic growth of the cyanobacterium *Synechocystis* sp. strain PCC 6803: a blue-light-requiring process. J Bacteriol 1991; 173(9): 2761-7. [PMID: 1902208]

[134] Zhang S, Bryant DA. Biochemical validation of the glyoxylate cycle in the cyanobacterium

Chlorogloeopsis fritschii strain PCC 9212. J Biol Chem 2015; 290(22): 14019-30. [http://dx.doi.org/10.1074/jbc.M115.648170] [PMID: 25869135]

[135] Stal LJ. Poly (hydroxyalkanoate) in cyanobacteria: an overview. FEMS Microbiol Lett 1992; 103(2-4): 169-80.
[http://dx.doi.org/10.1111/j.1574-6968.1992.tb05835.x]

[136] Ducat DC, Way JC, Silver PA. Engineering cyanobacteria to generate high-value products. Trends Biotechnol 2011; 29(2): 95-103.
[http://dx.doi.org/10.1016/j.tibtech.2010.12.003] [PMID: 21211860]

[137] Lau N-S, Matsui M, Abdullah AA-A. Cyanobacteria: Photoautotrophic microbial factories for the sustainable synthesis of industrial products. Biomed Res Int 2015.

[138] Zhang S, Liu Y, Bryant DA. Metabolic engineering of *Synechococcus* sp. PCC 7002 to produce poly-3-hydroxybutyrate and poly-3-hydroxybutyrate-co-4-hydroxybutyrate. Metab Eng 2015; 32: 174-83. [http://dx.doi.org/10.1016/j.ymben.2015.10.001] [PMID: 26474789]

[139] Woodger FJ, Bryant DA, Price GD. Transcriptional regulation of the CO_2-concentrating mechanism in a euryhaline, coastal marine cyanobacterium, *Synechococcus* sp. Strain PCC 7002: role of NdhR/CcmR. J Bacteriol 2007; 189(9): 3335-47.
[http://dx.doi.org/10.1128/JB.01745-06] [PMID: 17307862]

[140] Chen GQ, Wu Q. Microbial production and applications of chiral hydroxyalkanoates. Appl Microbiol Biotechnol 2005; 67(5): 592-9.
[http://dx.doi.org/10.1007/s00253-005-1917-2] [PMID: 15700123]

[141] Horvat P, Koller M, Braunegg G. Recent advances in elementary flux modes and yield space analysis as useful tools in metabolic network studies. World J Microbiol Biotechnol 2015; 31(9): 1315-28. [http://dx.doi.org/10.1007/s11274-015-1887-1] [PMID: 26066363]

[142] Yu Y, You L, Liu D, Hollinshead W, Tang YJ, Zhang F. Development of *Synechocystis* sp. PCC 6803 as a phototrophic cell factory. Mar Drugs 2013; 11(8): 2894-916.
[http://dx.doi.org/10.3390/md11082894] [PMID: 23945601]

[143] Taroncher-Oldenburg G, Nishina K, Stephanopoulos G. Identification and analysis of the polyhydroxyalkanoate-specific β-ketothiolase and acetoacetyl coenzyme A reductase genes in the cyanobacterium *Synechocystis* sp. strain PCC6803. Appl Environ Microbiol 2000; 66(10): 4440-8. [http://dx.doi.org/10.1128/AEM.66.10.4440-4448.2000] [PMID: 11010896]

[144] Panda B, Mallick N. Enhanced poly-β-hydroxybutyrate accumulation in a unicellular cyanobacterium, *Synechocystis* sp. PCC 6803. Lett Appl Microbiol 2007; 44(2): 194-8.
[http://dx.doi.org/10.1111/j.1472-765X.2006.02048.x] [PMID: 17257260]

[145] Sudesh K, Taguchi K, Doi Y. Effect of increased PHA synthase activity on polyhydroxyalkanoates biosynthesis in *Synechocystis* sp. PCC6803. Int J Biol Macromol 2002; 30(2): 97-104.
[http://dx.doi.org/10.1016/S0141-8130(02)00010-7] [PMID: 11911900]

[146] Hondo S, Takahashi M, Osanai T, *et al.* Genetic engineering and metabolite profiling for overproduction of polyhydroxybutyrate in cyanobacteria. J Biosci Bioeng 2015; 120(5): 510-7. [http://dx.doi.org/10.1016/j.jbiosc.2015.03.004] [PMID: 26055446]

[147] Okamura E, Tomita T, Sawa R, Nishiyama M, Kuzuyama T. Unprecedented acetoacetyl-coenzyme A

synthesizing enzyme of the thiolase superfamily involved in the mevalonate pathway. Proc Natl Acad Sci USA 2010; 107(25): 11265-70.
[http://dx.doi.org/10.1073/pnas.1000532107] [PMID: 20534558]

[148] Lau N-S, Foong CP, Kurihara Y, Sudesh K, Matsui M. RNA-Seq analysis provides insights for understanding photoautotrophic polyhydroxyalkanoate production in recombinant *Synechocystis* Sp. PLoS One 2014; 9(1): e86368.
[http://dx.doi.org/10.1371/journal.pone.0086368] [PMID: 24466058]

[149] Nakaya Y, Iijima H, Takanobu J, Watanabe A, Hirai MY, Osanai T. One day of nitrogen starvation reveals the effect of sigE and rre37 overexpression on the expression of genes related to carbon and nitrogen metabolism in *Synechocystis* sp. PCC 6803. J Biosci Bioeng 2015; 120(2): 128-34.
[http://dx.doi.org/10.1016/j.jbiosc.2014.12.020] [PMID: 25641577]

[150] Osanai T, Numata K, Oikawa A, *et al.* Increased bioplastic production with an RNA polymerase sigma factor SigE during nitrogen starvation in *Synechocystis* sp. PCC 6803. DNA Res 2013; 20(6): 525-35.
[http://dx.doi.org/10.1093/dnares/dst028] [PMID: 23861321]

[151] Kunasundari B, Sudesh K. Isolation and recovery of microbial polyhydroxyalkanoate. Express Polym Lett 2011; 5(7): 620-6.
[http://dx.doi.org/10.3144/expresspolymlett.2011.60]

[152] Koller M, Bona R, Chiellini E, Braunegg G. Extraction of short-chain-length poly-[(*R*)-hydroxyalkanoates] (*scl*-PHA) by the "anti-solvent" acetone under elevated temperature and pressure. Biotechnol Lett 2013; 35(7): 1023-8.
[http://dx.doi.org/10.1007/s10529-013-1185-7] [PMID: 23525946]

[153] Wang B, Pugh S, Nielsen DR, Zhang W, Meldrum DR. Engineering cyanobacteria for photosynthetic production of 3-hydroxybutyrate directly from CO_2. Metab Eng 2013; 16(1): 68-77.
[http://dx.doi.org/10.1016/j.ymben.2013.01.001] [PMID: 23333586]

[154] Balasundaram B, Harrison S, Bracewell DG. Advances in product release strategies and impact on bioprocess design. Trends Biotechnol 2009; 27(8): 477-85.
[http://dx.doi.org/10.1016/j.tibtech.2009.04.004] [PMID: 19573944]

[155] Herrero M, Ibáñez E. Green processes and sustainability: An overview on the extraction of high added-value products from seaweeds and microalgae. J Supercrit Fluids 2015; 96: 211-6.
[http://dx.doi.org/10.1016/j.supflu.2014.09.006]

[156] Balasundaram B, Skill SC, Llewellyn CA. A low energy process for the recovery of bioproducts from cyanobacteria using a ball mill. Biochem Eng J 2012; 69: 48-56.
[http://dx.doi.org/10.1016/j.bej.2012.08.010]

[157] Hahn SK, Chang YK, Kim BS, Chang HN. Optimization of microbial poly(3-hydroxybutyrate) recover using dispersions of sodium hypochlorite solution and chloroform. Biotechnol Bioeng 1994; 44(2): 256-61.
[http://dx.doi.org/10.1002/bit.260440215] [PMID: 18618692]

[158] Ryu HW, Cho KS, Lee EG, Chang YK. Recovery of poly(3-hydroxybutyrate) from coagulated Ralstonia eutropha using a chemical digestion method. Biotechnol Prog 2000; 16(4): 676-9.
[http://dx.doi.org/10.1021/bp000054c] [PMID: 10933846]

[159] de Morais MG, Stillings C, Dersch R, *et al.* Extraction of poly(3-hydroxybutyrate) from *Spirulina* LEB 18 for developing nanofibers. Polímeros 2015; 25(2): 161-7. [http://dx.doi.org/10.1590/0104-1428.1686]

[160] Ughy B, Nagy CI, Kós PB. Biomedical potential of cyanobacteria and algae. Acta Biol Szeged 2015; 59(2): 203-24.

[161] Matsui MS, Muizzuddin N, Arad S, Marenus K. Sulfated polysaccharides from red microalgae have antiinflammatory properties *in vitro* and *in vivo*. Appl Biochem Biotechnol 2003; 104(1): 13-22. [http://dx.doi.org/10.1385/ABAB:104:1:13] [PMID: 12495202]

[162] Mishima T, Murata J, Toyoshima M, *et al.* Inhibition of tumor invasion and metastasis by calcium spirulan (Ca-SP), a novel sulfated polysaccharide derived from a blue-green alga, *Spirulina platensis*. Clin Exp Metastasis 1998; 16(6): 541-50. [http://dx.doi.org/10.1023/A:1006594318633] [PMID: 9872601]

[163] Teixeira RE. Energy-efficient extraction of fuel and chemical feedstocks from algae. Green Chem. Royal Society of Chemistry 2012; 14(2): 419-27.

[164] Kim Y-H, Choi Y-K, Park J, *et al.* Ionic liquid-mediated extraction of lipids from algal biomass. Bioresour Technol 2012; 109: 312-5. [http://dx.doi.org/10.1016/j.biortech.2011.04.064] [PMID: 21601447]

[165] Kobayashi D, Fujita K, Nakamura N, Ohno H. A simple recovery process for biodegradable plastics accumulated in cyanobacteria treated with ionic liquids. Appl Microbiol Biotechnol 2015; 99(4): 1647-53. [http://dx.doi.org/10.1007/s00253-014-6234-1] [PMID: 25432673]

[166] Li W-J, Laurencin CT, Caterson EJ, Tuan RS, Ko FK. Electrospun nanofibrous structure: a novel scaffold for tissue engineering. J Biomed Mater Res 2002; 60(4): 613-21. [http://dx.doi.org/10.1002/jbm.10167] [PMID: 11948520]

[167] Yoshimoto H, Shin YM, Terai H, Vacanti JP. A biodegradable nanofiber scaffold by electrospinning and its potential for bone tissue engineering. Biomaterials 2003; 24(12): 2077-82. [http://dx.doi.org/10.1016/S0142-9612(02)00635-X] [PMID: 12628828]

[168] Fridrikh SV, Yu JH, Brenner MP, Rutledge GC. Controlling the fiber diameter during electrospinning. Phys Rev Lett 2003; 90(14): 144502. [http://dx.doi.org/10.1103/PhysRevLett.90.144502] [PMID: 12731920]

[169] Zeng J, Xu X, Chen X, *et al.* Biodegradable electrospun fibers for drug delivery. J Control Release 2003; 92(3): 227-31. [http://dx.doi.org/10.1016/S0168-3659(03)00372-9] [PMID: 14568403]

[170] Gungor-Ozkerim PS, Balkan T, Kose GT, Sarac AS, Kok FN. Incorporation of growth factor loaded microspheres into polymeric electrospun nanofibers for tissue engineering applications. J Biomed Mater Res A 2014; 102(6): 1897-908. [http://dx.doi.org/10.1002/jbm.a.34857] [PMID: 23852885]

[171] Said SS, Pickering JG, Mequanint K. Controlled delivery of fibroblast growth factor-9 from biodegradable poly(ester amide) fibers for building functional neovasculature. Pharm Res 2014; 31(12): 3335-47.

[http://dx.doi.org/10.1007/s11095-014-1423-2] [PMID: 24852896]

[172] Tseng Y-Y, Liu S-J. Nanofibers used for the delivery of analgesics. Nanomedicine (Lond) 2015; 10(11): 1785-800.
[http://dx.doi.org/10.2217/nnm.15.23] [PMID: 26080700]

[173] Fu R, Li C, Yu C, Xie H, Shi S, Li Z, *et al.* A novel electrospun membrane based on moxifloxacin hydrochloride/poly(vinyl alcohol)/sodium alginate for antibacterial wound dressings in practical application. Drug Deliv 2015; 1-12.
[PMID: 24870202]

[174] Ignatova M, Manolova N, Rashkov I. Electrospun antibacterial chitosan-based fibers. Macromol Biosci 2013; 13(7): 860-72.
[http://dx.doi.org/10.1002/mabi.201300058] [PMID: 23754600]

[175] Armentano I, Marinucci L, Dottori M, *et al.* Novel poly(L-lactide) PLLA/SWNTs nanocomposites for biomedical applications: material characterization and biocompatibility evaluation. J Biomater Sci Polym Ed 2011; 22(4-6): 541-56.
[http://dx.doi.org/10.1163/092050610X487873] [PMID: 20566045]

[176] Bianco A, Del Gaudio C, Baiguera S, *et al.* Microstructure and cytocompatibility of electrospun nanocomposites based on poly(epsilon-caprolactone) and carbon nanostructures. Int J Artif Organs 2010; 33(5): 271-82.
[PMID: 20593348]

[177] Migneco F, Huang Y-C, Birla RK, Hollister SJ. Poly(glycerol-dodecanoate), a biodegradable polyester for medical devices and tissue engineering scaffolds. Biomaterials 2009; 30(33): 6479-84.
[http://dx.doi.org/10.1016/j.biomaterials.2009.08.021] [PMID: 19712970]

[178] Kang IK, Xing ZC, Han SJ, Shin YS. Fabrication of biodegradable polyester nanocomposites by electrospinning for tissue engineering. J Nanomater 2011.

[179] Henry JA, Simonet M, Pandit A, Neuenschwander P. Characterization of a slowly degrading biodegradable polyester-urethane for tissue engineering scaffolds. J Biomed Mater Res A 2007; 82(3): 669-79.
[http://dx.doi.org/10.1002/jbm.a.31094] [PMID: 17323319]

[180] de Jesus Raposo MF, de Morais RM, de Morais AM. Health applications of bioactive compounds from marine microalgae. Life Sci 2013; 93(15): 479-86.
[http://dx.doi.org/10.1016/j.lfs.2013.08.002] [PMID: 23994664]

[181] de Morais MG, Stillings C, Dersch R, *et al.* Biofunctionalized nanofibers using Arthrospira (Spirulina) biomass and biopolymer. Biomed Res Int 2015.

[182] de Morais MG, Vaz B da S, de Morais EG, Costa JAV. Biological effects of *Spirulina* (*Arthrospira*) biopolymers and biomass in the development of nanostructured scaffolds. Biomed Res Int 2014.

[183] de Morais MG, Stillings C, Dersch R, *et al.* Preparation of nanofibers containing the microalga *Spirulina* (*Arthrospira*). Bioresour Technol 2010; 101(8): 2872-6.
[http://dx.doi.org/10.1016/j.biortech.2009.11.059] [PMID: 20056537]

[184] Suwantong O, Waleetorncheepsawat S, Sanchavanakit N, *et al. In vitro* biocompatibility of electrospun poly(3-hydroxybutyrate) and poly(3-hydroxybutyrate-co-3-hydroxyvalerate) fiber mats.

Int J Biol Macromol 2007; 40(3): 217-23.
[http://dx.doi.org/10.1016/j.ijbiomac.2006.07.006] [PMID: 16949148]

[185] Hufenus R, Reifler FA, Fernández-Ronco MP, Heuberger M. Molecular orientation in melt-spun poly(3-hydroxybutyrate) fibers: Effect of additives, drawing and stress-annealing. Eur Polym J 2015; 71: 12-26.
[http://dx.doi.org/10.1016/j.eurpolymj.2015.07.039]

[186] Asada Y, Miyake M, Miyake J, Kurane R, Tokiwa Y. Photosynthetic accumulation of poly-(hydroxybutyrate) by cyanobacteria--the metabolism and potential for CO_2 recycling. Int J Biol Macromol 1999; 25(1-3): 37-42.
[http://dx.doi.org/10.1016/S0141-8130(99)00013-6] [PMID: 10416648]

[187] Tan G-Y, Chen C-L, Li L, *et al.* Start a Research on Biopolymer Polyhydroxyalkanoate (PHA): A Review. Polymers (Basel) 2014; 6(3): 706-54.
[http://dx.doi.org/10.3390/polym6030706]

CHAPTER 4

Biopolymer Production by Mixed Microbial Cultures: Integrating Remediation with Valorization

Luísa Seuanes Serafim[1,*], **Diogo Queirós**[1], **Simona Rossetti**[2] and **Paulo Costa Lemos**[3]

[1] *CICECO-Aveiro Institute of Materials, Chemistry Department, University of Aveiro, Campus Universitário de Santiago, 3810-193 Aveiro, Portugal*

[2] *Water Research Institute, C.N.R., Via Salaria Km 29,300, 00015 Monterotondo, Italy*

[3] *LAQV, REQUIMTE, Chemistry Department, Faculty of Science and Technology, University Nova of Lisbon, 2829-516 Caparica, Portugal*

Abstract: Polyhydroxyalkanoates (PHA) production by mixed microbial cultures (MMC) became the central point of many research works. Firstly discovered in wastewater treatment systems, namely those related with biological phosphorus removal, it became a process to be considered for polymer production intended to replace petrochemical derived plastics. The ability to store PHA provides microbial communities a competitive advantage for survival under transient conditions typical of waste treatment systems. In this way, MMC can continuously adapt to the operational conditions increasing the number PHA-storing organisms with minimum requirements of sterility. MMC can produce PHA using waste or industrial by-products as substrates, allowing for their valorization. The use of waste, together with the lower requirements in sterility and process control, could signify a decrease on PHA production costs. The price is one of the main drawbacks that prevent the increase of world PHA market share. This chapter intends to contextualize the use of MMC for PHA production, by describing the different approaches and compare them with the traditional processes employing pure or genetically modified organisms. The main processes to select PHA-storing MMC are discussed, namely alternating of anaerobic and aerobic (AN/AE) or

*) **Corresponding author Luísa Seuanes Serafim:** Departamento de Química/CICECO, Universidade de Aveiro, Campus Universitário de Santiago, P-3810-193 Aveiro, Portugal; Tel: +351 234 370 360; Email: luisa.serafim@ua.pt.

Martin Koller (Ed.)

anoxic (AN/AO) conditions and aerobic dynamic feeding (ADF). The use of wastes as substrates, the operational conditions, the microorganisms involved as well as the molecular methods used for their identification are also the focus of this work.

Keywords: ADF, AN/AE, Mixed microbial cultures, Molecular methods, Operational conditions, Polyhydroxyalkanoates, Selection, Valorization, Waste, Waste treatment.

INTRODUCTION

Mixed Culture Definition

It is common knowledge that polyhydroxyalkanoates (PHA)-producing genera are quite ubiquitous in nature. So far, more than 300 were found and part of them cultivated but it is believed that a larger number still remains unknown [1, 2]. Until now, the search for pure and defined cultures with a high PHA-storing capacity has been the objective of most of the researchers working in this field. However, in nature, microorganisms live in community with different relationships among them. Most of the times, the isolation of bacteria with the desirable characteristics is quite difficult. It is believed that more than 99% of microorganisms have not been successfully cultivated yet and, surely, among this number many PHA-producing strains can be found [3]. For the majority of bacteria, the ability of accumulating internal reserves represents a major competitive advantage to survive in natural environments. Carbon-based internal reserves as PHA and glycogen or external like some polysaccharides, as well phosphorus (namely, polyphosphate - PolyP), nitrogen (cyanophycins) or sulfur-based reserves, are usually advantageous for the producing organisms and were found in many different species of bacteria, including those that constitute activated sludge populations, the most known and studied mixed microbial culture (MMC). Activated sludge is a complex consortium of several populations of microorganisms of undefined composition, which depends on feeding and process conditions. Other MMCs can be found in open biological systems that appear in natural environments, as soil, seawater, sediments, rumen, N_2-fixing bacteria, or, as activated sludge, in human created environments under non-sterile conditions, namely ponds, waste storage facilities, *etc.*

Historical Background

Due to its role in wastewater treatment plants (WWTPs), the first MMC to be studied was activated sludge. In early 70s, the presence of PHA granules inside cells belonging to an activated sludge community was observed for the first time [4]. Later on, PHA were found to play a key-role on the competition between polyphosphate-accumulating organisms (PAOs) and glycogen-accumulating organisms (GAOs) for the external carbon source in enhanced biological phosphorus removal (EBPR). In this type of systems, PHA are stored when the external carbon source, usually a short-chain organic acid (SCOA), is supplied in the absence of an external electron sink. The lack of electron sink prevents microbial growth but PHA-storing bacteria are able to store the external carbon source as intracellular reserves to be used as carbon and energy source in the absence of external substrate and/or an electron sink is available [5]. In this way, PAOs and GAOs were the first organisms present in MMC to be described as PHA-storing bacteria. Later on, a correlation between PHA storage and use and the phenotypic identity of PAOs and GAOs was established by using fluorescence *in situ* hybridization (FISH) and postFISH nile blue staining [6]. The first attempts to identify the microorganisms involved in EBPR, including both PAOs and GAOs, started almost 40 years ago, but so far no attempts were successful. Potential methods to successfully isolate these bacteria were tested through the use of *e.g.* micro-manipulation techniques from enriched cultures of these organisms [7]. In the following years, PHA production by MMC systems based on the alternation of availability of electron sink was developed, being the most successful those where GAOs were dominant [8].

Bulking was another aspect of activated sludge correlated with PHA storage observed in aerobic WWTP [9]. The process configuration with alternation of excess of carbon source periods with depletion periods was observed to disfavor the selection of filamentous bacteria responsible for the bulking phenomena and select floc-formers with enhanced PHA storage capacity [9]. The observations of Majone *et al.* (1996) confirmed the hypotheses raised by Daigger and Grady (1982) to explain the phenomenon. After a period of absence of external carbon source, during which the levels of enzymes and RNA required for growth decreased, a sudden increase of external substrate triggered the storage of internal reserves (faster response). Under such conditions, cell growth requires the synthesis of more enzymes and RNA which represents a more complex physiological adaptation resulting in a slower response [10]. This mechanism was the basis for the development of PHA production by MMC based on intermittent substrate supply. The concept of aerobic feast and famine process, later known as aerobic dynamic feeding (ADF), was first proposed by Majone *et al.* (1996) and was widely used in the subsequent years [11 - 16]. Also the previously developed system based on the alternation of availability of oxygen could be considered a feast and famine process, since the external carbon source is available only during the absence of oxygen [17].

Mixed Cultures *versus* Pure Cultures

In recent years, MMC processes gained marked importance and become competitive with pure cultures since they represent a clear alternative that can contribute to the decrease of PHA production costs. While pure culture processes, including those that employ recombinant organisms, rely on tight sterile conditions in order to guarantee the presence of bacteria of interest, this is not an issue for processes using MMC. In MMC processes, the dominance of PHA-storing organisms is imposed by the operational conditions. The operational conditions chosen, usually based on the alternation of periods of excess and lack of carbon substrate, should allow only for the survival of PHA-storing organisms. This situation requires that the selected PHA-storing bacteria should consume completely the external substrate in order to establish true feast and famine conditions. The famine period is relevant since it allows eliminating the non-storing bacteria. Moreover, during this period, the PHA-storing bacteria will use their internal reserves as carbon and energy source, and the decrease of these reserves will trigger a new storage process when the next feast period begins.

The use of complex feedstock as substrates is another advantage of mixed cultures over pure cultures, since a wide range of low value substrates was already successfully tested, including industrial waste of undefined composition or containing known microbial inhibitors [18 - 22]. Also, the lack in SCOAs of most of the feedstock, usually rich in carbohydrates, is not a drawback of MMC processes, since a preliminary acidification performed by anaerobic MMC can be included in the process, in order to convert the organic compounds into the desired substrates. The utilization of inexpensive feedstock and the possibility to work without sterile conditions can contribute to the decrease of PHA production costs when compared with pure culture processes.

Some weaknesses can be appointed to MMC processes, which are contributing to delay their implementation at industrial level. The first one is biomass concentration, which are way below than the values obtained in pure culture processes. The increase on the amount of biomass would improve volumetric productivities and may have a high impact on the downstream process costs, since the extraction yield is directly dependent on the amount of biomass collected [17].

Another drawback traditionally appointed to MMC processes is the composition of the microbial population selected. A wide range of bacteria was already identified in MMC production systems as recently reviewed by Queirós *et al.* (2015), many of them are bacteria traditionally related with pure culture systems [23]. Moreover, the diversity of populations is related with robustness of the systems to cope with situations where the feedstock suffers periodic variations in its composition [1, 8, 24, 25]. The hypothesis that this diversity could reflect negatively on the PHA characteristics is raised quite often, despite some works already showed the opposite, by analyzing molecular weights, polydispersities, melting and glass transition temperatures, melting enthalpies or crystallinity [26, 27]. More work in this field is still required since the characterization of the polymer produced is mandatory for the definition of the final application. Pure cultures allow for a better control of PHA characteristics, which is required by polymer processing industries. Nevertheless, PHA monomeric composition can be controlled by feeding the MMC with defined mixtures of SCOAs, as already showed by several researchers [26, 28]. Different mixtures of SCOAs can result in the production of copolymers of poly-(3-hydrobutyrate-*co*-3-hydroxyvalerate), PHBHV, with different composition and also other monomers as 3-hydroxy-2-methylbutyrate (3H2MB), 3-hydroxy-2-methylvalerate (3H2MV) or 3-hydroxyhexanoate (3HHx) were already obtained in MMC processes [8]. The variety of monomers that can be obtained by MMC is very low when compared with pure cultures, since monomers from 3 to 14 carbons were already obtained as well with a higher diversity of side groups [29 - 31].

The downstream processing is the one of the main drawbacks of PHA production by MMC. Despite the many methods tested [32], including some of those working with pure cultures, the first positive results obtained were recently published [33 - 35]. Cell wall and floc structure of MMC biomass seem to be much more complex and harder to disrupt than of pure cultures that become fragile when high PHA storage contents are attained. Still, researchers are looking for a low-cost, highly efficient, and environmentally friendly PHA recovery process that allows for high purity without polymer degradation.

Since the MMC process is based on the use of industrial waste or subproducts and microbial populations collected in WWTP facilities, the possibility of implementing the PHA production units on those locations should be explored as a cost-effective technology. However, only a few pilot plants were already being tested in the last five years [36 - 40] which is far from the situation of pure culture processes, already working at industrial scale. Moreover, the origin of PHA produced could raise some concerns about the final application of the polymer since some of the components of the feedstock used as substrate could be impurities hard to eliminate from the extracted product. Also, the use of microbial populations selected in sewage treatment systems could be a problem for the commercialization. In these situations, the final application of the polymer cannot be the traditional applications of PHA produced by pure cultures, as medical field and food and cosmetics packaging and disposable items, but less noble applications in agriculture and packaging of non-food items.

The objective of this chapter is to exploit the different aspects related with PHA production by MMC.

METABOLISM

Few specific works concerning the metabolism of PHA synthesis using MMCs are available in the literature. For most of the cases, it is assumed that MMCs perform in the same way as pure cultures, despite MMC preferred substrates for PHA production are SCOAs, as acetate, propionate, butyrate and valerate, resulting frequently in short-chain PHA.

Two types of MMC present in activated sludge processes involving AN/AE steps were characterized in terms of PHA metabolism. PAOs are the responsible for phosphorus sequestration from wastewaters, being PHA important intermediates in this process. GAOs also strive in such systems differing from the previous organisms mainly in the form of obtaining energy, where polyP is not involved, competing for the same carbon substrates. Both uptake SCOAs converting them to the respective hydroxyacyl-CoA and being polymerized to produce short-chain PHA. Under anaerobic conditions, glycogen plays a major role both as provider of part of the energy and reducing equivalents used, but also as a source of monomers for PHA synthesis, as acetyl or propionyl moieties [5, 41 - 43]. Both polymers are interconnected since while under anaerobic conditions PHA are produced with glycogen consumption, under aerobic condition carbon contained in the previously produced PHA are converted to glycogen.

The most well studied pathways involve the production of poly-3-hydroxyalkanoates. The consumption of acetate for the production of poly-3-hydroxybutyrate (PHB) by *Cupriavidus necator* is the most common example. Here acetate is converted to acetyl-CoA, two of such units condensed to acetoacetyl-CoA by the action of AY-ketothiolase, further processed to R-3-hydroxybutyryl-CoA (3HB) using acetoacetyl-CoA reductase, being this monomer added to the growing polymeric chain of PHB by PHA synthase. Other SCOAs, as butyrate and valerate, are also converted to their respective 3-hydroxyacyl-CoA (3HB and 3-hydroxyvaleryl-CoA (3HV), respectively). From propionate more than one monomer can be obtained, as combining two units of propionyl-CoA, 3H2MV can be obtained, while when one acetyl moiety combines with one propionyl moiety 3H2MB results.

The consumption of sugars and longer-chain organic acids can be performed in different ways. Sugars can be converted by glycolysis to pyruvate and further transformed into acetyl or propionyl moieties following the previously described scheme or the obtained acetyl-CoA introduced in the *de novo* fatty acids biosynthesis (Fig. **1**). The *R*-3-hydroxyacyl-ACP intermediates of this pathway can be processed by a transacylase followed by a reductase to the corresponding *R*-3-hydroxyacyl-CoAs. Fatty acids can also be consumed in this pathway for monomer production in PHA synthesis but can likewise be processed through the catabolic fatty acids pathway, β-oxidation. Concerning β-oxidation, several of its intermediates can be drifted for R-3-hydroxyacyl-CoA production using enzymes as reductases, hydratases and epimerases. From these two substrates both short-chain length (*scl*) and medium-chain length (*mcl*) monomers can result.

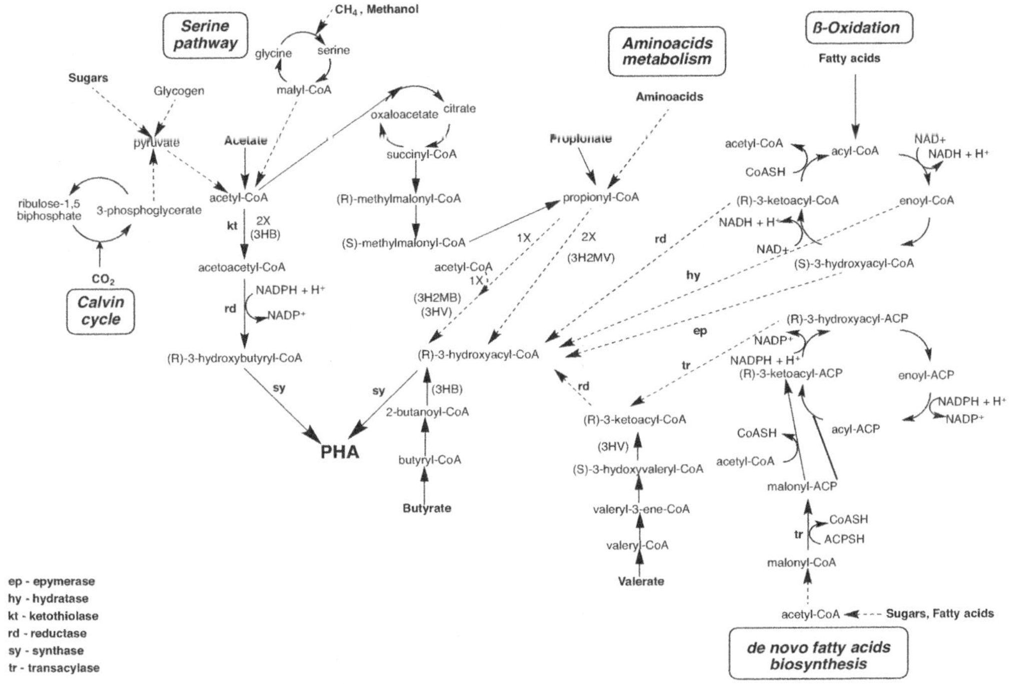

Fig. (1). Metabolic pathways for poly-3-hydroxyalkanoates production from different carbon sources (adapted from Serafim *et al.*(2008) and Tan *et al.*(2014) [17, 31]).

The type of PHA synthase of the microorganism determines the type of PHA produced, *scl*- or *mcl*-PHA. So far four different types of synthases are known, based on the nucleotide sequences of 59 PHA synthase genes obtained from 44 different bacteria [44]. This classification is based on the substrate specificity and subunit structure. The type organisms for PHA synthases Class I, II, III and IV are *Cupriavidus necator*, *Pseudomonas aeruginosa*, *Allochromatium vinosum* and *Bacillus megaterium*, respectively. As for substrate specificity Class I and III PHA synthases incorporate only short-chain-length hydroxyacyl monomers (C3 - C5), while class II PHA synthases utilize medium-chain-length hydroxyacyl monomers (C6 - C14). Class IV synthases are capable of polymerizing mainly short-chain-length monomer units but some works reported the ability of *B. megaterium* DSM 509, *Bacillus* sp. US163, and *Bacillus* sp. US177 to produce medium- or long-chain length PHA. Some diversity seems to exist among Class IV synthases since *B. cereus* and lineages have a broader substrate specificity producing *scl*-PHA from a wider variety of monomers than the class representative *B. megaterium* and its lineages [45]. Concerning the subunits structure Classes I and II have only one subunit (PhaC) whereas Class III is composed of two subunits, namely, a catalytic subunit PhaC and a second subunit PhaE. Class IV has also two subunits a catalytic subunit PhaC and a second subunit PhaR. Based on the differences on the nucleic acid sequences of Class IV PhaC Solaiman and Ashby suggested that this class could be divided into at least two subgroups one defined by *B. cereus* and another by *B. megaterium* [46].

Other carbon compounds were also investigated as possible substrates for PHA production. Aminoacids, as valine and isoleucine, were also shown to be involved in PHA production by a strain of *Alcaligenes eutrophus* (presently *Cupriavidus necator*) where the aminoacids provided the propionyl moiety present in the obtained copolymer of PHBHV [47]. Substrates with a single carbon have also been considered, namely for their low cost, as carbon dioxide, methane and methanol. Hydrogen-oxidizing bacteria, as *Cupriavidus necator* and *Ideonella* sp., can utilize H_2 as electron donor and O_2 as electron acceptor to fix CO_2 and incorporate into bacterial biomass [48, 49]. These autotrophs recur to the enzyme RUBISCO present in the Calvin cycle from which 3-phosphoglycerate can be drifted for central metabolism *via* pyruvate, or to the utilization of the reverse tricarboxylic acid cycle, providing acetyl-CoA moieties. Methane and methanol can be utilized by methylotrophs for PHA production using either the serine pathway or the ribulose monophosphate (RuMP) pathway. The serine pathway is the most utilized for PHB production namely by *Methylocystis* and *Methylosinus*, *Methylococcus*, and *Methylomonas* spp [50]. The utilization of CO_2 and CH_4 apart from a cheap substrate for PHA production could also help to mitigate the growing increase of greenhouse gases in the atmosphere.

Due to the high diversity of monomers, naturally occurring or chemically synthetized, PHA with monomers other than 3-hydroxyalkanoates can be produced. The presence of 4-hydroxybutyrate monomers (4HB) in PHA, with high utilization on biomedical applications, can be obtained from the use of 4-hydroxybutyric acid, γ-butyrolactone, 1,4-butanediol and other I, ω-alkanediols as carbon sources [51]. Other examples are the utilization of cyclohexanol to the production of 6-hydroxyhexanoyl-CoA and of 4,5-alkanolactone to obtain 4,5-hydroxyacyl-CoA (4,5-HA-CoA) [52].

STRATEGIES OF PHA PRODUCTION BY MMC

In the recent years, numerous approaches were studied for the production of PHA, making use of a variety of organisms such as bacteria, plants and fungi. Also, with the advent of genetic engineering, the use of several recombinant organisms become possible, like the well-known recombinant *Escherichia coli* or new transgenic plants [53]. With the aim of achieving the industrial production, several considerations can be taken into account: choosing cheap and suitable substrates in accordance with the metabolism of the producing organism; its storage capacity; the need for sterile conditions; type of reactor, feeding pattern; oxygen demand; behavior and evolution of the culture.

Over than 300 bacterial species were reported as PHA producers but only 75 were tested for large-scale production. Currently, industrial production processes rely on pure cultures of microorganisms in their wild form, or genetically modified, namely, *Cupriavidus necator*, *Alcaligenes latus*, *Comamonas acidovorans*, *Pseudomonas putida*, *Chromobacterium* sp., *Hydrogenophaga pseudoflava*, *Rhodococcus rubber*, *Azotobacter vinelandii*, *Pseudomonas oleovorans*, *Paracoccus denitrificans*, *Protomonas extorquens* and recombinant *Escherichia coli* [54]. Genetic engineering allowed for the construction of recombinant strains for a more cost-effective PHA production. By cloning PHA synthesis genes from several organisms, strains showing simultaneously fast growth and high cell density from inexpensive substrates and simple polymer purification, can be obtained. However, the use of pure culture and genetically modified organisms still represent high production costs and, as mentioned before, one way to overcome them is the use of MMC.

First Attempts

In PHA production by MMC, the selection of strains with high capacity to accumulate intracellular reserves results from the imposition of transient conditions. As mentioned before, the production of PHA by MMC was first verified in the EBPRs facilities [17]. Activated sludge processes with alternating anaerobic and aerobic conditions have been successfully used for EBPR from wastewater. Such EBPR processes are referred to as the anaerobic-aerobic (AN/AE) or anaerobic-oxic (AN/AO) process [55]. This process comprises two distinct phases. The anaerobic stage in which absolute anaerobic conditions are kept with neither oxygen nor NO_2^-/NO_3^- available as electron acceptor. During this stage, organic substrates from the phosphorus-enriched influent wastewater are taken up by PAOs with simultaneous phosphorus release, from stored PolyP, to the bulk medium. Energy from this hydrolysis is used to build PHA as a consequence of oxygen limitation in the presence of external organic carbon (Fig. **2**). The second stage is characterized by the presence of an external donor acceptor, oxygen or nitrate, and PAOs can utilize the stored PHA to grow and restore glycogen reserves and replenish PolyP from the soluble phosphate of wastewater. Depending on the conditions and substrates available in such systems, also GAOs can be enriched [7].

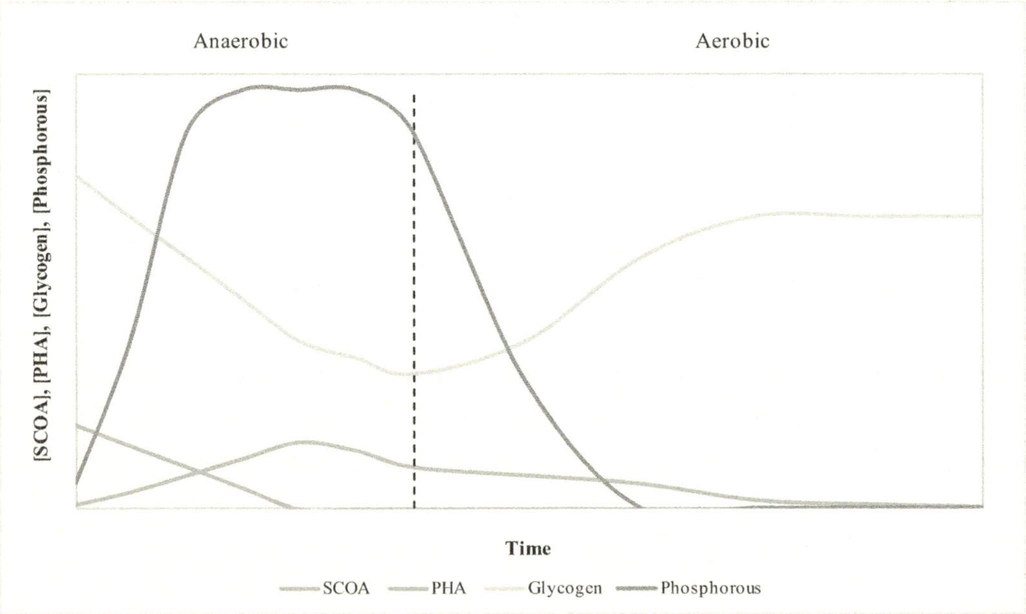

Fig. (2). Example of an anaerobic/aerobic cycle operation by PAOs.

In contrast, several low/zero Pi-content waste- and by-products can be used to enrich a culture in more efficient PHA-storing microorganisms. Depriving sludge from Pi, the community will channel the organic compounds towards PHA accumulation through glycogen consumption, maximizing its content in order to survive the aerobic stage without external substrate available (Fig. **3**). For this reason, GAOs gained importance as PHA-storing organisms in MMC processes [8, 56]. Nevertheless, the amount of PHA accumulated by these groups of microorganisms is generally lower than those selected by aerobic dynamic feeding conditions (ADF) further exploit in the next section [1].

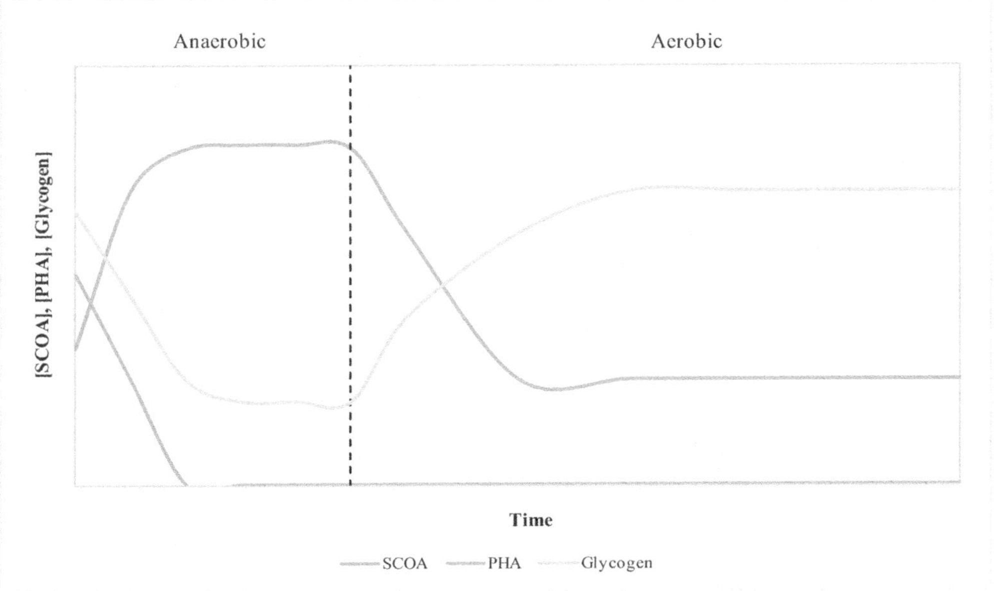

Fig. (3). Example of an anaerobic/aerobic cycle operation by GAOs.

Another process for PHA production is the microaerophilic-aerobic system, which was a modification of the AN/AE one. In this process an aerobic period alternates with microaerophilic conditions, where a very limited amount of oxygen is supplied. In this system, microorganisms consume organic substrates that allow for the achievement of energy through oxidative degradation. If the supply of oxygen is sufficient, microorganisms are able to get enough energy for assimilative activities such as the production of protein, glycogen, along with consuming of organic substrates. If the supply of oxygen is adequately controlled, the assimilative activity will be suppressed and the external substrate is drifted to PHA accumulation [57]. As PHA production requires less energy than needed for glycogen synthesis, the oxygen supplied should be around 0.51% of the chemical oxygen demand (COD) provided, in order to allow bacteria to store high amounts of PHA intracellularly [57]. Pratt *et al.* (2012) also observed that a higher fraction of substrate was accumulated as PHA in comparison to high dissolved oxygen (DO) conditions. However, the time required to accumulate the same PHA amount as in high DO conditions was three times longer [58].

An alternative to both processes described is the ADF system, also known as feast and famine, the MMC, under aerobic conditions, is submitted to alternating periods of external substrate availability (feast) and unavailability (famine) resulting in unbalanced growth. The ADF process will be described in more detail in the next sub-chapter, since it is the most studied and developed in the last years.

Three-Step Process

In order to become economically competitive with pure cultures, low-value substrates, such as agro-forestry residues, industrial by-products, and industrial and urban wastes, can be repurposed for PHA production by MMC. Nonetheless, these low-value substrates need to be converted into readily biodegradable carbon sources to be transformed into PHA by microorganisms in MMC. In 2004, Dionisi *et al.* proposed a three-step process for using waste or surplus-based feedstock: (1) Acidification of organic components of the feedstock in order to obtain readily biodegradable carbon sources, as SCOAs; (2) Enrichment in PHA-storing microorganisms; (3) PHA accumulation (Fig. **4**) [11].

The acidification step is required whenever the chosen feedstock is poor in SCOAs. This is a stage of anaerobic digestion, where carbon-based compounds from waste are transformed into SCOAs (*e.g.*, acetate, propionate or lactate) and alcohols (*e.g.*, ethanol). The anaerobic digestion is a complex process, which can be divided into four individual degradation stages: hydrolysis, acidogenesis, acetogenesis, and, finally, methanogenesis. Anaerobic digestion is a treatment-oriented technology employed within full-scale facilities worldwide for the treatment of industrial and urban wastewater and organic solid waste. This process is advantageous over aerobic systems because of high organic removal rates, low energy-input requirement, energy production, and low sludge production. However, for SCOAs production, it is essential to eliminate methanogenesis by manipulating parameters such pH, temperature, hydraulic retention time (HRT), sludge retention time (SRT), and organic loading rate (OLR) [59]. Therefore, the amount and type of SCOAs generated by the fermentation of organic constituents are crucial for the selection and accumulation steps, influencing the amount and type of PHA produced [26, 28, 60 - 62].

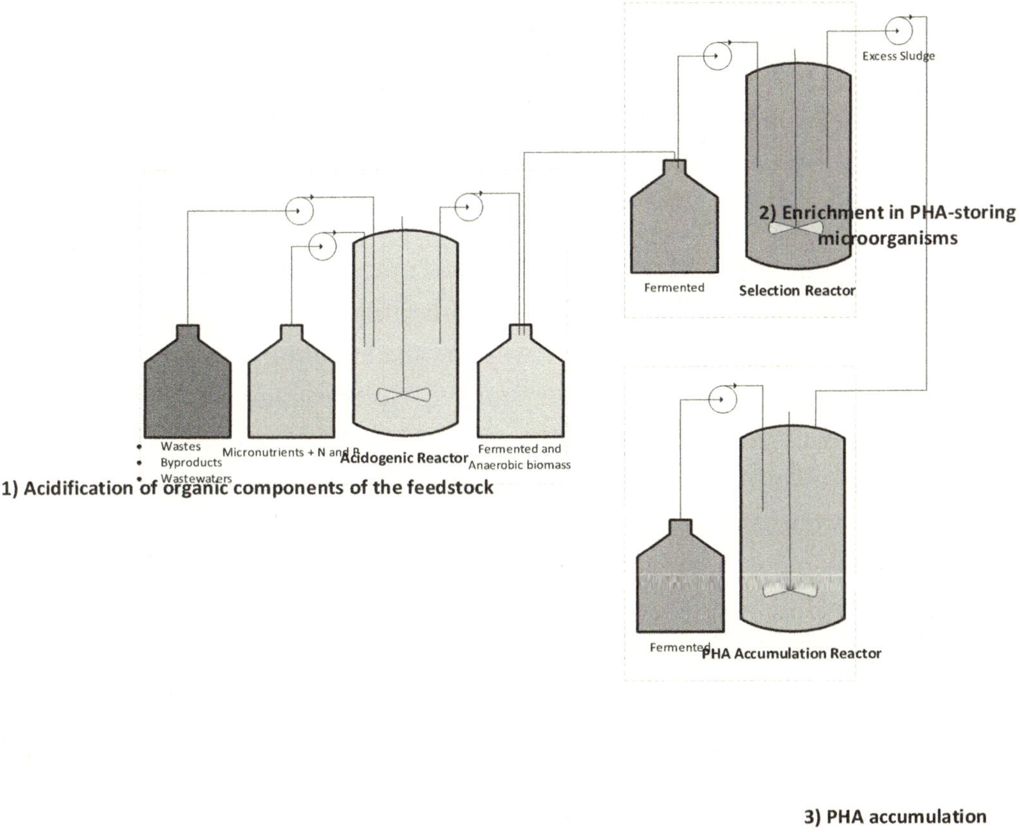

Fig. (4). Three-step PHA production process (Adapted from Dionisi *et al.* (2004) [11]).

The enrichment in PHA-storing organisms is considered one of the key-points of all PHA production processes by MMC. The objective of this step is to select the maximum number of organisms from the initial culture, with a high PHA storage capacity and eliminate, or at least reduce, non PHA-storing side populations. The system is operated in order to eliminate microorganisms presenting low or no storage capacity since they contribute negatively for the reduction of the average PHA cell content and storage yields by using the carbon source for growth. Very heterogeneous populations in terms of PHA storage ability have a negative impact on the downstream process. However, experimental results showed that by keeping a high storage capacity at this phase, an unstable population is usually obtained. Thus, the operation of the selection reactor should be optimized to obtain a homogeneous population with relatively high and stable storage capacity rather than to maximize the PHA cell content [17]. The accumulation step is usually projected to maximize the amount of polymers stored by cells harvested from the selection step using the substrate obtained from the acidification step.

In the three-step process, ADF is the most common methodology applied for biomass selection. In ADF systems the availability of an excess of external substrate after a long period of absence triggers the accumulation of intracellular reserves, due to the lower requirements of physiological adaptation by microorganisms (Fig. **5**). In this situation, the substrate is used almost entirely to the accumulation of intracellular reserves (about 70% of the substrate). However, cell growth can become dominant if the exposure time to substrate is long enough to allow physiological adaptation of microorganisms [63, 64]. Once more, parameters such pH, temperature, HRT, SRT, and OLR, among others, are the key to successfully steer the selection [65]. Despite the use of synthetic substrate, Johnson *et al.* (2009) selected a stable microbial community able to reach 89% cell dry mass (CDM) in the accumulation step, a value comparable to those obtained with pure cultures in terms of maximum PHA content [66]. Molecular techniques confirmed that the operational parameters imposed led to a virtual monoculture, highly competitive, later fully characterized and named *Plasticicumulans acidivorans* [67].

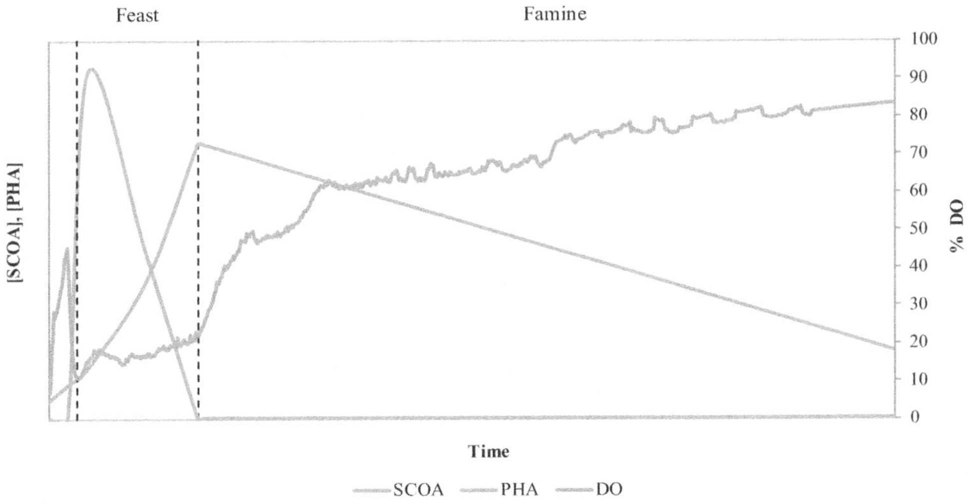

Fig. (5). Example of an ADF cycle operation by a fully aerobic culture.

A considerable number of works already began seeking for processes based on using complex substrates, namely olive mill wastewaters [68], paper mill wastewaters [16, 69], pulp and paper mill by-products [70], palm oil mill effluents [71], pyrolysis by-product and crude glycerol [21, 22] and sugar cane molasses [14]. Using fermented paper mill wastewaters, Jiang *et al.* obtained a MMC able to accumulate the highest PHA content so far reported with a complex substrate, namely 77% of CDM [16].

The mechanisms for PHA storage by PAOs and GAOs enriched under anaerobic/aerobic conditions are different from those observed in the ADF process. Nevertheless, in both cases, PHA storage occurs when growth is prevented. Contrary to ADF, where PHA storage occurs due to an internal growth limitation despite the presence of electron donor and acceptor, in the AN/AE process, storage is mainly caused by an external growth limitation due to absence of an electron sink (oxygen and/or nitrate). A few works employed this selection strategy, aiming the enrichment in GAO phenotypic behavior, reaching however lower PHA accumulations [8, 56, 72].

High accumulations are expected to be obtained at the third and last step of the process. The excess sludge produced in the selection reactor with a high storage response is exploited in a third reactor, operated at considerably higher organic load in order to saturate the sludge storage capacity. The accumulation stage, thus, aims the PHA accumulation in microbial cells and its efficiency strongly depends on the storage capacity of the selected MMC. A highly relevant parameter is the maximum attainable polymer content in the biomass that can decrease the polymer extraction cost. If the accumulation is performed under nutrient-limiting conditions the driving force for storage will be kept high until a PHA saturation level is reached. On the contrary, if the biomass is continuously exposed to nutrient rich conditions plus high organic loads, a growth response will progressively take place, overcoming, eventually, the storage response. Therefore, the maximum PHA content will be lower than the maximum storage capacity.

Finally, the PHA-rich biomass then flows to the downstream processing for extraction and purification of PHA.

Operation Modes

Selecting a culture with a high storage capacity is one of key stones in MMC processes in order to develop an economical and competitive PHA production process. The goal is to steer the population to a homogeneous microbial enrichment in terms of high PHA storage capacity [1, 23]. Selection leading to a heterogeneous community, in terms of PHA accumulation, will terminate into a negative impact on the overall productivity due to a reduction on the average PHA cell content, and feasibility of the process by increasing the PHA extraction costs. Nevertheless, the selection step should aim for a homogenous population with stable storage capacity, leaving the maximization of PHA cell content for the accumulation step [1].

During the selection period, the population shifts and dynamics should be followed in order to correlate the communities selected with the performance and results observed [23]. Several molecular methods, such FISH, denaturing gradient gel electrophoresis (DGGE), terminal restriction fragment length polymorphism (T-RFLP), pyrosequencing, and quantitative-PCR (qPCR) allow a periodically monitoring of the microbial community evolution, by identifying the different groups of microorganisms and handling the selection towards the best PHA producers [23].

Several strategies can be implemented to select PHA-storing microorganisms, from which stand out the AN/AE and ADF systems.

Anaerobic/Aerobic System

One of the main sources of pollution is the one caused by phosphorous accumulation due to industrial and agriculture intensive activity. In order to prevent a drastic decrease in water quality and aquatic ecosystems losses, EBPR systems were created and applied [73]. Based on this method, anaerobic/aerobic system was created. In this process, as mentioned before, PAOs and GAOs showed to be able to accumulate PHA [17].

A considerable number of works studied the use of synthetic mediums/synthetic wastewaters to select PHA-storing microorganisms under AN/AE conditions and understand its influence on MMC enrichment (Table **1**). Michinaka *et al.* (2007) aimed to select and identified the bacteria able to accumulate PHA containing 3H2MV by a MMC selected in an AN/AE SBR with synthetic wastewater containing propionate:acetate (molar ratio 14:3) as carbon sources. In fact, authors observed an evolution towards the increase on the amount of 3H2MV, from 6 to 38%. Through the construction of *phaC* clone libraries, 261 clones were found to have a putative lipase box, a conserved block, probably to be the catalyst center of PHA synthase [44], and applying T-RLPF the authors related the increase amount of 3H2MV produced with the presence of *Pseudomonas* sp [74]. Before moving towards complex substrates, synthetic ones were explored to select the best operational parameters. For 450 days, a stable reactor submitted to AN/AE was operated using only acetate [56]. The GAOs enriched culture accumulate up to 49% CDM (HB:HV = 73:27) under anaerobic conditions and up to 60% PHB of CDM under aerobic condition, in batch tests. Regarding microbial characterization *Candidatus* Competibacter phosphatis accounted to 57% and *Defluviicoccus*-related Cluster 1 to 2-6% during the steady-state, both communities were undetected in the beginning of the experiment [56]. Other parameters such the effect of COD/N ratio, were also studied using an AN/AE SBR fed with acetate. COD amount was gradual increased up to 1200 mg/L and

COD/N ratio fixed at 125, culminating in a PHA accumulation of 64.2% of CDM [72]. To follow the evolution, pyrosequencing allowed to identified a very rich community of nine phyla: *Proteobacteria*, *Acidobacteria*, *Firmicutes*, *Bacteroidetes*, *Gemmatimonadetes*, *Cyanobacteria/Chloroplast*, *Actinobacteria*, *Nitrospira*, TM7, and unclassified bacteria, representing approximately 36.34, 14.09, 4.41, 1.72, 0.22, 1.51, 19.68, 0.11, 0.75, and 21.18% of the sequences, respectively [72]. It was demonstrated that microbial communities selected could be quite diverse and still present considerable PHA accumulation capacities (92 taxa of PHA-accumulating bacteria were presented by the authors).

Under AN/AE conditions and using hydrolyzed molasses, supplemented with acetic acid, as substrate *Pseudomonas* was selected as the main PHA-storing population. The MMC was able to accumulate up to 24.8% CDM PHA after 30 minutes of beginning of anaerobic phase [75]. Simple variations such the origin of the inoculum and the matrix of the feed can lead to the selection of different microorganisms.

The final objective, after understanding how the several operational parameters could influence and guide the culture selection, is the use and valorization of residues and by-products from several industries. Several complex substrates were already tested to select a PHA-storing culture. Operated under AN/AE conditions, condensate wastewaters from pulp and paper mill and fermented municipal primary solids were proven to be adequate substrates for PHA production. The reactor fed with condensate wastewaters from pulp and paper mill was operated for four months with 24h cycles and a SRT of 4 days, showing an average PHB production of 6.6% of CDM [25]. The SBR fed with fermented municipal primary solids was operated under AN/AE conditions for nine months, with a PHA content fluctuating between 10 and 25% CDM [76]. Despite the long-term selection, both reactors recorded low PHA storing values and considerable fluctuations. Lack of oxygen to nearly absence could have led to complete inhibition of microbial community, once it was necessary to degrade some complex carbon sources in the condensate [25]. Fermented paper mill wastewater was used, alongside with anaerobic/aerobic conditions, to select GAOs for PHA production. The SCOAs-rich led to a population rich in GAOs related to *Defluviicoccus vanus* (56%) and *Candidatus* Competibacter phosphatis (22%).

Further accumulation tests, by applying a subsequent anaerobic period after an initial aerobic, led to a PHA content of 42% of CDM [77]. One last example falls on the use of food residues, such fermented sugar cane molasses. A culture already enriched in GAOs was submitted AN/AE conditions and reached 17% in CDM of PHA, being composed by 3HB, 3HV monomers in the first 81 days of reactor operation [8]. The microbial community maintained the GAO-phenotypic behavior and bared the capacity on handling short-term variations in the feeding composition, consequence of the acidogenic fermentation. Given the feed composition, an unforeseen medium chain length monomer of 3HHx, confirmed by ^{13}C-NMR, was observed further on reactor operation (137 days operation time) [8]. Microbial characterization by FISH analysis assigned the appearance of 3HHx to a shift in the community of GAO species, from *Candidatus* Competibacter phosphatis to *Defluviicoccus vanus*-related organisms [8].

Table 1. PHA-storing microorganisms selected under AN/AE conditions.

HRT	SRT	Substrate	Molecular Method	Main Organisms Selected	Study
8 h	10 days	Mixture of acetate and propionate	T-RFLP	*Pseudomonas* sp.	[74]
1.3 days	10 days	Acetate	FISH	*Candidatus* Competibacter phosphatis, *Defluviicoccus*-related Cluster 1 and 2	[56]
18 h	7 days	Synthetic wastewater	Pyrosequencing	*Proteobacteria, Acidobacteria* *	[72]
2 days	4 days	Hydrolyzed molasses supplemented with acetate	DGGE	*Pseudomonas*	[75]
3 days	10 days	Fermented paper mill wastewater	FISH	*Defluviicoccus vanus*, *Candidatus* Competibacter phosphatis	[77]
1.33 days	10 days	Fermented sugar cane molasses	FISH	*Defluvicoccus vanus-related organisms*	[8]

* - 92 different taxa were identified.

Aerobic Dynamic Feeding System

As previously stated, the most well studied process for the selection step is ADF. ADF process imposes, periodically, alternated availability of substrate creating periods of feast and famine (Fig. **5**).

Since its introduction by Majone *et al.* (1996), ADF became the best studied PHA storage process (Table **2**). Similar to AN/AE process, the selection of PHA-accumulating organisms under ADF conditions was performed with synthetic media containing single or mixed organic acids as sole carbon sources. Several reactor operational conditions were tested: OLR, HRT, SRT, carbon to nitrogen ratio (C/N), pH, oxygen concentration and temperature [1]. This was followed by the introduction of complex and low value substrates, such as agro-forestry residues, industrial byproducts and industrial and urban waste [14, 21, 25, 68 - 70].

Dionisi *et al.* (2005) studied the selection of a PHA-storing community feeding a SBR with a mixture of SCOAs, namely acetic, lactic and propionic acids. A high OLR was used as selective pressure, 12.75 g COD/L, that led to an accumulation of 14.7% on COD basis [78]. DGGE revealed a significant shift on the predominant microbial population throughout the reactor operation time. Authors observed a continuous narrowing on the number of microbial populations from the initial activated sludge inoculum. From the DGGE bands sequencing, genus such *Methylobacteriaceae*, *Flavobacterium* and *Thauera* were identified, all PHA accumulators [78].

Acetate and propionate are two of the most used SCOAs in culture selection. In order to evaluate the culture behavior in the presence of these SCOAs, two SBRs under ADF conditions, one fed with acetate (60 C-mmol/L·day) and other with propionate (30 C-mmol/L·day), with 1 and 10 days of HRT and SRT, respectively, and inoculated with an MMC from a stable phosphorus removal system, were studied [61]. The inoculum was dominated by *Candidatus* Accumulibacter phosphatis, *Candidatus* Competibacter phosphatis and tetrad-forming organisms (TFOs) affiliated to the *α-proteobacteria* class. After 5 SRTs, PAOs and GAOs disappeared because of the change from AN/AE to ADF conditions [61]. Despite the different feeding and OLR, the microbial populations selected were similar between both SBRs, presenting, however, different metabolism that resulted in diverse polymer compositions. The MMC from the acetate fed reactor produced a homopolymer of PHB and that from the propionate fed reactor, a copolymer of PHBHV (3HB:3HV = 28:72) [61]. In order to understand the microbial changes operated in the reactors under the influence of the applied SRT and OLR, the population was quantified by FISH [62]. In the reactor fed with acetate, 41.1 ± 2.2% of bacterial community belonged to *Thauera*, whereas it was only present at 1.9 ± 0.2% in the reactor fed with propionate. *Azoarcus* (23.3 ± 1.5%) and *Amaricoccus* (28.8 ± 1.8%) were the two main others genus present in the reactor fed with acetate. In the reactor fed with propionate, *Amaricoccus* was the dominant genus, 61.4 ± 1.9%, followed by *Azoarcus*, 3.9 ± 0.3%. A third SBR was introduced and operated under ADF conditions. Decreasing the SRT to 1 day and increasing the OLR of 120 C-mmol/L·day (acetate), faster-growing organisms were selected. In this way, *Amaricoccus* was washed out from the system and *Azoarcus* and *Thauera* accounted for 45.9 ± 1.5% and 49.4 ± 1.4%, respectively. This, consequently, also led to lower accumulations [62].

By using synthetic medium with acetic acid as sole carbon source the MMC that stored the highest amount of PHA accumulation, so far recorded, namely 89% of CDM, was selected. This content was obtained by submitting the MMC to a fed-batch experiment under growth limiting conditions. Through standard DGGE and 16S rRNA clone library construction, it was verified that the operational parameters imposed led to a virtual monoculture, highly competitive, without the need for sterilization [66]. The organism was later fully characterized and named *Plasticicumulans acidivorans* [67].

PHA can present several properties, manipulating its monomers constitution and side chains. *mcl*-PHA display particular physical properties making them more suitable for several applications. A MMC was enriched in *Pseudomonas aeruginosa* by feeding an ADF SBR nonanoic acid, as explained in the metabolism sub-chapter [79]. The culture was able to accumulate *mcl*-PHA up to 8.9% of PHA in CDM at the end of the feast phase, with a monomeric composition of 3-hydroxynonanoate (3HN) and 3-hydroxyheptanoate (3HHp) (79–84 mol-%, respectively) [79]. Regarding the monomeric constitution, a strategy that relies on feed composition must be adopted. Some experiments revealed that for obtaining the desired 3HV fraction in PHBHV, a proper selection of carbon sources for sludge acclimation and substrate composition for the accumulation step was necessary [80]. Therefore, the highest PHBHV content with MMC was obtained from a culture acclimated to acetate and propionate as substrate and further fed, in a batch test, with a mixture of acetate and propionate in a ratio of 1:1 (mg/mg). The MMC presented a PHBHV content of 47.7% of CDM, with an HV fraction of 48 mol-% [80]. The DGGE applied to the culture showed that *Sphingomonadales*, *Rhizobiales*, *Rhodobacterales*, *Rhodocyclales*, *Actinomycetales* and *Sphingobacteriales* dominated in both acclimation conditions, with acetate and propionate, which permitted to conclude that the addition of propionate for acclimatization did not dramatically shift the composition of the microbial community [80]. With the introduction of propionate in batch tests resulted in the disappearance of genera like *Thauera*, *Sphingomonas* and *Novosphingomonas*, selected at reactor fed with only acetate. The same was observed for *Flavobacterium* and *Diaphobacter*, although in this case the MMC was submitted to acclimatization with acetate and propionate. *Acidobacteria* and *Burkholderiales*, microorganisms that are able to consume propionate, were dominant when the highest PHBHV content was obtained. MMC, withdrawn from reactor fed with acetate and propionate, had a preference for propionate as substrate and utilized propionate before acetate, and the enrichment of these microorganisms may prompt PHBHV accumulation [80].

Besides feed composition, it is also important to explore the best operational parameters to select a culture. In a reactor fed with a mixture of acetate and propionate, the OLR and HRT [81], the pH-value and SCOAs mixture [15] and cycle length [82] were optimized. In this way, by applying ADF regimen, an OLR of 8.5 g COD/L and a pH-value of 7.5 were found to be the best conditions that led to higher PHA accumulations. DGGE analysis of 16S rRNA genes was done for every condition tested and revealed an enriched community dominated by *Lampropedia hyaline*. The best yields were observed with culture dominated by *L. hyaline* in opposition to a more heterogeneous culture which resulted in lower values of PHA production. Using the same OLR and pH-value studied for a cycle length of 2, 6 and 8 hours it was observed a clear evolution of the culture for the different cycles. However, *Lampropedia hyaline* outcompeted the remain population for the 6 and 2 hour cycles leading to accumulation up to 50% g COD PHA/g COD VSS in batch tests [82]. These findings proved that pH, OLR and cycle length play important roles in selecting the best PHA-storing community.

It is important to retain that despite the selection being, probably, the key factor in the PHA production by a MMC, not always a very narrow population is obtained specially with the use of complex substrates. In a different work, the enrichment and accumulation step led to a MMC enriched in PHA-accumulating organisms comprised in the *Dechloromonas*, *Thauera* and Compectibacter lineage [83]. It is important to stress that these results showed that different process operating conditions, mainly those applied and studied in the batch tests, led to different microbial community structures, which can be linked to different metabolic pathways activated under different operating conditions. A final analysis was done with nonmetric multidimensional scaling, which proved that microbial community structure had a close relationship with the process operating conditions also applied [83]. DGGE was applied to the original, acclimatized and PHB-accumulating MMCs. Contrary to what was described so far, it was observed that the number of DGGE bands from the acclimated sludge samples before or after PHB accumulation increased significantly in comparison to the non-acclimatized sludge sample [83].

Besides attempting to reduce PHA production costs, another very significant advantage of this strategy comes up, that fits in the concept of biorefinery: the valorization of industrial residues with high organic load and low value that generally represent a disposal problem to the industry, as stated earlier.

Fermented sugar molasses were proven to be a suitable substrate to enrich a MMC in PHA-storing microorganisms. Albuquerque *et al.* (2007) operated a SBR without pH control, 12h cycle, SRT of 10 days, with temperature control at 25 °C, ammonia limiting conditions and two different OLR (120 and 60 C-mmol/L.day) A stable PHA-accumulating culture was obtained when the lower OLR and higher ammonia concentration (2.5 mmol N/L) were used, accumulating up to 10% CDM PHA. It was also proven, this time for a complex substrate, that OLR and feast/famine ratio were a critical factor of the culture enrichment step, with the culture losing the PHA storage capacity at higher OLR for the same cycle length [14]. Later, Albuquerque *et al.* 2010 demonstrated that the SCOAs profiles on fermented sugar molasses and feeding regimen allowed to manipulated the PHA composition [28].

Food processing wastewaters can also be a potential substrate to be explored in

PHA production. In 2008, the possibility of using tomato cannery waste for simultaneous wastewater treatment and selection of PHA-storing microorganisms under ADF conditions was studied [84]. Authors adopted a two-stage process, where the selection bioreactor was operated with 4 days as HRT and SRT. The selected populations were able to remove 84% of the present COD and accumulate PHA within the range of 7% to 11% CDM in non-filtered wastewater and of 2% to 8% in filtered wastewater. A maximum PHA content of 20% of CDM was obtained in batch tests [84].

Moita and Lemos (2012) used bio-oil resulting from the fast-pyrolysis of chicken beds to acclimatize a MMC to produce PHA. An ADF system SBR, with 12 hours cycle, was operated with a HRT of 1 and a SRT of 10 days. It took 167 days to reach a pseudo-steady state where the sludge was able to accumulate up to 9% CDM of PHBHV with 30% of 3HV. Glycogen was also stored, 2.9% of CDM, probably due to the presence of sugars in bio-oil [21], decreasing the volumetric production of PHA. By day 61 of operation, the SRT was decreased to 5 days in order to favor the organisms with higher PHA accumulation capacity. The microbial community characterization revealed that *α-proteobacteria* class was the dominant one, over *α-proteobacteria* and *β-proteobacteria. Thauera* was the dominant genus during the operation, coming up, however, *Amaricoccus* and *Zooglea* with the change of the operational parameters [21].

A few works have been done regarding the valorization of pulp and paper side streams. Paper mill wastewater was fermented, reaching a 74% conversion of the soluble COD to SCOAs (acetate, propionate, butyrate and valerate). The SBR was operated at a pH-value of 7.3, HRT of 0.2 and 2.2 days and SRT of 7 days. The enriched culture was mainly composed of filaments and floc-forming bacteria, all of them showed PHA production as detected after Sudan Black or Nile Blue staining. The average PHA content was 11% of CDM. After batch accumulation test the maximum PHA content achieved was 48% of CDM, consisted of 31–47 mol-% 3HB and 53–69 mol-% 3HV [69].

By applying a three stage process in order to select a PHA-storing MMC using paper mill effluent as feedstock, an enrichment dominated by *Plasticicumulans acidivorans* was obtained, accounting for 56% of the microbial population [16].

Activated sludge from a nutrient sewage removal treatment plant was acclimatized, for 5 months, to fermented paper mill wastewater, resulting in the highest amount of PHA so far stored by a MMC fed with real substrate, 77% PHBHV (3HB:3HV = 86:14) in CDM [16].

Hardwood sulfite spent liquor (HSSL) is a by-product from the pulp and paper industry with a high COD content, namely lignosulphonates (60–80 g/L), xylose (35–45 g/L) and acetic acid (8–9 g/L). Queirós *et al.* (2014) studied the possibility of using the residue as substrate to select a PHA-storing enriched culture in a two-step process. An ADF SBR was operated to target the acetic acid consumption. HRT and SRT of 1 and 5 days, respectively, were imposed alongside with 4.2 g COD/LA day of OLR. Authors obtained a PHB accumulation of up to 65% of CDM, and observed that not only the acetic acid was involved in the PHA storage metabolic pathway, but also the xylose and other components. However, a stationary-state was not reached, revealed by the PHA accumulation fluctuations and the microbial evolution. By FISH, authors were able to identified *α-proteobacteria* as the dominant PHA-storing community, accounting for 72.74% after 65 days of operation. Due to the tendency observed, the culture was still adapting to the conditions [70]. Later, with a different batch of HSSL, more than 250 days were needed to reach a pseudo-steady state [20].

Biodiesel production originates high amounts of crude glycerol. Such by-product was already explored for PHA production by MMC. The selection reactor was operated under ADF conditions with 24 h cycles and 2 and 5 days of HRT and SRT, respectively, and pH value control between 7.2–8.2. The culture reached a steady-state after 8.5 SRTs, showing the ability to consume both glycerol and methanol fraction present in the crude and producing PHA around 6% of CDM [22]. The glycerol fraction was the only one contributing for the two biopolymers stored: PHB and glucose biopolymer. In batch tests, the enriched culture was able to accumulate 47% PHA in CDM at a productivity of 0.27 gX/L·day. However, authors believed that the productivity could be improved if the crude was completely redirected to PHA formation [22].

Campanari *et al.* (2014) assessed the feasibility of using olive oil mill effluents (OMEs) as a substrate in biodegradable polymer production. The enrichment of the culture was done in a SBR under feast and famine conditions with dephenolized and fermented OME. Authors aimed to optimize the OLR (ranging from 2.40 to 8.40. g COD/L·day). At an OLR of 4.70 g COD/L·day, the culture showed the highest values of storage rate and yield, 339.0 ± 48.0 mg mg COD/g COD·h and 0.56 ± 0.05 g COD/g COD, respectively [85]. As observed by Albuquerque *et al.*, the OLR applied to the SBR had influence on the performance of the PHA-accumulating reactor. Authors calculated that the polymer volumetric productivity, combining the SBR and the accumulation reactor, accounted for 1.50 g PHA/L·day [85].

Table 2. PHA-storing microorganisms selected under ADF conditions.

HRT	SRT	Substrate	Molecular Method	Main Microorganisms Selected	Study
1 day	1 day	Mixture of acetate, lactate and propionate	DGGE	*Methylobacteriaceae, Flavobacterium, Thauera*	[78]
1 day	10 days	Acetate	FISH	*Thauera, Azoarcus, Amaricoccus*	[61, 62]
1 day	10 days	Propionate		*Azoarcus, Amaricoccus*	
1 day	1 day	Acetate		*Thauera, Azoarcus*	
1 day	1 day	Acetate	DGGE and 16S rRNA gene clone library	*Plasticicumulans acidivorans*	[66, 67]
3 days	3 days	Nonanoic acid	DGGE	*Pseudomonas aeruginosa*	[79]
1.67 days	5 days	Acetate	DGGE	*Dechloromonas, Thauera,* Compectibacter lineage	[83]
1.67 days	5 days	Acetate	DGGE	*Sphingomonadales, Rhizobiales, Rhodobacterales, Rhodocyclales, Actinomycetales, Sphingobacteriales*	[80]
		Acetate:Propionate (3:1)			
1 day	1 day	Mixture of acetate and propionate	DGGE	*Lampropedia hyaline*	[82]
1 day	10 days	Fermented sugar molasses	FISH	*Azoarcus, Thauera*	[60]
1 day	10 days	Bio-oil from chicken beds fast-pyrolysis	FISH	*Thauera, Amaricoccus, Zoogloea*	[21]
2 days	2 days	Fermented paper mill wastewater	FISH	*Plasticicumulans acidivorans, Thauera*	[16]
1 day	5 days	Hardwood sulphite spent liquor	FISH	*I -Proteobacteria*	[70]

Other Possibilities

Despite the above presented processes, AN/AE and ADF, being the most common selection methods, a few others were studied in order to select the best PHA-storing microorganisms or even to eliminate this step.

A two-stage CSTR system (under ADF conditions) was used to enrich a MMC in PHA-storing microorganisms using fermented molasses as feedstock. The enrichment step was designed to impose feast and famine conditions in different reactors, by feeding the carbon substrate to the first reactor (feast reactor), and the resulting effluent passing by overflow to the second reactor (famine reactor). The imposed HRT and SRT were 1 and 10 days, respectively [86]. Two parallel systems allowed studying the effect of influent substrate concentration and feast-to-famine ratio (F/F) between reactors. Best results were obtained when high influent concentrations and low F/F ratio were introduced in the reactor, leading to a polymer yield on substrate of 0.59 C-mol PHA/C-mol SCOA and growth yield of 0.22 C-mol X/C-mol SCOA. Conversely, the famine growth yield reached higher values 0.42 C-mol X/C-mol SCOA. In this case a stable PHA content of 23% of CDM was recorded, whereas a content 61% PHA in CDM was attained in the batch test [86].

Recently, an attempt to eliminate the MMC enrichment step was done [87]. Waste activated sludge (WAS) was induced to accumulate PHB in a fed-batch reactor operated with growth limitation by phosphorus and using acetic acid as substrate. The behavior of WAS from three different sites by applying three defined proportions of carbon and phosphorus to each culture was investigated, resulting in PHA accumulations of 67, 57 and 45% of CDM, respectively [87]. Through pyrosequencing was possible to infer which bacterial groups were active during PHB accumulation during fed-batch operation of two reactors. Bacteria belonging to the genera *Acinetobacter*, *Zoogloea* and *Simplicispira*, already reported as PHA accumulators, were present in two of the three tests, representing about 40% of total active community. The relative abundance of each genus varied along the operation time. One of the reactors displayed less diversity of active bacteria, recording a strong metabolic activity from *Methylobacter*. In the other reactor, the culture kept on producing PHB after the first phase (dedicated to the depletion of residual phosphorus) and up to 71 hours of reaction [87]. This work pointed out that enrichment in PHA-producing microorganisms occurred, observing an increase in organisms already described as PHA accumulators.

Manipulation on the reactor configuration and feeding regimen imposed is also being explored, with an anoxic-aerobic dynamic feeding for enriching an MMC with denitrifying PHA-storing microorganisms. Two parallel bioreactors were set-up: one was operated as conventional activated sludge SBR (AS-SBR) and the second was an integrated biofilm activated sludge (IFAS), a hybrid moving bed-reactor operated in SBR mode (MBBR-SBR) [37]. The strategy aimed to mimetize the conditions imposed to the water treatment full-scale reactors in a beet sugar factory. Reactors were fed with condensate and wash waters and operated for 24 weeks, with 8 hours cycle and SRT of 16 days and HRT of 4-6 days for AS-SBR and MBBR-SBR, respectively. The highest PHA storage yields were reached at the end of the operations (20-24 weeks), 0.77 and 0.79 g COD/g COD, for AS- and MBBR-SBRs. Both SBRs presented, thus, biomass with PHA production potential while sustaining process water treatment for carbon, nitrogen and phosphorus for the factory process waters. PHA accumulation reached 60% of CDM for the MBBR-SBR which also presented better nitrification performance. A copolymer PHBHV (3HB:3HV = 85:15 on a molar basis) was produced from the beet tail press water [37].

APPLICATIONS

Due to the high diversity of monomers and the possibility to combine different ones into the final polymer, a wide range of properties can be obtained for this class of polymers, from thermoplastics to elastomers. Although PHA can replace many types of petrol-derived plastics their production price is not yet commercially competitive. PHB is stiff and brittle, limiting the number of useful applications. However, the utilization of copolymers with other monomers than 3HB, improve their properties. Copolymers of 3HB and 3HV from MMC do not differ from those obtained using pure cultures in terms of their properties [26, 27]. PHA produced by pure cultures is being used in many combinations with other biologically-derived polymers (polylatic acid, dextran, dextrin, amylose, other polysaccharides) or petrochemical-derived polymers (nylon, polyethylene oxide, polyvinyl acetate, polyvinyl chloride) and plasticizers resulting in the modification of other polymers properties in terms of processing and degradability, increasing PHA market share [88]. These combinations could be also possible for PHA obtained from MMC.

Despite the initial applications of PHA related with everyday consumer goods, particularly in the packaging area due to their good barrier properties, for food packaging and for making plastic beverage bottles [89], other possible applications were tested. They include shampoo bottles, packaging films used for containers and paper coatings, printing and photographic materials, disposable items such as razors, utensils, diapers, feminine hygiene products, cosmetic containers, and cups as well as medical surgical garments, upholstery, carpet, compostable bags and lids, or tubs for thermoformed articles [52]. Due to the origin of microorganisms and the substrates used, PHA by MMC should be targeted to less noble applications, where biomedical and pharmaceutical ones are not primary goals.

PHA can be used as a starting material in chemical industry, due to their chiral enantiomeric pure nature, or as raw material for the production of paints. PHA monomers after esterification, using 3-hydroxybutyrate methyl ester (3HBME) and *mcl* 3-hydroxyalkanoate methyl ester (3HAME) from PHB and *mcl*-PHA, can be used as biofuels [90].

In the agriculture area, PHA are used in the production of devices for controlled release of fertilizers, herbicides and insecticides [89]. They are also important growth promoters for plants, using bacterial PHA producers as inoculants, contributing to rhizosphere adaptation, root colonization, and plant growth promotion abilities. Their impact in symbiotic nitrogen-fixing is also acknowledge as well as contributing to a successful competitiveness in the nodulation process in different species of rhizobia [91]. Encapsulation of seeds, biodegradable films for crop protection or in the construction of greenhouses are other possibilities of utilization in this field [92].

For bioremediation purposes MMCs with PHA accumulation capacity can also be used, namely for oil-contaminated sites using different hydrocarbons such as benzene, toluene, and xylene (BTX) for PHA production, and providing the polymer a competitive advantage in order to adapt to stresses prevailing in the environment [93]. A natural occurring process where PHA also present an important role is on the treatment of wastewaters for the removal of organic and inorganic contaminants, as phosphorus and nitrogen.

CONCLUSION

Although being challenging, PHA production by MMC is a way of waste treatment through its valorization. Being PHA production a competitive advantage for microbial cultures, the enrichment in PHA-storing bacteria depends more on the operational conditions imposed to the process rather than the existence of sterility. PHA producing processes using MMC and waste as substrates could help to reduce the polymer costs but at the same time using the existing facilities of waste/wastewater treatment systems and residues on location could contribute to the overall profit of industries and WWTPs.

CONFLICT OF INTEREST

The authors confirm that they have no conflict of interest to declare for this publication.

ACKNOWLEDGEMENTS

This work was developed within the scope of the project CICECO-Aveiro Institute of Materials (Ref. FCT UID/CTM/50011/2013) and LAQV, REQUIMTE (Ref. FCT UID/QUI/50006/2013) financed by national funds through the FCT/MCTES and when appropriate co-financed by ERDF under the PT2020 Partnership Agreement (POCI-01-0145-FEDER-007265). Diogo Queirós thanks FCT/MEC for his PhD grant (SFRH/BD/87758/2012).

REFERENCES

[1] Dias JM, Lemos PC, Serafim LS, *et al.* Recent advances in polyhydroxyalkanoate production by mixed aerobic cultures: from the substrate to the final product. Macromol Biosci 2006; 6(11): 885-906. [http://dx.doi.org/10.1002/mabi.200600112] [PMID: 17099863]

[2] Chanprateep S. Current trends in biodegradable polyhydroxyalkanoates. 2010.

[3] Wintermute EH, Silver PA. Dynamics in the mixed microbial concourse. Genes Dev 2010; 24(23):

2603-14.
[http://dx.doi.org/10.1101/gad.1985210] [PMID: 21123647]

[4] Wallen LL, Rohwedder WK. Poly-b-hydroxyalkanoate from activated sludge. Environ Sci Technol 1974; 8: 576-9.
[http://dx.doi.org/10.1021/es60091a007]

[5] Pereira H, Lemos PC, Reis MA, *et al.* Model for carbon metabolism in biological phosphorus removal processes based on *in vivo* 13C-NMR labelling experiments. Water Res 1996; 30(9): 2128-38.
[http://dx.doi.org/10.1016/0043-1354(96)00035-8]

[6] Levantesi C, Serafim LS, Crocetti GR, *et al.* Analysis of the microbial community structure and function of a laboratory scale enhanced biological phosphorus removal reactor. Environ Microbiol 2002; 4(10): 559-69.
[http://dx.doi.org/10.1046/j.1462-2920.2002.00339.x] [PMID: 12366750]

[7] Oehmen A, Lemos PC, Carvalho G, *et al.* Advances in enhanced biological phosphorus removal: from micro to macro scale. Water Res 2007; 41(11): 2271-300.
[http://dx.doi.org/10.1016/j.watres.2007.02.030] [PMID: 17434562]

[8] Pisco AR, Bengtsson S, Werker A, Reis MA, Lemos PC. Community structure evolution and enrichment of glycogen-accumulating organisms producing polyhydroxyalkanoates from fermented molasses. Appl Environ Microbiol 2009; 75(14): 4676-86.
[http://dx.doi.org/10.1128/AEM.02486-08] [PMID: 19465533]

[9] Majone M, Massanisso P, Carucci A, Lindrea K, Tandoi V. Influence of storage on kinetic selection to control aerobic filamentous bulking. Water Sci Technol 1996; 34(5): 223-32.
[http://dx.doi.org/10.1016/0273-1223(96)00649-X]

[10] Daigger GT, Grady CP. An assessment of the role of physiological adaptation in the transient response of bacterial cultures. Biotechnol Bioeng 1982; 24(6): 1427-44.
[http://dx.doi.org/10.1002/bit.260240614] [PMID: 18546434]

[11] Dionisi D, Majone M, Papa V, Beccari M. Biodegradable polymers from organic acids by using activated sludge enriched by aerobic periodic feeding. Biotechnol Bioeng 2004; 85(6): 569-79.
[http://dx.doi.org/10.1002/bit.10910] [PMID: 14966798]

[12] Serafim LS, Lemos PC, Oliveira R, Reis MA. Optimization of polyhydroxybutyrate production by mixed cultures submitted to aerobic dynamic feeding conditions. Biotechnol Bioeng 2004; 87(2): 145-60.
[http://dx.doi.org/10.1002/bit.20085] [PMID: 15236243]

[13] Lemos PC, Serafim LS, Reis MA. Synthesis of polyhydroxyalkanoates from different short-chain fatty acids by mixed cultures submitted to aerobic dynamic feeding. J Biotechnol 2006; 122(2): 226-38.
[http://dx.doi.org/10.1016/j.jbiotec.2005.09.006] [PMID: 16253370]

[14] Albuquerque MG, Eiroa M, Torres C, Nunes BR, Reis MA. Strategies for the development of a side stream process for polyhydroxyalkanoate (PHA) production from sugar cane molasses. J Biotechnol 2007; 130(4): 411-21.
[http://dx.doi.org/10.1016/j.jbiotec.2007.05.011] [PMID: 17602776]

[15] Villano M, Beccari M, Dionisi D, *et al.* Effect of pH on the production of bacterial

polyhydroxyalkanoates by mixed cultures enriched under periodic feeding. Process Biochem 2010; 45(5): 714-23.
[http://dx.doi.org/10.1016/j.procbio.2010.01.008]

[16] Jiang Y, Marang L, Tamis J, van Loosdrecht MC, Dijkman H, Kleerebezem R. Waste to resource: Converting paper mill wastewater to bioplastic. Water Res 2012; 46(17): 5517-30.
[http://dx.doi.org/10.1016/j.watres.2012.07.028] [PMID: 22921584]

[17] Serafim LS, Lemos PC, Albuquerque MG, Reis MA. Strategies for PHA production by mixed cultures and renewable waste materials. Appl Microbiol Biotechnol 2008; 81(4): 615-28.
[http://dx.doi.org/10.1007/s00253-008-1757-y] [PMID: 19002455]

[18] Palmqvist E, Hahn-Hagerdal B. Fermentation of lignocellulosic hydrolysates. I: Inhibition and detoxification. Bioresour Technol 2000; 74(1): 17-24.
[http://dx.doi.org/10.1016/S0960-8524(99)00160-1]

[19] Palmqvist E, Hahn-Hagerdal B. Fermentation of lignocellulosic hydrolysates. II: Inhibitors and mechanisms of inhibition. Bioresour Technol 2000; 74(1): 25-33.
[http://dx.doi.org/10.1016/S0960-8524(99)00161-3]

[20] Queirós D, Fonseca A, Lemos PC, Serafim LS. Long-term operation of a two-stage polyhydroxyalkanoates production process from hardwood sulphite spent liquor. J Chem Technol Biotechnol 2016.

[21] Moita R, Lemos PC. Biopolymers production from mixed cultures and pyrolysis by-products. J Biotechnol 2012; 157(4): 578-83.
[http://dx.doi.org/10.1016/j.jbiotec.2011.09.021] [PMID: 21983233]

[22] Moita R, Freches A, Lemos PC. Crude glycerol as feedstock for polyhydroxyalkanoates production by mixed microbial cultures. Water Res 2014; 58: 9-20.
[http://dx.doi.org/10.1016/j.watres.2014.03.066] [PMID: 24731872]

[23] Queirós D, Lemos PC, Rossetti S, Serafim LS. Unveiling PHA-storing populations using molecular methods. Appl Microbiol Biotechnol 2015; 99(24): 10433-46.
[http://dx.doi.org/10.1007/s00253-015-7010-6] [PMID: 26428237]

[24] Carvalho G, Oehmen A, Albuquerque MG, Reis MA. The relationship between mixed microbial culture composition and PHA production performance from fermented molasses. N Biotechnol 2014; 31(4): 257-63.
[http://dx.doi.org/10.1016/j.nbt.2013.08.010] [PMID: 24025669]

[25] Coats ER, Loge FJ, Smith WA, Thompson DN, Wolcott MP. Functional stability of a mixed microbial consortium producing PHA from waste carbon sources. Appl Biochem Biotechnol 2007; 137-140(- -12): 909-25.
[http://dx.doi.org/10.1007/s12010-007-9107-6] [PMID: 18478444]

[26] Serafim LS, Lemos PC, Torres C, Reis MA, Ramos AM. The influence of process parameters on the characteristics of polyhydroxyalkanoates produced by mixed cultures. Macromol Biosci 2008; 8(4): 355-66.
[http://dx.doi.org/10.1002/mabi.200700200] [PMID: 18157852]

[27] Laycock B, Halley P, Pratt S, Werker A, Lant P. The chemomechanical properties of microbial

polyhydroxyalkanoates. Prog Polym Sci 2013; 38(3-4): 536-83.
[http://dx.doi.org/10.1016/j.progpolymsci.2012.06.003]

[28] Albuquerque MG, Martino V, Pollet E, AvA(c)rous L, Reis MA. Mixed culture polyhydroxyalkanoate (PHA) production from volatile fatty acid (VFA)-rich streams: effect of substrate composition and feeding regime on PHA productivity, composition and properties. J Biotechnol 2011; 151(1): 66-76.
[http://dx.doi.org/10.1016/j.jbiotec.2010.10.070] [PMID: 21034785]

[29] Chen G-Q, Hajnal I. The ?~PHAome ?(tm). Trends Biotechnol 2015; 33(10): 559-64.
[http://dx.doi.org/10.1016/j.tibtech.2015.07.006] [PMID: 26409775]

[30] Steinbiichel A, Steinbiichel A. Diversity of bacterial polyhydroxyalkanoic. FEMS Microbiol Lett 1995; 128: 219-28.
[http://dx.doi.org/10.1111/j.1574-6968.1995.tb07528.x]

[31] Tan G-Y, Chen C-L, Li L, *et al.* Start a Research on Biopolymer Polyhydroxyalkanoate (PHA): A Review. Polymers (Basel) 2014; 6(3): 706-54.
[http://dx.doi.org/10.3390/polym6030706]

[32] Madkour MH, Heinrich D, Alghamdi MA, Shabbaj II, SteinbA1/4chel A. PHA recovery from biomass. Biomacromolecules 2013; 14(9): 2963-72.
[http://dx.doi.org/10.1021/bm4010244] [PMID: 23875914]

[33] Jiang Y, Mikova G, Kleerebezem R, van der Wielen LA, Cuellar MC. Feasibility study of an alkaline-based chemical treatment for the purification of polyhydroxybutyrate produced by a mixed enriched culture. AMB Express 2015; 5(1): 5.
[http://dx.doi.org/10.1186/s13568-015-0096-5] [PMID: 25642402]

[34] Villano M, Valentino F, Barbetta A, Martino L, Scandola M, Majone M. Polyhydroxyalkanoates production with mixed microbial cultures: from culture selection to polymer recovery in a high-rate continuous process. N Biotechnol 2014; 31(4): 289-96.
[http://dx.doi.org/10.1016/j.nbt.2013.08.001] [PMID: 23954657]

[35] SamorA C, Abbondanzi F, Galletti P, *et al.* Extraction of polyhydroxyalkanoates from mixed microbial cultures: Impact on polymer quality and recovery. Bioresour Technol 2015; 189: 195-202.
[http://dx.doi.org/10.1016/j.biortech.2015.03.062] [PMID: 25889806]

[36] Tamisa J, Lu 3/4kov K, Jiang Y, van Loosdrecht MC, Kleerebezem R. Enrichment of Plasticicumulans acidivorans at pilot-scale for PHA production on industrial wastewater. J Biotechnol 2014; 192(Pt A): 161-9.
[http://dx.doi.org/10.1016/j.jbiotec.2014.10.022] [PMID: 25456060]

[37] Anterrieu S, Quadri L, Geurkink B, *et al.* Integration of biopolymer production with process water treatment at a sugar factory. N Biotechnol 2014; 31(4): 308-23.
[http://dx.doi.org/10.1016/j.nbt.2013.11.008] [PMID: 24361532]

[38] Morgan-Sagastume F, Hjort M, Cirne D, *et al.* Integrated production of polyhydroxyalkanoates (PHAs) with municipal wastewater and sludge treatment at pilot scale. Bioresour Technol 2015; 181: 78-89.
[http://dx.doi.org/10.1016/j.biortech.2015.01.046] [PMID: 25638407]

[39] Ntaikou I, Valencia Peroni C, Kourmentza C, *et al.* Microbial bio-based plastics from olive-mill

wastewater: Generation and properties of polyhydroxyalkanoates from mixed cultures in a two-stage pilot scale system. J Biotechnol 2014; 188: 138-47.
[http://dx.doi.org/10.1016/j.jbiotec.2014.08.015] [PMID: 25157746]

[40] Janarthanan OM, Laycock B, Montano-Herrera L, *et al.* Fluxes in PHA-storing microbial communities during enrichment and biopolymer accumulation processes. N Biotechnol 2016; 33(1): 61-72.
[http://dx.doi.org/10.1016/j.nbt.2015.07.007] [PMID: 26257140]

[41] Maurer M, Gujer W, Hany R, Bachmann S. Intracellular carbon flow in phosphorus accumulating organisms from activated sludge systems. Water Res 1997; 31(4): 907-17.
[http://dx.doi.org/10.1016/S0043-1354(96)00369-7]

[42] Lemos PC, Serafim LS, Santos MM, Reis MA, Santos H. Metabolic pathway for propionate utilization by phosphorus-accumulating organisms in activated sludge: 13C labeling and *in vivo* nuclear magnetic resonance. Appl Environ Microbiol 2003; 69(1): 241-51.
[http://dx.doi.org/10.1128/AEM.69.1.241-251.2003] [PMID: 12514001]

[43] Lemos PC, Dai Y, Yuan Z, Keller J, Santos H, Reis MA. Elucidation of metabolic pathways in glycogen-accumulating organisms with *in vivo* 13C nuclear magnetic resonance. Environ Microbiol 2007; 9(11): 2694-706.
[http://dx.doi.org/10.1111/j.1462-2920.2007.01382.x] [PMID: 17922754]

[44] Rehm BH. Polyester synthases: natural catalysts for plastics. Biochem J 2003; 376(Pt 1): 15-33.
[http://dx.doi.org/10.1042/bj20031254] [PMID: 12954080]

[45] Tsuge T, Hyakutake M, Mizuno K. Class IV polyhydroxyalkanoate (PHA) synthases and PHA-producing Bacillus. Appl Microbiol Biotechnol 2015; 99(15): 6231-40.
[http://dx.doi.org/10.1007/s00253-015-6777-9] [PMID: 26135986]

[46] Solaiman DK, Ashby RD. Rapid genetic characterization of poly(hydroxyalkanoate) synthase and its applications. Biomacromolecules 2005; 6(2): 532-7.
[http://dx.doi.org/10.1021/bm0493640] [PMID: 15762609]

[47] SteinbA1/4chel A, Pieper U. Production of a copolyester of 3-hydroxybutyric acid and 3-hydroxyvaleric acid from single unrelated carbon sources by a mutant of Alcaligenes eutrophus. Appl Microbiol Biotechnol 1992; 37: 1-6.
[http://dx.doi.org/10.1007/BF00174193]

[48] Tanaka K, Ishizaki A, Kanamaru T, Kawano T. Production of poly(D-3-hydroxybutyrate) from CO(2), H(2), and O(2) by high cell density autotrophic cultivation of Alcaligenes eutrophus. Biotechnol Bioeng 1995; 45(3): 268-75.
[http://dx.doi.org/10.1002/bit.260450312] [PMID: 18623147]

[49] Tanaka K, Miyawaki K, Yamaguchi A, Khosravi-Darani K, Matsusaki H. Cell growth and P(3HB) accumulation from CO2 of a carbon monoxide-tolerant hydrogen-oxidizing bacterium, Ideonella sp. O-1. Appl Microbiol Biotechnol 2011; 92(6): 1161-9.
[http://dx.doi.org/10.1007/s00253-011-3420-2] [PMID: 21695533]

[50] Khosravi-Darani K, Mokhtari Z-B, Amai T, Tanaka K. Microbial production of poly(hydroxybutyrate) from C1 carbon sources. Appl Microbiol Biotechnol 2013; 97(4): 1407-24.
[http://dx.doi.org/10.1007/s00253-012-4649-0] [PMID: 23306640]

[51] Doi Y, Segawa A, Nakamura S, Kunioka M. Production of biodegradable copolyesters by Alcaligenes eutrophus. In: Dawes EA, Ed. Novel biodegradable microbial polymers. Dordrecht, Netherlands: Kluwer Press 1990; pp. 37-48.
[http://dx.doi.org/10.1007/978-94-009-2129-0_4]

[52] Chen G-Q. Plastics completely synthesized by bacteria: Polyhydroxyalkanoates. In: Chen G-Q, Ed. Plastics from Bacteria: Natural Functions and Applications. Berlin/Heidelberg, Germany: Springer 2010; pp. 17-37.
[http://dx.doi.org/10.1007/978-3-642-03287-5_2]

[53] Koller M, Gasser I, Schmid F, Berg G. Linking ecology with economy: Insights into polyhydroxyalkanoate-producing microorganisms. Eng Life Sci 2011; 11(3): 222-37.
[http://dx.doi.org/10.1002/elsc.201000190]

[54] Magdouli S, Brar SK, Blais JF, Tyagi RD. How to direct the fatty acid biosynthesis towards polyhydroxyalkanoates production? Biomass Bioenergy 2015; 74: 268-79.
[http://dx.doi.org/10.1016/j.biombioe.2014.12.017]

[55] Mino T. Microbial selection of polyphosphate-accumulating bacteria in activated sludge wastewater treatment processes for enhanced biological phosphate removal. Biochemistry (Mosc) 2000; 65(3): 341-8.
[PMID: 10739477]

[56] Bengtsson S. The utilization of glycogen accumulating organisms for mixed culture production of polyhydroxyalkanoates. Biotechnol Bioeng 2009; 104(4): 698-708.
[PMID: 19530079]

[57] Satoh H, Iwamoto Y, Mino T, Matsuo T. Activated sludge as a possible source of biodegradable bioplastics. Water Sci Technol 1998; 38(2): 103-9.
[http://dx.doi.org/10.1016/S0273-1223(98)00435-1]

[58] Pratt S, Werker A, Morgan-Sagastume F, Lant P. Microaerophilic conditions support elevated mixed culture polyhydroxyalkanoate (PHA) yields, but result in decreased PHA production rates. Water Sci Technol 2012; 65(2): 243-6.
[http://dx.doi.org/10.2166/wst.2012.086] [PMID: 22233901]

[59] Lee WS, Chua AS, Yeoh HK, Ngoh GC. A review of the production and applications of waste-derived volatile fatty acids. Chem Eng J 2014; 235: 83-99.
[http://dx.doi.org/10.1016/j.cej.2013.09.002]

[60] Albuquerque MG, Torres CA, Reis MA. Polyhydroxyalkanoate (PHA) production by a mixed microbial culture using sugar molasses: effect of the influent substrate concentration on culture selection. Water Res 2010; 44(11): 3419-33.
[http://dx.doi.org/10.1016/j.watres.2010.03.021] [PMID: 20427069]

[61] Serafim LS, Lemos PC, Rossetti S, Levantesi C, Tandoi V, Reis MA. Microbial community analysis with a high PHA storage capacity. Water Sci Technol 2006; 54(1): 183-8.
[http://dx.doi.org/10.2166/wst.2006.386] [PMID: 16898151]

[62] Lemos PC, Levantesi C, Serafim LS, Rossetti S, Reis MA, Tandoi V. Microbial characterisation of polyhydroxyalkanoates storing populations selected under different operating conditions using a cell-

sorting RT-PCR approach. Appl Microbiol Biotechnol 2008; 78(2): 351-60. [http://dx.doi.org/10.1007/s00253-007-1301-5] [PMID: 18193421]

[63] Majone M, Dircks K, Beim JJ. Aerobic storage under dynamic conditions in activated sludge. The state of the art. Water Sci Technol 1999; 39(1): 61-73. [http://dx.doi.org/10.1016/S0273-1223(98)00776-8]

[64] van Loosdrecht MC, Pot MA, Heijnen JJ. Importance of bacterial storage polymers in bioprocesses. Water Sci Technol 1997; 35(1): 41-7. [http://dx.doi.org/10.1016/S0273-1223(96)00877-3]

[65] Reis MA, Albuquerque M, Villano M, Majone M. Mixed Culture Processes for Polyhydroxyalkanoate Production from Agro-Industrial Surplus / Wastes as Feedstocks. In: Mittal V, Ed. Rewable Polymers: Synthesis, Processing and Technology. Wiley, Inc. 2013.

[66] Johnson K, Jiang Y, Kleerebezem R, Muyzer G, van Loosdrecht MC. Enrichment of a mixed bacterial culture with a high polyhydroxyalkanoate storage capacity. Biomacromolecules 2009; 10(4): 670-6. [http://dx.doi.org/10.1021/bm8013796] [PMID: 19193058]

[67] Jiang Y, Sorokin DY, Kleerebezem R, Muyzer G, van Loosdrecht M. Plasticicumulans acidivorans gen. nov., sp. nov., a polyhydroxyalkanoate-accumulating gammaproteobacterium from a sequencing-batch bioreactor. Int J Syst Evol Microbiol 2011; 61(Pt 9): 2314-9. [http://dx.doi.org/10.1099/ijs.0.021410-0] [PMID: 20971833]

[68] Dionisi D, Carucci G, Papini MP, Riccardi C, Majone M, Carrasco F. Olive oil mill effluents as a feedstock for production of biodegradable polymers. Water Res 2005; 39(10): 2076-84. [http://dx.doi.org/10.1016/j.watres.2005.03.011] [PMID: 15913705]

[69] Bengtsson S, Werker A, Christensson M, Welander T. Production of polyhydroxyalkanoates by activated sludge treating a paper mill wastewater. Bioresour Technol 2008; 99(3): 509-16. [http://dx.doi.org/10.1016/j.biortech.2007.01.020] [PMID: 17360180]

[70] Queirós D, Rossetti S, Serafim LS. PHA production by mixed cultures: a way to valorize wastes from pulp industry. Bioresour Technol 2014; 157: 197-205. [http://dx.doi.org/10.1016/j.biortech.2014.01.099] [PMID: 24556373]

[71] Din MF, Mohanadoss P, Ujang Z, *et al.* Development of Bio-PORecA(r) system for polyhydroxyalkanoates (PHA) production and its storage in mixed cultures of palm oil mill effluent (POME). Bioresour Technol 2012; 124: 208-16. [http://dx.doi.org/10.1016/j.biortech.2012.08.036] [PMID: 22989648]

[72] Liu C, Wang H, Xing W, Wei L. Composition diversity and nutrition conditions for accumulation of polyhydroxyalkanoate (PHA) in a bacterial community from activated sludge. Appl Microbiol Biotechnol 2013; 97(21): 9377-87. [http://dx.doi.org/10.1007/s00253-013-5165-6] [PMID: 23974368]

[73] Ramasahayam SK, Guzman L, Gunawan G, Viswanathan T. A Comprehensive Review of Phosphorus Removal Technologies and Processes. J Macromol Sci Part A 2014; 51(6): 538-45. [http://dx.doi.org/10.1080/10601325.2014.906271]

[74] Michinaka A, Arou J, Onuki M, Satoh H, Mino T. Analysis of polyhydroxyalkanoate (PHA) synthase gene in activated sludge that produces PHA containing 3-hydroxy-2-methylvalerate. Biotechnol

Bioeng 2007; 96(5): 871-80.
[http://dx.doi.org/10.1002/bit.21085] [PMID: 16933327]

[75] Ciesielski S, Pokoj T, Klimiuk E. Molecular insight into activated sludge producing polyhydroxyalkanoates under aerobic-anaerobic conditions. J Ind Microbiol Biotechnol 2008; 35(8): 805-14.
[http://dx.doi.org/10.1007/s10295-008-0352-7] [PMID: 18418634]

[76] Coats ER, Loge FJ, Wolcott MP, Englund K, McDonald AG. Synthesis of polyhydroxyalkanoates in municipal wastewater treatment. Water Environ Res 2007; 79(12): 2396-403.
[http://dx.doi.org/10.2175/106143007X183907] [PMID: 18044356]

[77] Bengtsson S, Werker A, Welander T. Production of polyhydroxyalkanoates by glycogen accumulating organisms treating a paper mill wastewater. Water Sci Technol 2008; 58(2): 323-30.
[http://dx.doi.org/10.2166/wst.2008.381] [PMID: 18701781]

[78] Dionisi D, Beccari M, Di Gregorio S, Majone M, Papini MP, Vallini G. Storage of biodegradable polymers by an enriched microbial community in a sequencing batch reactor operated at high organic load rate. J Chem Technol Biotechnol 2005; 80(11): 1306-18.
[http://dx.doi.org/10.1002/jctb.1331]

[79] Lee SH, Kim JH, Mishra D, Ni Y-Y, Rhee YH. Production of medium-chain-length polyhydroxyalkanoates by activated sludge enriched under periodic feeding with nonanoic acid. Bioresour Technol 2011; 102(10): 6159-66.
[http://dx.doi.org/10.1016/j.biortech.2011.03.025] [PMID: 21463934]

[80] Wang Y, Cai J, Lan J, *et al.* Biosynthesis of poly(hydroxybutyrate-hydroxyvalerate) from the acclimated activated sludge and microbial characterization in this process. Bioresour Technol 2013; 148: 61-9.
[http://dx.doi.org/10.1016/j.biortech.2013.08.102] [PMID: 24035892]

[81] Villano M, Lampis S, Valentino F, Vallini G, Majone M, Beccari M. Effect of hydraulic and organic loads in Sequencing Batch Reactor on microbial ecology of activated sludge and storage polyhydroxyalkanoates. Chem Eng Trans 2010; 20: 187-92.

[82] Valentino F, Beccari M, Fraraccio S, Zanaroli G, Majone M. Feed frequency in a sequencing batch reactor strongly affects the production of polyhydroxyalkanoates (PHAs) from volatile fatty acids. N Biotechnol 2014; 31(4): 264-75.
[http://dx.doi.org/10.1016/j.nbt.2013.10.006] [PMID: 24184912]

[83] Liu Z, Wang Y, He N, *et al.* Optimization of polyhydroxybutyrate (PHB) production by excess activated sludge and microbial community analysis. J Hazard Mater 2011; 185(1): 8-16.
[http://dx.doi.org/10.1016/j.jhazmat.2010.08.003] [PMID: 20970922]

[84] Liu HY, Hall PV, Darby JL, *et al.* Production of polyhydroxyalkanoate during treatment of tomato cannery wastewater. Water Environ Res 2008; 80(4): 367-72.
[http://dx.doi.org/10.2175/106143007X221535] [PMID: 18536488]

[85] Campanari S, e Silva FA, Bertin L, Villano M, Majone M. Effect of the organic loading rate on the production of polyhydroxyalkanoates in a multi-stage process aimed at the valorization of olive oil mill wastewater. Int J Biol Macromol 2014; 71: 34-41.
[http://dx.doi.org/10.1016/j.ijbiomac.2014.06.006] [PMID: 24950311]

[86] Albuquerque MG, Concas S, Bengtsson S, Reis MA. Mixed culture polyhydroxyalkanoates production from sugar molasses: the use of a 2-stage CSTR system for culture selection. Bioresour Technol 2010; 101(18): 7123-33.
[http://dx.doi.org/10.1016/j.biortech.2010.04.019] [PMID: 20434907]

[87] CavaillA(c) L, Grousseau E, Pocquet M, *et al.* Polyhydroxybutyrate production by direct use of waste activated sludge in phosphorus-limited fed-batch culture. Bioresour Technol 2013; 149: 301-9.
[http://dx.doi.org/10.1016/j.biortech.2013.09.044] [PMID: 24121372]

[88] Lemos PC, Serafim LS, Ramos AM, Reis MA. Bioplastics from Waste Materials. In: Lens P, Hamelers B, Hoitink H, Bidlingmeie W, Eds. Resource Recovery and Reuse in Organic Waste Management. IWA Publis 2004; pp. 423-40.

[89] Philip S, Keshavarz T, Roy I. Polyhydroxyalkanoates: biodegradable polymers with a range of applications. J Chem Technol Biotechnol 2007; 82: 233-47.
[http://dx.doi.org/10.1002/jctb.1667]

[90] Zhang X, Luo R, Wang Z, Deng Y, Chen G-Q. Application of (R)-3-hydroxyalkanoate methyl esters derived from microbial polyhydroxyalkanoates as novel biofuels. Biomacromolecules 2009; 10(4): 707-11.
[http://dx.doi.org/10.1021/bm801424e] [PMID: 19249855]

[91] Lopez NI, Pettinari MJ, Nikel PI, MA(c)ndez BS. Polyhydroxyalkanoates: Much More than Biodegradable Plastics. Adv Appl Microbiol 2015; 93: 73-106.
[http://dx.doi.org/10.1016/bs.aambs.2015.06.001] [PMID: 26505689]

[92] Verlinden RA, Hill DJ, Kenward MA, Williams CD, Radecka I. Bacterial synthesis of biodegradable polyhydroxyalkanoates. J Appl Microbiol 2007; 102(6): 1437-49.
[http://dx.doi.org/10.1111/j.1365-2672.2007.03335.x] [PMID: 17578408]

[93] Nikodinovic J, Kenny ST, Babu RP, Woods T, Blau WJ, O'Connor KE. The conversion of BTEX compounds by single and defined mixed cultures to medium-chain-length polyhydroxyalkanoate. Appl Microbiol Biotechnol 2008; 80(4): 665-73.
[http://dx.doi.org/10.1007/s00253-008-1593-0] [PMID: 18629491]

CHAPTER 5

Ralstonia Eutropha and the Production of Value Added Products: Metabolic Background of the Wild-Type Strain and its Role as a Diverse, Genetically-Engineered Biocatalyst Organism

Christopher J. Brigham[*], **Abdulrahman A. Kehail** and **Jacob D. Palmer**

Department of Bioengineering, University of Massachusetts Dartmouth, 285 Old Westport Road, North Dartmouth, MA 02747, USA

Abstract: *Ralstonia eutropha* (a.k.a., *Cupriavidus necator*) is at the forefront of the research movement towards sustainable bioproductions. There are two principal reasons for the popularity of *R. eutropha* in this arena: 1) the organism has a versatile metabolism and can utilize a wide variety of carbon sources from sugars and lipids to carbon dioxide and aromatic compounds for growth and fermentative bioproductions, 2) it can store a large amount of carbon as intracellular polymer. *R. eutropha* is known as the model system for polymer (polyhydroxyalkanoate) biosynthesis and mobilization. Many valuable works have been published on this specific topic. In the recent years, however, *R. eutropha* is being used as a host organism for heterologous bioconversions, such as biofuel synthesis. Again, researchers are exploiting the metabolic versatility of the organism to create novel products and processes. This chapter chronicles the key discoveries of *R. eutropha* biology and presents the biomanufacturing context that inspired many research groups to tailor the organism as a novel biocatalyst.

Keywords: Autotrophy, Biofuel, Biopolymer, Bioremediation, *Cupriavidus necator*, Fed-batch cultivation, Fermentation, Isobutanol, Medical products, Metabolism, Polyhydroxyalkanoate (PHA), *Ralstonia eutropha.*

[*] **Corresponding author Christopher J. Brigham:** Department of Bioengineering, University of Massachusetts Dartmouth, Dartmouth, United States of America; Tel: 15089999149 Email: cbrigham@mit.edu.

Martin Koller (Ed.)

INTRODUCTION

In the early 1800s, the French gastronome Jean-Anthelme Brillat-Savarin said "Tell me what you eat, and I'll tell you what you are." This adage, when applied to the world of microorganisms, is most poignant, as many bacteria are defined and even named for the substrates from which they derive nutrition. Indeed, it is possible that Brillat-Savarin was familiar with the 1676 discoveries by Antonie van Leeuwenhoek of "animalcules" [*i.e.,* bacteria] in many different environmental niches. Nowadays, we understand that these microbes populate their respective niches due to the availability of usable nutrients. Since the advent of microbial ecology and subsequent physiological characterization, microbiologists seek to classify bacteria in part based on the compounds they are capable of eating.

Ralstonia eutropha, also named *Cupriavidus necator,* is a Gram negative bacterium that is found in soil and aquatic environments. The species name "*eutropha*" can be defined loosely as "good eating." Indeed, nutrition and carbon conversion have defined the bacterium almost since its discovery in 1952 [1]. *R. eutropha* is a master carbon-utilizer and carbon-converter, and this attribute has defined its role as a model organism for producing biotechnologically relevant products like polyhydroxyalkanoate (PHA) polymers. This bacterium is capable of growth in many different niches: in the presence or absence of O_2, in a chemoautotrophic environment, and, obviously, in a heterotrophic environment. It is this versatility that has made *R. eutropha* a microbial chassis for bioconversion of cheap carbon (waste) to value added compounds.

A Brief History of *R. eutropha* (*C. necator*): What's in a Name?

In 1952, Albert Schatz and Carlton Bovell isolated a bacterium from "lawn soil" that exhibited hydrogenase activity. The isolated bacterium was also shown to assimilate a variety of organic acids [*e.g.,* acetate, α-ketoglutarate]. Most notable about this find was the highly active hydrogenases expressed by the organism and, to Schatz and Bovell, the ease of which the organism could be manipulated. To the latter end, the newly isolated organism was named *Hydrogenomonas facilis*, "because of the ease with which it lent itself to experimentation" [1, 2]. The same organism was being characterized in Europe by Wilde and coworkers and was named *Hydrogenomonas eutropha* because of its ability to grow robustly on a variety of carbon sources, including CO_2 [3]. In fact, the prototypical wild-type strain, H16, was isolated by Wilde. Almost a decade later, *H. eutropha* was renamed *Alcaligenes eutrophus* because of its physiological indicators as a rod-shaped denitrifying bacterium that produces no pigment [4]. It had been suggested that the name *Hydrogenomonas* was illegitimately applied and *A. eutrophus* was listed in the Approved List of Bacterial Names [4, 5]. In 1995, the name was changed again to *Ralstonia eutropha.* It was determined that select species from the genus *Burkholderia* and *Alcaligenes* were phenotypically similar to each other and, with 16S rRNA/rDNA sequence comparison data, were determined to fit best in the genus *Ralstonia* [5]. This classification stood until 2004, when a new genus *Wautersia* was established and *R. eutropha* was then classified as *Wautersia eutropha.* The *Wautersia* lineage was established mainly as a result of re-analysis of 16S rRNA/rDNA sequences, and *R. eutropha*'s renaming was a result of a previous observation of two distinct sub-classes of *Ralstonia* [6]. The *W. eutropha* classification lasted less than a year when it was renamed *Cupriavidus necator* as a result of DNA-DNA hybridization experiments. The authors of the study, Vandamme and Coenye, cited Rules 15, 17, 23a and 37a [1] of the International Code of Nomenclature of Bacteria [7]. It should be noted that members of the genus *Cupriavidus* tend to exhibit resistance to metals (Hence the first part of the name *Cupriavidus*, as in copper]. Indeed, several strains of *R. eutropha* (*C. necator*) have demonstrated resistance to metals like nickel, cadmium and others [8]. There have also been found putative metal tolerance genes in the genome sequence of *R. eutropha* strain H16 [9]. Vandamme and Coenye state that there can be only one correct name for a bacterial species, according to the International Code of Nomenclature of Bacteria [7, 10]. They go on to state that the correct genus and species name for *R. eutropha* should be *Cupriavidus necator*, since the latter was established in 1987, well before either the *R. eutropha* or *W. eutropha* names were designated [7]. Thus, *C. necator* has become a generally accepted name for the organism, and, to date, no further name changes appear to have been suggested. However, it should be noted that several pioneering researchers, namely the Sinskey and Steinbüchel groups and offshoots of those groups, have elected to continue use of the *R. eutropha* name and still publish findings based on that designation. For the remainder of the chapter, we will refer to the organism as *Ralstonia eutropha*, although it should be noted that we acknowledge the accepted name of *Cupriavidus necator*.

R. eutropha Research, the Early Years

A schematic timeline of important discoveries pertaining to *R. eutropha* is shown in (Fig. **1**). When *R. eutropha* was discovered almost 50 years ago, there was a keen interest among researchers in hydrogen bacteria and autotrophic microbial growth. It is because of this attribute, the ability to produce biomass from CO_2, H_2, and O_2, that *R. eutropha* became an organism of great significance. To this day, the mechanisms of *R. eutropha* autotrophy remain an area where many advances are taking place, leading to a greater understanding of the physiology and the biochemistry behind this phenotype.

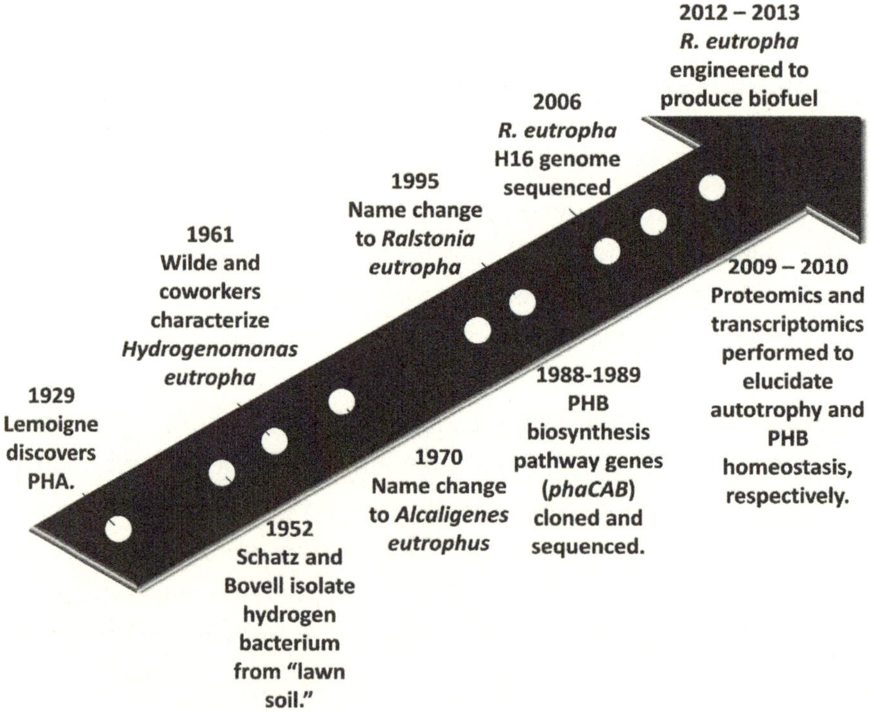

Fig. (1). A timeline indicating some of the important events of *R. eutropha*. Naturally, there are many more important events than could fit on this timeline, but here we indicate the key events where important metabolic pathways were discovered (*e.g.*, autotrophic growth, PHB biosynthesis) and exploited (*e.g.*, biofuel synthesis by recombinant organisms).

Early studies of cellular metabolism examined oxygen deficient growth and production. It is well known that *R. eutropha* can also utilize nitrate and nitrite as terminal electron acceptors [11]. Vollbrecht, *et al.* [11, 12] demonstrated the secretion of ethanol, propanol, 2,3-butanediol, succinate and *cis*-aconitate, among other metabolites, when respiration was restricted. In restricted oxygen growth conditions, it was found that polyhydroxybutyrate (PHB) could still be produced, totaling up to 50% of the cell's dry weight [11]. This phenomenon suggested that PHB can be considered "an inert cellular fermentation product" [13].

One of the early motivations for study of autotrophic bacteria like *R. eutropha* was its potential use as a sustainable food source [14]. In fact, the nutritional quality of proteins and lipids in *R. eutropha* was measured by feeding dried biomass to rats [15, 16]. It was determined that the "single-cell protein" provided by *R. eutropha* could be used as a nutritional supplement. Given the autotrophic growth capabilities of the bacteria, it was suggested that *R. eutropha* be used as both a CO_2 mitigation mechanism in space crafts and concomitantly as nutritional supplements for astronauts in space flight [14]. However, it was determined that the lipid content in *R. eutropha* cells was poorly absorbed by rats and did not promote robust growth [16]. The "lipid content" of the cells as described in these nutritional studies is the intracellular polyester PHB. Since *R. eutropha* cells are capable of storing up to 90% of their dry weight as this poorly-digestible PHB [17], the idea of using these bacteria as single-cell protein factories was quickly shelved. Interestingly, a recent study proposed turning the initial disadvantage of the *R. eutropha* single-cell protein approach into an advantage for a polymer recovery scheme. Rats were fed lyophilized cells containing $<$40 wt% PHB and $>$40 wt% protein. Fecal pellets recovered from rats contained over 95 wt% PHB. Recovery yield of PHB was 74%, suggesting a potential biological method for recovering polymer [18].

Since these early nutritional studies, research on the physiology of *R. eutropha* has gone in two often-overlapping directions: 1) the study of intracellular PHA biosynthesis and mobilization, and 2) the study of autotrophic growth. In both arenas, *R. eutropha* has proven to be a model organism and has facilitated numerous discoveries that have confirmed its versatility of metabolism and proclivity for production of value-added compounds.

POLYHDROXYALKANOATE HOMEOSTASIS

PHAs are important storage compounds found in a variety of prokaryotes that promote their survivability in times of carbon starvation and exist in the form of an intracellular inclusion, known as a granule. In organisms like *R. eutropha*, these granular polymers are stored as excess carbon during times of nutrient (*e.g.*, nitrogen, phosphate) starvation. The lack of nutrients generally suggests that the cells are in stasis, *i.e.,* they do not divide into daughter cells, due to the lack of specific raw materials. However, this does not prevent the cells from storing excess carbon as PHA to be used when the environmental conditions become more favorable. The ability of *R. eutropha* to store up to ~90% of its dry weight as PHAs, as well as its genetic manipulability, makes it a model organism in which to study PHA homeostasis [13, 19, 20]. The interest in PHA, and moreover the interest in *R. eutropha* as a model organism, stems from the observations that PHA polymers themselves are quite remarkable due to their biological nature. While, in general, they exhibit thermal and mechanical properties that are similar to petroleum-based plastics, PHAs are also biodegradable and biocompatible. These attributes make the polymers an attractive option as a replacement for the aforementioned petroleum-based thermoplastics, to be used potentially in both commercial and medical applications. PHAs can be broken down, theoretically, into carbon dioxide and water and will degrade in soil, sewage, or within a biological host [21, 22].

PHAs are a large family of polymers, with many potential permutations, as over 150 monomer types have been found to be incorporated [23]. Wild-type *R. eutropha* synthesizes PHA with short chain length (*scl*) monomers like 3-hydroxybutyrate (3HB) and 3-hydroxyvalerate (3HV). In the presence of most types of carbon source, wild-type *R. eutropha* will synthesize the homopolymer PHB [24, 25]. The synthesis of PHB starting from the central metabolite acetyl-CoA is a very simple, straightforward pathway and is accomplished by a process involving three enzymatic activities, starting with acetoacetyl-CoA catalysis by β-ketothiolase (PhaA). The resulting acetoacetyl-CoA is converted to 3-hydroxybutyryl-CoA (3HB-CoA) by the acetoacetyl-CoA dehydrogenase enzyme (PhaB). Lastly, the 3-hydroxybutyryl-CoA is used as the substrate in the polymerization of PHB by poly(3-hydroxyalkanoate) synthase (PhaC). Fig. **2** illustrates multiple pathways that can lead to PHA biosynthesis in wild-type and recombinant *R. eutropha*.

The simplicity of the biochemical pathways does not do justice to the complex nature of PHB granules and complex regulation of PHB homeostasis. To date, researchers have still not assigned concrete roles to every PHA-associated protein or enzyme in *R. eutropha*. The structure of PHB granules is highly organized, covered by an array of 19 different proteins. These proteins and others involved in PHA homeostasis are discussed below.

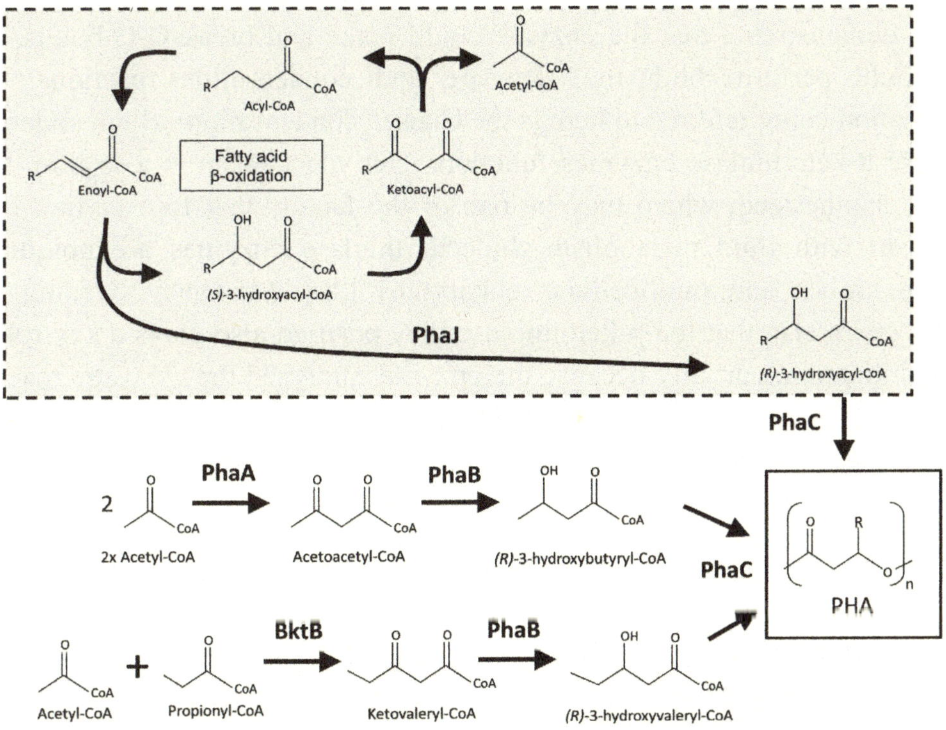

Fig. (2). Pathways to PHA. The most common pathway, the PHB pathway, proceeds via the PhaA, PhaB, and PhaC enzymes as described in the text. PHBHV can be produced in the presence of odd-chain-length carbon substrates starting with the thiolase BktB catalyzing the condensation of acetyl-CoA and propionyl-CoA to ketovaleryl-CoA. In gray is the recombinant pathway of P(HB-co-HHx) biosynthesis used by some organisms involving the native fatty acid β-oxidation pathway and recombinant PhaJ and PhaC enzymes. R. eutropha does, however, possess PhaJ enzymes that can be overexpressed and used in this engineered pathway.

PHA Homeostasis - Key Proteins

β-Ketothiolase (PhaA, BktB)

In the late 1980's, Professor Satoru Masamune of Massachusetts Institute of Technology (MIT) kindled the interest of his lab group about how organisms make carbon-carbon (C-C) bonds. Roughly 15 years earlier, Oeding and Schlegel had begun to examine this very phenomenon in *Hydrogenomonas* (*Ralstonia*) *eutropha* by partially purifying a β-ketothiolase enzyme from strain H16. Not only was the β-ketothiolase enzyme, now known as PhaA, purified, but Oeding and Schlegel demonstrated that the enzyme could make and break C-C bonds, *i.e.*, PhaA could perform both the cleavage and condensation reactions. The "condensation being referred to here is the Claisen condensation, which underpins how these β-ketothiolase enzymes function. This mechanism is also how fatty acids are synthesized, which may be one of the factors that forever link PHA metabolism with lipid metabolism. Indeed, thiolase enzymes are ubiquitous among microbes and multicellular eukaryotes [26]. Curiously, Oeding and Schlegel claim also that the β-ketothiolase they purified also plays a key role in PHB mobilization, partially because the enzyme catalyzed the cleavage reaction *in vitro* [27]. Of course, Masamune and colleagues investigated the PhaA activity, first in *Zoogloea ramigera*, where active site cysteine residues were identified [28], and then *R. eutropha*.

In 1988, Schubert *et al.*, documented the transfer of β-ketothiolase activity from *R. eutropha* to *E. coli*, which corresponded to the transfer of the ability to produce PHB [29]. This suggested 2 things: 1) β-ketothiolase activity was required for PHB synthesis, and 2) the genes for *phaA* and the PHB synthase were in somewhat close proximity in *R. eutropha*. Less than a year later, predicted amino acid sequences of the *Z. ramigera* and *R. eutropha* PhaA and PhaB enzymes were published. The DNA sequence showed that *phaA* and *phaB* genes were adjacent to each other [30]. Recently, it has been shown that *R. eutropha* contains multiple genes (~15) that potentially encode ketothiolases [9, 31]. These genes were systematically deleted from the *R. eutropha* genome and effect on physiology was examined. Individually, the deletion of *phaA* was shown to have the greatest impact on PHB biosynthesis [31]. Multiple ketothiolase gene deletions that included *phaA, bktB* [see below] and locus tag H16_A0170 saw dramatic decreases in PHB production [31, 32]. Recently, the crystal structure of the *R. eutropha* PhaA enzyme has been reported [33].

In 1987, Doi, *et al.* demonstrated the biosynthesis of poly(hydroxybutyrate-*co*-hydroxyvalerate) copolymer [PHBHV] in *R. eutropha* by feeding ^{13}C-labeled acetate and propionate [34]. This suggested that wild-type *R. eutropha* can synthesize more than just the homopolymer PHB. This finding also established that carbon source can influence the type of PHA polymer synthesized. A ketothiolase, termed BktB, was identified in *R. eutropha*, which possesses different substrate specificity compared to PhaA [35]. Using thiolytic cleavage as the assay for activity, BktB was shown to be able to use C5 carbon substrates as well as C4, whereas PhaA exhibited a preference only for C4 substrates. Furthermore, engineered *E. coli* systems containing only BktB as the functional ketothiolase were shown to produce PHBHV in the presence of propionate. A system using PhaA as the sole ketothiolase was shown to produce mainly PHB in the presence of propionate. Another ketothiolase, BktC, was also hypothesized to exist in R. eutropha [35], but little characterization of this gene was performed and no follow up work has been done.

Acetoacetyl-CoA Reductase (PhaB)

Schubert *et al.* demonstrated both NADH-dependent and NADPH-dependent acetoacetyl-CoA reductase activity in *R. eutropha* H16 [29]. Peoples and Sinskey also demonstrated acetoacetyl-CoA reductase activity in an *E. coli* system expressing genes from a DNA fragment originating from *R. eutropha* H16 [30]. This group also reported the gene and putative protein sequence of what would later be called *phaB1* (PhaB1). In a similar case to PhaA, 17 other isologs of PhaB are predicted in *R. eutropha* [9]. The deletion of genes encoding three isologs, termed *phaB1*, *phaB2* and *phaB3*, was investigated and the roles of each reductase in PHB homeostasis in *R. eutropha* H16 were further elucidated. No single deletion of a *phaB* gene was observed to result in an adverse effect on PHB biosynthesis. However, deletion of both *phaB1* and *phaB3* genes resulted in a dramatically decreased production of PHB under nutrient limiting conditions when cells were grown on fructose as the carbon source. When palm oil was used as the carbon source, the *phaB1* and *phaB2* double deletion strain produced only small amounts of PHB, in contrast to the phenotype of this mutant strain in fructose cultures [36]. Examination of global gene expression of *R. eutropha* in the presence of fructose as compared to a triacylglycerol, it was shown that growth on triacylglycerols decreased *phaB3* gene expression [24, 36], which would explain the phenotype observed with the *phaB1* and *phaB2* double mutant strain. These observations also suggest that *R. eutropha* has ways of modulating PHB biosynthesis based on the types of carbon source present in the extracellular milieu. Examination of enzyme activity of whole cell extracts of wild-type and *phaB* mutant strains suggests that NADPH is the cofactor that supplies the reducing power in the PhaB reaction [36]. The crystal structure of *R. eutropha* PhaB has been solved, and this is presumably the structure of PhaB1 though this is unclear based on the published report [37].

PHA Synthases (PhaC)

Arguably the most important protein of the entire polymer synthesis process is the PHA synthase, PhaC. Four separate classes (Class I – IV) of PHA synthase have been characterized [38], with the class with the most individual constituents being Class I, which are homodimeric enzymes that polymerize *scl*-PHAs. The active PhaC enzyme expressed by wild-type *R. eutropha* is a Class I enzyme that is responsible for using 3-hydroxybutyryl-CoA as a substrate to produce PHB. This protein is present in the cell cytoplasm as both monomeric and homodimeric forms in equilibrium, but only the homodimeric form is active [38, 39]. Schubert *et al.* demonstrated PhaC activity in *R. eutropha* H16 [29]. Less than a year following, the gene sequence of *phaC* and predicted amino acid sequence of the gene product were reported. It was shown that *phaA1*, *phaB1* and *phaC* formed a PHA synthesis gene cluster on the chromosome of *R. eutropha* [40].

PhaC is constitutively expressed in the cell [24, 25], and the amount of synthase in the host plays an important role in controlling the molecular weight and polydispersity of the resulting polymer where the polymers molecular weight is inversely related to the concentration of PhaC. This PHB synthase, like many other proteins discussed in this section, is located on the surface of PHB granules [41].

There has been much debate about the exact molecular mechanism of PhaC-mediated PHB polymerization. Early models for the polymerization mechanism using 3HB-CoA as the precursor, postulated the existence of two thiols in the active site of the enzyme that would pass the nascent PHB chain back and forth between them as a new precursor molecule made its way to the active site and ultimately resulted in elongation of the polymer [42, 43]. The active site thiol residue in PhaC was identified to be C319. Indeed, all PHA synthase enzymes have been shown to possess a "lipase box" in which the active site serine [traditional for lipase activity] is replaced by a cysteine (essential for PHA synthase activity). Such a lipase box is centered around this essential C319 residue in *R. eutropha* PhaC [23, 44]. However, to satisfy the original model of the nascent PHA chain being passed back and forth between thiol residues, it was predicted that this unknown 2nd thiol would have to be added to the enzyme by posttranslational modification [42]. Zhang and coworkers suggested that the homodimeric nature of PhaC could provide the two thiol groups with two C319 residues available in two active sites of the active PhaC homodimer [45]. This model was offered, even though an earlier study by the same group indicated that the active PhaC homodimer appears to only have one accessible active site with only one accessible C319 residue [46]. Another important residue, H508, has been demonstrated to act as a general base catalyst to activate the C319 in the active site of PhaC for nucleophilic attack. Alteration of either C319 or H508 results in severe reduction of PHB synthase activity in purified enzyme preparations. Completing the catalytic triad is D480. This residue does not seem to be essential for activity, according to mutagenesis studies, but it was shown using a specially formulated PHA synthase probe that deacylation activity is reduced in a D480N mutant, as compared to the wild-type PhaC [47].

Yuan and coworkers purified an epitope tagged PhaC heterologously from *E. coli.* Enzyme activity and substrate specificity were determined, but other essential kinetic parameters, such as K_m and V_{max}, were not reported [48]. The main reason for these kinetic parameters not being reported for *R. eutropha* PhaC and other class I PHA synthase enzymes is because these enzymes display a lag in enzyme activity *in vitro*. Hoping to alleviate this issue, an epitope tagged PhaC was purified from engineered *R. eutropha.* Initial enzyme activity studies showed a lack of lag phase, but further inspection revealed the presence of three specific subgroups of enzyme present in the preparation isolated from *R. eutropha* cells: 1) monomers, 2) dimers and 3) a high molecular weight fraction. The high molecular weight fraction was shown to exhibit PHB synthase activity with no lag phase, but was also found to contain other factors, such as oligo-HB and granule associated proteins like PhaP1. Monomer and dimer fractions did still exhibit a lag in activity [39]. For all PhaC enzymes, the rate of initiation (3HB initially binding to C319) is much slower than the rate of elongation, and this partially explains the differences in activity profiles seen between the monomer/dimer and high molecular weight fractions of the purified PhaC preparations [39]. Ignoring the lag phase, a recent study has demonstrated a K_m of 3300 μM 3HB-CoA for *R. eutropha* PhaC [49].

The genome sequence of *R. eutropha* strain H16 reveals the presence of another gene, *pha*C2, encoding a putative PHA synthase [9]. Little is known about *pha*C2, and thus it remains little-studied. The putative protein sequence of *pha*C2 shows the presence of a lipase box similar to that of *pha*C1, suggesting *pha*C2 might be a functional PHA synthase. However, no additional PHA synthase activity can be seen in *pha*C1 mutant strains of *R. eutropha* [47]. It was suggested that *pha*C2 may not be expressed at all under conditions tested [50]. This was confirmed by a whole-genome transcriptome analysis that showed the quantity of *pha*C2 transcript to be below the reasonable lower limit of expression [25]. *pha*C2 was demonstrated to actively produce PHB when overexpressed in the PHB-negative strain of *R. eutropha*, $PHB^{-}4$ [50]. However, *pha*C2 was shown not to facilitate polymer synthesis when expressed from a plasmid in other strains [50]. This contrast between results is puzzling, and no further investigation has been performed to determine if additional factors are needed that were present in $PHB^{-}4$, but not other strains of *R. eutropha*. A recent study has shown that *pha*C2 was present on purified PHB granule preparations, along with *pha*C1 and other granule associated proteins. The relative abundance of *pha*C2, compared to *pha*C1 in this case, was 100-200X less [51]. Still, the role of *pha*C2 in PHA homeostasis in *R. eutropha* cells remains a mystery.

Phasins (PhaP)

In 1995, a 24-kDa protein was discovered to be associated with the PHA granule surface and not freely soluble in the cell cytoplasm. This "phasin" protein was termed PhaP [52]. Nowadays, of course, this protein is called PhaP1, as there are now 7 known phasins, granule-associated proteins that accumulate onto the PHA granule surface [53 - 59]. The most abundant phasin on the surface of the PHB granule, and therefore the most notable, is PhaP1, which contributes substantially to the proteinaceous makeup of the granule surface [57, 60]. PhaP1 expression is has been shown to be induced by PHB in *R. eutropha*, and its expression levels increase linearly with PHB production [61, 62]. This is due to the regulation of PhaP1 expression by PhaR, the regulatory protein discussed below. The roles of PhaP1 in granule formation are controlling the amount, size, and number of granules within the cell, and likely preventing PHB crystallization [60, 63]. Phasin synthesis occurs only in the stationary phase of growth, when PHA is synthesized [60].

The role of phasins on the PHA granule surface is, at first glance, very simple. However, like PHA homeostasis in *R. eutropha*, the roles of phasins are very complex indeed. Whole genome transcriptome analysis showed that, while most phasin genes reached peak expression during PHB biosynthesis [nitrogen limitation], two were shown to have unusual gene expression pattern, with *phaP2* expression highest during PHB utilization and *phaP3* expression the highest during growth [25]. Recently, a two-hybrid analysis was performed of *R. eutropha* granule associated proteins. There was demonstrated to be a strong interaction between PhaP2 and PhaP4 [55], though the nature of this interaction is not yet well understood. Several phasin proteins (PhaP1-4) were also shown to interact with the intracellular PHB depolymerase enzyme, *phaZ*1 [55]. In fact, PHB mobilization has been demonstrated to involve PhaP1, according to *in vitro* studies [64]. Two new phasin proteins, PhaP6 and PhaP7, have been discovered and characterized. These proteins have been shown to be associated with the PHB granule, but no role has yet been uncovered for them in PHB homeostasis. However, overexpression of PhaP6 and PhaP7 results in localization of PHA granules at the cell poles [54].

Phasin Regulation by PhaR

Accumulation of the phasin PhaP1 has been shown to be dependent on the presence of PHA in the *R. eutropha* cell [65]. The regulation of PhaP1 expression has been shown to be controlled by an autoregulated repressor encoded by PhaR. Deletion of the *phaR* gene results in constitutive expression of PhaP1 [61]. This regulatory protein binds upstream of the *phaP1* gene at three different sites, according to DNaseI footprinting studies [58]. PhaR also appears to bind upstream of the *phaR* gene near the translational start codon [58]. Binding upstream of both *phaP1* and *phaR* allows PhaR to regulate its own expression along with the expression of PhaP1. Thus, PhaR is acting as a negative regulator for both *phaR* and *phaP1* expression. Quartz crystal microbalance studies show that PhaR binds more readily to the upstream region of the *phaP*1 gene than to the promoter of *phaR* [66]. This can be explained in part by the presence of three PhaR binding sites upstream of *phaP1* compared to one upstream of *phaR* [58]. However, the PhaR protein also binds to PHA granules, and PHA in this case acts as the inducer molecule of PhaP1 expression. Two separate domains of the PhaR protein have been postulated, one to modulate the protein's binding of DNA at the requisite regulatory regions and the other for binding PHA. Mutant PhaR proteins that lack DNA binding capabilities were shown to still be able to bind PHB [67]. The function of PhaR seems to be the repression of PhaP1 production when PHB is not being produced. When nascent PHA granules are formed, PhaR binds to the surface. This frees the upstream regions of both *phaR* and *phaP1* genes, so that transcription is activated. The newly transcribed PhaP1 then binds to the granule surface. If the granule continues to grow, the PhaR will continue to bind to it. However, when the intracellular granule ceases to grow, the released PhaR again binds to the upstream regions of both *phaR* and *phaP1* genes. In this sense, PhaR regulates the transcription of PhaP1, thus preventing the synthesis of PhaP1 when PHA synthesis is not occurring [59, 61, 68].

RT-qPCR studies on PHB-related gene expression demonstrated that *phaR* expression does not change significantly over the course of PHB homeostasis in *R. eutropha* [62]. Similarly, whole genome transcriptome studies showed that *phaR* gene expression levels remained at a constantly high level throughout [25]. While PhaR has been demonstrated to bind to target DNA sequences and to PHB, two-hybrid studies have shown that no significant interaction with other PHA homeostasis proteins has been determined [55].

Multifunctional Granule Associated Protein (PhaM)

Since PHA biosynthesis was first studied in *R. eutropha*, there has been much debate over how polymer granule formation is initiated. There have been three main models of PHA granule formation offered. The micelle model depicts PHA granule formation occurring analogous to the way detergent micelles expand in aqueous solution. Several active PhaC dimers are in close proximity to each other and polymer synthesis is initiated into a space between the enzymes, with a hydrophobic center of PHA and a hydrophilic surface consisting of PhaC enzymes and an ever-increasing amount of PhaP proteins as the granules enlarge. The membrane budding model depicts PhaC enzymes associating with the inner leaflet of the cytoplasmic membrane of the bacterial cell. PHA is synthesized and fills the space between the membrane leaflets until the PHA granule, surrounded by proteins and membrane lipids, buds off of the cell membrane to form its own intracellular, almost organelle-like compartment. Lastly, the scaffolding model is similar to the micelle model in terms of mechanism of granule formation initiation, except that in the scaffolding model, the nascent PHA granule is attached by some as-yet-to-be-defined intermolecular force to a scaffold made by an unknown substance [39]. The scaffold model has its roots in a transmission electron microscopy study that showed PHB granules forming around a dense body in the *R. eutropha* cell. This dense body was hypothesized to be a scaffold of some kind, on which PHB granules will anchor themselves [69]. A cryo-electron tomography study of PHB granules in *R. eutropha* suggests that the membrane budding model is not an accurate representation. In this study, no lipids were observed surrounding the intracellular polymer granules [70]. Other recent endeavors seem to support the conclusion of Beeby and coworkers with regards to the inaccuracy of the membrane budding model (unpublished data).

We are therefore left with two PHA granule biogenesis models, both similar to each other, yet with one crucial difference. Without characterization of a "scaffold substance," it would seem that the micelle model was the most accurate model for PHA granule formation. A new PHA granule associated protein has been characterized that plays an interesting and critical role in granule biogenesis and stability. The protein itself has been termed PhaM, and the history of its discovery and characterization is almost as complex as its role in PHA homeostasis. An epitope-tagged PhaC enzyme was isolated to near purification from engineered *R. eutropha* cells. The enzyme preparation was shown to be heterogeneous, including the presence of a high molecular weight fraction that demonstrated polymerase activity with no lag phase, but also contained oligo-HB and some associated proteins, including PhaP1. Another protein associated with this high molecular weight fraction was noted to be a ~32 kDa "hypothetical protein" [39]. This protein was subjected to *N*-terminal peptide sequencing and identified. However, there was insufficient interest to pursue study of this and the other uncharacterized proteins in the high molecular weight fraction of the PhaC preparation. Independent of the studies of Cho and coworkers, a novel and as-ye--unidentified granule associated protein, PhaM, was characterized [71]. This PhaM was found to be the same protein as the ~32 kDa hypothetical mentioned above.

PhaM was shown to interact with PHA synthase, *pha*C1 [71]. Examination of the amino acid sequence of the PhaM protein revealed similarities to TadA, a lipid body associated protein in the oleaginous bacterium *Rhodococcus opacus* [72]. Indeed, PhaM was shown to bind to PHB granules. PhaM was also shown to bind non-specifically to genomic DNA through its positively charged cytoplasmic surface. In conditions where PHB was produced and concomitant cell division was still taking place, PhaM was shown to distribute granules in daughter cells, presumably based on interaction with the *R. eutropha* chromosome during segregation [71].

Depolymerases (phaZ)

For PHA to play its role in the *R. eutropha* cell as a carbon and energy storage molecule, the polymer must be able to be readily mobilized and broken down into its core monomer, or some similar form, in order to be assimilated for growth. In the laboratory, *R. eutropha* cells replete with intracellular PHB are incubated in minimal medium lacking exogenous carbon source, but containing all other essential nutrients. These conditions allow for intracellular PHB to be mobilized as a carbon source in the cell [21, 22, 73, 74]. PHA depolymerase enzymes play the role of mobilization enzymes. Some *phaZ* enzymes are granule associated, while some are predicted to be secreted extracellularly in order to break down PHA present in the environment. *phaZ*1 is the most notable and well-understood PHA depolymerase expressed in *R. eutropha.* Deletion of the *phaZ*1 gene was shown to diminish the overall PHB depolymerase activity in the cell [74]. The *phaZ*1 protein only degrades amorphous PHB, and fails to have any effect on crystalline PHB [22, 75]. When *phaZ*1 was co-expressed with *phaCAB* in a heterologous *E. coli* system, secretion of 3-hydroxybutyrate [3HB] was demonstrated at high pH, indicating the activity of *phaZ*1 depolymerizing PHB, further suggesting that *phaZ*1 is granule associated [76]. *In vitro* thiolysis of PHB in the presence of free coenzyme A [CoA] by *phaZ*1 has also been demonstrated. This *phaZ*1-mediated thiolysis did not occur in heterologously synthesized PHB granules (*i.e.*, not a full complement of granule associated proteins) without PhaP1 being present. This suggests that PhaP1 and *phaZ*1 act together, by an unknown mechanism, to mobilize PHB [77]. A recent study has shown that PHB degradation by *phaZ*1 ultimately produces crotonyl-CoA. The compound *(S)*-3-hydroxybutyryl-CoA, originally thought to have been the ultimate degradation product, is a transient intermediate in the PHB mobilization process. In this study, it was shown that PhaP1 must also be present for optimal activity. Other phasin proteins were substituted for PhaP1, but were not as effective in modulating PHB degradation in the system studied [64]. Site-directed mutagenesis studies determined the key catalytic residues for *phaZ*1. The residues are a catalytic triad: C183, D355, and H388 [78].

Another depolymerase, *phaZ*2, has been shown to play a role in PHB degradation. Like *phaZ*1, *phaZ*2 is an intracellular, presumably granule associated enzyme [79]. Deletion of both the *phaZ*1 and *phaZ*2 genes resulted in negligible PHB mobilization in *R. eutropha* cells [74]. It is interesting to note that intracellular *phaZ* proteins typically share no similarity in sequence to extracellular depolymerases. Strangely, a *phaZ*2 single mutant does not seem to have any noticeable defect in PHB mobilization, when compared to wild-type *R. eutropha*. Thus, *phaZ*2 appears to have its greatest effect on the mobilization of PHB only when *phaZ*1 is present [74]. However, this claim was challenged in a more recent study where *phaZ* genes were introduced individually *in trans* to complement a mutant *R. eutropha* strain deleted for all intracellular *phaZ* genes. Expression of plasmid-borne *phaZ*2 alone exhibited a wild-type-like PHB degradation phenotype under the conditions studied, showing that *phaZ*2 could act in the absence of *phaZ*1 [73]. Even less is known about other intracellular PHA depolymerases in *R. eutropha*, *phaZ*3 and *phaZ*5, though recent research suggests *phaZ*3 and *phaZ*5 have roles in altering the granule architecture inside the cell. Deletion of *phaZ*3 and *phaZ*5 genes does not affect mobilization of PHB significantly. However, electron microscopy studies were performed on *phaZ* mutant *R. eutropha* cells. Data from this study suggested that *phaZ*3 may help control intracellular PHB granule numbers and *phaZ*5 may play a role in control of granule volume [73]. It is possible these and other depolymerases may also affect the crystallinity of intracellular PHB, but this has yet to be experimentally demonstrated.

More depolymerase genes are present in the *R. eutropha* genome, specifically *phaZ*4, *phaZ*6 and *phaZ*7 [9]. According to the PHA Depolymerase Engineering Database (http://www.ded.uni-stuttgart.de/) [80], *phaZ*4 is classified as an intracellular depolymerase that acts on native PHB, while *phaZ*6 and *phaZ*7are designated as extracellular depolymerases, acting on denatured [crystalline] PHA [73]. None of these proteins seem to have any noticeable effect on the mobilization of PHBs in conditions studied thus far. Interestingly, *phaZ*6 (also known as *phaZ*d1) and *phaZ*7 (also known as *phaZ*d2) seem to have a more mysterious function. Mutant strains with deleted *phaZ*6 and *phaZ*7 genes appeared to have no variation in PHB accumulation and mobilization compared to the wild type strain [81], and there was no detectable level of these proteins either in purified PHB granules or in the soluble (cell extract) fraction [51]. However, these proteins are active with native PHB *in vitro*. Even more interesting is the observation from *in vitro* studies that demonstrated *phaZ*6 and *phaZ*7 did not act on denatured PHB. When artificially expressed [*i.e.*, overexpressed] *in vivo*, *phaZ*6 and *phaZ*7 reduced and even prevented PHB accumulation under PHB permissive conditions. Translational fusion studies showed that *phaZ*6 and *phaZ*7 are not expressed in the *R. eutropha* cell [81], but whole genome transcriptome analysis showed that *phaZ*6 and *phaZ*7 transcripts are detectable and variable throughout the course of PHB homeostasis [25]. These results suggest that there exists some form of translation inhibition of these two genes. It remains to be determined whether this inhibition is a result of negative regulation of gene expression, or some other phenomenon.

R-Specific Enoyl-CoA Hydratase (PhaJ)

In many species of *Pseudomonas* that synthesize *mcl*-PHA polymers, 3-hydroxyacyl-CoA (3HA-CoA) precursors are synthesized from fatty acid degradation (β-oxidation) intermediates, namely enoyl-CoA molecules [82]. The enzyme enoyl-CoA hydratase (PhaJ) performs a hydration reaction on the enoyl-CoA molecule, producing an *(R)*-3HACoA [83]. Typically, when the PHA precursor pool is broadened in engineered *R. eutropha* strains, a heterologous *phaJ* gene is expressed. This will be discussed in greater detail below. Kawashima and coworkers have shown that *R. eutropha* possesses its own enoyl-CoA hydratase genes, *phaJ4a, phaJ4b,* and *phaJ4c*. These genes were expressed heterologously in *E. coli* and the resulting enzymes' substrate specificity was examined. Each PhaJ enzyme was shown to have substrate specificity for C4, C6, and C8 substrates, producing 3HB-CoA, 3-hydroxyhexanoyl-CoA (3HHx-CoA) and 3-hydroxyoctanoyl-CoA (3HO-CoA), respectively. The PhaJ enzymes were shown to be active in PHA precursor synthesis from β-oxidation of fatty acids from soybean oil in *R. eutropha* strains where the *phaA* gene was deleted. Overexpression of *phaJ* genes was shown to more effectively incorporate hydroxyhexanoate monomer (3HHx) into PHA [84].

3-Hydroxybutyrate Oligomer Hydrolase (PhaY)

It has been suggested that 3HB oligomers are one of the main PHB breakdown products. As such, a gene encoding 3-hydroxybutyrate oligomer hydrolase was isolated and expressed heterologously. The enzyme, PhaY, was shown to be active in hydrolyzing 3HB dimers and trimers. Site-directed mutagenesis revealed a unique active site sequence, SVSNG, with the central serine residue as the catalytic residue [85]. The PhaY enzyme was shown to be intracellular, and was later shown to hydrolyze 3HB tetramers and pentamers, as well. PhaY showed a very slight ability to hydrolyze artificial, amorphous PHB granule, and displayed no activity against native PHB granules [86]. Two *phaY* genes have been found in the *R. eutropha* H16 genome [73].

THE BUSINESS OF PHA: *R. EUTROPHA* AS A PLATFORM ORGANISM FOR INDUSTRIAL CULTIVATIONS

Given the nature of PHA, a bio-based, biodegradable, biocompatible plastic-like polymer, it is not a surprise that many corporate ventures have been launched with PHA as the principal product. PHA production has been constantly fine-tuned towards high yields of different polymers over the past 30 years. Imperial Chemical Industries, Inc. (ICI) had originally developed a single-cell protein program with *R. eutropha* as a potential biocatalyst of choice. After the single-cell protein program was scrapped, in 1982 ICI started using *R. eutropha* as a biocatalyst to produce PHA under the trade name Biopol. In 1990, ICI terminated their Biopol venture, and it was continued by Monsanto for approximately 4 years. During this time, Metabolix (www.metabolix.com) was founded based on work performed in Anthony Sinskey's laboratory at MIT. Currently, however, Metabolix uses mostly heterologous organisms as biocatalysts for PHA productions. Around the same time as the founding of Metabolix, a Brazilian firm began development of PHA under the trade name Biocycle, produced by "an *Alcaligenes* species" using sugarcane as the feedstock (www.biocycle.com.br). Because of its amenability to genetic manipulation, its recognition as a model system for PHA homeostasis, and its ability to utilize a wide variety of carbon sources, *R. eutropha* remains an industrially relevant organism suitable for large scale bioconversions.

R. eutropha Productions: Fermentative PHA Production

Researchers argue the importance of *R. eutropha* as a model organism and have backed up this argument with the publication of dozens of classic works that characterize PHB homeostasis and autotrophic growth. Some corporate enterprises have even been founded based on the production of biopolymer, discovered and characterized over 30 years ago in *R. eutropha.* While Metabolix, Biocycle, and a few other firms are developing commodity products fabricated using PHA as a raw material, many are using recombinant organisms like *E. coli,* based on the rapid rate of biomass accumulation. Whither the role of *R. eutropha* as an industrial producer of such biodegradable commodities? Studies of controlled fermentations of PHA biosynthesis remind the reader of what this bacterium can accomplish as a biocatalyst for an industrial production platform. Indeed, many controlled fermentations have been performed where R. eutropha is the biocatalyst, and the data from many of these is reviewed below.

PHAs are polymers that are commonly comprised of one or more monomers, as homopolymers (single monomer) or copolymers (two or more monomers), respectively [87 - 89]. The classification of PHAs depends on the number of carbon atoms present in their requisite monomers as either short-chain-length (*scl*-PHA; 3-5 C atoms) or medium-chain-length PHAs, (*mcl*-PHA; ≥6 C atoms) [90]. The family of PHAs include a broad number of polymers due to the diverse monomeric composition, but only a small number of them have been produced in the large-scale [pilot scale or greater] productions. The polymers that have been examined in controlled fermentations at larger scale are: PHB, PHBHV and poly(hydroxybutyrate-*co*-hydroxyhexanoate) [P(HB-*co*-HHx)] [91, 92]. The chemical composition of the PHA product, as well as the thermal and mechanical properties, can be adjusted by metabolic engineering of the host organism and other biotechnological strategies of polymer production [1]. PHB is a semi-crystalline polymer, characterized by a polymorphic crystallization, that is able to crystallize into two forms, α and β [90]. The most common conformation for PHB crystals is the α-form which consists in lamellar crystals [90, 93] and the β-form characterized as a planar zigzag conformation that can be used in films and fibers with high tensile strength [90, 93, 94]. It has been reported that the crystallization process affects the thermal and mechanical properties, as well as the biodegradability of polymers [90]. The PHB films or fibers produced by cold, melt, or solution crystallization are brittle, and their mechanical properties change due to a secondary crystallization that occurs under ambient conditions. Furthermore, PHB homopolymer has a high melting temperature, limiting its processability. Several researchers have attempted to improve the mechanical performance of PHB polymeric materials by altering the processing and crystallization conditions. High-strength PHB fibers or films have been prepared by the drawing process, an extrusion type of manufacturing that involves pulling the material in order to achieve a desired shape and thickness. The strength and tensile modulus of these fibers and films increase with the β-form content that can be controlled by changing the drawing temperature, as well as the annealing time and temperature. The formation of the β-form crystals by drawing leads to the improvement in the mechanical strength of PHB [95 - 98].

As mentioned above, the metabolic pathway for PHB biosynthesis in *R. eutropha* starts with the condensation of two molecules of acetyl-CoA by the PhaA enzyme to form acetoacetyl-CoA. The PhaB enzyme then converts acetoacetyl-CoA to the PHB precursor, 3HB-CoA, using NADPH as the source of reducing equivalents. Lastly, the enzyme PHA synthase, PhaC, polymerizes the 3HB-CoA monomers to make PHB (Fig. **2**) releasing a molecule CoA with each monomer added [38, 99]. In some microorganisms, such as *R. eutropha* and its close relatives, the PHB biosynthetic genes *phaA*, *phaB* and *phaC* are clustered and presumably organized in a single operon [30, 40]. This has been most well-studied in *R. eutropha* [100]. Given that transcriptomic data has shown that the PHB biosynthetic enzymes are expressed constitutively throughout the PHB cycle, we can infer that regulation of the pathways occurs at the enzyme level, specifically monomer supply and enzyme activity [25, 100]. It has been observed that nitrogen limitation results in an increase in the intracellular acetyl-CoA pool [17, 100, 101]. This is certainly a scenario that would favor the production of PHB, given that acetyl-CoA is the principal precursor. The PhaA β-ketothiolase enzymes have been shown to be inhibited by free coenzyme A (CoA) [27, 100, 102]. During growth, it can be predicted that CoA levels are high in *R. eutropha* cells, thus restricting the production of PHB. However, when nutrient limitation sets in, intracellular CoA levels are likely to be significantly lower, thus switching the focus of *R. eutropha* cellular metabolism to polymer biosynthesis [100].

The ability of microbial strains to incorporate a variety of monomers into PHA is due in part to the broad substrate range exhibited by the PHA synthase enzymes [99]. The nature of the carbon source provided to the cells can also dictate, to a certain extent, the monomer variety in the synthesis of PHAs, *via* the metabolic routes that convert the carbon source into the hydroxyacyl-CoA monomers [103]. In many organisms that harbor the biosynthetic pathways for the synthesis of PHAs composed of C6–C16 hydroxyalkanoic acids (*mcl*-PHAs), the CoA thioester precursors are derived from fatty acid metabolism. These precursors can be obtained from β-oxidation of alkanes, alkanols or alkanoic acids [104, 105], by an enantioselective enoyl-CoA hydratase, encoded by the *phaJ* gene [106], or from fatty acid *de novo* biosynthesis using an (*R*)-3-hydroxyacyl-ACP:CoA transacylase, encoded by *phaG*, to produce the substrates for polymer synthesis from an unrelated carbon source, such as carbohydrates [107 - 110]. It has been demonstrated that heterocopolymeric PHA containing *scl*- and *mcl*-3HA-CoA monomers possesses favorable thermal and mechanical properties [87, 111]. Thus the heterologous expression of a *phaG* and/or *phaJ* genes, or the enhancement of *R. eutropha*'s native PhaJ activity will result in production of useful biopolymer.

Metabolic Pathways and Genetics Involved in Production of PHBHV

The PHBHV biosynthetic pathway is illustrated in Fig. (**2**). The first study of copolymer production in *R. eutropha* was PHBHV [112]. The 3HV-CoA precursor is formed by condensation of propionyl-CoA with acetyl-CoA by β-ketoacyl-CoA thiolase, BktB, followed by reduction to 3HV-CoA. This route was confirmed by use of ^{13}C-labeled carbon sources. By adjusting the ratios of acetate and propionate in the substrate, *R. eutropha* H16 can accumulates PHBHV up to 50% of the cell dry mass, with a 3HV monomer percentage varying between 0 and 45% [34]. The use of even-numbered carbon chain growth substrates, *e.g.*, organic acids, fatty acids, *n*-alkanoates (within C_2-C_{22}) as the sole carbon source for *R. eutropha* H16 cultures results in synthesis of PHB homopolymer, whereas use of odd-numbered carbon chain substrates, such as propionate and valerate, produced copolymers of PHBHV [113].

Production of PHB from glucose and production of PHBHV from mixtures of glucose and propionic acid have both been produced by fed-batch cultures of *R. eutropha* [114]. In an 800-mL culture of PHB production culture, cells were grown in phosphate-limiting medium and encountered nutrient limitation after about 60 h of culture time, and accumulated PHB during the course of the next 40–60 h from supplied glucose. By controlling the glucose concentration at 10 g/L during fed-batch culture, a final cell mass, a PHB concentration and a PHB content of 164 g/L, 121 g/L, and 76% of cell dry mass (CDM), respectively, were obtained, resulting in a high productivity of 2.42 g/(Lh) with the yield of 0.3 g

PHB per one gram of glucose [115].

The addition of propionic acid or valeric acid to culture media already containing glucose leads to the production of the copolymer PHBHV [114]. When valerate was supplied as the carbon source to a culture of *R. eutropha* strain NCIMB11599, the 3HV fraction in the polymer was 85wt% [116]. Also, butyrate and valerate were used as the main carbon sources for *R. eutropha* H16 and *R. eutropha* NCIMB11599. The result was NCIMB11599 was able to synthesize PHBHV with a greater percentage of 3HV monomer [90 wt%] than compared to H16 [75 wt%] [117]. As has been shown in multiple studies, the 3HV monomer fraction in PHBHV can be adjusted by feeding of specific carbon sources, most typically propionate and valerate [32, 118 - 120]. Metabolic fluxes in *R. eutropha* cells have been examined during PHA biosynthesis with feeding of mixed chain-length organic acids [acetic, propionic and butyric acids]. Propionate is taken up by cells and converted to propionyl-CoA, which is one of the substrates for the BktB β-ketothiolase enzyme. However, a relatively small portion (<10 mol-%) of propionyl-CoA flux is directed towards BktB for the purpose of PHA biosynthesis. The majority of propionyl-CoA in *R. eutropha* cells was shown to be converted to methylcitrate [121].

Yang *et al.* 2010 used an organic acid mixture analysis model to optimize *R. eutropha* culture conditions, in order to enhance PHBHV production, including the 3HV monomer content in the copolymer. Systematic mixture analysis was employed to determine the optimal levels of acetate, propionate, and butyrate for cell growth, PHA production, and 3HV content by *R. eutropha* H16 [118]. It was found that butyrate is the preferred acid for increasing cell growth to result in high PHA production; cell growth and PHA production increased about 1.5-fold over initial conditions when the proportion of butyrate was increased. These studies also determined that the 3HV content in the resulting copolymer depends on the proportion of the propionate initially present in the growth medium.

There are toxic effects of organic acids in cultures of *R. eutropha*, which can be easily observed at high concentrations. Therefore, a specially designed fed-batch cultivation strategy is necessary. Huschner *et al.* [122] designed a pH-stat fed-batch culture technique, using acetic, butyric, and propionic acids as the main carbon source, in combination with an additional pO_2-dependent feed for delivering organic acid salts. The developed strategy results in robust growth and PHA production, showing a high productivity of more than 2 g PHA/L/h. After 44 h of fermentation, 112.4 ± 2.3 g/L CDM with 83.3 ± 1.1% PHBHV in CDM containing 5.6 ± 0.4 mol-% 3HV were achieved. This technique showed the highest results in productivity, CDM and PHBHV in CDM values for *R. eutropha* growth on mixed organic acids as the main carbon source.

R. eutropha H16 has been shown to improve the production of PHB and PHBHV when placed under oxidative stress. Growth and PHB biosynthesis in the presence of the oxidant molecules hydrogen peroxide and methyl viologen was shown to stimulate the flux of NADPH to the PHA biosynthetic pathway. The oxidants significantly activated the glucose 6-phosphate dehydrogenase (G6PD) of *R. eutropha*, increasing the NADPH levels used for the reduction reaction of acetoacetyl-CoA to β-hydroxybutyryl-CoA, thus, enhancing the biosynthesis of PHB and PHBHV [123].

Engineered *R. eutropha* and P(HB-*co*-HHx) Synthesis

Metabolic Pathways and Genetics Involved in Production of P(HB-co-HHx)

As mentioned previously, *R. eutropha* strain H16 has been engineered to produce copolymers with favorable thermal and mechanical properties. *R. eutropha* strains have been modified to produce the copolymer P(HB-*co*-HHx) with different 3-hydroxyhexanoate (3HHx) content, based on the genotype of the adjusted strains and the carbon sources used [124]. Additionally, *R. eutropha* does grow efficiently using plant oils as the main carbon source, which can affect the content and concentration of copolymers associated in the PHB polymer [24]. As previously mentioned, the wild-type strain produces only *scl*-PHA, which limits its utilization as an industrial PHA production organism, due to the thus-far limited applications of the *scl*-polymers. *Aeromonas caviae* was one of the first wild-type organisms identified that synthesizes P(HB-*co*-HHx) [125]. When grown on plant oils and fatty acids, this bacterium was shown to accumulate low levels of P(HB-*co*-HHx) [126]. Researchers investigated the ability of *A. caviae* to synthesize P(HB-*co*-HHx) and found that the PHA synthase from *A. caviae*

($PhaC_{Ac}$) was able to accept 3HB-CoA and 3HHx-CoA as substrates [127]. Additionally, *A. caviae* expresses a gene encoding an (*R*)-specific enoyl-CoA hydratase (*phaJ*$_{Ac}$), which allows for conversion of fatty acid β-oxidation intermediates to PHA precursors (Fig. **2**) [128, 129]. Thus, researchers have been able to utilize these genes and others from similar organisms in the metabolic engineering of *R. eutropha* in order to produce increased quantities of P(HB-*co*-HHx) copolymer, which demonstrates improved thermomechanical properties over PHB produced by wild-type strains, for potential production of industrially relevant polymers. In one study [127], the *phaC*$_{Ac}$ gene was incorporated into *R. eutropha* strain PHB$^-$4, a PHA-negative mutant derived from the wild-type strain, H16 [130], utilizing a broad-host-range expression vector. Insertion of *phaJ*$_{Ac}$ into a synthetic *pha* operon significantly increased 3HHx content without interfering with the cell growth and polyester accumulation. These strains were able to accumulate high levels of PHA [up to 75% of cell dry mass] containing up to 10 mol-% HHx monomer in cultures with soybean oil as the sole carbon source.

Plant oils and fatty acids are attractive feedstocks for industrial PHA production due to their high carbon content. The yield production of PHA from plant oils and fatty acids (0.65–0.98g PHA/g carbon source) is shown to be higher than that from conventional sugars (0.32–0.48g PHA/g glucose) [131]. In a recent work, high PHB productivity was reached by *R. eutropha* using soybean oil in fed-batch culture [261], where a peak cell concentration 83 g/L was achieved, and the PHB productivity was observed to be 2.5 g/(Lh). The PHB content was 81% (w/w), with a maximal PHB concentration of 67.2 g/L. The metabolism of lipids and fatty acids can influence the monomer composition of the produced PHA [132]. Several groups have produced P(HB-*co*-HHx) by using recombinant *R. eutropha* expressing *phaC*$_{Ac}$ [133 - 136]. However, the production of PHA in these cases with high 3HHx monomer content (5 mol-% or greater) require feeding the cells medium chain-length fatty acids (<12 carbons), which tend to be more expensive as feedstocks than raw plant oil. Fukui *et al.* 1998 reported that PHA can be produced from plant oils using strain H16 of *R. eutropha* or its recombinant strain, PHB$^-$4/pJRDEE32d13 harboring a plasmid-borne PHA synthase gene from *A. caviae* (*phaC*$_{Ac}$). In these cultivations, a cellular content of about 80% PHA (w/w) was produced in flask-scale experiments [137]. Kahar *et al.* [134] also demonstrated an efficient production of PHA with a high yield from soybean oil. PHA was obtained ranging from 0.72 to 0.76 g PHA per gram soybean oil used. These workers have successfully demonstrated both PHB homopolymer production and the P(HB-*co*-HHx) copolymer production with high yields by wild-type and engineered *R. eutropha* strains, respectively, using soybean oil as the main carbon source. Yield coefficient and volumetric productivity were 0.83 g PHA per one gram oil and 1.46 g/(Lh), respectively, higher than what Kahar *et al.* [134] obtained, moreover, using a cheaply available waste oil. Propanol at 1 vol.-% enhanced both PHA and biomass formation significantly, and resulted in incorporation of 3HV units into PHA structure. Therefore, propanol was used as an effective precursor of 3HV for production of PHBHV copolymer by *R. eutropha.* Although the highest content of 3HV in PHA was using 0.75 vol.-% propanol, the highest growth and PHA production was with 1 vol.-% propanol, and as a result, during the fed-batch cultivation, propanol concentration was maintained at 1 vol.-%, which resulted in 8 mol-% content of 3HV in the copolymer.

An engineered *R. eutropha* strain was reported in which genes encoding acetoacetyl-CoA reductases (*phaB1-3*) were deleted, and this strain synthesizes significantly less PHB than the wild type [36]. In order to engineer this strain to produce P(HB-*co*-HHx), a series of strains were constructed in which plasmid-borne PHA synthase genes from *A. caviae* and *Rhodococcus aetherivorans* I24 [138] were expressed by engineered *R. eutropha* strains with different levels of acetoacetyl-CoA reductase activity. *Rhodococcus* species have been shown to synthesize PHA copolymers [139], and analysis of a draft genome of *R. aetherivorans* revealed two putative PHA synthase genes. When these PHA synthase genes from *R. aetherivorans* strain I24 were expressed by a *phaC* deletion strain of *R. eutropha*, the recombinant strains accumulated P(HB-*co*-HHx) when grown on even-chain-length fatty acids (*e.g.*, octanoate), with the strain expressing *pha*$C2_{Ra}$ synthesizing polymer containing a significantly higher amount of 3HHx monomer (>18mol-%) compared to a strain expressing the *pha*$C1_{Ra}$ synthase gene (11.5mol-%) [140]. These strains expressing $phaC_{Ra}$ genes also accumulated P(HB-*co*-HHx) when grown on palm oil, but the 3HHx content of the PHA was significantly lower (1.5mol-% 3HHx, compared to >10mol-%

with octanoate). Further improvements were made to the strains by incorporating *phaJ* genes from *A. caviae* or *Pseudomonas aeruginosa*, and by increasing gene dosage with a stable plasmid expression system [140]. PhaJ enzymes from *A. caviae* and *P. aeruginosa* hydrate crotonyl-CoA and 2-hexenoyl-CoA at similar rates, which leads to the synthesis of 3HB-CoA and 3HHx-CoA precursors [128, 141]. It was shown that the expression of either $phaJ1_{Pa}$ by the recombinant strains led to higher PHA accumulation (≥63% PHA per cell dry mass), as well as higher 3HHx concentration (up to 24 mol-%). Two notable P(HB-*co*-HHx) synthesizing strains of *R. eutropha* were produced with these methods, Re2058/pCB113 that accumulated 71 wt% PHA with 17 mol-% 3HHx, and Re2160/pCB113 that accumulated 66 wt% PHA with a high 3HHx content of 30 mol-%, both using palm oil as the main carbon source for growth and PHA biosynthesis [140].

Fermentation Strategies

A schematic of inputs and outputs in *R. eutropha* fermentations to produce PHA is shown in Fig. (**3**). Very few species of bacteria capable of producing PHB have been used in industrial-scale fermentations (*i.e.*, ≥ 10 tons annually) [91, 103]. Selection of the microorganism for PHA production should be based on several elements including the ability of the cell to utilize several low cost carbon sources, growth rate, polymer synthesis rate, and the maximum extent of polymer accumulation [114]. *R. eutropha* is a good organism to select, according to these parameters, as it accumulates a large amount of polymer (up to 90% of its cell dry mass, in some cases) when nitrogen or phosphorous is completely depleted [17, 115, 133, 142, 143].

PHA has been produced by *R. eutropha* either in batch, fed-batch or continuous cultures. Batch fermentation for PHB production has been a well-studied process, due to its relatively low operation costs. However, the disadvantages of batch cultures include low yield and productivity. Thus, fed-batch cultures have been employed to achieve high cell densities and a high concentration of PHA [122, 143]. Fed-batch cultivations are systems where one or more nutrients are intermittently supplied to the bioreactor, and the products and other components are kept within the system until the end of fermentation. Usually, one of the intermittently supplied nutrients is the carbon substrate that is being used for growth and production. This means that there is an inflow but no outflow [144]. *R. eutropha* requires the limitation of an essential nutrient such as nitrogen or phosphorous for the synthesis of PHA from an excess of carbon source. High productivity of PHA is indeed obtained by fed-batch fermentations with well-designed feeding strategies as can be seen in recent works using *R. eutropha* H16 and derivatives as biocatalysts [122, 142].

In fed-batch cultures of microorganisms involving biosynthesis of a value-added product, there are two main possible cases: the production of a growth-associated product and the production of a non-growth-associated product. In some cases, mixed production (growth- and non-growth-associated) is seen. These types of fermentations can be illustrated by the Luedeking-Piret equation (1):

$$q_P = \alpha\mu + \beta \qquad \textbf{(1)}$$

where q_P is the specific rate of product formation, α and β are growth-associated and non-growth-associated coefficients and μ is the specific rate of growth (also termed μ_g). When determining the specific rate of product formation for a growth-associated product (based on Eq. 1), $\beta = 0$ and α is non-zero. Alternatively, if a non-growth-associated product is being considered, then $\alpha = 0$ and β is non-zero. In the case of mixed production, both coefficients are non-zero [145].

During PHA production in *R. eutropha* and related organisms, a non-growt--associated product synthesis is observed. Thus, the growth-associated coefficient of the Luedeking-Piret equation would be 0, resulting in a simple equation (2):

$$q_P = \beta \qquad \textbf{(2)}$$

where PHA production would not be correlated to growth rate. This observed phenomenon makes sense, as PHA biosynthesis occurs during nutrient depletion, a condition that results in growth cessation, as observed by non-growth-associated synthesis of PHA in fed-batch cultivation by *R. eutropha* H16 [146]. For other fed-batch cultures of *R. eutropha*, a distinct two-step cultivation method has been employed: the first step is a growth phase where the culture is fed with plentiful nutrients, and the second step is a production phase (when concentrations of a nutrient are limited) in which excess carbon source is fed to the bioreactor [147]. Initially, a desired concentration of biomass is obtained in a robust nutritive environment, followed by nutrient limitation in order to allow for efficient PHA synthesis. In the nutrient limiting stage, the residual cell concentration (the total biomass concentration subtracting the PHA concentration) remains almost constant and the total biomass concentration increases due to intracellular accumulation of PHA [142, 143].

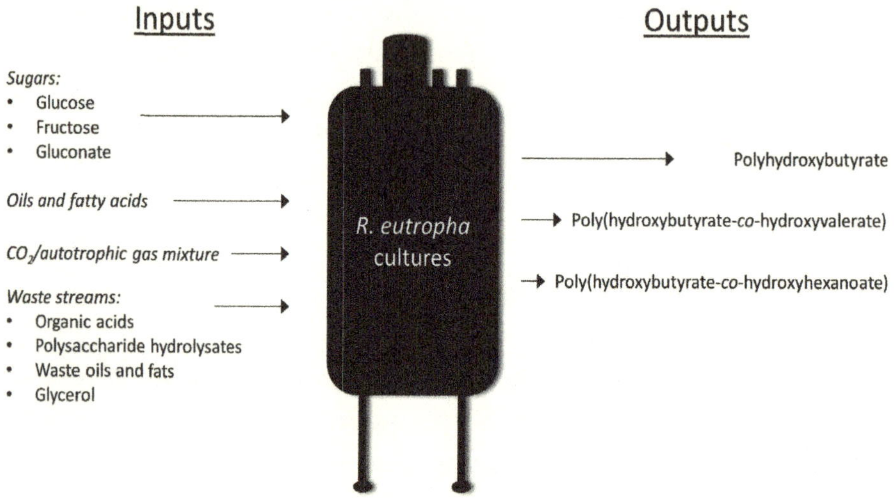

Fig. (3). Many controlled fermentations have been performed using wild-type or recombinant R. eutropha as a biocatalyst. One critical need for controlling production costs is the choice of inputs (carbon sources). One critical attribute required in determining applications of polymer is the choice of strain and feeding platform to control outputs.

PHA Biosynthesis Strategy - Nitrogen vs. Phosphorus Limitation

Many controlled fermentative PHA productions utilizing *R. eutropha* as a biocatalyst have focused on nitrogen limitation to usher in PHA production [122, 142, 143, 148]. Ho Nam Chang and coworkers have examined, in detail, phosphate limitation in *R. eutropha* cultures [149, 150]. In many microbiological minimal media, a phosphate buffer system is used to control the pH-value of the culture. Thus, if phosphorus is to be the limiting ingredient in PHA productions, a different buffer system will need to be used. As an alternative, Ryu *et al.* [150]

used citric acid and its conjugate base as a buffer for phosphate-limited PHB production media. HEPES (4-(2-hydroxyethyl)-1-piperazineethanesulfonic acid) can also serve as an appropriate alternative buffering system in phosphate-limited conditions [25]. It was demonstrated that PHB productivity is higher in phosphate-limited cultures (maximum productivity = 3.75 g/(Lh)) as compared to nitrogen-limited cultures (maximum productivity = 2.42 g/(Lh)) [150]. In general, both nitrogen and phosphate limitation provide suitable conditions for high productivity PHA productions using *R. eutropha* as the biocatalyst.

Waste Streams as Carbon Feedstocks

Primary expenses in the production of PHA biopolymers include the cost of the fermentation substrate, polymer extraction from the cells and treatment of extraction waste [151]. The cost of the carbon source indeed has a significant influence on the price of polymer production, thus, inexpensive, alternative carbon sources have been studied in hope of developing a process capable of high-yield PHA production at a decreased cost.

Crude glycerol, a by-product of the large-scale production of biodiesel, has been evaluated for its potential use as an inexpensive feedstock for PHB production [152, 153]. Using *R. eutropha* strains, yield of PHB synthesized from glycerol was 62.5% of cell dry mass, with a maximal biomass and PHB concentration of 104.7 g/L and 65.6 g/L, respectively [153].

The laboratory of Stanislav Obruca at Brno Institute of Technology has examined PHA biosynthesis by *R. eutropha* using a variety of waste carbon sources. In fed-batch fermentation, the group obtained biomass and PHA yields of 138 g/L and 105 g/L, respectively, by using waste rapeseed oil [154]. In a follow-up work, the production of PHB from the waste rapeseed oil was enhanced by strain engineering of *R. eutropha.* A strain uncovered by a mutagenesis screen, EO1, was shown to synthesize ~2.5 g/L more PHB in waste rapeseed oil cultures, compared to wild-type strain H16. When propionate was added to the culture media, there was also a slight increase in 3HV monomer content in the resulting PHBHV synthesized by EO1, compared to H16. The EO1 strain was partially characterized, and it is notable that enzyme activity assays show a doubling of the specific activity of malic enzyme in EO1 compared to wild type. Other enzymes, such as NADP-dependent isocitrate dehydrogenase, glucose-6 phosphate dehydrogenase and glutamate dehydrogenase, were also shown to exhibit increased activity in *R. eutropha* EO1. It was suggested that the likely alteration of the reducing environment of the cell due to the higher activity of the aforementioned enzymes, results in more robust PHA biosynthesis [155]. Obruca's group also extracted oils from spent coffee grounds [SCG] and demonstrated their utility as a carbon source for growth and PHA production by *R. eutropha*. The oils extracted from SCG contained mostly palmitic acid and linoleic acid. Using oil from SCG as the sole carbon source, *R. eutropha* produced 55.4 g/L of dry cell mass, containing 49.4 g/L PHB [89.1 wt% PHB], in fed-batch fermentations. The biomass yield for these cultivations was calculated to be 0.92 g per g oil, and the PHB yield was calculated to be 0.82 g per g of oil [156].

Another interesting study involving the repurposing of waste carbon as feedstock for PHB production by *R. eutropha* was performed by Zahari *et al.* [157, 158], who utilized the sugar-rich juice derived from crushed palm fronds. Palm fronds are routinely obtained during harvesting of the fruit bunches, and have been underutilized as a waste stream, with ~83 million tons (wet weight) fronds generated per year [157]. The juice extracted from palm fronds using a sugarcane pressing machine contained mostly glucose, fructose and sucrose. In batch bioreactor cultures, ~5 g/L PHB was produced using *C. necator* strain CCUG52238^T. The cells were shown to utilize most of the glucose and fructose in the palm frond juice mixture, yet the sucrose proved to not be a preferred carbon source [158].

Verlinden *et al.* [159] set up a small-scale batch fermentation studies using different concentrations of pure vegetable oil, heated vegetable oil and waste frying oil. They were able to achieve a PHB concentration from waste frying oil of 1.2 g/L, which is similar to the concentration that can be obtained from fructose, while the PHB harvest from pure oil and heated oil was 0.62 g/L and 0.9 g/L respectively. These findings recommend the recycling of waste frying oil, since PHB production yields using the waste oil as the main carbon source is higher than using pure or heated vegetable oil. It should be noted that overall PHB yields from this work are quite low, even for batch cultivations.

One of the most robust studies of high cell density *R. eutropha* cultivation and concurrent P(HB-*co*-HHx) copolymer production was performed by Riedel *et al.* [143]. In this work, batch, extended batch and fed-batch cultivations were examined for their PHA yields and productivities. Palm oil was chosen as the main carbon source for cultures of engineered *R. eutropha*. Nitrogen sources, ammonium chloride and urea, were also compared. Fed batch cultivations using urea as the nitrogen source were shown to provide the best P(HB-*co*-HHx) yield and productivity. In these high cell density palm oil fermentations, the final output consisted of >139 g/L of biomass with a PHA content of 74% and an average 3HHx monomer content of 19 mol-%. The productivity of P(HB-*co*-HHx) biosynthesis was shown to increase from 0.2 g/[L h] in batch cultures [140] to 1.1 g/(Lh) in fed batch fermenter cultivations with urea used as the nitrogen source [143].

In a follow-up study, Riedel and coworkers also examined the recovery of P(HB-*co*-HHx) copolymer from wet and dry *R. eutropha* biomass using non-halogenated solvents [262]. Works from the academic and patent literature [160 - 163] have identified many non-halogenated solvents that can serve as alternatives to chloroform for PHA recovery. The solvents methyl isobutyl ketone and ethyl acetate (MIBK and EA, respectively) were examined for P(HB-*co*-HHx) produced from palm oil cultures [143]. These solvents were found to dissolve the polymer most effectively, compared to tested alternatives. PHA extractions were performed at volumes ranging from 2 mL to 3 L using a PHA to solvent ratio of 2% [wt./vol.). EA showed high recovery levels [up to 94%] and high product purities [up to 99%] when using dry cells, while MIBK was shown to be a better solvent for PHA recovery from wet cells; purities of up to 99% and total recovery yields of up to 84% were reached. Recovery from wet cells eliminates a biomass drying step during the downstream process, thus, saving time and cost. All PHA in solution was precipitated by addition of threefold volumes of *n*-hexane [262].

Extraction, separation, and washing steps are required to isolate the polymer from the cells. MIBK and EA have lower densities than water, which allows for simple decantation of PHA solutions after extraction and centrifugation. Also, the residual biomass is retained in the aqueous phase, allowing for easier separation from the polymer solution. This phenomenon is an advantage over chloroform, as chloroform has higher density than water, causing PHA-chloroform solutions to form the bottom phase along with the residual cell material, thus making the mixture difficult to separate.

MORE THAN JUST A GOOD IDEA: UTILITY OF PHA BIOPOLYMERS

Regardless of the discrepancy in price of microbially-produced PHAs compared to petrochemical polymers, the potential is there for PHA to replace traditional polymers as raw material in many specific applications. PHAs have a wide range of applications due to their novel features [114]. Initial applications for PHA included everyday articles, such as shampoo bottles and other packaging materials. Also, packaging films, shopping bags and paper coatings have previously been targeted for PHA application [91]. Companies like Procter and Gamble (P&G), Biomer, and Metabolix have pushed the envelope of what is possible for PHA applications. PHA has been examined for items like disposable razors, diapers, carpets, upholstery, *etc*. It should be noted that not all of these applications are necessarily ideal for a biodegradable polymer. However, the sheer number of different potential applications that have been examined for PHA should give the reader an idea of how versatile this biopolymer is, based on the similarities of its properties with conventional plastics.

One type of application where a biodegradable polymer can be a boon is in medical and surgical devices (*e.g.*, sutures, implants and meshes). Many studies have shown PHAs to be biocompatible, indicating that these polymers do not cause adverse reactions in the mammalian body when they are implanted nor when they are broken down by action of non-specific enzymes in the body [164, 165]. Indeed, sutures fabricated from PHA of *R. eutropha* origin have been implanted into mice with no adverse effects [166]. A study has also shown that PHA, if purified from bacterial cells in such a way that allows for the almost complete removal of residual lipopolysaccharide (LPS) and other contaminants, can be used in contact with human blood. These findings suggest that, with rigorous recovery and purification procedures, PHA can be produced for medical applications. Many other works in the literature confirm this, as PHA has been implanted in rats, mice, dogs and rabbits [165].

Given the biodegradability of PHA, the polymer is also considered for use in bioactive compound release. The most obvious application for this would be drug delivery from a surgically implanted PHA matrix. Studies have demonstrated release of antibiotic from various types of matrix. The most common polymer tested was PHBHV, which has shown sustained release of antibiotic over the course of 60 days [165]. Implantation of drug-impregnated PHA particles has been shown to treat carcinoma *in vivo*. PHB with rubomycin was shown to release the drug and inhibit proliferation of Ehrlich's carcinoma, concomitant with degradation, in mice [167]. Tepha, Inc. (www.tepha.com) is currently producing drug delivery matrices fabricated from poly-4hydroxybutyrate (P4HB). Release of bioactive compound in PHA has also been exploited for pesticide and herbicide delivery in agricultural applications [168].

While many companies are producing and fabricating products from PHA, the biocatalyst that most firms use for fermentative production is recombinant *E. coli*. The principal reason for this is the short doubling time of the organism; *R. eutropha* possesses a significantly longer doubling time in culture, compared to *E. coli*. However, with the broad carbon substrate range of *R. eutropha* and its ability to store carbon as PHA in up to 90% of its cell dry mass, the hope is that the model organism will someday have its big payday in the biomanufacturing industry.

An Outlook of *R. eutropha* and the PHA Industry

R. eutropha is an interesting organism for the biotechnological production of polyesters, and the results discussed here demand this bacterium's consideration as an industrial biocatalyst. There are many advantages of using *R. eutropha* in PHA production: it can convert a wide range of renewable feedstocks and heterotrophic resources into various value-added products, it is genetically tractable, versatile, and high cell density fermentations of this bacterium can be easily performed. This bacterium is certainly amenable to metabolic engineering by conventional molecular genetic approaches. Lastly, the complete genome sequence of *R. eutropha* H16, including the pHG1 megaplasmid, is sequenced, annotated and readily available.As a result, a variety of subsequent proteomic, metabolomic and transcriptomic studies have provided a large quantity of valuable data that explains the complex metabolism of *R. eutropha*. Based on these advantages, it is clear that this organism provides and possesses all the necessary characteristics required for a biotechnological production strain [169]. Despite these advantages, *R. eutropha* usage in the PHA production industry is limited, chiefly due to its high production costs for large-scale production of bacterial biomass and value-added products. The use of inexpensive raw materials could significantly reduce the production expenses. In this regard, conversion of a variety of feedstocks to PHA by *R. eutropha* is a feasible production method. Industrial by-products, waste and residual compounds like glycerol, waste frying oil, and substrates containing a high carbon content per weight (*e.g.*, palm, soybean, and corn oils), are promising carbon substrates for large-scale fermentative PHA biosynthesis. In addition, raw sugar-rich materials like molasses and lignocellulose are a potential untapped renewable source of carbon for polymer productions. While modifications will be required for using *R. eutropha* as a production strain, these are also promising renewable substrates for fermentative PHA productions.

Biosynthesis of Other Unique Biopolymer Compounds by *R. eutropha*

Several groups have demonstrated *R. eutropha*'s metabolic diversity by engineering the organism to produce alternative compounds to PHA, such as biofuel molecules and unique non-PHA polymers. Polythioesters (PTEs) have been produced by engineered strains as a result of substrate feeding in the growth and production medium. PTEs are non-biodegradable in nature, but they are still thought to be a durable and bio-based alternative to petroleum-based polymers. The principal challenge with PTE production in *R. eutropha* strains is the toxicity of the precursors that are fed to cells in the media. This finding suggests that feeding strategies would need to be perfected in fermenter cultures for significant production of PTEs by *R. eutropha*. Biosynthesis of the copolymer poly(3-hydroxybutyrate-*co*-3-mercaptopropionate) [P(3HB-*co*-3MP)] was demonstrated by feeding *R. eutropha* strains 3-mercaptopropionic acid, 3,3'-dithiopropionic acid, or 3,3'-thiodipropionic acid [170, 171].

Engineered *R. eutropha* strains have been examined as biocatalysts for the production of cyanophycins. These polymers, which are produced at low levels in some cyanobacteria (hence the name), are thought to be a biodegradable alternative to polyacrylate [172]. Cyanophycin accumulation of up to 32% of the cell dry mass has been demonstrated in engineered strains of *R. eutropha* [173]. Production of this polymer has also been attempted in a high-cell-density fermentation setting [174]. Recently, three cyanophycin biosynthesis gene homologs were found in the *R. eutropha* H16 sequenced genome. Functional analysis of the genes and their products showed activity consistent with cyanophycin synthesis when the genes were expressed heterologously, but no cyanophycin production was observed in *R. eutropha* cells [175]. Thus, the function of the three genes remains a mystery. It is possible, however, that the conditions necessary for cyanophycin production in wild-type *R. eutropha* have not been examined in the laboratory thus far.

Autotrophy

R. eutropha, like many industrially relevant bacteria, has been defined in the scientific and patent literature by its metabolic capabilities, with the majority of studies focusing on PHA biosynthesis and autotrophic growth. One can easily imagine an ideal production system where *R. eutropha* utilizes waste CO_2 (power plant and factory emissions, *etc.*) to synthesize a useful, sustainable biopolymer. Large scale production of any compound using autotrophic cultures presents very interesting challenges. The typical autotrophic gas mixture is H_2, O_2 and CO_2, with the major component being H_2. Using a mixture of this type, operators must be wary of gas leakage and consequently, explosion risks. Furthermore, H_2 is very weakly soluble in aqueous solutions (*i.e.*, microbial growth media), which may partially explain the vast excess used in the autotrophic gas mixture. In 1979, Sonnleitner, *et al.*, demonstrated modest autotrophic cell growth and PHB production [176]. Other groups investigating synthesis of PHB by *R. eutropha* using CO_2 as the carbon source was the laboratory of Kenji Tanaka in the early 1990s. In a closed culture system, the group initially demonstrated that growth rate was higher under lower partial pressures of O_2 [177]. Using a recycled gas closed circuit culture system, Tanaka *et al.* were able to grow 90 g/dm^3 of *R. eutropha* biomass, producing 68% PHB per cell dry mass, which up to that point were the highest yields attained [178]. Furthermore, the stoichiometric equations (3) and (4) for autotrophically growing *R. eutropha* cells and PHB biosynthesis were shown to be:

$$\text{Cells: } 21.36\ H_2 + 6.21\ O_2 + 4.09\ CO_2 + 0.76\ NH3 \rightarrow C_{4.09}H_{7.13}O_{1.89}N_{0.76} \quad (3)$$

$$\text{PHB: } 33\ H_2 + 12\ O_2 + 4\ CO_2 \rightarrow C_4H_6O_2 + 30\ H_2O \quad (4)$$

It should be noted that $C_4H_6O_2$ in equation #2 represents the chemical formula of a hydroxybutyryl monomer in the PHB polymer [179].

Regarding the challenges of autotrophic growth of organisms, assuring a high-density culture and prevention of an explosive situation are concomitant. As mentioned above, the substrate gas mixture is poorly soluble in aqueous solution. Thus, it is imperative that gas mass transfer be improved in order to supply more nutrients to the cells for growth and production. In the case of PHB accumulation in the cells, dissolved hydrogen levels must always be high enough to facilitate polymer synthesis [177, 179]. Improvement of gas mass transfer is also critical to avoid an explosion in an autotrophic culture system. The key to preventing an explosive situation is to keep the oxygen levels in the gas phase below the lower limit for explosion, 6.9 vol.-%. In such an environment, the cellular productivity decreases significantly, and a high K_La value must be maintained for effective autotrophic PHA production [179]. One strategy to increase gas mass transfer in liquid culture is to generate smaller gas bubbles. The liquid film surrounding the gas bubble in an aqueous solution is one of the principal barriers to mass transfer. Generation of bubbles with smaller diameter will decrease the thickness of this surrounding layer, which will result in better gas mass transfer. One method for production of microbubbles is the use of a microbubble generator [180]. Gas bubble coalescence must also be avoided in these situations, although microbubbles are less likely to rapidly coalesce in aqueous growth medium due to the presence of solutes [181].

A key enzyme in CO_2 metabolism is carbonic anhydrase. The carbonic anhydrase enzyme catalyzes the interconversion of CO_2 to bicarbonate, as equilibrium of these two different forms of the compound is necessary for proper physiological function [263]. There are five main classes of carbonic anhydrase (CA) found in a variety of different organisms and tissues. *R. eutropha* has been shown to express representatives of three of the classes of this enzyme, α-,β- and γ-CA. It should be noted that *R. eutropha* possesses two different β class enzymes. One of the β-CA enzymes, Can, was demonstrated to be essential for growth of *R. eutropha* in atmospheric CO_2 concentrations [182]. However, addition of CO_2 bubbled into the growth medium restored growth of a *can* deletion strain [263]. These results suggest that the deletion of the *can* gene upsets the CO_2:bicarbonate equilibrium in the cell under normal atmospheric conditions. Deletion of the gene encoding an α-CA, *caa,* was shown to severely inhibit growth of *R. eutropha* on atmospheric CO_2. However, the growth phenotype was not complemented by the addition of excess CO_2 in the media. A quadruple carbonic anhydrase gene deletion mutant exhibited severe growth defects, even in the presence of excess CO_2 [263]. Indeed, the phenotype of the quadruple carbonic anhydrase mutant is potentially due to the deletion of *caa*. The equilibrium of Caa lies in the direction of bicarbonate, whereas the equilibrium of the other carbonic anhydrase enzymes lies towards CO_2. Caa is predicted to be localized in the periplasm of the *R. eutropha* cell [263]. This prediction agrees with the observations of the *caa* deletion phenotype observed above, if indeed Caa is the only periplasmic carbonic anhydrase enzyme. Indeed the model of carbonic anhydrase interplay offered by Gai *et al.* [263] is that Caa converts CO_2 to bicarbonate in the periplasm, which is then channeled across the cytoplasmic membrane to be converted back to CO_2 by one of the other carbonic anhydrase enzymes for carbon fixation. Precise roles of each of the enzymes would become further elucidated with continued study, but the presence of multiple carbonic anhydrase enzymes is yet another example of the interesting metabolic feats that *R. eutropha* can perform.

R. eutropha assimilates carbon *via* the Calvin-Benson-Bassham (CBB) cycle. The key enzyme in this cycle is ribulose-1,5 bisphosphate carboxylase/oxygenase (RuBisCO). The CBB cycle is one of the most common and most well-studied carbon fixation cycles. The cycle is comprised of 13 separate enzymatic reactions. *R. eutropha* harbors two *cbb* operons, one present on Chromosome 2 and the other on the megaplasmid pHG1. The gene encoding the regulator for the CBB cycle enzyme expression, *cbbR*, is present in both operons. Interestingly, the chromosomally encoded CbbR seems to be responsible for regulation of both the chromosomal and the plasmid-borne *cbb* operons [182]. The expression of the *cbb* operons is tightly controlled, as operation of the CBB cycle during heterotrophic growth would be a tremendous waste of energy for the cell. Proteomic data show that CBB enzymes are highly expressed in autotrophic culture as compared to heterotrophic culture [183, 184]. It was determined that phosphoenolpyruvate (PEP) can act as an inducer molecule for *cbb* gene expression. No other metabolite has as yet been uncovered as affecting *cbb* gene expression in *R. eutropha* [182, 185].

When the CBB cycle is operational in *R. eutropha* cells, once a molecule of CO_2 is fixed, along with the co-substrate ribulose-1,5 bisphosphate, by RuBisCO to form two 3-phosphoglycerate molecules, seven ATP equivalents and five reductants will be used in the production of one pyruvate [186, 187]. Needless to say, there is a net ATP gain with CO_2 fixation performed *via* the CBB cycle, albeit a small one. Reducing equivalents are replenished by the presence of hydrogen in the autotrophic gas mixture. Hydrogen and electrons are transferred from H_2 to NADH or NADPH by the action of hydrogenases.

R. eutropha is capable of utilizing H_2 as an energy source, and its H_2-utilization machinery are well-characterized. In all hydrogenase enzymes studied thus far, the active sites contain protein-bound metal complexes. In many hydrogenase examples, the active site is connected to the primary electron acceptor or donor through a relay of iron-sulfur (Fe-S) clusters [188]. Ni-Fe hydrogenases, which contain a complex of nickel and iron in the active site, are common in bacteria and archaea. *R. eutropha* utilizes Ni-Fe hydrogenases during autotrophic growth. *R. eutropha* is known as a "Knallgas bacterium," which means that it can drive its energy metabolism *via* a controlled oxidation of H_2 with O_2 as the terminal electron acceptor [9, 188]. The two main hydrogenases present in *R. eutropha* are the cytoplasmic "soluble" hydrogenase (SH, Fig. **4**), shown to be related to the mitochondrial NADH:ubiquinone reductase. The other main hydrogenase that powers autotrophic growth is the membrane-bound hydrogenase (MBH, Fig. **4**), which, perhaps unsurprisingly, is linked to the electron transport chain [189]. MBH is an O_2-tolerant hydrogenase, unlike similar complexes present in anaerobic bacteria that are completely inactivated in the presence of ~0.05 mM O_2. As has been experimentally shown, the MBH of *R. eutropha* is transiently inhibited by increasing concentrations of O_2 (0.05 – 0.3 mM), but when the O_2 is flushed from the system, the MBH once again will operate at peak activity [190].

The SH complex is made up of two "modules:" (1) the hydrogenase comprised of the polypeptides HoxH and HoxY, and their requisite cofactors, and (2) the NADH-dehydrogenase comprised of HoxF and HoxU and cofactors, including Flavin mononucleotide-b (FMN-b). An additional component of this complex, HoxI, has also been discovered. HoxI is hypothesized to provide a binding site for NADPH, allowing for activity similar to NADH:ubiquinone reductase [191]. The MBH consists of the periplasmic proteins HoxG and HoxK, along with their requisite cofactors, linked to the respiratory chain by HoxZ, a di-heme cytochrome *b* [192]. Strains containing mutations in SH or MBH were isolated. SH mutants were bradytrophic on hydrogen, likely due to difficulty in providing cells with sufficient NADH. MBH mutants did not display any H_2 oxidation activities, but growth using H_2 as the energy source was not impaired significantly, suggesting that a functional SH compensated for the loss of MBH [189]. There are two modes of regulation for hydrogenase expression and activity. Both hydrogenases are dependent on σ^{54} and the response regulator HoxA for expression [193]. In every two-component system, a response regulator must have a partner histidine protein kinase. In that respect, HoxA is phosphorylated by the histidine protein kinase HoxJ, and thus inactivated. This inactivation of the response regulator by phosphorylation is perhaps counter-intuitive to the role played by other protein kinases in two-component systems. In the presence of H_2, either HoxJ-mediated phosphorylation would be inhibited or dephosphorylation of phosphor-HoxA would be initiated. HoxJ is complexed with a regulatory hydrogenase (RH, Fig. **4**) that includes the proteins HoxB and HoxC. It is believed that strong interaction of the histidine protein kinase HoxJ with RH is necessary for H_2-sensing and response.

The hydrogenases of *R. eutropha* are complex metabolic machines. It can be imagined that these complexes require exact and intricate construction in order to function properly in the cell. Indeed, the numerous different maturation proteins (termed "Hyp") of *R. eutropha* are required for proper maturation of both SH and MBH [192]. Furthermore, for MBH, the twin-arginine translocase (TAT) system is necessary for properly timed export of the HoxK/HoxG dimer into the periplasm [194].

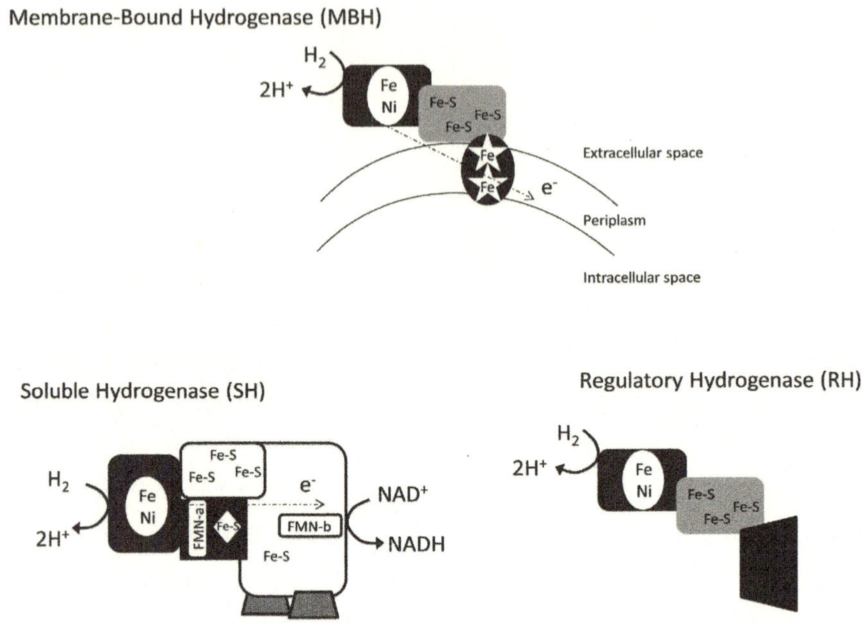

Fig. (4). Schematic drawings of *R. eutropha* hydrogenases. The membrane-bound hydrogenase (MBH) oxidizes hydrogen with electrons feeding into the electron transport chain. The soluble hydrogenase (SH) oxidizes hydrogen to reduce NAD^+ to NADH. The regulatory hydrogenase (RH) senses the amount of available hydrogen and activates HoxA, the response regulator responsible for activating expression of hydrogenase genes.

Growth in the Presence of Carbon Monoxide [CO]

Combining the discussion of autotrophic growth by *R. eutropha* with that of waste carbon source utilization, a dream bioconversion would be the use of syngas, a product of coal-fired power plants, as a carbon source for value added product biosynthesis. Syngas is a mixture of H_2, CO_2 and CO gases. The laboratory Tatiana Volova in the Siberian Federal University, Krasnoyarsk, has been studying growth and productions by *R. eutropha* using a syngas mixture. The key physiological concept here is the tolerance of the organism to carbon monoxide (CO), which can be toxic to cells in most living systems. Volova and coworkers have been able to demonstrate growth of *R. eutropha*, as well as synthesis of 70-75% of cell dry mass in PHB using syngas as a feedstock. Since autotrophic growth gas mixtures require the presence of large ratios of hydrogen compared to CO_2 and O_2, syngas was treated with water vapor at high temperature to increase the concentration of H_2 to make the gas mixture more suitable for *R. eutropha* growth and polymer production [19]. The presence of CO did not affect the polymer yield and productivity [195 - 197]. In fact, CO did not appear to be utilized by the *R. eutropha* culture, and built up in the culture vessel to the point where it could mostly be flushed from the system in favor of replacement by fresh syngas substrate [196]. In controlled fermentations in the presence of up to 20% CO, *R. eutropha* was shown to produce 17.5 g/L cell mass and 12.7 g/L PHB, with hydrogenase and PHA biosynthesis enzyme activities relatively stable [196].

Ralstonia eutropha JMP134 – Master Bioremediator

Thus far in this chapter, we have discussed the myriad of compounds that *R. eutropha* can make. In doing so, we have also discussed carbon sources that *R. eutropha* can break down, many of which are common, some of which are cheaply available. However, there is a strain of *R. eutropha* that is best known for what it can break down.

Ralstonia eutropha strain JMP134 is, like its relative strain H16, both a soil and water dwelling betaproteobacterium, capable of utilizing a wide range of carbon sources for energy. Isolation of *R. eutropha* JMP134 was originally performed based on its ability to utilize 2,4 dichlorophenoxyacetic acid (2, 4-D), a common herbicide [198]. The *R. eutropha* JMP134 genome is comprised of two chromosomes, one megaplasmid, and one plasmid (pJP4), and the genes for 2,4-D utilization have been identified as belonging to the tfd_{I} and tfd_{II} gene clusters of pJP4 [199]. Additionally, *R. eutropha* JMP134 has demonstrated the ability to utilize a variety of other aromatic and chloroaromatic compounds, including: 4-chloro-2-methylphenoxyacetate (MCPA) [200], 2-methylphenoxyacetic acid [200], 3-chlorobenzoate [201], 2,4,6-trichlorophenol [202], 4-fluorobenzoate [203], 2,6-dinitrophenol [204], 3-nitrophenol [205], 4-methylcatecol [206], and various other aromatic substrates [207]. As such, *R. eutropha* JMP134 can be thought of as a master bioremediator of environmental pollutants.

Aromatic Degradation Pathways

The industrial and environmental relevance of *R. eutropha* JMP134 stems primarily from its catabolic degradation of environmental pollutants, specifically those with aromatic structures, of which it is known to utilize 60 unique aromatic compounds as a sole carbon and energy source. The genome of *R. eutropha* JMP134 has been sequenced and pathways for aromatic catabolism can be differentiated into central and peripheral pathways, with peripheral pathways converting a variety of substrates into key intermediate molecules, which can then be processed *via* the central pathways [207]. *R. eutropha* JMP134 contains 10 out of the 11 central metabolic pathways for ring-cleavage known to be present in Proteobacteria, with the sole pathway excluded from its genome being the homoprotocatechuate pathway [207]. The various aerobic ring-cleavage pathways tend to follow the same general two-step degradation process. First, an oxygenation step transforms an aromatic structure into a dihydroxylated benzene ring. Second, deoxygenases induce ring-fission in generally either an *ortho* or *meta* cleavage, or, if the hydroxyl groups are in a *para* formation on the ring, a gentisate pathway cleavage will take place [208].

β-Ketoadipate Pathway

The β-ketoadipate pathway (also known as: 3-oxoadipate pathway) is a multi-branch metabolic pathway that is widely distributed in soil microorganisms, with all branches transforming substrates into β-ketoadipate [also: 3-oxoadipate] for further metabolic processing. Many funneling pathways are able to convert various plant-based aromatics and industrial pollutants into two primary ring cleavage substrates: catechol and protocatechuate [208]. From these substrates, *ortho* cleavage of the aromatic ring proceeds in a 4-step enzymatic process, resulting in β-ketoadipate. Catechol is converted *via* genes of the *cat* gene cluster, while protocatechuate is converted *via* genes of the *pca* gene cluster. From β-ketoadipate, two more enzymatic steps are able to catalyze the transformation into the common metabolites succinyl-CoA and acetyl-CoA. A generic example of conversion from an aromatic environmental molecule into β-ketoadipate *via* the precursor protocatechuate is demonstrated in Fig. (**5**).

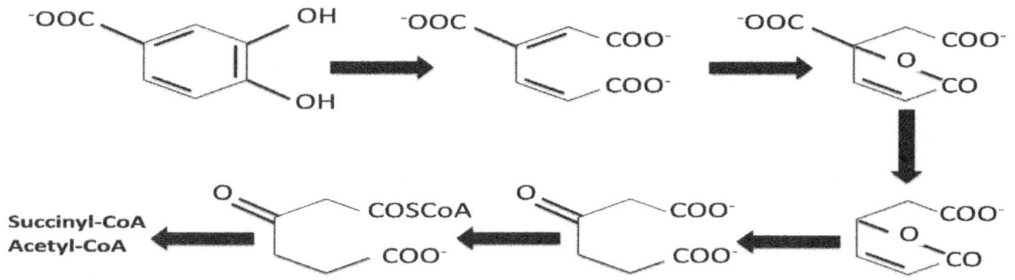

Fig. (5). Conversion of protocatechuate to common Krebs Cycle intermediates, succinyl CoA and acetyl CoA. Intermediate molecules pictured are β-carboxy-*cis,cis*-muconate, γ-carboxymuconolactone, β-ketoadipate enol-lactone, β-ketoadipate and β-ketoadipate:succinyl-CoA.

Common initial substrates for funneling into the β-ketoadipate pathway include natural byproducts of plant decay, shikimate and quinone [209], a narrow range of chlorinated substrates including 4-chlorobenzoate [210], and toxic byproducts of industrial processes, such as *p*-cresol and naphthalene [211, 212].

2,4-dichlorophenoxyacetate

Metabolism of 2,4-dichlorophenoxyacetate (2,4-D) is in a ring-cleavage pathway that does not necessarily fit within the scope of the other pathways, as it is a common environmental aromatic contaminant and, while numerous chloroaromatics can be degraded by the Tfd enzymes, its presence is required in order to activate transcription of gene products for degradation *via* the 2,4-D cleavage pathway [213]. Enzymes of this pathway are contained within the *tfd* gene cluster [named for 2,4-dichlorophenoxyacetate], and encode for a complete set of enzymes required for the degradation of chloroaromatics to β-ketoadipate, the common intermediate of aromatic degradation of the catechol and protocatechuate pathways described in the previous section. Interestingly, *R. eutropha* JMP134 encodes a second set of *tfd* genes. The original *tfd*$_I$ *ABCDEF* gene cluster has a second copy of all genes, except for *tfd*$_I$*A*, and has the nomenclature *tfd*$_{II}$*BCDEF*. Additionally, studies have shown that some the *tfd*$_{II}$ genes demonstrate differing kinetic properties to that of their *tfd*$_I$ homologs [214], and that overall quantity of *tfd* genes can play a distinct role in growth rate of JMP134 on chloroaromatic substrates [215].

4-Methylcatechol *Ortho* Ring-Cleavage Pathway

The 4-methylcatechol *ortho* ring-cleavage pathway is a thus-far rare aromatic degradation pathway. 4-Methyl aromatics are capable of mineralization *via meta* cleavage pathways in numerous prokaryotes [216], though these pathways generally result in "dead-end" metabolites. The *ortho* cleavage pathway for 4-methylcatechol is known, to date, in just 3 organisms: *R. eutropha* JMP134 [206], *Rhodococcus rhodochrous* N75 [217], and *Pseudomonas reinekei* MT1 [218]. A key enzyme of this process is a first-step isomerization of 4-methylcatechol into 3-methylcatechol by 4-methyl-muconolactone methylisomerase [219]. In *R. eutropha* JMP134, enzymes produced by the *mml* gene cluster appear fully responsible for the ultimate conversion of 4-methylcatechol to 4-methyl-3-oxoadipate, though all of the enzymes and genes responsible for this conversion are yet to be elucidated [218]. 4-methyl-3-oxoadipate may then be further metabolized in a manner congruent with the traditional β-ketoadipate pathway (as shown in Fig. **5**), where 4-methyl-3-oxoadipate is converted into 4-methyl-3-oxoadipyl-CoA, a substrate that is then converted into methylsuccinyl-CoA and acetyl-CoA [218]. However, in contrast to the β-ketoadipate pathway, which results in standard central metabolites succinyl-CoA and acetyl-CoA, the eventual fate of methylsuccinyl-CoA, along with the genes and enzymes responsible, is not well established [218].

Gentisate Ring-Cleavage Pathway

Gentisate is a similar initial substrate to both catechol and protochuate, with the primary variance being the *para* position of the two hydroxyl groups on the benzene ring. However, due to this *para* positioning, the enzymes encoded by the *pca* and *cat* gene clusters are not equipped for ring-fission, requiring gentisate to be processed *via* the aptly named gentistate ring-cleavage pathway. Initial substrates of the gentisate cleavage pathway include molecules such as ethyl salicylate and 3-hydroxybenzyl alcohol, which are enzymatically transformed into the precursor molecules of the gentisate pathway, salicylate and 3-hydroxybenzoate, which, interestingly, do not have *para* positioned hydroxyl groups (Fig. **6**).

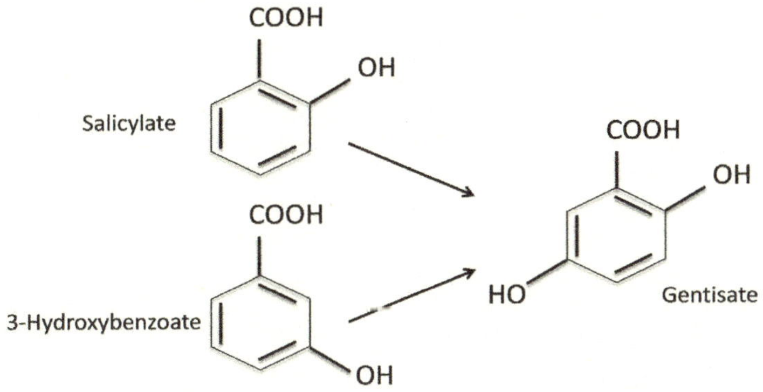

Fig. (6). Salicylate and 3-hydroxybenzoate, the starting substrates of the gentisate pathway. Note the *ortho* hydroxyl group of salicylate and the *meta* hydroxyl group of 3-hydroxybenzoate.

In *Sphingomonas*, *Acinetobacter* and *Pseudomonas* species, salicylate has been shown to be converted to catechol by salicylate-1-hydroxylases, making it available for degradation by the β-ketoadipate pathway [220 - 223]. However, most salicylate-1-hydroxylases known to date are active with 4- and 5-chlorosalicylate, which, coupled with the *tfd* gene products described above, are able to degrade chlorosalicylates, a trait which *R. eutropha* JMP134 does not possess [223]. Additionally, genome searches for homology in *R. eutropha* JMP134 for any known salicylate-1-hydroxylase genes yielded open reading frames [ORFs] generating gene products with a maximum amino acid homology of just 30% [223], providing further evidence against the conversion of salicylate to catechol. An alternative salicylate degradation pathway was described in *Ralstonia* sp. U2 (formerly *Pseudomonas*), catalyzed by the formation of gentisate from salicylate by a salicylate-5-hydroxylase, with a second confirmed salicylate-5-hydroxylase in *Pseudomonas aeruginosa* JB2 [224, 225]. A putative gene cluster, *hybRBCDA*, was identified in strain *R. eutropha* JMP134, with the HybR, HybB, HybC and HybD gene products having relatively high amino acid homology (56-78%) with components of these two confirmed multicomponent salicylate-5-hydroxylases, providing strong evidence for the gentisate pathway being the method by which *R. eutropha* JMP134 catabolizes salicylate.

The second gentisate pathway precursor molecule, 3-hydroxybenzoate, can be converted into gentisate by 3-hydroxybenzoate-6-hydroxylase [226], or into protocatechuate by 3-hydroxybenzoate-4-hydroxylase [227]. In genome analysis of *R. eutropha* JMP134, no open reading frames (ORFs) were found to have any significant homology with 3-hydroxybenzoate-4-hydroxylase genes, but two separate ORFs had 85% amino acid homology with the 3-hydroxybenzoate-6-hydroxylases of *Pseudomonas alcaligenes* NCIB 9867 [228] and *Klebsiella pneumoniae* M5a1 [229], lending evidence towards the degradation of 3-hydroxybenzoate by the gentisate pathway, a suggestion originally made in 1971 by Johnson and Stanier relating to *Ralstonia eutropha* strain 335, while it was known as *Alcaligenes eutrophus* [230].

Generally, once precursor molecules are converted to gentisate, aromatic cleavage is initiated by gentisate-1,2-dioxygenase and then proceeds *via* glutathione-dependent or glutathione-independent steps to form common metabolic intermediates: pyruvate and maleate [223]. In *R. eutropha* JMP134, amino acid homology of gene products of the cluster *mhb* show strong correlation with known enzymes of the glutathione-dependent pathway, yielding evidence for glutathione-dependent pathway utilization by JMP134 [224].

Phenylacetyl-CoA Ring-Cleavage Pathway

The phenylacetyl-CoA ring-cleavage pathway is widespread in nature, as phenylacetate is a relatively common source of carbon for various bacteria, and the general pathway for degradation of this substrate has been characterized in a number of organisms, including *Escherichia coli* [231], and *Azoarcus evansii* [232, 233], which is also a betaproteobacterium. Indeed, genes of the phenylacetate degradation pathway have been found in the genome of *R. eutropha* H16 [9]. There are a few interesting deviations from the established *E. coli* and *A. evansii* pathways by *R. eutropha* JMP134 that are worth mentioning, but for a detailed analysis of the full pathway, please consult the referenced works above, the pathway as characterized in *Pseudomonas putida* U, by Olivera *et al.*, 1998 [234][1], and (Fig. 7).

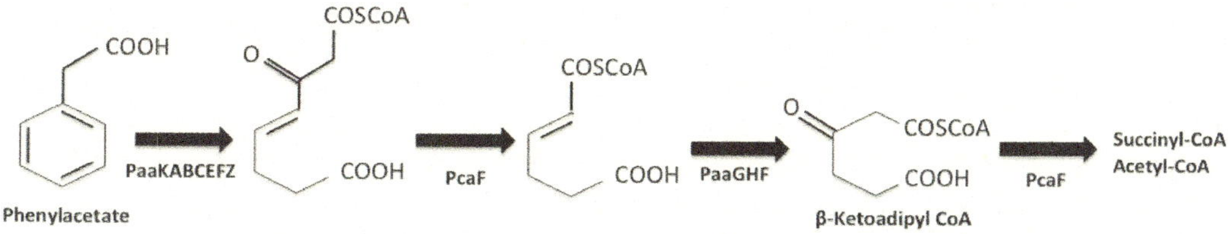

Fig. (7). Phenylacetate catabolic pathway demonstrating key intermediates and enzymes involved in process of conversion from phenylacetate to succinyl-CoA and acetyl-CoA. Not all intermediates are shown as individual steps, but all relevant enzymes are present. Note the necessity of PcaF enzymes in order to complete the catabolic conversion.

In the final enzymatic steps of (Fig. **7**), one may notice similarities to the final steps of (Fig. **5**), of the β-ketoadipate pathway. In *E. coli*, *R. eutropha*, and other phenylacetate-utilizing organisms, genes relating to the phenylacetyl-CoA ring cleavage pathway are generally categorized within the *paa* gene cluster, and appropriately named as such. The PaaJ/PaaE[1] (now simply termed PaaE) gene product in *E. coli*, *P. putida*, and other organisms was shown to catalyze the final step of the pathway: conversion of β -ketoadipyl-CoA into succinyl-CoA and acetyl-CoA [235]. However, *R. eutropha* JMP134 very notably lacks homology for any *paaJ*/*paaE* gene in the *paa* gene cluster of its genome. Yet, Nogales *et al.*, 2007, were able to demonstrate that *E. coli* PaaJ/PaaE was able to successfully complete the final step of the β-ketoadipate in *P. putida* U *pcaF* mutants [235]. Therefore, it seems reasonable to assume that the *pcaF* gene product of the β-ketoadipate pathway performs the function of a β-ketoadipyl-CoA thiolase (*i.e.* PaaJ/PaaE in *E. coli*) to produce succinyl-CoA and acetyl-CoA in *R. eutropha* JMP134, demonstrating an interesting overlap in the two aromatic degradation pathways.

Homogentisate Ring-Cleavage Pathway

Tyrosine and phenylalanine are two of the three aromatic amino acids, and are both degraded *via* the homogentisate pathway. As such, it is well distributed throughout nature in both eukaryotic and prokaryotic organisms, and has been well characterized in a host of organisms including proteobacteria *Pseudomonas putida* [236] and *Burkholderia xenovorans* LB400 [237]. For comprehensive detail of the entire pathway, please consult the previously listed works. There are, however, a few interesting deviations from the standard homogentisate pathway in *R. eutropha* JMP134 worth mentioning.

Most aromatic amino acid hydroxylases bind iron at the catalytic site, including all known mammalian amino acid hydroxylases [238]. Interestingly, the most homologous known phenylalanine hydroxylase to the phenylalanine hydroxylase of JMP134 is from *Chromobacterium violaceum*, with an amino acid homology of 69%, yet this enzyme is known to bind copper at its active site, a rare feature for a hydroxylase of this nature [239]. This has led researchers to speculate that *R. eutropha* JMP134 may, too, bind copper at its catalytic core [223]. This speculation has yet to be determined experimentally.

Secondly, all three isomers of hydroxylphenylacetate are metabolized by *R. eutropha* JMP134, despite demonstrating no significant matches from a genomic search for enzymes involved in the homoprotocatechuate pathway, the standard route for 3- and 4-hydroxyphenylacetate [223]. However, an alternative 4-hydroxyphenylacetate pathway has been reported in a number of Betaproteobacteria [233, 240]. 3-Hydroxyphenylacetate degradation by the homogentisate pathway has been shown in *P. putida* U [236], and speculation of homogentisate pathway utilization in bacteria for 2-hydroxylphenylacetate has been published [241], leading researchers to believe that it is reasonable to add *R. eutropha* JMP134 to the speculative list of organisms utilizing this pathway for hydroxylphenylacetate catabolism [223]. Degradation of hydroxyphenylacetate by *R. eutropha* JMP134 has yet to be demonstrated.

2,3-dihydroxyphenylpropionate Meta Ring-Cleavage Pathway

Initial substrates for catabolism *via* the 2,3-dihydroxyphenylpropionate *meta* ring-cleavage pathway include both 3-hydroxyphenylpropionate and 3-hydroxycinnamate (common plant-derived aromatic compounds synthesized from phenylalanine), but not 2-hydroxyphenylpropinoate or 2-hydroxycinnamate. Detailed analysis of substrates, steps, and final products of this pathway may be found in numerous *E. coli* studies [242, 243]. While some enzymes of the pathway in *R. eutropha* JMP134 show higher amino acid homology for enzymes derived from other organisms (*i.e.*, *Comamonas testosteroni* strain TA441 [244]), the actual pathway and production of intermediates in JMP134 and *E. coli* are believed to be the same, allowing 3-hydroxyphenylpropionate to be transformed into succinate, pyruvate, and acetyl-CoA, and 3-hydroxycinnamate transformed into fumarate, pyruvate, and acetyl-CoA [223, 243].

Catechol *Meta* Ring-Cleavage Pathway

The catechol *meta* ring-cleavage pathway is funneled primarily from starting substrates of benzene and toluene, which are then converted to catechol, just as they would be for central metabolism by the β-ketoadipate pathway. However, the β-ketoadipate pathway proceeds *via* an *ortho* cleavage of catechol by benzoate-1,2-dioxygenase, while the *meta* pathway proceeds *via* a *meta* cleavage by catechol 2,3-dioxygenase and does not utilize β-ketoadipate as an intermediate [245]. Interestingly, this pathway was shown to be activated (*i.e*., production of catechol 2,3-dioxygenase) in the presence of high concentrations of benzoate as a sole carbon source [~40-55mM], in *R. eutropha* strain 335 [245], yet this particular activation phenomenon has not been demonstrated in JMP134. It is an interesting phenomenon and potentially arose as a mechanism for continued catabolism in very high levels of aromatic substrates which would otherwise be toxic, as 55 mM concentrations of benzene as sole carbon source have been shown to inactivate benzoate-1,2-dioxygenase, thus halting catabolism *via* the β-ketoadipate pathway [245].

The final products of this pathway are pyruvate and acetyl-CoA, yet *R. eutropha* JMP134 arrives at these products *via* production of enzymes produced outside of the general catechol *meta* cleavage pathway operon (*phl* genes), a feature currently known to be unique to JMP134. These two enzymes, a 4-hydroxy-2-ketovalerate aldolase and an aldehyde dehydrogenase show a high homology to the *xylQ* and *xylK* genes of *P. putida* [223], though JMP134 does not contain an analogous *xyl* operon.

Chlorohydroxyquinol *Ortho* Ring-Cleavage Pathway

In *R. eutropha* JMP134, 2,4,6-trichlorophenol is the sole chlorinated starting substrate for metabolism *via* the chlorohydroxyquinol *ortho* ring-cleavage pathway [223], and is only one of two chlorinated phenols able to be metabolized as the sole carbon source, along with 2,4-dichlorophenol [202]. Hydroxyquinol in also catabolized in this general pathway, though this metabolism occurs *via* a different set of enzymes. The gene cluster *tcpRXABCYD* encodes the enzymes responsible for catabolism of 2,4,6-trichlorophenol [246], while the *hxqC* and *hxqD* genes are responsible for hydroxyquinol metabolism [223]. Both starting substrates are eventually converted to maleylacetate, which is then converted to β-ketoadipate Fig. (**8**) and eventually succinyl-CoA and acetyl-CoA by the Pca enzymes of the β-ketoadipate pathway [223].

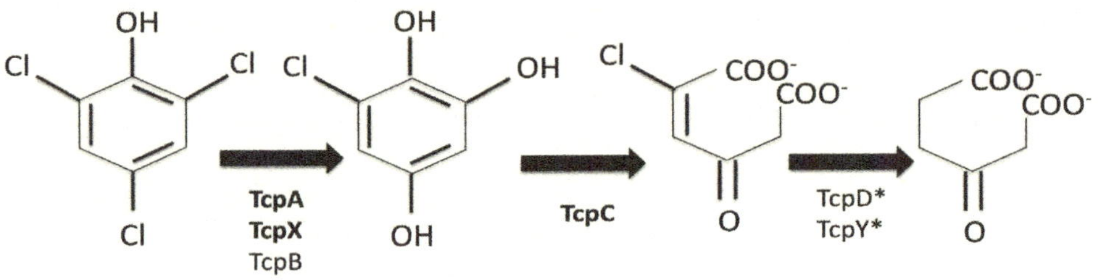

Fig. (8). 2,4,6-Trichlorophenol and key [but not all] intermediates along catabolic pathway to β-ketoadipate. Enzymes in **bold** are required complete catabolism to β-ketoadipate. Enzymes not in bold are not required for complete catabolism, yet may provide increased catabolic rates or increased substrate concentration tolerance. * mutant strains lacking these enzymes were still able to metabolize 2,4,6-trichlorophenol intermediates, likely by putative gene products elsewhere in the genome (genes: *tfdF1* and ReutB4129) [247].

The *tcpRXABCYD* gene cluster has been proposed as a putative catabolic operon, with TcpR assuming the role of the transcriptional regulator, activated in the presence of 2,4,6-trichlorophenol [247]. Knockout strains of *R. eutropha* JMP134 have been developed for each of the genes comprising this cluster, yet the exact role that each gene product plays in the complete transformation of 2,4,6-trichlorophenol to β-ketoadipate has not been unanimously agreed upon, based on the relevant literature [223, 246, 247].

Aminohydroquinone Ring-Cleavage Pathway

The aminohydroquinone ring-cleavage pathway can be thought of as the second of 2 branches of a broader, "hydroquinone" pathway, as initial substrates that can be utilized by either branch are converted eventually to maleylacetate and eventually β-ketoadipate [223]. 3-hydroxylaminophenol, 2-chloro-5-nitrophenol,

3-nitrophenol and aminobenzoquinone are all capable of being utilized as sole carbon and energy sources for *R. eutropha* JMP134 [223, 248, 249]. These initial substrates are converted into aminohydroquinone as a central metabolite by a hydroxyaminophenol mutase, which may then be further transformed into β-ketoadipate by the products of the *mnpC*, *mnpD*, and *mnpE* genes, which encode an aminohydroquinone dioxygenase, an amidase, and a maleylacetate reductase, respectively [223]. The gene cluster relating to aminohydroquinone ring-cleavage, *mnpCARED*, is located on the megaplasmid pJPL [223].

2-Aminobenzoyl-CoA Ring-Cleavage Pathway

L-Tryptophan can serve as a sole carbon and energy source for *R. eutropha* JMP134, yet the strain appears to accomplish this *via* utilization of *abm* genes, homologous to the 2-aminobenzoyl-CoA ring cleavage pathway discovered in 2001 in *Azoarcus evansii* [250], and *kyn* genes, originally documented in *Cupriavidus metallidurans* CH34 [251], rather than a more traditional anthranilate catabolism pathway, based on genome searches for relevant homologs [223]. However, all relevant molecular intermediates of this pathway in *R. eutropha* JMP134 have yet to be elucidated.

R. eutropha JMP134 – A Question

Many readers might be wondering: "Does *R. eutropha* JMP134 synthesize PHA?" A glance of the genome of this strain shows that it possesses the PHA production operon *phaCAB* similar in makeup to that of strain H16 [207]. However, not many published works have focused on PHA biosynthesis in *R. eutropha* JMP134. One report did demonstrate poly(hydroxybutyrate-*co*-hydroxypropionate) [P(HB-*co*-HP)] production by JMP134. In this work, recombinant *R. eutropha* JMP134 synthesized a similar yield of P(HB-*co*-HP) as did strain H16 in low-cell-density cultures. No examination of controlled fermenter cultivations have been performed using JMP134 as the biocatalyst. A question indeed remains for the future: "Does *R. eutropha* JMP134 synthesize PHA when grown on aromatic compounds as a carbon source?"

The Dawning of a New Era – Biofuels Production in *R. eutropha*

As laid out extensively, *R. eutropha* directs excess carbon under nutrient limiting conditions to synthesize PHA as an intracellular molecule of carbon and energy storage. This, as well as the ability of *R. eutropha* to utilize a myriad of carbon sources for growth and polymer production, leads researchers to wonder if carbon flux could be redirected to produce other value added products of interest. Indeed, this question has been partially answered by the works summarized in the previous section on non-PHA polymer synthesis by *R. eutropha*. Given the uncertainty surrounding petroleum prices, recent decades have spawned a tremendous interest in the synthesis and production of bio-based fuels. These biofuels can take many forms, such as alcohols, fatty acids to be converted to biodiesel, or methyl/ethyl esters of long chain organic acids. Thus, in order for biofuel researchers to redirect the cellular machinery of *R. eutropha* from carbon storage to the production of other types of molecules, it is often necessary to remove or otherwise inactivate the operon containing the genes that encode the PHA biosynthesis pathway (*phaCAB*).

R. eutropha strains containing inactivated *phaC* genes have been shown to produce no detectable PHA and secrete pyruvate and possibly acetate into the growth medium [25, 252]. It is for these reasons that *phaC* mutant strains make excellent parental strains for the development of a biofuel synthesis platform. Alternatively, other researchers have opted to leave the *phaCAB* operon intact, yet overexpress introduced or innate genes for redirection of carbon and cellular energy for the production of non-PHA related, high carbon content molecules [253]. However, while recombinant *R. eutropha* can be created to produce a variety of fuel molecules, researchers have expressed reservations regarding the method of both reorganizing "front-end" metabolism by muting or directing carbon away from the PHA biosynthetic pathway, as well as "downstream" metabolism in regards to producing large amounts of a non-native molecule that is not naturally produced by *R. eutropha* and stored intracellularly or secreted into the extracellular milieu [20].

Short-Chained Alcohol Production

Many studies of biofuel production using recombinant microbes involve the synthesis of short-chain alcohol products. Generally, short-chain alcohols have energy densities similar to that of gasoline and possess reasonable research octane numbers (RON) indicating their suitability for use in automobile engines. Methanol is not suitable and as it has a rather a low energy density. Ethanol is more energy dense than methanol, but cannot be used in high concentrations in most countries' existing fuel delivery infrastructures. The biofuel molecules of choice for most researchers tend to be isobutanol and butanol, both C4 alcohols that can be used as "drop in" fuels for existing delivery infrastructures of most nations [254]. In one work, Lu *et al.* demonstrated the ability for recombinant *R. eutropha* to utilize fructose as a sole carbon and energy source to produce isobutanol and 3-methyl-1-butanol from standard intermediates α-ketoisovalerate and α-ketoisocaproate, respectively [252]. While these intermediates are standard cellular metabolic intermediates, wild-type *R. eutropha* is unable to convert them to the key precursors isobutyraldehyde and 3-methyl-1-butyraldehyde, respectively. Therefore, Lu *et al.*, expressed the heterologous *kivD* gene of *Lactococcus lactis* in *R. eutropha* in order to complete this pathway [252]. The KivD gene product is a ketoisovalerate decarboxylase that converts ketoisovalerate and ketoisocaproate to isobutyraldehyde and 3-methyl-1-butyrate, respectively [252]. Simply adding a pathway for the production of branched-chain alcohol precursors is not enough, however. Short chain alcohol dehydrogenase activity is needed, and at ambient oxygen temperatures, *R. eutropha* lacks the necessary dehydrogenase activity required to convert these precursors into their alcohol state. Thus, Lu *et al.* utilized previously constructed strains with constitutive alcohol dehydrogenase activity [255] as parental strains for construction of *R. eutropha* capable of synthesizing isobutanol as the main alcohol product (Fig. **9**), in order to ensure the conversion into alcohols [252]. The isobutanol and 3-methyl-1-butyraldehyde production pathways are initially fed by pyruvate, which, in this case, is utilized by amino acid biosynthetic pathways for the production of valine and leucine. Introduction of the *L. lactis kivD* gene and use of a constitutively expressed alcohol dehydrogenase gene allows for the production of alcohol biofuels from unrelated carbon substrates in engineered *R.*

eutropha. Therefore, while modest production values were achieved using the aforementioned strain modifications, increased production levels were achieved as a result of overexpression of the valine biosynthesis genes and the deletion of several genes allowing carbon to be diverted away from fuel molecule pathways (*e.g.*, an *ilvE* gene that encodes the enzyme to convert 2-ketoisovalerate to valine), with values of isobutanol reaching 200 mg/L, and 3-methyl-1-butanol reaching 30 mg/L [252]. In a follow-up work, a branched-chain alcohol yield of 790 mg/L was achieved in fed-batch cultivation, again with fructose as the carbon source [254]. In this work, and in this field, the challenge of purifying and separating valuable bio-produced components is almost completely avoided, as *R. eutropha* secretes isobutanol extracellularly during metabolism. However, while the fuel molecules are secreted from the *R. eutropha* cell, their presence in the extracellular environment can still be toxic. Therefore, it is necessary to remove the isobutanol or 3-methyl-1-butanol frequently in order to avoid growth inhibitory concentrations (~0.8 vol.-%) [252]. Of course, separation of the two fuel molecules to pure streams of isobutanol and 3-methyl-1-butanol is not needed, since the product is to be used as fuel for combustion.

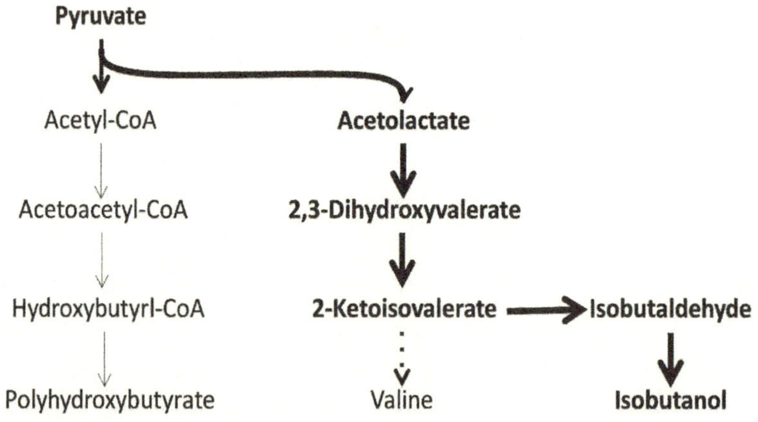

Fig. (9). Primary pathway of carbon flux away from processing *via* the polyhydroxybutyrate (PHB) synthesis pathway and through the altered amino acid synthesis pathway for isobutanol production. Carbon flux is directed towards isobutanol production, even when phaCAB operon fully intact. Thin arrows denote the typical PHB production pathway, despite lack of usage in isobutanol producing strains. The dotted arrow denotes decreased expression of a gene required to catalyze transformation.

Another strain of *R. eutropha* unable to synthesize PHA was used for isopropanol production. The authors of this study took advantage of a different metabolic pathway to divert carbon flux to make alcohol. A *phaB* deletion strain that is incapable of synthesizing 3HB-CoA was used as the parental strain for this endeavor. A maximum observed yield of 3.44 g/L was produced from batch cultures using fructose as the main carbon source. The overall isopropanol production yield was 0.24 Cmol/Cmol [256].

Utilization of CO_2 and Electricity

While *R. eutropha* can utilize numerous refined carbon sources, naturally utilizable waste carbon sources, or be genetically engineered to utilize other waste carbon sources, perhaps the most intriguing metabolic aspect of *R. eutropha* is its ability to utilize CO_2 and H_2 as carbon and energy sources in order to produce isobutanol and 3-methyl-1-butanol. In fact, this process has been done utilizing essentially the same genetic concepts as Lu *et al.*, but instead of utilizing fructose for carbon and energy, Li *et al.* drove CO_2 fixation by electrochemically creating formic acid in a medium with 15% CO_2. This simple process of growing genetically engineered *R. eutropha* solely on CO_2 and electricity produced 140 mg/L of biofuel molecules (isobutanol and 3-methyl-1-butanol combined) [257]. This allows one to imagine that it may be possible to utilize man-made photo-voltaic devices in order to generate electricity to power branched chain alcohol fuel production, an interesting cohesion of microbial and mechanical expertise leading to a common goal of a liquid transportation fuel developed from renewable resources.

A novel hybrid electrical/microbial water splitting and biofuel production system was constructed by Torella *et al.*, to produce isopropanol from CO_2. Using a cobalt phosphate (CoP) anode and a nickel molybdenum zinc (NiMoZn) or stainless steel (SS) cathode, water was split electrically into H_2 and O_2, which were then used by the *R. eutropha* culture in the presence of CO_2. Using this method, 0.2 mg/L isopropanol was produced by an engineered strain of *R. eutropha* [258]. While this yield is significantly lower than that of Grousseau, *et al.* [256], the means by which autotrophic growth and production were established shows great promise for CO_2 conversion into bioproducts.

Medium-Chain Length Alcohol Production

While isobutanol and 3-methyl-1-butanol serve as bio-based blending alcohols with gasoline or drop-in fuels for gasoline engines, they are not compatible with diesel engines, which require a higher energy-density molecule. Medium chain length hydrocarbons, and specifically, medium chain methyl ketones, produced from genetically engineered *R. eutropha*, may provide a possible solution. Wild-type *R. eutropha* does not produce detectable levels of methyl ketones, nor does it contain a high concentration of methyl ketone precursor molecules (fatty acids). Therefore, Müller *et al.* first increased fatty acid production in strains *via* a tripartite approach. First, based on studies in *E. coli* [259], the gene product of '*tesA*, a thioesterase, was overexpressed, which led to a 180-fold increase in fatty acid production over the wild-type strain [253]. Second, a β-oxidation mutant was developed (Δ*fadE*); however this demonstrated no further increase in fatty acid production, in contrast to what had been originally theorized [253]. And finally, as is a common theme in non-PHA production pathways, in order to avoid lost carbon to PHA biosynthesis, the *phaCAB* operon was deleted, leading to an overall 17-fold increase in fatty acid production over the wild-type strain [253].

With a strain overproducing methyl ketone precursors, Müller and coworkers were then able to manipulate the pathways associated with β-oxidatation during heterotrophic growth in order to produce maximum levels of β-keto acids, which can spontaneously decarboxylate to form methyl ketones. This was done *via* a bipartite method. First, overproduction of β-ketoacyl-CoA is achieved *via* a truncated β-oxidation pathway that utilizes FadE expressed from a heterologous *Micrococcus lateus* gene as an acyl-CoA oxidase, FadB from a heterologous *E. coli* gene, and the native *fadA* is deleted. Second, overexpression of FadM, a thioesterase, from *E. coli* converts the aforementioned β-ketoacyl-CoA into β-keto acid [253]. The highest yields of methyl ketone production came from strains which included both a β-oxidatation (*fadE*) deletion in partnership with the engineered truncated version, designed to produce methyl ketones. The strain with these characteristics produced more methyl ketones than even the strain containing the *fadE* and the *phaCAB* gene deletions, and data suggests that methyl ketone production through the truncated β-oxidatation pathway will divert carbon flux away from PHA synthesis and through the truncated pathway. In the studies of this strain grown on fructose as the main carbon source, production levels were able to reach >50mg/L of total methyl ketones [253].

As with the previous work in isobutanol and 3-methyl-1-butanol production, chemolithoautotrophic production is of particular interest based on its inherent sustainable nature. Here, Müller *et al.* were able to produce methyl ketones under chemolithoautotrophic conditions, with CO_2 as the sole carbon source, in titers ranging between 50-70 mg/L after 48 h of incubation, which are very comparable levels to that observed under heterotrophic conditions [253].

Eutrophanol? Future Directions for Fuel Production

Researchers have demonstrated the ability of *R. eutropha* to utilize refined sugars, waste carbon, and CO_2 as carbon sources for production of short chain fuel molecules, isobutanol and 3-methyl-1-butanol, and medium-chain fuel molecules, methyl ketones. In all cases, it is notable that the yield and productivity of fuel is very low. The productivity in all cases is too low to be considered a viable alternative fuel production method. The reason for this is likely due to the fact that these are foreign products for *R. eutropha.* Indeed, PHA is an intrinsic product of *R. eutropha* metabolism, and thus manipulation of the bacterium and/or culture conditions to achieve greater yields and productivities is seemingly limitless.

It is true that improving production efficiency of these fuel molecules will be key to any future industrial applications, though these early results show intrigue based on the robust degradative properties and ease of cultivation of *R. eutropha*, as high cell-densities are achievable at ambient room temperatures. Research on fuel development with *R. eutropha* is in its infancy, and so future research may focus on production efficiency *via* improved flux at metabolic bottlenecks, utilization of novel carbon feedstocks, genetic engineering for enhanced metabolic properties, or some other yet undiscovered avenue. One of the principal considerations for engineered, fuel-producing *R. eutropha* strains is how the cells react to their new product. Studies of gene expression and product tolerance are underway in *R. eutropha* (Brigham *et al.*, unpublished data) to aid in understanding the direction future strain engineering must take to construct a robust alcohol-producing organism.

CONCLUDING REMARKS: *RALSTONIA* WRAP UP

"The idea was, could you take a bug like *Ralstonia eutropha* and mess around with its guts…" [260]

-Daniel Nocera, Harvard University

We believe that several research groups have started conversations in much the same way as Professor Nocera, who is discussing the solar-mediated splitting of water to make H_2 and O_2 for autotrophic *R. eutropha* cultures. The soil has yielded many such "bioconverters" as *R. eutropha* and its close relatives. Specifically, *R. eutropha* has been a bacterium of interest for many years because of what it offers: autotrophic growth ability, genetic manipulability, high concentration carbon storage ability, and versatility of carbon substrate utilization. It has become a bacterium of industrial interest because of the ease of metabolic engineering and its relatively rapid growth and product accumulation rate.

There are, however, still challenges to be faced and overcome if we are to develop a bioprocess with *R. eutropha* as the biocatalyst. First, the costs of PHA production must be lowered significantly to compete with the cost of petroleum-based plastic as raw material. This is indeed an area of focus for research groups. Second, to engineer heterologous biosynthesis pathways in *R. eutropha*, the path forward to high yield productions must be clear. This will undoubtedly involve genomic, proteomic and metabolomics efforts to maximize carbon flux to the proper product while still maintaining cell viability. The future is bright for *R. eutropha* because challenges abound.

CONFLICT OF INTEREST

The authors confirm that they have no conflict of interest to declare for this publication.

ACKNOWLEDGEMENTS

The authors would like to thank the editors of "Microbial Polyesters – Production, Performance and Processing Aspects," especially Dr. Martin Koller, for the opportunity to write this chapter. We would also like to thank Mr. Nicholas

Phillips for his help in preparing our manuscript. CJB wishes to thank Prof. Anthony Sinskey and Prof. ChoKyun Rha for the opportunity to work on the exciting challenges associated with PHA homeostasis and metabolic engineering in *R. eutropha*.

NOTES

[1] PaaJ/PaaE is a single protein, now referred to solely as PaaE, but was previously referred to as PaaJ. It is the gene product of a single gene, *paaE*, which was previously known as *paaJ*.

REFERENCES

[1] Schatz A, Bovell C Jr. Growth and hydrogenase activity of a new bacterium, *Hydrogenomonas facilis*. J Bacteriol 1952; 63(1): 87-98.
[PMID: 14927551]

[2] Bowien B, Schlegel HG. Physiology and biochemistry of aerobic hydrogen-oxidizing bacteria. Annu Rev Microbiol 1981; 35: 405-52.
[http://dx.doi.org/10.1146/annurev.mi.35.100181.002201] [PMID: 6271040]

[3] Wilde E. Untersuchungen über Wachstum und Speikerstoffsynthese von Hydrogenomonas. Arch Mikrobiol 1962; 43: 109-37.
[http://dx.doi.org/10.1007/BF00406429] [PMID: 14042378]

[4] Davis DH, Stanier RY, Doudoroff M, Mandel M. Taxonomic studies on some gram negative polarly flagellated "hydrogen bacteria" and related species. Arch Mikrobiol 1970; 70(1): 1-13.
[http://dx.doi.org/10.1007/BF00691056] [PMID: 4987616]

[5] Yabuuchi E, Kosako Y, Yano I, Hotta H, Nishiuchi Y. Transfer of two Burkholderia and an Alcaligenes species to Ralstonia gen. Microbiol Immunol 1995; 39(11): 897-904.

[6] Vaneechoutte M, Kämpfer P, De Baere T, Falsen E, Verschraegen G. Wautersia gen. nov., a novel genus accommodating the phylogenetic lineage including Ralstonia eutropha and related species, and proposal of Ralstonia [Pseudomonas] syzygii (Roberts et al. 1990) comb. nov. Int J Syst Evol Microbiol 2004; 54(Pt 2): 317-27.
[http://dx.doi.org/10.1099/ijs.0.02754-0] [PMID: 15023939]

[7] Vandamme P, Coenye T. Taxonomy of the genus *Cupriavidus*: a tale of lost and found. Int J Syst Evol Microbiol 2004; 54(Pt 6): 2285-9.
[http://dx.doi.org/10.1099/ijs.0.63247-0] [PMID: 15545472]

[8] Tibazarwa C, Wuertz S, Mergeay M, Wyns L, van Der Lelie D. Regulation of the cnr cobalt and nickel resistance determinant of *Ralstonia eutropha* (*Alcaligenes eutrophus*) CH34. J Bacteriol 2000; 182(5): 1399-409.
[http://dx.doi.org/10.1128/JB.182.5.1399-1409.2000] [PMID: 10671464]

[9] Pohlmann A, Fricke WF, Reinecke F, *et al.* Genome sequence of the bioplastic-producing "Knallgas"

bacterium *Ralstonia eutropha* H16. Nat Biotechnol 2006; 24(10): 1257-62. [http://dx.doi.org/10.1038/nbt1244] [PMID: 16964242]

[10] Lapage SP, Sneath PH, Lessel EF, Skerman VB, Seeliger HP, Clark WA, Eds. International Code of Nomenclature of Bacteria: Bacteriological Code, 1990 Revision. Washington, DC: ASM Press 1992.

[11] Vollbrecht D, Schlegel HG, Stoschek G, Janczikowski A. Excretion of metabolites by hydrogen bacteria: IV. Respiration rate-dependent formationof primary metabolites and poly-3-hydroxybutanoate. Eur J Appl Microbiol Biotechnol 1979; 7: 267-76. [http://dx.doi.org/10.1007/BF00498021]

[12] Vollbrecht D, El Nawawy MA, Schlegel HG. Excretion of metabolites by hydrogen bacteria: I. Autotrophic and heterotrophic fermentations. Eur J Appl Microbiol Biotechnol 1978; 6: 145-55. [http://dx.doi.org/10.1007/BF00504426]

[13] Reinecke F, Steinbüchel A. *Ralstonia eutropha* strain H16 as model organism for PHA metabolism and for biotechnological production of technically interesting biopolymers. J Mol Microbiol Biotechnol 2009; 16(1-2): 91-108. [http://dx.doi.org/10.1159/000142897] [PMID: 18957865]

[14] Schlegel H, Lafferty R. Novel Energy and Carbon Sources. Advances in Biochemical Engineering 1971; 1: 143-68. [http://dx.doi.org/10.1007/BFb0044733]

[15] Calloway DH, Kumar AM. Protein quality of the bacterium *Hydrogenomonas eutropha.* Appl Microbiol 1969; 17(1): 176-8. [PMID: 4975453]

[16] Waslien CI, Calloway DH. Nutritional value of lipids in *Hydrogenomonas eutropha* as measured in the rat. Appl Microbiol 1969; 18(2): 152-5. [PMID: 4979940]

[17] Steinbüchel A. Polyhydroxyalkanoic Acids. In: Byrom D, Ed. Biomaterials London: MacMillan. 1991; pp. 123-213. [http://dx.doi.org/10.1007/978-1-349-11167-1_3]

[18] Kunasundari B, Murugaiyah V, Kaur G, Maurer FH, Sudesh K. Revisiting the single cell protein application of *Cupriavidus necator* H16 and recovering bioplastic granules simultaneously. PLoS One 2013; 8(10): e78528. [http://dx.doi.org/10.1371/journal.pone.0078528] [PMID: 24205250]

[19] Brigham CJ, Zhila N, Shishatskaya E, Volova TG, Sinskey AJ. Manipulation of *Ralstonia eutropha* carbon storage pathways to produce useful bio-based products. Subcell Biochem 2012; 64: 343-66. [http://dx.doi.org/10.1007/978-94-007-5055-5_17] [PMID: 23080259]

[20] Riedel SL, Lu J, Stahl U, Brigham CJ. Lipid and fatty acid metabolism in *Ralstonia eutropha*: relevance for the biotechnological production of value-added products. Appl Microbiol Biotechnol 2014; 98(4): 1469-83. [http://dx.doi.org/10.1007/s00253-013-5430-8] [PMID: 24343766]

[21] Gebauer B, Jendrossek D. Assay of poly(3-hydroxybutyrate) depolymerase activity and product determination. Appl Environ Microbiol 2006; 72(9): 6094-100.

[http://dx.doi.org/10.1128/AEM.01184-06] [PMID: 16957234]

[22] Jendrossek D, Handrick R. Microbial degradation of polyhydroxyalkanoates. Annu Rev Microbiol 2002; 56: 403-32.
[http://dx.doi.org/10.1146/annurev.micro.56.012302.160838] [PMID: 12213937]

[23] Steinbüchel A, Hustede E, Liebergesell M, Pieper U, Timm A, Valentin H. Molecular basis for biosynthesis and accumulation of polyhydroxyalkanoic acids in bacteria. FEMS Microbiol Rev 1992; 9(2-4): 217-30.
[http://dx.doi.org/10.1016/0378-1097(92)90313-D] [PMID: 1476773]

[24] Brigham CJ, Budde CF, Holder JW, *et al.* Elucidation of beta-oxidation pathways in *Ralstonia eutropha* H16 by examination of global gene expression. J Bacteriol 2010; 192(20): 5454-64.
[http://dx.doi.org/10.1128/JB.00493-10] [PMID: 20709892]

[25] Brigham CJ, Speth DR, Rha C, Sinskey AJ. Whole-genome microarray and gene deletion studies reveal regulation of the polyhydroxyalkanoate production cycle by the stringent response in *Ralstonia eutropha* H16. Appl Environ Microbiol 2012; 78(22): 8033-44.
[http://dx.doi.org/10.1128/AEM.01693-12] [PMID: 22961894]

[26] Masamune S, Walsh CT, Sinskey AJ, Peoples OP. Poly-(R)-3-hydroxybutyrate (PHB) biosynthesis: mechanistic studies on biological Claisen condensation catalyzed by (beta)-ketoacyl thiolase. Pure Appl Chem 1989; 61(3): 303-12.
[http://dx.doi.org/10.1351/pac198961030303]

[27] Oeding V, Schlegel HG. Beta-ketothiolase from *Hydrogenomonas eutropha* H16 and its significance in the regulation of poly-beta-hydroxybutyrate metabolism. Biochem J 1973; 134(1): 239-48.
[http://dx.doi.org/10.1042/bj1340239] [PMID: 4198758]

[28] Thompson S, Mayerl F, Peoples OP, Masamune S, Sinskey AJ, Walsh CT. Mechanistic studies on beta-ketoacyl thiolase from *Zoogloea ramigera*: identification of the active-site nucleophile as Cys89, its mutation to Ser89, and kinetic and thermodynamic characterization of wild-type and mutant enzymes. Biochemistry 1989; 28(14): 5735-42.
[http://dx.doi.org/10.1021/bi00440a006] [PMID: 2775734]

[29] Schubert P, Steinbüchel A, Schlegel HG. Cloning of the *Alcaligenes eutrophus* genes for synthesis of poly-beta-hydroxybutyric acid (PHB) and synthesis of PHB in *Escherichia coli*. J Bacteriol 1988; 170(12): 5837-47.
[PMID: 2848014]

[30] Peoples OP, Sinskey AJ. Poly-beta-hydroxybutyrate biosynthesis in *Alcaligenes eutrophus* H16. Characterization of the genes encoding beta-ketothiolase and acetoacetyl-CoA reductase. J Biol Chem 1989; 264(26): 15293-7.
[PMID: 2670935]

[31] Lindenkamp N, Peplinski K, Volodina E, Ehrenreich A, Steinbüchel A. Impact of multiple beta-ketothiolase deletion mutations in *Ralstonia eutropha* H16 on the composition of 3-mercaptopropionic acid-containing copolymers. Appl Environ Microbiol 2010; 76(16): 5373-82.
[http://dx.doi.org/10.1128/AEM.01058-10] [PMID: 20601511]

[32] Lindenkamp N, Volodina E, Steinbüchel A. Genetically modified strains of Ralstonia eutropha H16 with β-ketothiolase gene deletions for production of copolyesters with defined 3-hydroxyvaleric acid

contents. Appl Environ Microbiol 2012; 78(15): 5375-83.
[http://dx.doi.org/10.1128/AEM.00824-12] [PMID: 22636005]

[33] Kim EJ, Kim KJ. Cloning, expression, purification, crystallization and X-ray crystallographic analysis of PhaA from *Ralstonia eutropha.* Acta Crystallogr F Struct Biol Commun 2014; 70(Pt 11): 1566-9.
[http://dx.doi.org/10.1107/S2053230X14022778] [PMID: 25372833]

[34] Doi Y, Kunioka M, Nakamura Y, Soga K. Biosynthesis of copolyesters in *Alcaligenes eutrophus* H16 from ^{13}C-labeled acetate and propionate. Macromolecules 1987; 20(12): 2988-91.
[http://dx.doi.org/10.1021/ma00178a006]

[35] Slater S, Houmiel KL, Tran M, *et al.* Multiple beta-ketothiolases mediate poly(beta-hydroxyalkanoate) copolymer synthesis in *Ralstonia eutropha.* J Bacteriol 1998; 180(8): 1979-87.
[PMID: 9555876]

[36] Budde CF, Mahan AE, Lu J, Rha C, Sinskey AJ. Roles of multiple acetoacetyl coenzyme A reductases in polyhydroxybutyrate biosynthesis in *Ralstonia eutropha* H16. J Bacteriol 2010; 192(20): 5319-28.
[http://dx.doi.org/10.1128/JB.00207-10] [PMID: 20729355]

[37] Kim EJ, Kim J, Ahn JW, Kim YJ, Chang JH, Kim KJ. Crystal structure of (S)-3-hydroxybutyryl-CoA dehydrogenase from Clostridium butyricum and its mutations that enhance reaction kinetics. J Microbiol Biotechnol 2014; 24(12): 1636-43.
[http://dx.doi.org/10.4014/jmb.1407.07027] [PMID: 25112316]

[38] Rehm BH. Polyester synthases: natural catalysts for plastics. Biochem J 2003; 376(Pt 1): 15-33.
[http://dx.doi.org/10.1042/bj20031254] [PMID: 12954080]

[39] Cho M, Brigham CJ, Sinskey AJ, Stubbe J. Purification of polyhydroxybutyrate synthase from its native organism, *Ralstonia eutropha*: implications for the initiation and elongation of polymer formation *in vivo.* Biochemistry 2012; 51(11): 2276-88.
[http://dx.doi.org/10.1021/bi2013596] [PMID: 22369488]

[40] Peoples OP, Sinskey AJ. Poly-beta-hydroxybutyrate (PHB) biosynthesis in *Alcaligenes eutrophus* H16. Identification and characterization of the PHB polymerase gene (*phbC*). J Biol Chem 1989; 264(26): 15298-303.
[PMID: 2670936]

[41] Gerngross TU, Reilly P, Stubbe J, Sinskey AJ, Peoples OP. Immunocytochemical analysis of poly-beta-hydroxybutyrate (PHB) synthase in *Alcaligenes eutrophus* H16: localization of the synthase enzyme at the surface of PHB granules. J Bacteriol 1993; 175(16): 5289-93.
[PMID: 8349571]

[42] Gerngross TU, Snell KD, Peoples OP, *et al.* Overexpression and purification of the soluble polyhydroxyalkanoate synthase from *Alcaligenes eutrophus*: evidence for a required posttranslational modification for catalytic activity. Biochemistry 1994; 33(31): 9311-20.
[http://dx.doi.org/10.1021/bi00197a035] [PMID: 8049232]

[43] Kawaguchi Y, Doi Y. Kinetics and mechanism of synthesis and degradation of poly(3-hydroxybutyrate) in *Alcaligenes eutrophus.* Macromolecules 1992; 25(9): 2324-9.
[http://dx.doi.org/10.1021/ma00035a007]

[44] Rehm BH, Antonio RV, Spiekermann P, Amara AA, Steinbüchel A. Molecular characterization of the

poly(3-hydroxybutyrate) (PHB) synthase from *Ralstonia eutropha*: *in vitro* evolution, site-specific mutagenesis and development of a PHB synthase protein model. Biochim Biophys Acta 2002; 1594(1): 178-90.
[http://dx.doi.org/10.1016/S0167-4838(01)00299-0] [PMID: 11825620]

[45] Zhang S, Kolvek S, Lenz RW, Goodwin S. Mechanism of the polymerization reaction initiated and catalyzed by the polyhydroxybutyrate synthase of *Ralstonia eutropha.* Biomacromolecules 2003; 4(3): 504-9.
[http://dx.doi.org/10.1021/bm020099x] [PMID: 12741763]

[46] Zhang S, Yasuo T, Lenz RW, Goodwin S. Kinetic and mechanistic characterization of the polyhydroxybutyrate synthase from *Ralstonia eutropha.* Biomacromolecules 2000; 1(2): 244-51.
[http://dx.doi.org/10.1021/bm005513c] [PMID: 11710107]

[47] Jia Y, Yuan W, Wodzinska J, Park C, Sinskey AJ, Stubbe J. Mechanistic studies on class I polyhydroxybutyrate (PHB) synthase from *Ralstonia eutropha*: class I and III synthases share a similar catalytic mechanism. Biochemistry 2001; 40(4): 1011-9.
[http://dx.doi.org/10.1021/bi002219w] [PMID: 11170423]

[48] Yuan W, Jia Y, Tian J, *et al.* Class I and III polyhydroxyalkanoate synthases from *Ralstonia eutropha* and *Allochromatium vinosum*: characterization and substrate specificity studies. Arch Biochem Biophys 2001; 394(1): 87-98.
[http://dx.doi.org/10.1006/abbi.2001.2522] [PMID: 11566031]

[49] Zhang W, Shrestha R, Buckley RM, *et al.* Mechanistic insight with $HBCH_2CoA$ as a probe to polyhydroxybutyrate (PHB) synthases. ACS Chem Biol 2014; 9(8): 1773-9.
[http://dx.doi.org/10.1021/cb5002735] [PMID: 24896226]

[50] Peplinski K, Ehrenreich A, Döring C, Bömeke M, Reinecke F, Hutmacher C, *et al.* Genome-wide transcriptome analyses of the "Knallgas" bacterium *Ralstonia eutropha* H16 with regard to PHA metabolism. Microbiology 2010; 156(Pt. 7): 2136-52.
[http://dx.doi.org/10.1099/mic.0.038380-0] [PMID: 20395272]

[51] Sznajder A, Pfeiffer D, Jendrossek D. Comparative proteome analysis reveals four novel polyhydroxybutyrate (PHB) granule-associated proteins in *Ralstonia eutropha* H16. Appl Environ Microbiol 2015; 81(5): 1847-58.
[http://dx.doi.org/10.1128/AEM.03791-14] [PMID: 25548058]

[52] Wieczorek R, Pries A, Steinbüchel A, Mayer F. Analysis of a 24-kilodalton protein associated with the polyhydroxyalkanoic acid granules in *Alcaligenes eutrophus.* J Bacteriol 1995; 177(9): 2425-35.
[PMID: 7730274]

[53] Jendrossek D, Pfeiffer D. New insights in the formation of polyhydroxyalkanoate granules (carbonosomes) and novel functions of poly(3-hydroxybutyrate). Environ Microbiol 2014; 16(8): 2357-73.
[http://dx.doi.org/10.1111/1462-2920.12356] [PMID: 24329995]

[54] Pfeiffer D, Jendrossek D. Localization of poly(3-hydroxybutyrate) (PHB) granule-associated proteins during PHB granule formation and identification of two new phasins, PhaP6 and PhaP7, in *Ralstonia eutropha* H16. J Bacteriol 2012; 194(21): 5909-21.
[http://dx.doi.org/10.1128/JB.00779-12] [PMID: 22923598]

[55] Pfeiffer D, Jendrossek D. Interaction between poly(3-hydroxybutyrate) granule-associated proteins as revealed by two-hybrid analysis and identification of a new phasin in *Ralstonia eutropha* H16. Microbiology 2011; 157(Pt 10): 2795-807.
[http://dx.doi.org/10.1099/mic.0.051508-0] [PMID: 21737497]

[56] Pötter M, Müller H, Reinecke F, *et al.* The complex structure of polyhydroxybutyrate (PHB) granules: four orthologous and paralogous phasins occur in *Ralstonia eutropha.* Microbiology 2004; 150(Pt 7): 2301-11.
[http://dx.doi.org/10.1099/mic.0.26970-0] [PMID: 15256572]

[57] Pötter M, Steinbüchel A. Poly(3-hydroxybutyrate) granule-associated proteins: impacts on poly(3-hydroxybutyrate) synthesis and degradation. Biomacromolecules 2005; 6(2): 552-60.
[http://dx.doi.org/10.1021/bm049401n] [PMID: 15762612]

[58] Pötter M, Müller H, Steinbüchel A. Influence of homologous phasins (PhaP) on PHA accumulation and regulation of their expression by the transcriptional repressor PhaR in *Ralstonia eutropha* H16. Microbiology 2005; 151(Pt 3): 825-33.
[http://dx.doi.org/10.1099/mic.0.27613-0] [PMID: 15758228]

[59] Pötter M, Madkour MH, Mayer F, Steinbüchel A. Regulation of phasin expression and polyhydroxyalkanoate (PHA) granule formation in *Ralstonia eutropha* H16. Microbiology 2002; 148(Pt 8): 2413-26.
[http://dx.doi.org/10.1099/00221287-148-8-2413] [PMID: 12177335]

[60] York GM, Stubbe J, Sinskey AJ. New insight into the role of the PhaP phasin of *Ralstonia eutropha* in promoting synthesis of polyhydroxybutyrate. J Bacteriol 2001; 183(7): 2394-7.
[http://dx.doi.org/10.1128/JB.183.7.2394-2397.2001] [PMID: 11244085]

[61] York GM, Stubbe J, Sinskey AJ. The *Ralstonia eutropha* PhaR protein couples synthesis of the PhaP phasin to the presence of polyhydroxybutyrate in cells and promotes polyhydroxybutyrate production. J Bacteriol 2002; 184(1): 59-66.
[http://dx.doi.org/10.1128/JB.184.1.59-66.2002] [PMID: 11741844]

[62] Lawrence AG, Schoenheit J, He A, *et al.* Transcriptional analysis of *Ralstonia eutropha* genes related to poly-(R)-3-hydroxybutyrate homeostasis during batch fermentation. Appl Microbiol Biotechnol 2005; 68(5): 663-72.
[http://dx.doi.org/10.1007/s00253-005-1969-3] [PMID: 15924243]

[63] Kuchta K, Chi L, Fuchs H, Pötter M, Steinbüchel A. Studies on the influence of phasins on accumulation and degradation of PHB and nanostructure of PHB granules in *ralstonia eutropha* H16. Biomacromolecules 2007; 8(2): 657-62.
[http://dx.doi.org/10.1021/bm060912e] [PMID: 17291089]

[64] Eggers J, Steinbüchel A. Poly(3-hydroxybutyrate) degradation in *Ralstonia eutropha* H16 is mediated stereoselectively to (S)-3-hydroxybutyryl coenzyme A (CoA) *via* crotonyl-CoA. J Bacteriol 2013; 195(14): 3213-23.
[http://dx.doi.org/10.1128/JB.00358-13] [PMID: 23667237]

[65] York GM, Junker BH, Stubbe JA, Sinskey AJ. Accumulation of the PhaP phasin of *Ralstonia eutropha* is dependent on production of polyhydroxybutyrate in cells. J Bacteriol 2001; 183(14): 4217-26.
[http://dx.doi.org/10.1128/JB.183.14.4217-4226.2001] [PMID: 11418562]

[66] Yamada M, Takahashi S, Okahata Y, Doi Y, Numata K. Monitoring and kinetic analysis of the molecular interactions by which a repressor protein, PhaR, binds to target DNAs and poly-((R)-3-hydroxybutyrate. AMB Express 2013; 3(1)

[67] Yamada M, Wakuda A, Taguchi S. Morphological change in cellular granule formation of poly[(R)-3-hydroxybutyrate] caused by DNA-binding-related mutations of an autoregulated repressor PhaR. Biosci Biotechnol Biochem 2007; 71(6): 1572-6.
[http://dx.doi.org/10.1271/bbb.70038] [PMID: 17587694]

[68] Yamada M, Yamashita K, Wakuda A, *et al.* Autoregulator protein PhaR for biosynthesis of polyhydroxybutyrate [P(3HB)] possibly has two separate domains that bind to the target DNA and P(3HB): Functional mapping of amino acid residues responsible for DNA binding. J Bacteriol 2007; 189(3): 1118-27.
[http://dx.doi.org/10.1128/JB.01550-06] [PMID: 17122335]

[69] Tian J, Sinskey AJ, Stubbe J. Kinetic studies of polyhydroxybutyrate granule formation in *Wautersia eutropha* H16 by transmission electron microscopy. J Bacteriol 2005; 187(11): 3814-24.
[http://dx.doi.org/10.1128/JB.187.11.3814-3824.2005] [PMID: 15901706]

[70] Beeby M, Cho M, Stubbe J, Jensen GJ. Growth and localization of polyhydroxybutyrate granules in *Ralstonia eutropha.* J Bacteriol 2012; 194(5): 1092-9.
[http://dx.doi.org/10.1128/JB.06125-11] [PMID: 22178974]

[71] Pfeiffer D, Wahl A, Jendrossek D. Identification of a multifunctional protein, PhaM, that determines number, surface to volume ratio, subcellular localization and distribution to daughter cells of poly(3-hydroxybutyrate), PHB, granules in *Ralstonia eutropha* H16. Mol Microbiol 2011; 82(4): 936-51.
[http://dx.doi.org/10.1111/j.1365-2958.2011.07869.x] [PMID: 22023320]

[72] MacEachran DP, Prophete ME, Sinskey AJ. The *Rhodococcus opacus* PD630 heparin-binding hemagglutinin homolog TadA mediates lipid body formation. Appl Environ Microbiol 2010; 76(21): 7217-25.
[http://dx.doi.org/10.1128/AEM.00985-10] [PMID: 20851968]

[73] Brigham CJ, Reimer EN, Rha C, Sinskey AJ. Examination of PHB Depolymerases in Ralstonia eutropha: Further Elucidation of the Roles of Enzymes in PHB Homeostasis. AMB Express 2012; 26;2(1)

[74] York GM, Lupberger J, Tian J, Lawrence AG, Stubbe J, Sinskey AJ. Ralstonia eutropha H16 encodes two and possibly three intracellular Poly[D-(-)-3-hydroxybutyrate] depolymerase genes. J Bacteriol 2003; 185(13): 3788-94.
[http://dx.doi.org/10.1128/JB.185.13.3788-3794.2003] [PMID: 12813072]

[75] Saegusa H, Shiraki M, Kanai C, Saito T. Cloning of an intracellular Poly[D(-)-3-Hydroxybutyrate] depolymerase gene from Ralstonia eutropha H16 and characterization of the gene product. J Bacteriol 2001; 183(1): 94-100.
[http://dx.doi.org/10.1128/JB.183.1.94-100.2001] [PMID: 11114905]

[76] Uchino K, Saito T, Jendrossek D. Poly(3-hydroxybutyrate) (PHB) depolymerase *phaZ*a1 is involved in mobilization of accumulated PHB in *Ralstonia eutropha* H16. Appl Environ Microbiol 2008; 74(4): 1058-63.
[http://dx.doi.org/10.1128/AEM.02342-07] [PMID: 18156336]

[77] Uchino K, Saito T, Gebauer B, Jendrossek D. Isolated poly(3-hydroxybutyrate) (PHB) granules are complex bacterial organelles catalyzing formation of PHB from acetyl coenzyme A (CoA) and degradation of PHB to acetyl-CoA. J Bacteriol 2007; 189(22): 8250-6.
[http://dx.doi.org/10.1128/JB.00752-07] [PMID: 17720797]

[78] Kobayashi T, Saito T. Catalytic triad of intracellular poly(3-hydroxybutyrate) depolymerase (*phaZ*1) in *Ralstonia eutropha* H16. J Biosci Bioeng 2003; 96(5): 487-92.
[http://dx.doi.org/10.1016/S1389-1723(03)70136-4] [PMID: 16233560]

[79] Kobayashi T, Shiraki M, Abe T, Sugiyama A, Saito T. Purification and properties of an intracellular 3-hydroxybutyrate-oligomer hydrolase (*phaZ*2) in *Ralstonia eutropha* H16 and its identification as a novel intracellular poly(3-hydroxybutyrate) depolymerase. J Bacteriol 2003; 185(12): 3485-90.
[http://dx.doi.org/10.1128/JB.185.12.3485-3490.2003] [PMID: 12775684]

[80] Knoll M, Hamm TM, Wagner F, Martinez V, Pleiss J. The PHA Depolymerase Engineering Database: A systematic analysis tool for the diverse family of polyhydroxyalkanoate (PHA) depolymerases. BMC Bioinformatics 2009.

[81] Sznajder A, Jendrossek D. To be or not to be a poly(3-hydroxybutyrate) (PHB) depolymerase: *phaZ*d1 (*phaZ*6) and *phaZ*d2 (*phaZ*7) of *Ralstonia eutropha*, highly active PHB depolymerases with no detectable role in mobilization of accumulated PHB. Appl Environ Microbiol 2014; 80(16): 4936-46.
[http://dx.doi.org/10.1128/AEM.01056-14] [PMID: 24907326]

[82] Chung MG, Rhee YH. Overexpression of the (R)-specific enoyl-CoA hydratase gene from *Pseudomonas chlororaphis* HS21 in *Pseudomonas* strains for the biosynthesis of polyhydroxyalkanoates of altered monomer composition. Biosci Biotechnol Biochem 2012; 76(3): 613-6.
[http://dx.doi.org/10.1271/bbb.110871] [PMID: 22451412]

[83] Tsuge T, Sato S, Hiroe A, Ishizuka K, Kanazawa H, Shiro Y, *et al.* Contribution of the distal pocket residue to the acyl-chain-length substrate specificity of (R)-specific enoyl-CoA hydratases from *Pseudomonas*: site-directed mutagenesis, chimeragenesis, X-ray crystallographic analysis, and homology modeling. Appl Environ Microbiol 2015; 81(23): 8076-83.
[http://dx.doi.org/10.1128/AEM.02412-15] [PMID: 26386053]

[84] Kawashima Y, Cheng W, Mifune J, Orita I, Nakamura S, Fukui T. Characterization and functional analyses of R-specific enoyl coenzyme A hydratases in polyhydroxyalkanoate-producing *Ralstonia eutropha.* Appl Environ Microbiol 2012; 78(2): 493-502.
[http://dx.doi.org/10.1128/AEM.06937-11] [PMID: 22081565]

[85] Saegusa H, Shiraki M, Saito T. Cloning of an intracellular D(-)-3-hydroxybutyrate-oligomer hydrolase gene from *Ralstonia eutropha* H16 and identification of the active site serine residue by site-directed mutagenesis. J Biosci Bioeng 2002; 94(2): 106-12.
[http://dx.doi.org/10.1016/S1389-1723(02)80128-1] [PMID: 16233278]

[86] Kobayashi T, Uchino K, Abe T, Yamazaki Y, Saito T. Novel intracellular 3-hydroxybutyrate-oligomer hydrolase in *Wautersia eutropha* H16. J Bacteriol 2005; 187(15): 5129-35.
[http://dx.doi.org/10.1128/JB.187.15.5129-5135.2005] [PMID: 16030206]

[87] Sudesh K, Abe H, Doi Y. Synthesis, structure, and properties of polyhydroxyalkanoates: biological polyesters. Prog Polym Sci 2000; 25: 1503-55.

[http://dx.doi.org/10.1016/S0079-6700(00)00035-6]

[88] Kessler B, Witholt B. Factors involved in the regulatory network of polyhydroxyalkanoate metabolism. J Biotechnol 2001; 86(2): 97-104. [http://dx.doi.org/10.1016/S0168-1656(00)00404-1] [PMID: 11245898]

[89] Brigham CJ, Sinskey AJ. Polyhydroxyalkanoate Production Enzymes: a Survey and Biological Perspective. JSiberian Fed Univ 2012; 3: 220-42.

[90] Pan P, Inoue Y. Polymorphism and isomorphism in biodegradable polyesters. Polym Degrad Stabil 2009; 34: 605-40.

[91] Chen GQ. A microbial polyhydroxyalkanoates (PHA) based bio- and materials industry. Chem Soc Rev 2009; 38(8): 2434-46. [http://dx.doi.org/10.1039/b812677c] [PMID: 19623359]

[92] Chanprateep S. Current trends in biodegradable polyhydroxyalkanoates. J Biosci Bioeng 2010; 110(6): 621-32. [http://dx.doi.org/10.1016/j.jbiosc.2010.07.014] [PMID: 20719562]

[93] Kabe T, Tsuge T, Kasuya K, Takemura A, Hikima T, Takata M, *et al.* Physical and structural effects of adding ultra-high-molecular-weight poly((R)-3-hydroxybutyrate) to wild type poly((R)-3-hydroxybutyrate). Macromolecules 2012; 45: 1858-65. [http://dx.doi.org/10.1021/ma202285c]

[94] Iwata T. Strong fibers and films of microbial polyesters. Macromol Biosci 2005; 5(8): 689-701. [http://dx.doi.org/10.1002/mabi.200500066] [PMID: 16052600]

[95] Aoyagi Y, Doi Y, Iwata T. Mechanical properties and highly ordered structure of ultra-hig--molecular-weight poly((R)-3- hydroxybutyrate) films: effects of annealing and two-step drawing. Polym Degrad Stabil 2003; 79: 209-16. [http://dx.doi.org/10.1016/S0141-3910(02)00273-2]

[96] Iwata T, Aoyagi Y, Fujita M, Yamane H, Doi Y, Suzuki Y, *et al.* Processing of a strong biodegradable poly((R)-3-hydroxybutyrate) fiber and a new fiber structure revealed by micro-beam X-ray diffraction with synchrotron radiation. Macromol Rapid Commun 2004; 25: 1100-4. [http://dx.doi.org/10.1002/marc.200400110]

[97] Iwata T, Tsunoda T, Aoyagi Y, Kusaka S, Yonezawa N, Doi Y. Mechanical properties of uniaxially cold-drawn films of poly(*(R)*- 3-hydroxybutyrate). Polym Degrad Stabil 2003; 79: 217-24. [http://dx.doi.org/10.1016/S0141-3910(02)00274-4]

[98] Yamane H, Terao K, Hiki S, Kimura Y. Mechanical properties and higher order structure of bacterial homo poly(3-hydroxybutyrate) melt spun fibers. Polymer (Guildf) 2001; 42: 3241-8. [http://dx.doi.org/10.1016/S0032-3861(00)00598-X]

[99] Stubbe J, Tian J, He A, Sinskey AJ, Lawrence AG, Liu P. Nontemplate-dependent polymerization processes: polyhydroxyalkanoate synthases as a paradigm. Annu Rev Biochem 2005; 74: 433-80. [http://dx.doi.org/10.1146/annurev.biochem.74.082803.133013] [PMID: 15952894]

[100] Steinbüchel A, Schlegel HG. Physiology and molecular genetics of poly(beta-hydroxy-alkanoic acid) synthesis in *Alcaligenes eutrophus.* Mol Microbiol 1991; 5(3): 535-42. [http://dx.doi.org/10.1111/j.1365-2958.1991.tb00725.x] [PMID: 2046547]

[101] Zinn M, Witholt B, Egli T. Occurrence, synthesis and medical application of bacterial polyhydroxyalkanoate. Adv Drug Deliv Rev 2001; 53(1): 5-21.
[http://dx.doi.org/10.1016/S0169-409X(01)00218-6] [PMID: 11733115]

[102] Anderson AJ, Haywood GW, Dawes EA. Biosynthesis and composition of bacterial poly(hydroxyalkanoates). Int J Biol Macromol 1990; 12(2): 102-5.
[http://dx.doi.org/10.1016/0141-8130(90)90060-N] [PMID: 2078525]

[103] Peña C, Castillo T, García A, Millán M, Segura D. Biotechnological strategies to improve production of microbial poly-(3-hydroxybutyrate): a review of recent research work. Microb Biotechnol 2014; 7(4): 278-93.
[http://dx.doi.org/10.1111/1751-7915.12129] [PMID: 24898500]

[104] Lageveen RG, Huisman GW, Preusting H, Ketelaar P, Eggink G, Witholt B. Formation of Polyesters by *Pseudomonas oleovorans*: Effect of Substrates on Formation and Composition of Poly-(R)-3-Hydroxyalkanoates and Poly-(R)-3-Hydroxyalkenoates. Appl Environ Microbiol 1988; 54(12): 2924-32.
[PMID: 16347790]

[105] Brandl H, Gross RA, Lenz RW, Fuller RC. *Pseudomonas oleovorans* as a Source of Poly(beta-Hydroxyalkanoates) for Potential Applications as Biodegradable Polyesters. Appl Environ Microbiol 1988; 54(8): 1977-82.
[PMID: 16347708]

[106] Tsuge T, Taguchi K, Seiichi T, Doi Y. Molecular characterization and properties of (R)-specific enoyl-CoA hydratases from *Pseudomonas aeruginosa*: metabolic tools for synthesis of polyhydroxyalkanoates *via* fatty acid beta-oxidation. Int J Biol Macromol 2003; 31(4-5): 195-205.
[http://dx.doi.org/10.1016/S0141-8130(02)00082-X] [PMID: 12568928]

[107] Rehm BH, Krüger N, Steinbüchel A. A new metabolic link between fatty acid de novo synthesis and polyhydroxyalkanoic acid synthesis. The PHAG gene from *Pseudomonas putida* KT2440 encodes a 3-hydroxyacyl-acyl carrier protein-coenzyme a transferase. J Biol Chem 1998; 273(37): 24044-51.
[http://dx.doi.org/10.1074/jbc.273.37.24044] [PMID: 9727022]

[108] Hoffmann N, Steinbüchel A, Rehm BH. Homologous functional expression of cryptic *phaG* from *Pseudomonas oleovorans* establishes the transacylase-mediated polyhydroxyalkanoate biosynthetic pathway. Appl Microbiol Biotechnol 2000; 54(5): 665-70.
[http://dx.doi.org/10.1007/s002530000441] [PMID: 11131392]

[109] Hoffmann N, Steinbüchel A, Rehm BH. The *Pseudomonas aeruginosa phaG* gene product is involved in the synthesis of polyhydroxyalkanoic acid consisting of medium-chain-length constituents from non-related carbon sources. FEMS Microbiol Lett 2000; 184(2): 253-9.
[http://dx.doi.org/10.1111/j.1574-6968.2000.tb09023.x] [PMID: 10713430]

[110] Matsumoto K, Matsusaki H, Taguchi S, Seki M, Doi Y. Cloning and characterization of the *Pseudomonas* sp. 61-3 *phaG* gene involved in polyhydroxyalkanoate biosynthesis. Biomacromolecules 2001; 2(1): 142-7.
[http://dx.doi.org/10.1021/bm005604+] [PMID: 11749165]

[111] Wong YM, Brigham CJ, Rha C, Sinskey AJ, Sudesh K. Biosynthesis and characterization of polyhydroxyalkanoate containing high 3-hydroxyhexanoate monomer fraction from crude palm kernel

oil by recombinant *Cupriavidus necator*. Bioresour Technol 2012; 121: 320-7.
[http://dx.doi.org/10.1016/j.biortech.2012.07.015] [PMID: 22858502]

[112] Holmes PA, Collins SH, Wright LF. inventors. 3-hydroxybutyrate polymers. US patent 4,477,654 1984.

[113] Akiyama M, Taima Y, Doi Y. Production of poly(3-hydroxyalkanoates) by a bacterium of the genus *Alcaligenes* utilizing long chain fatty acids. Appl Microbiol Biotechnol 1992; 37: 698-701.
[http://dx.doi.org/10.1007/BF00174830]

[114] Khanna S, Srivastava AK. Recent advances in microbial polyhydroxyalkanoates. Process Biochem 2005; 40: 607-19.
[http://dx.doi.org/10.1016/j.procbio.2004.01.053]

[115] Kim BS, Lee SC, Lee SY, Chang HN, Chang YK, Woo SI. Production of poly(3-hydroxybutyric acid) by fed-batch culture of *Alcaligenes eutrophus* with glucose concentration control. Biotechnol Bioeng 1994; 43(9): 892-8.
[http://dx.doi.org/10.1002/bit.260430908] [PMID: 18615882]

[116] Doi Y, Tamaki A, Kunioka M, Soga K. Biosynthesis of terpolyesters of 3-hydroxybutyrate, 3-hydroxyvalerate, and 5-hydroxyvalerate, in *Alcaligenes eutrophus* from 5-chloropentanoic and pentanoic acids. Macromol Rapid Commun 1987; 8: 631-5.
[http://dx.doi.org/10.1002/marc.1987.030081209]

[117] Madison LL, Huisman GW. Metabolic engineering of poly(3-hydroxyalkanoates): from DNA to plastic. Microbiol Mol Biol Rev 1999; 63(1): 21-53.
[PMID: 10066830]

[118] Yang YH, Brigham CJ, Budde CF, *et al.* Optimization of growth media components for polyhydroxyalkanoate (PHA) production from organic acids by *Ralstonia eutropha.* Appl Microbiol Biotechnol 2010; 87(6): 2037-45.
[http://dx.doi.org/10.1007/s00253-010-2699-8] [PMID: 20535466]

[119] Shimizu H, Kozaki Y, Kodama H, Shioya S. Maximum production strategy for biodegradable copolymer P(HB-*co*-HV) in fed-batch culture of *Alcaligenes eutrophus.* Biotechnol Bioeng 1999; 62(5): 518-25.
[http://dx.doi.org/10.1002/(SICI)1097-0290(19990305)62:5<518::AID-BIT3>3.0.CO;2-6] [PMID: 10099560]

[120] Choi JI, Lee SY. High-level production of poly(3-hydroxybutyrate-*co*-3-hydroxyvalerate) by fed-batch culture of recombinant *Escherichia coli.* Appl Environ Microbiol 1999; 65(10): 4363-8.
[PMID: 10508061]

[121] Yu J, Si Y. Metabolic carbon fluxes and biosynthesis of polyhydroxyalkanoates in *Ralstonia eutropha* on short chain fatty acids. Biotechnol Prog 2004; 20(4): 1015-24.
[http://dx.doi.org/10.1021/bp034380e] [PMID: 15296425]

[122] Huschner F, Grousseau E, Brigham CJ, Plassmeier J, Popovic M, Rha CK, *et al.* Development of a feeding strategy for high cell and PHA densityfed-batch fermentation of *Ralstonia eutropha* H16 from organic acidsand their salts. Process Biochem 2015; 50(2): 165-72.
[http://dx.doi.org/10.1016/j.procbio.2014.12.004]

[123] Jung YM, Lee YH. Utilization of oxidative pressure for enhanced production of poly-bet--hydroxybutyrate and poly(3-hydroxybutyrate-3-hydroxyvalerate) in *Ralstonia eutropha*. J Biosci Bioeng 2000; 90(3): 266-70.
[http://dx.doi.org/10.1016/S1389-1723(00)80080-8] [PMID: 16232855]

[124] Budde CF, Riedel SL, Hübner F, *et al.* Growth and polyhydroxybutyrate production by *Ralstonia eutropha* in emulsified plant oil medium. Appl Microbiol Biotechnol 2011; 89(5): 1611-9.
[http://dx.doi.org/10.1007/s00253-011-3102-0] [PMID: 21279345]

[125] Shimamura E, Kasuya K, Kobayashi G, Shiotani T, Shima Y, Doi Y. Physical properties and biodegradability of microbial poly. Macromolecules 1994; 27: 878-80. [3-hydroxybutyrate-*co*-3-hydroxyhexanoate].
[http://dx.doi.org/10.1021/ma00081a041]

[126] Doi Y, Kitamura S, Abe H. Microbial Synthesis and Characterization of Poly(3-hydroxybutyrate-*co*-3-hydroxyhexanoate). Macromolecules 1995; 28: 4822.
[http://dx.doi.org/10.1021/ma00118a007]

[127] Fukui T, Doi Y. Cloning and analysis of the poly(3-hydroxybutyrate-*co*-3-hydroxyhexanoate) biosynthesis genes of *Aeromonas caviae*. J Bacteriol 1997; 179(15): 4821-30.
[PMID: 9244271]

[128] Fukui T, Shiomi N, Doi Y. Expression and characterization of (R)-specific enoyl coenzyme A hydratase involved in polyhydroxyalkanoate biosynthesis by *Aeromonas caviae*. J Bacteriol 1998; 180(3): 667-73.
[PMID: 9457873]

[129] Hisano T, Tsuge T, Fukui T, Iwata T, Miki K, Doi Y. Crystal structure of the (R)-specific enoyl-CoA hydratase from *Aeromonas caviae* involved in polyhydroxyalkanoate biosynthesis. J Biol Chem 2003; 278(1): 617-24.
[http://dx.doi.org/10.1074/jbc.M205484200] [PMID: 12409309]

[130] Schlegel HG, Lafferty R, Krauss I. The isolation of mutants not accumulating poly-bet--hydroxybutyric acid. Arch Mikrobiol 1970; 71(3): 283-94.
[http://dx.doi.org/10.1007/BF00410161] [PMID: 4097070]

[131] Yamane T. Cultivation engineering of microbial bioplastics production. FEMS Microbiol Rev 1992; 103: 257-64.
[http://dx.doi.org/10.1111/j.1574-6968.1992.tb05846.x]

[132] Akiyama M, Tsuge T, Doi Y. Environmental life cycle comparison of polyhydroxyalkanoates produced from renewable carbon resources by bacterial fermentation. Polym Degrad Stabil 2003; 80: 183-94.
[http://dx.doi.org/10.1016/S0141-3910(02)00400-7]

[133] Chen GQ, Zhang G, Park SJ, Lee SY. Industrial scale production of poly(3-hydroxybutyrate-*co*-3-hydroxyhexanoate). Appl Microbiol Biotechnol 2001; 57(1-2): 50-5.
[http://dx.doi.org/10.1007/s002530100755] [PMID: 11693933]

[134] Kahar P, Tsuge T, Taguchi K, Doi Y. High yield production of polyhydroxyalkanoates from soybean oil by *Ralstonia eutropha* and its recombinant strain. Polym Degrad Stabil 2004; 83: 79-86.

[http://dx.doi.org/10.1016/S0141-3910(03)00227-1]

[135] Loo CY, Lee WH, Tsuge T, Doi Y, Sudesh K. Biosynthesis and characterization of poly(3-hydroxybutyrate-*co*-3- hydroxyhexanoate) from palm oil products in a *Wautersia eutropha* mutant. Biotechnol Lett 2005; 27(18): 1405-10.
[http://dx.doi.org/10.1007/s10529-005-0690-8] [PMID: 16215858]

[136] Mifune J, Nakamura S, Fukui T. Targeted engineering of *Cupriavidus necator* chromosome for biosynthesis of poly(3-hydroxybutyrate-*co*-3-hydroxyhexanoate) from vegetable oil. Can J Chem 2008; 86: 621-7.
[http://dx.doi.org/10.1139/v08-047]

[137] Fukui T, Doi Y. Efficient production of polyhydroxyalkanoates from plant oils by *Alcaligenes eutrophus* and its recombinant strain. Appl Microbiol Biotechnol 1998; 49(3): 333-6.
[http://dx.doi.org/10.1007/s002530051178] [PMID: 9581296]

[138] Buckland BC, Drew SW, Connors NC, *et al.* Microbial conversion of indene to indandiol: a key intermediate in the synthesis of CRIXIVAN. Metab Eng 1999; 1(1): 63-74.
[http://dx.doi.org/10.1006/mben.1998.0107] [PMID: 10935755]

[139] Williams DR, Anderson AJ, Dawes EA, Ewing DF. Production of a co-polyester of 3-hydroxybutyric acid and 3-hydroxyvaleric acid from succinic acid by *Rhodococcus ruber*: biosynthetic considerations. Appl Microbiol Biotechnol 1994; 40: 717-23.
[http://dx.doi.org/10.1007/BF00173334]

[140] Budde CF, Riedel SL, Willis LB, Rha C, Sinskey AJ. Production of poly(3-hydroxybutyrate-*co*-3-hydroxyhexanoate) from plant oil by engineered Ralstonia eutropha strains. Appl Environ Microbiol 2011; 77(9): 2847-54.
[http://dx.doi.org/10.1128/AEM.02429-10] [PMID: 21398488]

[141] Tsuge T, Fukui T, Matsusaki H, *et al.* Molecular cloning of two (R)-specific enoyl-CoA hydratase genes from *Pseudomonas aeruginosa* and their use for polyhydroxyalkanoate synthesis. FEMS Microbiol Lett 2000; 184(2): 193-8.
[http://dx.doi.org/10.1111/j.1574-6968.2000.tb09013.x] [PMID: 10713420]

[142] Riedel SL, Jahns S, Koenig S, *et al.* Polyhydroxyalkanoates production with *Ralstonia eutropha* from low quality waste animal fats. J Biotechnol 2015; 214: 119-27.
[http://dx.doi.org/10.1016/j.jbiotec.2015.09.002] [PMID: 26428087]

[143] Riedel SL, Bader J, Brigham CJ, *et al.* Production of poly(3-hydroxybutyrate-co-3-hydroxyhexanoate) by Ralstonia eutropha in high cell density palm oil fermentations. Biotechnol Bioeng 2012; 109(1): 74-83.
[http://dx.doi.org/10.1002/bit.23283] [PMID: 21809332]

[144] Mejía MA, Segura D, Espín G, Galindo E, Peña C. Two-stage fermentation process for alginate production by *Azotobacter vinelandii* mutant altered in poly-beta-hydroxybutyrate (PHB) synthesis. J Appl Microbiol 2010; 108(1): 55-61.
[http://dx.doi.org/10.1111/j.1365-2672.2009.04403.x] [PMID: 19583796]

[145] Shuler M, Kargi F. Bioprocess Engineering: Basic concepts. 2nd ed., Upper Saddle River, NJ: Prentice Hall 2002.

[146] Ienczak JL, Schmidell W, Aragão GM. High-cell-density culture strategies for polyhydroxyalkanoate production: a review. J Ind Microbiol Biotechnol 2013; 40(3-4): 275-86. [http://dx.doi.org/10.1007/s10295-013-1236-z] [PMID: 23455696]

[147] McNeil B, Harvey LM, Eds. Fermentation, a practical approach. Tokyo: IRL Press 1990.

[148] Verlinden RA, Hill DJ, Kenward MA, Williams CD, Radecka I. Bacterial synthesis of biodegradable polyhydroxyalkanoates. J Appl Microbiol 2007; 102(6): 1437-49. [http://dx.doi.org/10.1111/j.1365-2672.2007.03335.x] [PMID: 17578408]

[149] Shang L, Jiang M, Chang HN. Poly(3-hydroxybutyrate) synthesis in fed-batch culture of *Ralstonia eutropha* with phosphate limitation under different glucose concentrations. Biotechnol Lett 2003; 25(17): 1415-9. [http://dx.doi.org/10.1023/A:1025047410699] [PMID: 14514042]

[150] Ryu HW, Hahn SK, Chang YK, Chang HN. Production of poly(3-hydroxybutyrate) by high cell density fed-batch culture of *Alcaligenes eutrophus* with phospate limitation. Biotechnol Bioeng 1997; 55(1): 28-32. [http://dx.doi.org/10.1002/(SICI)1097-0290(19970705)55:1<28::AID-BIT4>3.0.CO;2-Z] [PMID: 18636441]

[151] Chen GQ. Plastics completely synthesized by bacteria: Polyhydroxyalkanoates. In: Steinbüchel A, Chen GQ, Eds. Plastics From Bacteria Berlin. Springer-Verlag 2010; pp. 17-37. [http://dx.doi.org/10.1007/978-3-642-03287-5_2]

[152] Cavalheiro JM, de Almeida MC, Grandfils C, da Fonseca MM. Poly (3-hydroxybutyrate) production by *Cupriavidus necator* using waste glycerol. Process Biochem 2009; 44: 509-15. [http://dx.doi.org/10.1016/j.procbio.2009.01.008]

[153] Mozumder MS, De Wever H, Voicke EI, Garcia-Gonzalez L. A robust fed-batch feeding strategy independent of the carbon source for optimal polyhydroxybutyrate production. Process Biochem 2014; 155: 272-80.

[154] Obruca S, Marova I, Snajdar O, Mravcova L, Svoboda Z. Production of poly(3-hydroxybutyrate-c--3-hydroxyvalerate) by Cupriavidus necator from waste rapeseed oil using propanol as a precursor of 3-hydroxyvalerate. Biotechnol Lett 2010; 32(12): 1925-32. [http://dx.doi.org/10.1007/s10529-010-0376-8] [PMID: 20814716]

[155] Obruca S, Snajdar O, Svoboda Z, Marova I. Application of random mutagenesis to enhance the production of polyhydroxyalkanoates by *Cupriavidus necator* H16 on waste frying oil. World J Microbiol Biotechnol 2013; 29(12): 2417-28. [http://dx.doi.org/10.1007/s11274-013-1410-5] [PMID: 23801326]

[156] Obruca S, Petrik S, Benesova P, Svoboda Z, Eremka L, Marova I. Utilization of oil extracted from spent coffee grounds for sustainable production of polyhydroxyalkanoates. Appl Microbiol Biotechnol 2014; 98(13): 5883-90. [http://dx.doi.org/10.1007/s00253-014-5653-3] [PMID: 24652066]

[157] Zahari MA, Zakaria MR, Ariffin H, *et al.* Renewable sugars from oil palm frond juice as an alternative novel fermentation feedstock for value-added products. Bioresour Technol 2012; 110: 566-71. [http://dx.doi.org/10.1016/j.biortech.2012.01.119] [PMID: 22342083]

[158] Zahari MA, Ariffin H, Mokhtar MN, Salihon J, Shirai Y, Hassan MA. Factors Affecting Poly(3-hydroxybutyrate) Production from Oil Palm Frond Juice by *Cupriavidus necator* (CCUG52238^T). J Biomed Biotechnol 2012; 125865.

[159] Verlinden RA, Hill DJ, Kenward MA, Williams CD, Piotrowska-Seget Z, Radecka IK. Production of polyhydroxyalkanoates from waste frying oil by *Cupriavidus necator*. AMB Express 2011; 1(1)

[160] Noda I, Green PR, Satkowski MM, Schechtman LA. Preparation and properties of a novel class of polyhydroxyalkanoate copolymers. Biomacromolecules 2005; 6(2): 580-6. [http://dx.doi.org/10.1021/bm049472m] [PMID: 15762616]

[161] Noda I. Solvent extraction of polyhydroxyalkanoates from biomass facilitated by the use of marginal nonsolvent 1998.

[162] Kinoshida K, Oskada Y, Ueda Y, Narasimhan K, Cearly AC, Yee K, *et al.* Method for producing polyhydroxyalkanoate crystal 2005.

[163] Van Walsem J, Zhong L, Shih S. Polymer extraction methods 2007.

[164] Shishatskaya EI, Volova TG, Gordeev SA, Puzyr AP. Degradation of P(3HB) and P(3HB-co-3HV) in biological media. J Biomater Sci Polym Ed 2005; 16(5): 643-57. [http://dx.doi.org/10.1163/1568562053783678] [PMID: 16001722]

[165] Brigham CJ, Sinskey AJ. Applications of Polyhydroxyalkanoates in the Medical Industry. Int J Biotechnol Wellness Ind 2012; 1: 53-60.

[166] Shishatskaya EI, Volova TG, Puzyr AP, Mogilnaya OA, Efremov SN. Tissue response to the implantation of biodegradable polyhydroxyalkanoate sutures. J Mater Sci Mater Med 2004; 15(6): 719-28. [http://dx.doi.org/10.1023/B:JMSM.0000030215.49991.0d] [PMID: 15346741]

[167] Shishatskaya EI, Goreva AV, Voinova ON, Inzhevatkin EV, Khlebopros RG, Volova TG. Evaluation of antitumor activity of rubomycin deposited in absorbable polymeric microparticles. Bull Exp Biol Med 2008; 145(3): 358-61. [http://dx.doi.org/10.1007/s10517-008-0091-9] [PMID: 19039944]

[168] Philip S, Keshavarz T, Roy I. Polyhydroxyalkanoates: biodegradable polymers with a range of applications. J Chem Technol Biotechnol 2007; 82: 233-47. [http://dx.doi.org/10.1002/jctb.1667]

[169] Volodina E, Raberg M, Steinbüchel A. Engineering the heterotrophic carbon sources utilization range of *Ralstonia eutropha* H16 for applications in biotechnology. Crit Rev Biotechnol 2015; •••: 1-14. [http://dx.doi.org/10.3109/07388551.2015.1079698] [PMID: 26329669]

[170] Kawada J, Lütke-Eversloh T, Steinbüchel A, Marchessault RH. Physical properties of microbial polythioesters: characterization of poly(3-mercaptoalkanoates) synthesized by engineered *Escherichia coli*. Biomacromolecules 2003; 4(6): 1698-702. [http://dx.doi.org/10.1021/bm0341327] [PMID: 14606898]

[171] Lütke-Eversloh T, Steinbüchel A. Novel precursor substrates for polythioesters (PTE) and limits of PTE biosynthesis in *Ralstonia eutropha*. FEMS Microbiol Lett 2003; 221(2): 191-6. [http://dx.doi.org/10.1016/S0378-1097(03)00185-X] [PMID: 12725926]

[172] Aboulmagd E, Voss I, Oppermann-Sanio FB, Steinbüchel A. Heterologous expression of cyanophycin synthetase and cyanophycin synthesis in the industrial relevant bacteria *Corynebacterium glutamicum* and *Ralstonia eutropha* and in *Pseudomonas putida.* Biomacromolecules 2001; 2(4): 1338-42. [http://dx.doi.org/10.1021/bm010075a] [PMID: 11777412]

[173] Voss I, Steinbüchel A. Application of a KDPG-aldolase gene-dependent addiction system for enhanced production of cyanophycin in *Ralstonia eutropha* strain H16. Metab Eng 2006; 8(1): 66-78. [http://dx.doi.org/10.1016/j.ymben.2005.09.003] [PMID: 16266816]

[174] Lin K, Elbahloul Y, Steinbüchel A. Physiological conditions conducive to high cell density and high cyanophycin content in *Ralstonia eutropha* strain H16 possessing a KDPG aldolase gene-dependent addiction system. Appl Microbiol Biotechnol 2012; 93(5): 1885-94. [http://dx.doi.org/10.1007/s00253-011-3685-5] [PMID: 22080348]

[175] Adames K, Euting K, Bröker A, Steinbüchel A. Investigations on three genes in *Ralstonia eutropha* H16 encoding putative cyanophycin metabolizing enzymes. Appl Microbiol Biotechnol 2013; 97(8): 3579-91. [http://dx.doi.org/10.1007/s00253-012-4599-6] [PMID: 23224585]

[176] Sonnleitner B, Heinzle E, Braunegg G, Lafferty RM. Formal kinetics of poly-(beta)-hydroxybutyric acid (PHB) production in *Alcaligenes eutrophus* H16 and *Mycoplasma rubra* R14 with respect to the dissolved oxygen tension in ammonium-limited batch cultures. Eur J Appl Microbiol Biotechnol 1979; 7: 1-10. [http://dx.doi.org/10.1007/BF00522473]

[177] Ishizaki A, Tanaka K. Production of poly-[beta]-hydroxybutyric acid from carbon dioxide by *Alcaligenes eutrophus* ATCC 17697^{T}. J Ferment Bioeng 1991; 70: 254-7. [http://dx.doi.org/10.1016/0922-338X(91)90277-N]

[178] Tanaka K, Ishizaki A, Kanamaru T, Kawano T. Production of poly(D-3-hydroxybutyrate) from CO(2), H(2), and O(2) by high cell density autotrophic cultivation of *Alcaligenes eutrophus.* Biotechnol Bioeng 1995; 45(3): 268-75. [http://dx.doi.org/10.1002/bit.260450312] [PMID: 18623147]

[179] Ishizaki A, Tanaka K, Taga N. Microbial production of poly-D-3-hydroxybutyrate from CO_2. Appl Microbiol Biotechnol 2001; 57(1-2): 6-12. [http://dx.doi.org/10.1007/s002530100775] [PMID: 11693935]

[180] Worden RM. Liu YCCatalytic bioreactors and methods of using same 2014.

[181] Arjunwadkar SJ, Sarvanan K, Kulkarni PR, Pandit AB. Gas-liquid mass transfer in dual impeller bioreactor. Biochem Eng J 1998; 1: 99-106. [http://dx.doi.org/10.1016/S1385-8947(97)00083-1]

[182] Bowien B, Kusian B. Genetics and control of CO(2) assimilation in the chemoautotroph *Ralstonia eutropha.* Arch Microbiol 2002; 178(2): 85-93. [http://dx.doi.org/10.1007/s00203-002-0441-3] [PMID: 12115053]

[183] Schwartz E, Voigt B, Zühlke D, *et al.* A proteomic view of the facultatively chemolithoautotrophic lifestyle of *Ralstonia eutropha* H16. Proteomics 2009; 9(22): 5132-42. [http://dx.doi.org/10.1002/pmic.200900333] [PMID: 19798673]

[184] Kohlmann Y, Pohlmann A, Otto A, *et al.* Analyses of soluble and membrane proteomes of *Ralstonia eutropha* H16 reveal major changes in the protein complement in adaptation to lithoautotrophy. J Proteome Res 2011; 10(6): 2767-76.
[http://dx.doi.org/10.1021/pr101289v] [PMID: 21561103]

[185] Grzeszik C, Jeffke T, Schäferjohann J, Kusian B, Bowien B. Phosphoenolpyruvate is a signal metabolite in transcriptional control of the *cbb* CO_2 fixation operons in *Ralstonia eutropha.* J Mol Microbiol Biotechnol 2000; 2(3): 311-20.
[PMID: 10937440]

[186] Berg IA. Ecological aspects of the distribution of different autotrophic CO_2 fixation pathways. Appl Environ Microbiol 2011; 77(6): 1925-36.
[http://dx.doi.org/10.1128/AEM.02473-10] [PMID: 21216907]

[187] Berg IA, Kockelkorn D, Ramos-Vera WH, *et al.* Autotrophic carbon fixation in archaea. Nat Rev Microbiol 2010; 8(6): 447-60.
[http://dx.doi.org/10.1038/nrmicro2365] [PMID: 20453874]

[188] Fritsch J, Lenz O, Friedrich B. Structure, function and biosynthesis of O□-tolerant hydrogenases. Nat Rev Microbiol 2013; 11(2): 106-14.
[http://dx.doi.org/10.1038/nrmicro2940] [PMID: 23321533]

[189] Kömen R, Schmidt K, Friedrich B. Hydrogenase mutants of Alcaligenes eutrophus H16 show alterations in the electron transport system. FEMS Microbiol Lett 1992; 75(2-3): 173-8.
[http://dx.doi.org/10.1111/j.1574-6968.1992.tb05412.x] [PMID: 1398034]

[190] Lenz O, Ludwig M, Schubert T, *et al.* H2 conversion in the presence of O2 as performed by the membrane-bound [NiFe]-hydrogenase of Ralstonia eutropha. ChemPhysChem 2010; 11(6): 1107-19.
[http://dx.doi.org/10.1002/cphc.200901002] [PMID: 20186906]

[191] Burgdorf T, van der Linden E, Bernhard M, *et al.* The soluble NAD+-Reducing [NiFe]-hydrogenase from Ralstonia eutropha H16 consists of six subunits and can be specifically activated by NADPH. J Bacteriol 2005; 187(9): 3122-32.
[http://dx.doi.org/10.1128/JB.187.9.3122-3132.2005] [PMID: 15838039]

[192] Burgdorf T, Lenz O, Buhrke T, *et al.* [NiFe]-hydrogenases of Ralstonia eutropha H16: modular enzymes for oxygen-tolerant biological hydrogen oxidation. J Mol Microbiol Biotechnol 2005; 10(2-4): 181-96.
[http://dx.doi.org/10.1159/000091564] [PMID: 16645314]

[193] Schwartz E, Gerischer U, Friedrich B. Transcriptional regulation of *Alcaligenes eutrophus* hydrogenase genes. J Bacteriol 1998; 180(12): 3197-204.
[PMID: 9620971]

[194] Schubert T, Lenz O, Krause E, Volkmer R, Friedrich B. Chaperones specific for the membrane-bound [NiFe]-hydrogenase interact with the Tat signal peptide of the small subunit precursor in Ralstonia eutropha H16. Mol Microbiol 2007; 66(2): 453-67.
[http://dx.doi.org/10.1111/j.1365-2958.2007.05933.x] [PMID: 17850259]

[195] Volova TG, Voĭnov NA. [Study of *Ralstonia eutropha* culture producing polyhydroxyalkanoates on products of coal processing]. Prikl Biokhim Mikrobiol 2004; 40(3): 296-300.

[PMID: 15283331]

[196] Volova TG, Voinov NA. Kinetic Parameters of a Culture of the Hydrogen-Oxidizing Bacterium *Ralstonia eutropha* Grown under Conditions Favoring Polyhydroxybutyrate Biosynthesis. Appl Biochem Microbiol 2003; 39(2): 166-70.
[http://dx.doi.org/10.1023/A:1022585912873]

[197] Volova TG, Voĭnov NA. [Kinetic parameters of a culture of the hydrogen-oxidizing Ralstonia eutropha, grown under the regimen of biosynthesis of polyhydroxybutyrate]. Prikl Biokhim Mikrobiol 2003; 39(2): 189-93.
[PMID: 12722652]

[198] Don RH, Pemberton JM. Properties of six pesticide degradation plasmids isolated from Alcaligenes paradoxus and *Alcaligenes eutrophus*. J Bacteriol 1981; 145(2): 681-6.
[PMID: 6257648]

[199] Leveau JH, König F, Füchslin H, Werlen C, Van Der Meer JR. Dynamics of multigene expression during catabolic adaptation of *Ralstonia eutropha* JMP134 (pJP4) to the herbicide 2, 4-dichlorophenoxyacetate. Mol Microbiol 1999; 33(2): 396-406.
[http://dx.doi.org/10.1046/j.1365-2958.1999.01483.x] [PMID: 10411755]

[200] Pieper DH, Reinecke F, Engesser KH, Knackmuss HJ. Metabolism of 2,4-dichlorophenoxyacetic acid, 4-chloro-2-methylphenoxy- acetic acid and 2-methylphenoxyacetic acid by *Alcaligenes eutrophus* JMP 134. Arch Microbiol 1998; 150: 95-102.
[http://dx.doi.org/10.1007/BF00409724]

[201] Ghosal D, You IS, Chatterjee DK, Chakrabarty AM. Genes specifying degradation of 3-chlorobenzoic acid in plasmids pAC27 and pJP4. Proc Natl Acad Sci USA 1985; 82(6): 1638-42.
[http://dx.doi.org/10.1073/pnas.82.6.1638] [PMID: 3856842]

[202] Clément P, Matus V, Cárdenas L, González B. Degradation of trichlorophenols by *Alcaligenes eutrophus* JMP134. FEMS Microbiol Lett 1995; 127(1-2): 51-5.
[http://dx.doi.org/10.1111/j.1574-6968.1995.tb07449.x] [PMID: 7737484]

[203] Schlömann M, Schmidt E, Knackmuss HJ. Different types of dienelactone hydrolase in 4-fluorobenzoate-utilizing bacteria. J Bacteriol 1990; 172(9): 5112-8.
[PMID: 2394679]

[204] Ecker S, Widmann T, Lenke H, *et al.* Catabolism of 2,6-dinitrophenol by *Alcaligenes eutrophus* JMP 134 and JMP 222. Arch Microbiol 1992; 158: 149-54.
[http://dx.doi.org/10.1007/BF00245219]

[205] Schenzle A, Lenke H, Fischer P, Williams PA, Knackmuss H. Catabolism of 3-Nitrophenol by *Ralstonia eutropha* JMP 134. Appl Environ Microbiol 1997; 63(4): 1421-7.
[PMID: 16535572]

[206] Pieper DH, Engesser KH, Don RH, Timmis KN, Knackmuss HJ. Modified ortho-cleavage pathway in *Alcaligenes eutrophus* JMP134 for the degradation of 4-methylcatechol. FEMS Microbiol Lett 1985; 29: 63-7.
[http://dx.doi.org/10.1111/j.1574-6968.1985.tb00836.x]

[207] Lykidis A, Pérez-Pantoja D, Ledger T, *et al.* The complete multipartite genome sequence of

Cupriavidus necator JMP134, a versatile pollutant degrader. PLoS One 2010; 5(3): e9729. [http://dx.doi.org/10.1371/journal.pone.0009729] [PMID: 20339589]

[208] Harwood CS, Parales RE. The beta-ketoadipate pathway and the biology of self-identity. Annu Rev Microbiol 1996; 50: 553-90. [http://dx.doi.org/10.1146/annurev.micro.50.1.553] [PMID: 8905091]

[209] Hawkins AR, Lamb HK, Moore JD, Charles IG, Roberts CF. The pre-chorismate (shikimate) and quinate pathways in filamentous fungi: theoretical and practical aspects. J Gen Microbiol 1993; 139(12): 2891-9. [http://dx.doi.org/10.1099/00221287-139-12-2891] [PMID: 8126417]

[210] Dunaway-Mariano D, Babbitt PC. On the origins and functions of the enzymes of the 4-chlorobenzoate to 4-hydroxybenzoate converting pathway. Biodegradation 1994; 5(3-4): 259-76. [http://dx.doi.org/10.1007/BF00696464] [PMID: 7765837]

[211] Barnsley EA. Role and regulation of the ortho and meta pathways of catechol metabolism in pseudomonads metabolizing naphthalene and salicylate. J Bacteriol 1976; 125(2): 404-8. [PMID: 1245462]

[212] Hewetson L, Dunn HM, Dunn NW. Evidence for a transmissible catabolic plasmid in *Pseudomonas putida* encoding the degradation of p-cresol *via* the protocatechuate ortho cleavage pathway. Genet Res 1978; 32(3): 249-55. [http://dx.doi.org/10.1017/S0016672300018747] [PMID: 751853]

[213] Pieper D, Engesser K, Knackmuss H. Regulation of catabolic pathways of phenoxyacetic acids and phenols in *Alcaligenes eutrophus* JMP134. Arch Microbiol 1989; 151(4): 365-71. [http://dx.doi.org/10.1007/BF00406566]

[214] Plumeier I, Pérez-Pantoja D, Heim S, González B, Pieper DH. Importance of different *tfd* genes for degradation of chloroaromatics by *Ralstonia eutropha* JMP134. J Bacteriol 2002; 184(15): 4054-64. [http://dx.doi.org/10.1128/JB.184.15.4054-4064.2002] [PMID: 12107121]

[215] Trefault N, Clément P, Manzano M, Pieper DH, González B. The copy number of the catabolic plasmid pJP4 affects growth of *Ralstonia eutropha* JMP134 (pJP4) on 3-chlorobenzoate. FEMS Microbiol Lett 2002; 212(1): 95-100. [http://dx.doi.org/10.1111/j.1574-6968.2002.tb11250.x] [PMID: 12076793]

[216] Sala-Trepat JM, Murray K, Williams PA. The metabolic divergence in the meta cleavage of catechols by *Pseudomonas putida* NCIB 10015. Physiological significance and evolutionary implications. Eur J Biochem 1972; 28(3): 347-56. [http://dx.doi.org/10.1111/j.1432-1033.1972.tb01920.x] [PMID: 4342908]

[217] Bruce N, Cain R. Methylmuconolactone, a key intermediate in the dissimilation of methylaromatic compounds by a modified 3-oxoadipate pathway evolved in nocardioform actinomycetes. FEMS Microbiol Lett 1988; 50: 233-9.

[218] Marín M, Pérez-Pantoja D, Donoso R, Wray V, González B, Pieper DH. Modified 3-oxoadipate pathway for the biodegradation of methylaromatics in *Pseudomonas reinekei* MT1. J Bacteriol 2010; 192(6): 1543-52. [http://dx.doi.org/10.1128/JB.01208-09] [PMID: 20061479]

[219] Pieper DH, Stadler-Fritzsche K, Knackmuss HJ, Engesser KH, Bruce NC, Cain RB. Purification and characterization of 4-methylmuconolactone methylisomerase, a novel enzyme of the modified 3-oxoadipate pathway in the gram-negative bacterium *Alcaligenes eutrophus* JMP 134. Biochem J 1990; 271(2): 529-34.
[http://dx.doi.org/10.1042/bj2710529] [PMID: 2241929]

[220] Demanèche S, Meyer C, Micoud J, Louwagie M, Willison JC, Jouanneau Y. Identification and functional analysis of two aromatic-ring-hydroxylating dioxygenases from a *sphingomonas* strain that degrades various polycyclic aromatic hydrocarbons. Appl Environ Microbiol 2004; 70(11): 6714-25.
[http://dx.doi.org/10.1128/AEM.70.11.6714-6725.2004] [PMID: 15528538]

[221] Jones RM, Pagmantidis V, Williams PA. *sal* genes determining the catabolism of salicylate esters are part of a supraoperonic cluster of catabolic genes in *Acinetobacter* sp. strain ADP1. J Bacteriol 2000; 182(7): 2018-25.
[http://dx.doi.org/10.1128/JB.182.7.2018-2025.2000] [PMID: 10715011]

[222] You IS, Ghosal D, Gunsalus IC. Nucleotide sequence analysis of the *Pseudomonas putida* PpG7 salicylate hydroxylase gene (*nahG*) and its 3′-flanking region. Biochemistry 1991; 30(6): 1635-41.
[http://dx.doi.org/10.1021/bi00220a028] [PMID: 1993181]

[223] Pérez-Pantoja D, De la Iglesia R, Pieper DH, González B. Metabolic reconstruction of aromatic compounds degradation from the genome of the amazing pollutant-degrading bacterium *Cupriavidus necator* JMP134. FEMS Microbiol Rev 2008; 32(5): 736-94.
[http://dx.doi.org/10.1111/j.1574-6976.2008.00122.x] [PMID: 18691224]

[224] Zhou NY, Al-Dulayymi J, Baird MS, Williams PA. Salicylate 5-hydroxylase from *Ralstonia* sp. strain U2: a monooxygenase with close relationships to and shared electron transport proteins with naphthalene dioxygenase. J Bacteriol 2002; 184(6): 1547-55.
[http://dx.doi.org/10.1128/JB.184.6.1547-1555.2002] [PMID: 11872705]

[225] Hickey WJ, Sabat G, Yuroff AS, Arment AR, Pérez-Lesher J. Cloning, nucleotide sequencing, and functional analysis of a novel, mobile cluster of biodegradation genes from *Pseudomonas aeruginosa* strain JB2. Appl Environ Microbiol 2001; 67(10): 4603-9.
[http://dx.doi.org/10.1128/AEM.67.10.4603-4609.2001] [PMID: 11571162]

[226] Poh CL, Bayly RC. Evidence for isofunctional enzymes used in *m*-cresol and 2,5-xylenol degradation *via* the gentisate pathway in *Pseudomonas alcaligenes*. J Bacteriol 1980; 143(1): 59-69.
[PMID: 6995451]

[227] Michalover JL, Ribbons DW, Hughes H. 3-Hydroxybenzoate 4-hydroxylase from *Pseudomonas testosteroni*. Biochem Biophys Res Commun 1973; 55(3): 888-96.
[http://dx.doi.org/10.1016/0006-291X(73)91227-8] [PMID: 4148586]

[228] Gao X, Tan CL, Yeo CC, Poh CL. Molecular and biochemical characterization of the *xlnD*-encoded 3-hydroxybenzoate 6-hydroxylase involved in the degradation of 2,5-xylenol *via* the gentisate pathway in *Pseudomonas alcaligenes* NCIMB 9867. J Bacteriol 2005; 187(22): 7696-702.
[http://dx.doi.org/10.1128/JB.187.22.7696-7702.2005] [PMID: 16267294]

[229] Liu DQ, Liu H, Gao XL, Leak DJ, Zhou NY. Arg169 is essential for catalytic activity of 3-hydroxybenzoate 6-hydroxylase from *Klebsiella pneumoniae* M5a1. Microbiol Res 2005; 160(1): 53-9.

[http://dx.doi.org/10.1016/j.micres.2004.09.003] [PMID: 15782938]

[230] Johnson BF, Stanier RY. Dissimilation of aromatic compounds by *Alcaligenes eutrophus.* J Bacteriol 1971; 107(2): 468-75.
[PMID: 5113598]

[231] Ferrández A, Miñambres B, García B, *et al.* Catabolism of phenylacetic acid in *Escherichia coli.* Characterization of a new aerobic hybrid pathway. J Biol Chem 1998; 273(40): 25974-86.
[http://dx.doi.org/10.1074/jbc.273.40.25974] [PMID: 9748275]

[232] Rost R, Haas S, Hammer E, Herrmann H, Burchhardt G. Molecular analysis of aerobic phenylacetate degradation in *Azoarcus evansii.* Mol Genet Genomics 2002; 267(5): 656-63.
[http://dx.doi.org/10.1007/s00438-002-0699-9] [PMID: 12172805]

[233] Mohamed Mel-S, Ismail W, Heider J, Fuchs G. Aerobic metabolism of phenylacetic acids in *Azoarcus evansii.* Arch Microbiol 2002; 178(3): 180-92.
[http://dx.doi.org/10.1007/s00203-002-0438-y] [PMID: 12189419]

[234] Olivera ER, Miñambres B, García B, *et al.* Molecular characterization of the phenylacetic acid catabolic pathway in *Pseudomonas putida* U: the phenylacetyl-CoA catabolon. Proc Natl Acad Sci USA 1998; 95(11): 6419-24.
[http://dx.doi.org/10.1073/pnas.95.11.6419] [PMID: 9600981]

[235] Nogales J, Macchi R, Franchi F, *et al.* Characterization of the last step of the aerobic phenylacetic acid degradation pathway. Microbiology 2007; 153(Pt 2): 357-65.
[http://dx.doi.org/10.1099/mic.0.2006/002444-0] [PMID: 17259607]

[236] Arias-Barrau E, Olivera ER, Luengo JM, *et al.* The homogentisate pathway: a central catabolic pathway involved in the degradation of L-phenylalanine, L-tyrosine, and 3-hydroxyphenylacetate in *Pseudomonas putida.* J Bacteriol 2004; 186(15): 5062-77.
[http://dx.doi.org/10.1128/JB.186.15.5062-5077.2004] [PMID: 15262943]

[237] Méndez V, Agulló L, González M, Seeger M. The homogentisate and homoprotocatechuate central pathways are involved in 3- and 4-hydroxyphenylacetate degradation by *Burkholderia xenovorans* LB400. PLoS One 2011; 6(3): e17583.
[http://dx.doi.org/10.1371/journal.pone.0017583] [PMID: 21423751]

[238] Zhao G, Xia T, Song J, Jensen RA. *Pseudomonas aeruginosa* possesses homologues of mammalian phenylalanine hydroxylase and 4 alpha-carbinolamine dehydratase/DCoH as part of a three-component gene cluster. Proc Natl Acad Sci USA 1994; 91(4): 1366-70.
[http://dx.doi.org/10.1073/pnas.91.4.1366] [PMID: 8108417]

[239] Onishi A, Liotta LJ, Benkovic SJ. Cloning and expression of *Chromobacterium violaceum* phenylalanine hydroxylase in *Escherichia coli* and comparison of amino acid sequence with mammalian aromatic amino acid hydroxylases. J Biol Chem 1991; 266(28): 18454-9.
[PMID: 1655752]

[240] Hareland WA, Crawford RL, Chapman PJ, Dagley S. Metabolic function and properties of 4-hydroxyphenylacetic acid 1-hydroxylase from *Pseudomonas acidovorans.* J Bacteriol 1975; 121(1): 272-85.
[PMID: 234937]

[241] Baggi G, Boga MM, Catelani D, Galli E, Treccani V. Styrene Catabolism by a Strain of *Pseudomonas fluorescens*. Syst Appl Microbiol 1983; 4(1): 141-7.
[http://dx.doi.org/10.1016/S0723-2020(83)80042-3] [PMID: 23196308]

[242] Díaz E, Ferrández A, Prieto MA, García JL. Biodegradation of aromatic compounds by *Escherichia coli*. Microbiol Mol Biol Rev 2001; 65(4): 523-69.
[http://dx.doi.org/10.1128/MMBR.65.4.523-569.2001] [PMID: 11729263]

[243] Ferrández A, Garciá JL, Díaz E. Genetic characterization and expression in heterologous hosts of the 3-(3-hydroxyphenyl)propionate catabolic pathway of *Escherichia coli* K-12. J Bacteriol 1997; 179(8): 2573-81.
[PMID: 9098055]

[244] Arai H, Yamamoto T, Ohishi T, Shimizu T, Nakata T, Kudo T. Genetic organization and characteristics of the 3-(3-hydroxyphenyl)propionic acid degradation pathway of *Comamonas testosteroni* TA441. Microbiology 1999; 145: 2813-20.

[245] Ampe F, Lindley ND. Flux limitations in the ortho pathway of benzoate degradation of *Alcaligenes eutrophus*: metabolite overflow and induction of the meta pathway at high substrate concentrations. Microbiology 1996; 142(Pt 7): 1807-17.
[http://dx.doi.org/10.1099/13500872-142-7-1807] [PMID: 8757743]

[246] Matus V, Sánchez MA, Martínez M, González B. Efficient degradation of 2,4,6-Trichlorophenol requires a set of catabolic genes related to tcp genes from Ralstonia eutropha JMP134(pJP4). Appl Environ Microbiol 2003; 69(12): 7108-15. [pJP4].
[http://dx.doi.org/10.1128/AEM.69.12.7108-7115.2003] [PMID: 14660355]

[247] Sánchez MA, González B. Genetic characterization of 2,4,6-trichlorophenol degradation in *Cupriavidus necator* JMP134. Appl Environ Microbiol 2007; 73(9): 2769-76.
[http://dx.doi.org/10.1128/AEM.02584-06] [PMID: 17322325]

[248] Schenzle A, Lenke H, Spain JC, Knackmuss HJ. Chemoselective nitro group reduction and reductive dechlorination initiate degradation of 2-chloro-5-nitrophenol by *Ralstonia eutropha* JMP134. Appl Environ Microbiol 1999; 65(6): 2317-23.
[PMID: 10347008]

[249] Schenzle A, Lenke H, Spain JC, Knackmuss HJ. 3-Hydroxylaminophenol mutase from *Ralstonia eutropha* JMP134 catalyzes a Bamberger rearrangement. J Bacteriol 1999; 181(5): 1444-50.
[PMID: 10049374]

[250] Schühle K, Jahn M, Ghisla S, Fuchs G. Two similar gene clusters coding for enzymes of a new type of aerobic 2-aminobenzoate (anthranilate) metabolism in the bacterium Azoarcus evansii. J Bacteriol 2001; 183(18): 5268-78.
[http://dx.doi.org/10.1128/JB.183.18.5268-5278.2001] [PMID: 11514509]

[251] Kurnasov O, Jablonski L, Polanuyer B, Dorrestein P, Begley T, Osterman A. Aerobic tryptophan degradation pathway in bacteria: novel kynurenine formamidase. FEMS Microbiol Lett 2003; 227(2): 219-27.
[http://dx.doi.org/10.1016/S0378-1097(03)00684-0] [PMID: 14592712]

[252] Lu J, Brigham CJ, Gai CS, Sinskey AJ. Studies on the production of branched-chain alcohols in

engineered *Ralstonia eutropha.* Appl Microbiol Biotechnol 2012; 96(1): 283-97.
[http://dx.doi.org/10.1007/s00253-012-4320-9] [PMID: 22864971]

[253] Müller J, MacEachran D, Burd H, *et al.* Engineering of *Ralstonia eutropha* H16 for autotrophic and heterotrophic production of methyl ketones. Appl Environ Microbiol 2013; 79(14): 4433-9.
[http://dx.doi.org/10.1128/AEM.00973-13] [PMID: 23686271]

[254] Fei Q, Brigham CJ, Lu J, Fu R, Sinskey AJ. Production of branched-chain alcohols by *Ralstonia eutropha* in fed batch cultivation. Biomass Bioenergy 2013; 56: 334-41.
[http://dx.doi.org/10.1016/j.biombioe.2013.05.009]

[255] Jendrossek D, Krüger N, Steinbüchel A. Characterization of alcohol dehydrogenase genes of derepressible wild-type *Alcaligenes eutrophus* H16 and constitutive mutants. J Bacteriol 1990; 172(9): 4844-51.
[PMID: 2144274]

[256] Grousseau E, Lu J, Gorret N, Guillouet SE, Sinskey AJ. Isopropanol production with engineered *Cupriavidus necator* as bioproduction platform. Appl Microbiol Biotechnol 2014; 98(9): 4277-90.
[http://dx.doi.org/10.1007/s00253-014-5591-0] [PMID: 24604499]

[257] Li H, Opgenorth PH, Wernick DG, *et al.* Integrated electromicrobial conversion of CO_2 to higher alcohols. Science 2012; 335(6076): 1596.
[http://dx.doi.org/10.1126/science.1217643] [PMID: 22461604]

[258] Torella JP, Gagliardi CJ, Chen JS, *et al.* Efficient solar-to-fuels production from a hybrid microbial-water-splitting catalyst system. Proc Natl Acad Sci USA 2015; 112(8): 2337-42.
[http://dx.doi.org/10.1073/pnas.1424872112] [PMID: 25675518]

[259] Steen EJ, Kang Y, Bokinsky G, *et al.* Microbial production of fatty-acid-derived fuels and chemicals from plant biomass. Nature 2010; 463(7280): 559-62.
[http://dx.doi.org/10.1038/nature08721] [PMID: 20111002]

[260] Casey M. Scientists using sunlight to make liquid fuel. Available from: http://www.cbsnews.com/news/scientists-using-sunlight-to-make-liquid-fuel/ 2015.

[261] Pradella JG, Ienczak JL, Delgado CR, Taciro MK. Carbon source pulsed feeding to attain high yield and high productivity in poly(3-hydroxybutyrate) (PHB) production from soybean oil using *Cupriavidus necator*. Biotechnol Lett 2012; 34(6): 1003-7.
[http://dx.doi.org/10.1007/s10529-012-0863-1] [PMID: 22315097]

[262] Riedel SL, Brigham CJ, Budde CF, *et al.* Recovery of poly(3-hydroxybutyrate-c--3-hydroxyhexanoate) from Ralstonia eutropha cultures with non-halogenated solvents. Biotechnol Bioeng 2013; 110(2): 461-70.
[http://dx.doi.org/10.1002/bit.24713] [PMID: 22903730]

[263] Gai CS, Lu J, Brigham CJ, Bernardi AC, Sinskey AJ. Insights into bacterial CO_2 metabolism revealed by the characterization of four carbonic anhydrases in *Ralstonia eutropha* H16. AMB Express 2014; 4(1)

Recent Advances in Biotechnology, Vol. 1, 2016, 348-379

CHAPTER 6

PHBHV Biosynthesis by *Haloferax mediterranei*: from Genetics, Metabolism, and Engineering to Economical Production

Hua Xiang*

State Key Laboratory of Microbial Resources, Institute of Microbiology, Chinese Academy of Sciences, Beijing, China

Abstract: The halophilic archaeon *Haloferax mediterranei*, has shown promise for the production of poly(3-hydroxybutyrate-*co*-3-hydroxyvalerate) (PHBHV), a desirable bioplastic, from many low-cost carbon sources. To obtain a full understanding of PHBHV biosynthesis and to provide novel strategies for polyhydroxyalkanaote (PHA) production by haloarchaea, the complete genome of *H. mediterranei* has been determined, and a genome-wide investigation of the genetics and metabolism of this haloarchaeon involved in PHBHV biosynthesis was performed in the past few years. This exploration covers the identification of key genes for PHBHV biosynthesis, the specific precursor supplying pathways, the PHA granule modulation, and PHBHV mobilization, as well as its linkage to haloarchaeal-specific acetyl-CoA and propionyl-CoA assimilation. These advances have not only revealed many haloarchaeal-specific enzymes and pathways for PHBHV biosynthesis, such as the archaeal-type PHA synthases, the haloarchaeal-type β-ketothiolases with two distinct subunits, and the novel 3-hydroxypropionate pathway coupling CO_2 fixation into PHBHV, but have also provided strategies for bioengineering haloarchaea for economical production of PHBHV or tailor-made higher-value PHA. This chapter will summarize these current progresses of PHBHV biosynthesis by *H. mediterranei*, thus providing a perspective for economical production of industry-scale or higher-value PHAs by haloarchaea in the near future.

* **Corresponding author Hua Xiang:** State Key Laboratory of Microbial Resources, Institute of Microbiology, Chinese Academy of Sciences, NO.1 Beichen West Road, Chaoyang District, Beijing 100101, People's Republic of China; Tel: 8610-64807472; Fax: 8610-64807472; Email: xiangh@im.ac.cn.

Martin Koller (Ed.)

Keywords: Acetoacetyl-CoA reductase, Acetyl-CoA, β-ketothiolase, β-oxidation, Carbon sources, Copolyester, Economical production, Genes, Genome, *Haloferax mediterranei*, Halophilic archaea, PHA biosynthesis, PHA depolymerase, PHA granule, PHA mobilization, PHA synthase, Polyhydroxyalkanoates (PHA), Poly(3-hydroxybutyrate-*co*-3-hydroxyvalerate) (PHBHV), Propionyl-CoA, Regulation, Tailor-made PHA, Terpolyester.

INTRODUCTION

Haloferax mediterranei, previously known as *Halobacterium mediterranei*, was first isolated from seawater evaporation ponds near Alicante, Spain [1, 2]. *H. mediterranei* is a pleomorphic rod-shaped, slightly motile halophile belonging to the newly proposed *Haloferacaceae* family (moved from the *Halobacteriaceae* family) of the domain Archaea [3, 4]. The extracellular polysaccharides produced by *H. mediterranei* display xanthan-like properties and give the carotenoid-rich red colonies a typical mucous character on agar plates [5]; the strong coloration, based on C50-carotenoids pigments, is also easily visible in liquid cultures of this organism (Fig. **1**).

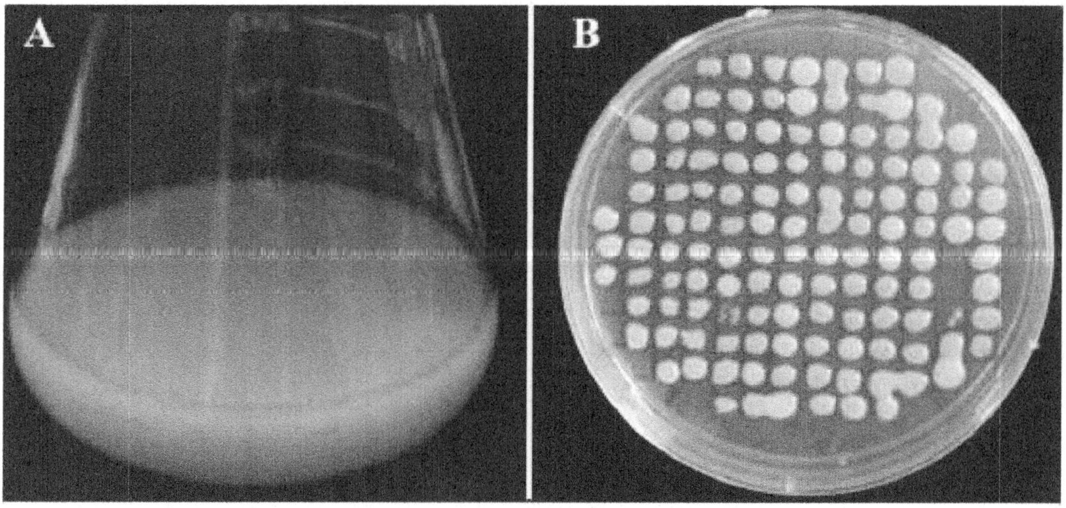

Fig. (1). Culturing of *H. mediterranei* in liquid media (A) and on agar plate (B).

H. mediterranei is strongly amylolytic and lipolytic, and it can use a variety of carbohydrates, polyalcohols, organic acids, or amino acids as a sole source of carbon and energy. It grows faster than most other members of *Halobacteria* (1, 2). It can grow aerobically over a wide range of salt concentrations (optimum at 20% and survival above 3% total salts, including 20-40 mM Mg^{2+}) with an optimal temperature of 37-51 °C and pH-value of 6.5 to 7.2. *H. mediterranei* can also grow anaerobically by reducing nitrate and nitrites to N_2 [2, 6]. *H. mediterranei* was observed to fix CO_2 in the form of HCO_3^-, which might also be an important carbon source for *H. mediterranei* at high salt concentration [7 - 10].

In stationary phase or upon other stress conditions, *H. mediterranei* can produce halocins [11], gas vesicles [12, 13], and accumulate polyhydroxyalkanoate (PHA) [14]. This makes it an interesting model haloarchaeon for the research of haloarchaeal physiology. The PHA produced by *H. mediterranei* was determined as poly(3-hydroxybutyrate-*co*-3-hydroxyvalerate) (PHBHV) [15], a desirable bioplastic with multiple applications. The genome of *H. mediterranei* has already been determined, and the genetic manipulation tools have also been well-established [16, 17]; thus, *H. mediterranei* is a potential novel microbial cell factory for PHBHV production.

In this chapter, we will summarize recent progress from genetics, metabolism, and bioengineering of *H. mediterranei* for PHBHV biosynthesis to tailor their properties, and to make the production of PHBHV by this haloarchaeal cell factory more economic.

THE GENOME OF *H. MEDITERRANEI*

The complete genome sequence (3,904,707 bp) of *H. mediterranei* was determined in 2012 by a combination of 454 pyrosequencing and Sanger sequencing, which revealed four circular replicons of the genome [17], including one chromosome and three mega-plasmids named pHM100, pHM300 and pHM500 (Table **1**). The genome encodes 3,864 putative proteins, 57 tRNAs, and two 5S-16S-23S rRNA operons. Many important genes involved in PHBHV biosynthesis are located in the mega-plasmid pHM300 (Fig. **2**).

Table 1. General features of *H. mediterranei* ATCC 33500 genome [17].

Feature	Chromosome	pHM100	pHM300	pHM500
Length (bp)	2,948,884	129,210	321,908	504,705
GC Content	61.1%	57.6%	57.9%	57.5%
Number of Encoded Proteins	2,951	142	291	480
Number of rRNA Operons (16S,23S,5S)	2	0	0	0
Number of tRNAs	56	0	1	0
Number of 7S RNA	1	0	0	0
Average pI of Proteins	4.77	5.09	4.94	4.81

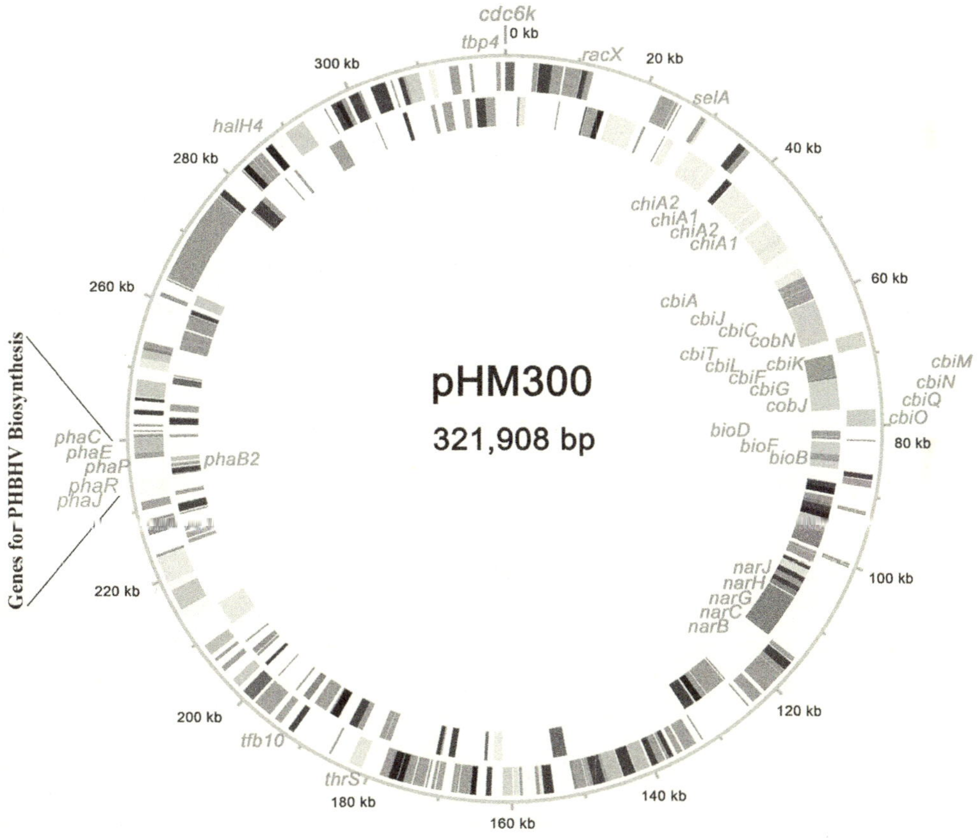

Fig. (2). The mega-plasmid pHM300 of *H. mediterranei* and the location of selected PHA biosynthesis genes (*phaB*, *phaC*, *phaE*, *phaJ*, *phaP* and *phaR*, see below for details), modified from [19].

There are multiple replication origins in the chromosome of *H. mediterranei* [18], and the copy number of the chromosome can be up to ~50 copies per cell during the exponential growth phase. Notably, the copy ratio of pHM300 to the chromosome increases from the exponential to stationary phase, indicating that the copy number control of the pHM300 is independent from that of the chromosome, which may be favorable for expression of the PHA synthesis genes that are located in pHM300 during stationary phase [19].

THE KEY ENZYMES FOR PHBHV BIOSYNTHESIS

PHA Synthase (PhaEC)

PHA synthases, the key enzymes that catalyze the polymerization of 3-hydroxyacyl-CoA into PHAs, have been well-characterized in bacteria and are grouped into four classes according to their substrate specificity and the subunit compositions. Class III and IV bacterial synthases are composed of two subunits. The catalytic subunit is PhaC (~40 kDa), while the other subunit is PhaE (~40 kDa) for class III synthases and PhaR (~22 kDa) for class IV synthases [20]. In the domain of *Archaea*, although the presence of PHA granules in haloarchaeal cells was observed forty years ago [21], the genes involved in PHA biosynthesis were unknown until 2004, when a homologue of a *phaC* gene was annotated in the *Haloarcula marismortui* genome [22]. With this information, the first haloarchaeal PHA synthase was then experimentally determined in both *Haloarcula hispanica* and *H. marismortui* in 2007 [23]. The *Haloarcula* PHA synthases contain two subunits, as gene knockout and complementation clearly indicated that an additional gene (named *phaE* after this work) located immediately upstream of the *phaC* gene is indispensable for the function of the PHA synthase in *H. hispanica.* Although the *Haloarcula* PhaCs are homologous (37 to 43% amino acid identity) to the PhaC subunits of the bacterial class III PHA synthases, and the *Haloarcula* PhaEs also containes a typical "PhaE box", these haloarchaeal PhaCs (~53 kDa) are much larger, while the PhaEs (~21 kDa) are much smaller than their counterparts of the bacterial Class III synthases of bacteria [23].

Based on the highly conserved regions of the known haloarchaeal PhaCs, the gene cluster (*phaEC*) encoding a PHBHV synthase in *H. mediterranei* was cloned *via* thermal asymmetric interlaced PCR [24] and confirmed by gene knockout and complementation. The *phaE* and *phaC* genes of *H. mediterranei* are constitutively expressed, which may be one of the reasons that *H. mediterranei* can accumulate considerable PHBHV in either the nutrient-rich AS-168 or nutrient-limited (with an excessive carbon source) MST media (Table **2**). Western blotting analysis indicated that both PhaE and PhaC proteins are strongly bound to the PHBHV granules, and both subunits are essential for the synthase activity either *in vivo* or *in vitro* [24]. All haloarchaeal PhaCs have a longer C-terminus that is absent in their bacterial counterparts. Deletion of this region in PhaC sharply reduces the PHA synthesis ability, suggesting the important function of this C-terminal part for the haloarchaeal-type PhaCs [24]. This additional domain that is not found in mesohalic PhaCs may be important for correct protein folding at high salt concentrations or may offer partial function that is supplied by bacterial PhaE because all of the known haloarchaeal PhaE subunits (~21 kDa) are much smaller than their bacterial counterparts (~40 kDa).

Table 2. The nutrient-rich AS-168 and nutrient-limited MST (with excessive starch) or MG (with excessive glucose) media*.

AS-168				MST or MG	
Bacto casamino acids	5	g	Bacto yeast extract	1	g
Bacto yeast extract	5	g	Starch (MST) or Glucose (MG)	10	g
Sodium glutamate	1	g	Sodium glutamate	1	g
Trisodium citrate	3	g	$MgSO_4 \cdot 7H_2O$	20	g
$MgSO_4 \cdot 7H_2O$	20	g	KH_2PO_4	37.5	mg
KCl	2	g	KCl	2	g
NaCl	200	g	NaCl	200	g
$FeSO_4 \cdot 7H_2O$	50	mg	$FeSO_4 \cdot 7H_2O$	50	mg
$MnCl_2 \cdot 4H_2O$	0.036	mg	$MnCl_2 \cdot 4H_2O$	0.036	mg
ddH_2O	to 1	L	ddH_2O	to 1	L
pH=7.0~7.2			pH=7.0~7.2		

* PIPS (15 g/L) may be included as a stable pH-buffer in flask culturing.

Screening of the PHA synthase genes in the PHA producers of haloarchaea (up to

12 genera) clearly indicated that the *phaEC* genes are widely distributed among these haloarchaea. Phylogenetic analysis shows that all the type III haloarchaeal PHA synthases are clustered separately from those of bacteria. As the haloarchaeal PHA synthases also differ considerably in both molecular weight and certain conserved motifs from their bacterial counterparts, we propose that the haloarchaeal PHA synthases represent a subtype IIIA (A for archaea), while the bacterial type III PHA synthases are proposed as the subtype IIIB (B for bacteria) (Fig. **3**) [25].

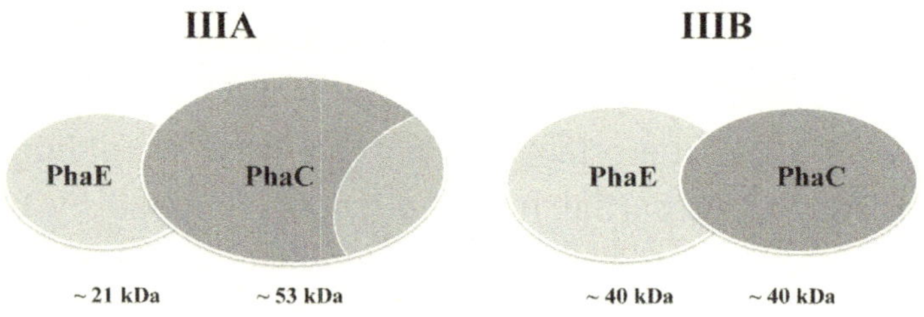

Fig. (3). Comparison of the archaeal subtype (IIIA, A for archaea) and the bacterial subtype (IIIB, B for bacteria) of the type III PHA synthases.

PHA Biosynthetic Ketothiolases (PhaA and BktB)

The PHA biosynthetic ketothiolases (PhaA and BktB), which condense two acetyl coenzyme A (acetyl-CoA) molecules to acetoacetyl-CoA (by PhaA) or one acetyl-CoA and one propionyl-CoA to 3-ketovaleryl-CoA (by BktB), were identified in haloarchaea in 2013 [26]. Whole-genome analysis has revealed eight potential ketothiolase genes in the haloarchaeon *H. mediterranei*. Gene knockout or gene disruption indicates that only the β-ketothiolases encoded by HFX_6004-HFX_6003 (located in mega-plasmid pHM500) and HFX_1023-HFX_1022 (located in the chromosome) are involved in PHBHV biosynthesis. Knockout of HFX_6004-6003 leads to the accumulation of PHB only without the 3HV monomer. Although knockout of the HFX_1023-1022 had no significant effect on PHBHV accumulation, the simultaneous knockout of HFX_6004-6003 and HFX_1023-1022 caused the strain to lose the ability to accumulate PHA. These results indicate that HFX_6004-6003 is able to catalyze the synthesis of acetoacetyl-CoA from two acetyl-CoA molecules and also produce 3-ketovalery-

-CoA from acetyl-CoA and propionyl-CoA, thus encoding the enzyme of BktB. In contrast, HFX_1023-1022 is restricted to the biosynthesis of acetoacetyl-CoA from two acetyl-CoA molecules, thus encoding PhaA [26].

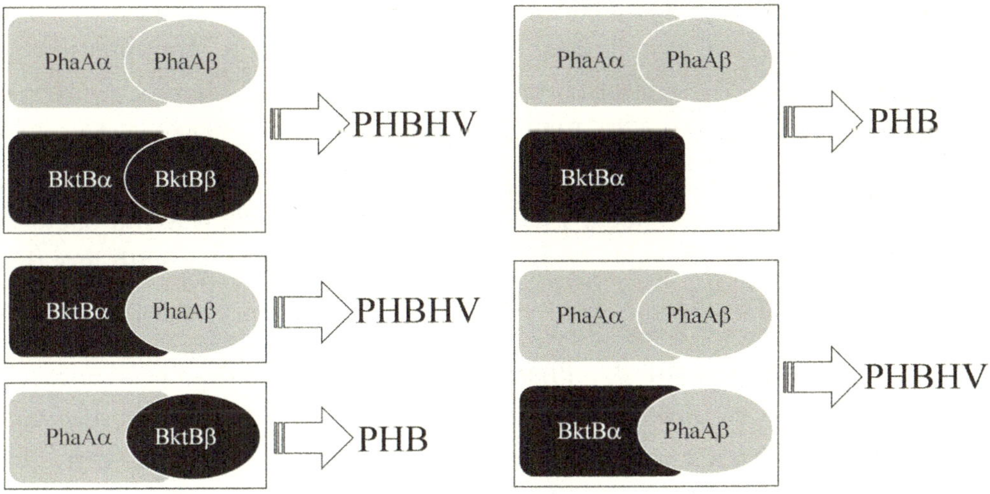

Fig. (4). The two-subunit β-ketothiolases (PhaA and BktB) of *H. mediterranei*, and the interchange of the small subunits (BktBβ and PhaAβ) in PHB or PHBHV production.

Unlike all known bacterial counterparts encoded by a single gene, the haloarchaeal PhaA and BktB are both composed of two different types of subunits. The PhaAα and PhaAβ subunits of PhaA are encoded by the co-transcribed HFX_1023 and HFX_1022 genes, whereas the BktBα and BktBβ subunits of BktB are encoded by the co-transcribed genes HFX_6004 and HFX_6003, respectively. BktBα and PhaAα are the catalytic subunits and determine substrate specificities. Their catalytic triad "Ser-His-His" is distinct from the bacterial "Cys-His-Cys." BktBβ and PhaAβ both contain an oligosaccharide-binding fold domain, which is essential for the ketothiolase activity. BktBβ and PhaAβ were functionally interchangeable, although PhaAβ prefer functioning with PhaAα (Fig. 4). This is the first report on the haloarchaeal type of PHBHV-specific β-ketothiolases, which are distinct from their bacterial counterparts in both subunit composition and catalytic residues [26].

PHA-Specific Acetoacetyl-CoA Reductase (PhaB)

Before the PHA-specific acetoacetyl-CoA reductase (PhaB) of *H. mediterranei* was identified, the PhaB was first discovered in *H. hispanica* in 2009 [27]. This PhaB (previously annotated as FabG1 in *Haloarcula*) was distinguished from all other FabG (ketoacyl-ACP reductase) paralogs through a combination of bioinformatics and gene knockout-complementation analyses. It was confirmed by enzymatic activity assays, and a PHA biosynthetic pathway was reconstructed in *Haloferax volcanii*, a PHA-defective haloarchaeon, with this *phaB* gene as well as *phaEC* genes. This haloarchaeal PhaB is NADPH-dependent as its bacterial counterparts, but it is only distantly related to the bacterial PhaBs, with less than 38% amino acid identity [27].

With the gene information of the PhaB in *H. hispanica*, two PhaB enzymes (PhaB1 and PhaB2) were identified in *H. mediterranei* [28]. Knockout of *phaB1* gene did not affect the accumulation of PHBHV, but when *phaB2* gene was knocked out, the yield of PHBHV and the fraction of the 3HV monomer decreased significantly. Notably, when the *phaB1* and *phaB2* were knocked out in combination, the mutant strain (Δ*phaB1phaB2*) no longer produced PHBHV, indicating both PhaB1 and PhaB2 were involved in PHBHV biosynthesis in *H. mediterranei*. The PhaB2 (encoded by HFX_5215, located in pHM300) plays a key role in supplying the precursors of (*R*)-3-hydroxybutyrate-CoA (3HB-CoA) and (*R*)-3-hydroxyvalerate-CoA (3HV-CoA) for PHBHV biosynthesis. [28].

Taken together, after a few years' exploration, the key enzymes for PHBHV biosynthesis from acetyl-CoA and propionyl-CoA *via* a three-step process catalyzed by ketothiolase (PhaA/BktB), ketoacyl-CoA reductase (PhaB1 and B2), and PHA synthase (PhaEC) sequentially, were all identified and characterized in *H. mediterranei.*

PHA Granule-Associated PhaP and PhaR

Polyhydroxyalkanoates (PHAs) are synthesized and assembled as PHA granules that undergo well-regulated formation in many microorganisms. Although PHA granule-associated proteins (PGAPs) are important for PHA synthesis and granule formation, most haloarchaeal PGAP gene sequences could not be found through homology searches of the haloarchaeal genomes. To find the specific haloarchaeal PGAPs, two-dimensional gel electrophoresis (2-DE) and a matrix-assisted laser desorption ionization tandem time-of-flight mass spectrometric (MALDI-TOF/TOF MS) analyses were used to separate and identify the proteins located on the PHA granule of *H. mediterranei*. Five PGAPs were identified, and notably, the genes encoding these PGAPs were unexpectedly located adjacent to each other, forming a gene cluster (HFX_5217-HFX_5221, encoding *phaJRPEC*) on a mega-plasmid pHM300 of *H. mediterranei* (Figs. **2** and **5a**). The HFX_5220 and HFX_5221 genes encode the two subunits of the PHA synthase (PhaE and PhaC) as described above. HFX_5219 encodes the predominant (more than 90% abundance) PGAP and was named as PhaP, the major phasin of *H. mediterranei* [29]. The HFX_5218 encodes a small protein (12 kDa) containing a DNA binding motif at the C-terminal, implying a regulatory function and thus was named PhaR [30]. The HFX_5217 gene is located upstream of but oriented away from the HFX_5218. HFX_5217 is annotated as a putative enoyl-CoA hydratase (MaoC) and has high identities (45 and 36%, respectively) to PhaJ1 and PhaJ4 of *Pseudomonas aeruginosa* [29]; thus, it was renamed PhaJ and was found to be involved in PHA mobilization (see below).

The function of PhaP involved in PHA accumulation in *H. mediterranei* was demonstrated through an integrative approach of gene knockout/complementation and fermentation analyses. The number of PHA granules in the Δ*phaP* mutant cells was significantly decreased [29]. In most of the Δ*phaP* cells there is only a single large granule, while no granule was observed in some cells (Fig. **5b**). These results demonstrate that the *H. mediterranei* PhaP is the main structural protein (phasin) on the PHA granules involved in PHA accumulation and granule formation, and this may be the major reason that a significant decrease in PHA production was also observed in the Δ*phaP* strain of *H. mediterranei* [29]. Notably, the amino acid sequence of the *H. mediterranei* PhaP is distinct from all known PhaPs in the domain of *Bacteria*. Bioinformatic and phylogenetic analyses indicate that this type of PhaP is exclusively conserved in haloarchaea; thus, it is a representative of the haloarchaeal type PHA phasin [29].

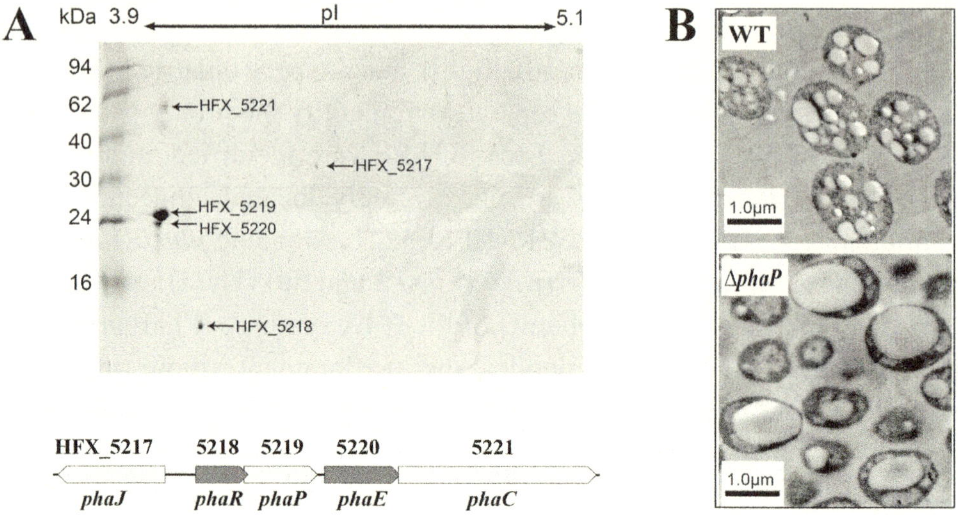

Fig. (5). Identification of the PHA granule-associated proteins of *H. mediterranei* by 2-DE and MALDI-TOF/TOF analyses (**a**) and characterization of the PhaP function by gene deletion (**b**).

The PhaR encoded by HFX_5218 is the only known PGAP of *H. mediterranei* that specifically binds both DNA and PHA granules. It has a putative DNA-binding domain consisting of a swapped-hairpin barrel fold, which shows low homology to the corresponding domain of the AbrB (antibiotic resistance protein B) superfamily of regulators [30]. The *phaR-phaP* constitutes an operon and is co-transcribed. PhaR can directly bind to the promoter of the *phaRP* gene cluster and negatively regulate this operon and thus regulate the PHA accumulation through PhaP [30]. As the PhaR can bind onto the PHA granules, the free amount of PhaR for *phaRP* expression regulation was also modulated by PHA biosynthesis. Although a similar PHA-sensing regulation strategy by PhaR is reported in bacteria [31], *H. mediterranei* has developed a different regulation pattern to achieve such a general regulation strategy (Fig. **6**). First, although *phaR* and *phaP* are co-transcribed from a single promoter P.*phaRP* located upstream of *phaR*, the translation efficiency for PhaP and the binding efficiency of PhaP on the PHA granules were likely much higher than those of PhaR. Therefore, at the beginning of the PHA biosynthesis, the PhaR would have space on the PHA granules (a function of the PHA granules in titration of the PhaR) and thus no free

PhaR binding onto the promoter, so the *phaRP* gene expression would not be inhibited by PhaR, and PhaP is actively expressed. This will accelerate PHA accumulation and increase PHA granule formation. After a large amount of PhaP and PhaR are available due to the high expression of the *phaRP* operon, and when the PHA granules reach a certain number, there would be no more space on the granules for PhaR. Therefore, there would be more free PhaR available to bind to the promoter of *phaRP* and thus shut down the expression of the *phaRP*. Without more PhaP, new PHA granules will not form. By this feedback regulation of PhaRP expression, the size and number of the PHA granules could be maintained at a reasonable range in *H. mediterranei* (Fig. **6**) [30].

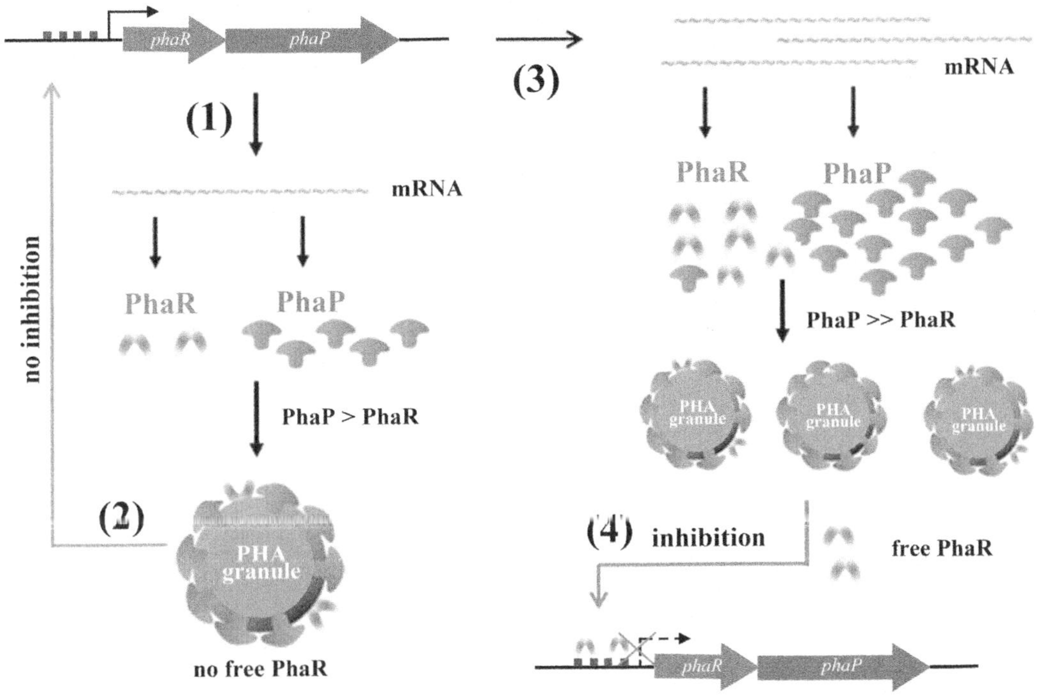

Fig. (6). Regulation of the PhaP-dependent PHBHV granule formation by PhaR in *H. mediterranei*.

H. mediterranei PhaR is likely a bifunctional protein. When both of the two major PGAPs, PhaR and PhaP, are absent, there would be a deficiency in the protection layer between the PHA granule and the cytoplasm; therefore, the cells of the *phaRP*-deletion strain could only accumulate a small amount of PHA. When *phaR* was overexpressed in the *phaRP*-deleted strain, PHA accumulation returned to a wild-type level, and the cells synthesized granules larger than those of the *phaRP*-deleted strain (Fig. **7a**). It has been hypothesized that PhaR might facilitate PHA synthesis in a PhaP-independent mechanism, such as by promoting PHA synthase activity or by acting as a phasin protein [30]. Alternatively, when PhaP is overexpressed in the *phaRP*-deleted strain, a large number of PHA granules with a much smaller size were observed while the PHA amounts were not increased (Fig. **7b**). This indicates that the PhaP is the main phasin, but the proper amount of PhaP is critical to the formation of regular PHA granules. Therefore, *H. mediterranei* PhaR, which is essential for the expressional control of *phaP*, plays a central role in maintaining proper granule formation and PHA accumulation.

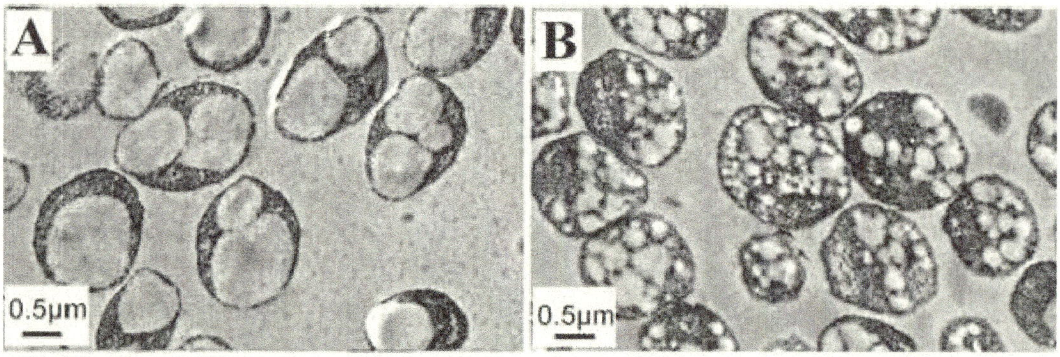

Fig. (7). Overexpression of PhaR (**a**) or PhaP (**b**) in the *phaRP*-deleted strain of *H. mediterranei*, (modified from [30]).

MULTIPLE PATHWAYS FOR PROPIONYL-COA SUPPLYING

H. mediterranei is able to accumulate PHBHV with more than 10 mol-% 3-hydroxyvalerate (3HV) from 3HV-unrelated carbon sources, and the highly efficient propionyl-CoA (the precursor of 3HV monomer) supplying pathways of this strain have long been one of the most attractive questions in the PHA field. Through bioinformatic analysis of the *H. mediterranei* genome sequence, we have predicted four pathways for propionyl-CoA biosynthesis in this haloarchaeon, which were recently further confirmed by genetic and molecular experiments [9].

The first two pathways for the propionyl-CoA supply are the citramalate/2-oxobutyrate pathway and the aspartate/2-oxobutyrate pathway. These two pathways share the reaction steps of transforming 2-oxobutyrate to propionyl-CoA, but they differ from each other in how they produce 2-oxobutyrate. The citramalate/2-oxobutyrate pathway generates 2-oxobutyrate from pyruvate and acetyl-CoA, catalyzed by CimA-LeuBCD, while the aspartate/2-oxobutyrate pathway generates 2-oxobutyrate from threonine and methionine, catalyzed by threonine dehydratase and methionine γ-lyase, respectively (Fig. **8**). The third route is the methylmalonyl-CoA pathway. Briefly, succinyl-CoA could be isomerized into methylmalonyl-CoA, which was then decarboxylated to propionyl-CoA by methylmalonyl-CoA decarboxylase (Fig. **8**). The 3HV fraction increase in PHBHV was observed when co-feeding the pathway intermediates, and 3HV fraction decreases occurred upon knockout/disruption of the pathway genes, indicating the involvement of these pathways in propionyl-CoA supplying for PHBHV biosynthesis [9].

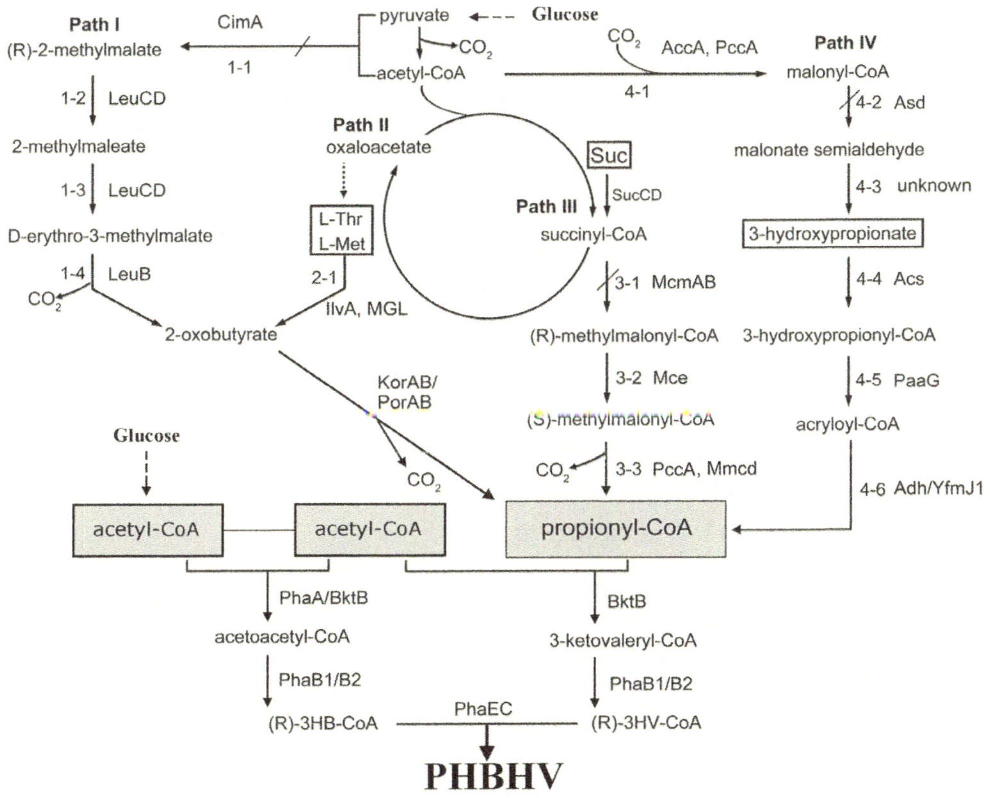

Fig. (8). Multiple pathways for propionyl-CoA synthesis; updated and modified from [9].

The fourth pathway proposed for the propionyl-CoA supply is the first part of the 3-hydroxypropionate/4-hydroxybutyrate cycle that was recently identified in an autotrophic member of the archaeal order, *Sulfolobales* [32], and is named the 3-hydroxypropionate pathway. In this pathway, acetyl-CoA is first carboxylated by the biotin-dependent acetyl-CoA carboxylase to malonyl-CoA, and then reduced *via* malonate semialdehyde to 3-hydroxypropionate that is usually catalyzed by two enzymes named malonyl-CoA reductase (MCR) and malonate semialdehyde reductase (MSR). 3-Hydroxypropionate is further reductively converted to propionyl-CoA in three steps by acyl-CoA synthetase, enoyl-CoA hydratase, and acryloyl-CoA reductase (Fig. **8**). Although the genes encoding the MSR are still not well understood, the 3-hydroxypropionate pathway was confirmed by co-feeding 3-hydroxypropionate and inactivating the key genes [9]. The fixation of CO_2 into PHBHV by this pathway was further confirmed by analysis of ^{13}C positional enrichment in 3HV of the PHBHV in ^{13}C-labelling experiments [9].

Notably, ^{13}C metabolic flux analysis showed that the citramalate/2-oxobutyrate pathway (53.0% flux) and the 3-hydroxypropionate pathway (30.6% flux) were the two main generators of propionyl-CoA from glucose [9].

Identification of these multiple propionyl CoA-supplying pathways for PHBHV biosynthesis has not only answered the long-standing question why *H. mediterranei* is able to accumulate PHBHV from 3HV-unrelated carbon sources, but it also may provide more strategies for the production of PHBHV with a controllable 3HV molar fraction in *H. mediterranei.*

ENZYMES AND PATHWAYS FOR PHBHV MOBILIZATION

PhaZh1-Drived PHBHV Degradation

While there have been important advances in the haloarchaeon *H. mediterranei* PHBHV biosynthesis in the past several years, no molecular clue was available for PHA mobilization in haloarchaea until recently, when a PHA depolymerase PhaZh1 was reported in *H. mediterranei* [33]. The PhaZh1 is also located on the PHA granules, but it was not identified previously [29], indicating a smaller amount of this protein on the PHA granules at the stage of PHA biosynthesis. However, when the *H. mediterranei* cells containing PHA were transferred to a freshly prepared carbon-free medium, this novel protein emerged on the granules [33]. To isolate this potential PHA depolymerase, the native PHA (nPHA) granules were collected and purified from cells in which PHA was degraded. SDS-PAGE and MALDI-TOF/TOF MS analysis revealed that this granule-associated protein is encoded by HFX_6464 (on the mega-plasmid pHM500) and is a patatin-like protein. Patatin is a major storage protein found in potatoes that also has lipid acyl hydrolase and acyl transferase activities [34]. A classical lipase box (G-T-S47-G-G), which is important for the hydrolysis activity of many bacterial PHA depolymerases (PhaZs), was also found in the amino acid sequence encoded by HFX_6464. This indicates that HFX_6464 is a potential haloarchaeal PHA depolymerase, and thus it was named PhaZh1 [33].

To determine the function of PhaZh1 on nPHA granule degradation, the *phaZh1* gene was knocked out in *H. mediterranei.* When the nPHA granules were isolated from the *H. mediterranei* cells and the *in vitro* degradation was monitored, it was clear that nPHA granules isolated from the *H. mediterranei* wild-type strain were gradually degraded *in vitro*, while degradation of the nPHA granules isolated from the Δ*phaZh1* mutant strain was not detectable. In contrast, complementation of *phaZh1* in the Δ*phaZh1* strain recovered the auto-degradation of the nPHA granules *in vitro*. This result indicated that PhaZh1 is a key enzyme involved in nPHA granule degradation *in vitro* [33].

To further biochemically determine the hydrolysis activity of PhaZh1 with the nPHA granules, PhaZh1 was overexpressed and purified. The purified PhaZh1 efficiently hydrolyzed the nPHA granules isolated from either *H. mediterranei* or bacteria in high salt concentration, with a 3-hydroxybutyrate (3HB) monomer as the primary product. The site-directed mutagenesis of PhaZh1 indicated that Gly16, Ser47 (in a classical lipase box, G-X-S47-X-G), and Asp195 of this depolymerase were essential for its activity in nPHA granule hydrolysis [33] (**Fig. 9**).

When the effect of *phaZh1* deletion on intracellular PHA degradation in *H. mediterranei* was evaluated, unexpectedly we found that the PhaZh1 mutation (Δ*phaZh1*) had no significant effect on intracellular PHA mobilization in *H. mediterranei,* as the PHA in both the wild-type and Δ*phaZh1* strains was degraded to a similar level after 5 days when cultivated in a freshly prepared carbon-free medium. This indicated that the PhaZh1 is likely not the key enzyme for PHA mobilization *in vivo*, and there would be alternative PHA mobilization pathway(s) that function more effectively within the cells of *H. mediterranei*.

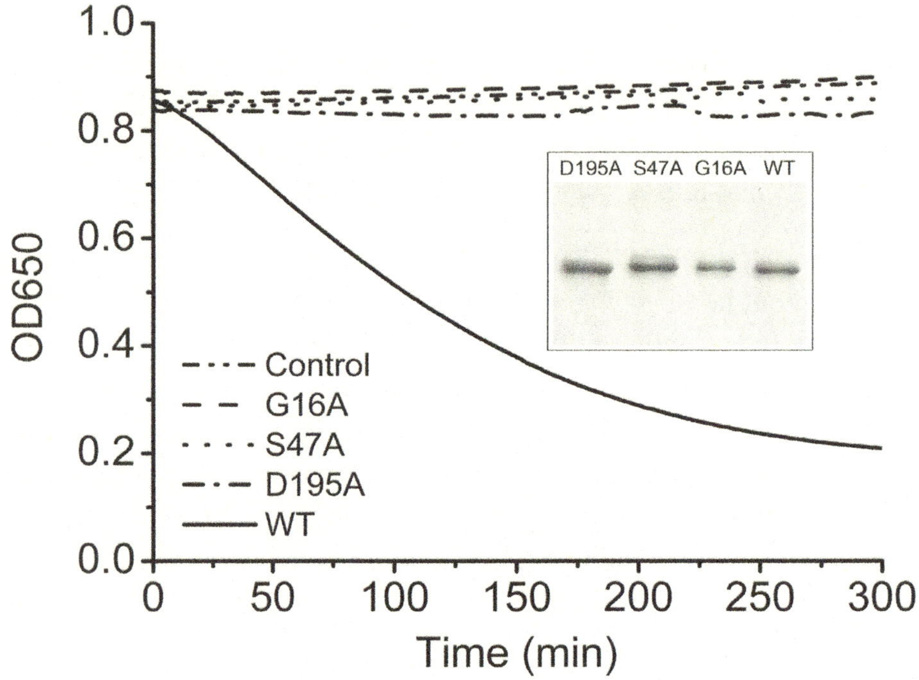

Fig. (9). Degradation of nPHA *in vitro* by PhaZh1 (WT) and PhaZh1 mutants (S16A, S47A and D195A).

PhaJ-Mediated PHA Mobilization

To reveal the efficient intracellular PHA mobilization pathway(s) in *H. mediterranei*, we went back to check the function of another PAGP encoded by HFX_5217 (Fig. **5**), the only PAGP whose function was still not clear. HFX_5217 is annotated as MaoC or PhaJ, a putative (*R*)-specific enoyl-CoA hydratase (R-ECH). As the hydration reaction catalyzed by R-ECHs is a reversible process, we proposed that R-ECHs of *H. mediterranei* may be involved in PHA degradation as well as PHA biosynthesis. Although five putative R-ECHs were found in the *H.*

mediterranei genome, gene knockout and complementation showed that only the R-ECH encoded by HFX_5217 was associated with PHA granules and is involved in PHA mobilization. As this R-ECH can catalyze the conversion between (*R*)-3 - hydroxyacyl-CoA and enoyl-CoA, we named it PhaJ in *H. mediterranei*. Given the products of the *in vivo* degradation of PHA could be (*R*)-3-hydroxyacyl-CoA, the involvement of PhaJ in PHA mobilization would be its function in conversion of the PHA degradation product (3-hydroxyacyl-CoA) to the intermediate (enoyl-CoA) of the β-oxidation cycle. Therefore, PhaJ plays the key role in linking the PHA mobilization to the β-oxidation pathway in *H. mediterranei*. We recently found that PhaJ-mediated PHA mobilization might be the key pathway for *in vivo* PHBHV degradation in *H. mediterranei*, as PHA degradation was not obvious in *H. mediterranei ΔphaJ* when the cells containing PHA were transferred to a freshly prepared carbon-free medium (unpublished data)[1].

As acetyl-CoA and propionyl-CoA would be the final products of PHBHV degradation through either β-oxidation cycle or other unknown pathways, the assimilation of both acetyl-CoA and propionyl-CoA in haloarchaea was addressed recently [35, 36]. The propionyl-CoA carboxylation pathway has been demonstrated as the only pathway for propionyl-CoA assimilation in *H. mediterranei* [35], and the methylaspartate cycle was suggested to assimilate the PHA-generated acetyl-CoA in most haloarchaea [36]. However, as the methylaspartate cycle was not found in *H. mediterranei,* the alternative anaplerotic pathways for acetyl-CoA assimilation (*e.g.*, glyoxylate cycle bypass) may be involved in PHBHV utilization in *H. mediterranei*.

Taken together, the current studies indicate that at least two PHA degradation pathways exist in *H. mediterranei*. PhaZh1-invovled PHA degradation functions well *in vitro* but not *in vivo* and thus may play a role in carbon recycling when PHA granules are released from the dead cells of a population. In contrast, the PhaJ-mediated PHA mobilization is likely the key intracellular PHA degradation pathway in *H. mediterranei*, which may be important for a flexible regulatio between PHA mobilization and PHA biosynthesis. Further investigation of the mechanisms of these pathways will provide novel insights for understanding the adaptation of haloarchaea in hypersaline environments and may also provide novel insights for engineering *H. mediterranei* for high-level PHBHV production.

PRODUCTION OF PHBHV FROM INEXPENSIVE CARBON RESOURCES

One attractive feature of *H. mediterranei* in the economic production of PHA is that this haloarchaeon can produce PHBHV from 3HV-unrelated carbon resources, such as glucose and soluble starch [24, 37]. To make the PHBHV production more cost effective, many other low cost carbon sources have been investigated as alternative substrates for *H. mediterranei*, including extruded rice bran [38], whey [39 - 42], vinasse [43], stillage [44, 45], glycerol [46], as well as chitin [47] (Table **3**). The use of such industrial or agricultural by-products for PHBHV production by *H. mediterranei* has now been considered as an important strategy to reduce PHBHV production costs.

Rice Bran and other Agricultural By-Products

As *H. mediterranei* cannot use native rice bran or corn starch as a carbon source, extruded rice bran (ERB) and extruded corn starch (ECS) with α-amylase treatment were used as carbon sources for PHBHV production by *H. mediterranei*. By employing a pH-stat control strategy to maintain pH-value at 6.9-7.1 and using ERB:ECS (1:8 g/g) as the major carbon source, Chen CW's research group has obtained a cell concentration of 140 g/L, PHBHV concentration of 77.8 g/L and PHBHV content of 55.6 wt.-% in a repeated fed-batch fermentation [38]. In contrast, when ECS (without ERB) was used as the major carbon source, they obtained 62.6 g/L cell concentration, 24.2 g/L PHA concentration and 38.7 wt.-% PHBHV content. This indicates that ERB is important for high cell density cultivation as well as for PHBHV accumulation of *H. mediterranei*. Although large amount of yeast extracts (YE) and oxygen were also used in this PHBHV production (the mass ratio of carbon material and YE was kept at 1:1.7 within 118 h of culture) [38], which might raise the cost for PHBHV production, the PHBHV concentration of 77.8 g/L and volumetric productivity of 0.659 g/(Lh) are likely the highest level of PHBHV biosynthesis for *H. mediterranei* reported to date (Table **3**). It would be interesting to investigate the factors (specific carbon sources, vitamins, or inorganic ions, *etc.*) in ERB that are likely important for high cell-density cultivation of *H. mediterranei* for PHBHV production.

Whey Hydrolysates from Dairy Industry

Whey is a major by-product in the dairy industry from cheese and casein production, but is also a cheap carbon source as it contains 81% of the original lactose in milk. To address the waste problem of whey and to develop a novel strategy for economical PHBHV production, Koller *et al.* first investigated the potential of *H. mediterranei* as a producer of PHBHV from whey lactose [39]. The whey lactose was hydrolyzed by β-galactosidase to generate glucose and galactose for PHBHV production, as lactose is not a favorable carbon source for *H. mediterranei.* Among various archaeal and eubacterial strains as industrial PHA producers from whey, Koller *et al.* found that *H. mediterranei* is the most promising candidate considering all of the fermentation conditions, kinetic data in PHBHV production as well as the characteristics of the produced biopolymers. *H. mediterranei* can accumulate 50 wt.-% of PHBHV from hydrolyzed whey without addition of 3HV precursors but with a low volumetric productivity of 0.05 g/(Lh) [39]. When *H. mediterranei* was cultivated in a 42-liter bioreactor and the medium was supplemented with 10 g/L of hydrolyzed whey and 5 g/L of yeast extract, the PHBHV content can reach 72.8 wt.-%, and productivity can reach 0.09 g/(Lh) [40]. Recently, chemically hydrolyzed cheese whey for the cultivation and PHBHV production of *H. mediterranei* was also reported [41]. The acid (HCl) hydrolysis provided a less expensive method to obtain cheese whey hydrolysates. Batch bioreactor cultivation of *H. mediterranei* with this chemically hydrolyzed cheese sugars and supplementation of micronutrients resulted in the production of PHBHV content of 53 wt.-% and a volumetric productivity of 0.16 g/(Lh) [41]. Although the volumetric productivity is still relatively low (Table **3**), successful production of PHA from whey hydrolysates makes it possible to integrate the industrial scale production of PHBHV into the production lines of large dairies, which provides a solution for both the waste whey treatment of dairy industry and an economically competitive strategy for PHBHV production [39, 42].

Table 3. Selected reports on PHBHV production by *H. mediterranei* from low cost carbon sources.

Carbon Source	PHBHV [g/L]	PHBHV Content* [wt.-%]	PHBHV Volumetric Productivity [g/(Lh)]	3HV in PHBHV [mol-%]	Ref.
Extruded rice bran and corn starch (1:8 w/w), and Yeast extract	77.8	55.6	0.659	-	[38]
Enzymatically hydrolyzed whey	5.5-12.2	50-73	0.05-0.09	8-10	[39, 40]
Chemically hydrolyzed whey	7.92	53	0.16	1.5	[41]
Glycerol from biodiesel industry	13.4–16.2	75-76	0.12	10	[42]
Vinasse from the sugarcane-based ethanol industry	17.4-19.7	66-70	0.18-0.21	12.36-14.09	[43]
Stillage from the rice-based ethanol industry	16.4	71	0.17	15.4	[45]
Starch (with strain ES1)	21.28	48	0.296	10	[47]

*Because cell dry mass (CDM) were measured by different methods in different laboratories, the PHBHV content (PHBHV/CDM, wt.-%) is listed for reference but not for direct comparisons.

Vinasse and Stillage from Ethanol Industry

To utilize the sugarcane molasses-based distillery vinasse for commercial production of PHAs, the raw vinasse was pretreated with activated carbon and was used as a low cost carbon source for *H. mediterranei* PHBHV production [43]. Therefore, 25 vol.-% pre-treated vinasse replaced the glucose in the MST medium [24], leading to a maximum PHA concentration of 19.7 g/L with a volumetric productivity of 0.21 g/(Lh) [43]. The higher PHA productivity and low-cost carbon source, as well as other advantages of PHBHV production by *H. mediterranei* indicate its potential to integrate this process into ethanol manufacturing. However, a pre-treatment process for the removal of growth-inhibitory compounds in vinasse is cumbersome for the overall process, so Mukherjee's lab has also endeavored to utilize stillage derived from the rice-based ethanol industry as a free-of-cost raw material for the production of PHBHV by *H. mediterranei* [44, 45]. With simultaneous recovery and re-use of medium salts, 99.3% of the medium salts were recovered and re-used for PHBHV production. Technology and cost analysis (2.05 US-$/kg) demonstrate that PHBHV production by *H. mediterranei* integration with ethanol plants might be feasible in the near future [44, 45].

Surplus Glycerol from the Biodiesel Industry and Chitin

The crude glycerol phase from biodiesel production has also been used as carbon source for *H. mediterranei* PHBHV production, and a comparable volumetric productivity (0.12 g/(Lh)) is achieved [46]. A recent study demonstrated that *H. mediterranei* could also utilize colloidal or even powdered chitins for growth and PHBHV accumulation, and the gene cluster (HFX_5025-5039 on pHM300) encoding the chitin catabolism pathway has already been experimentally identified [47]. Chitin is the second most abundant natural polysaccharide after cellulose. Discovery of the capability of *H. mediterranei* for chitin utilization may provide an alternative method for producing PHBHV from cheap carbon sources.

ENGINEERING AND TAILOR-MADE PRODUCTION OF PHBHV

Genetic Engineering of *H. mediterranei* for PHA Production

A specific and highly efficient gene knock-out system (*H. mediterranei* DF50/pHFX), based upon the uracil auxotrophic host *H. mediterranei* DF50 (Δ*pyrF*) and the suicide plasmid vector pHFX (with *pyrF* as a selectable, as well as a counter-selectable marker), were developed in *H. mediterranei* in 2011 [16]. The effectiveness of this system has already been confirmed by knocking out dozens of genes involved in the PHBHV biosynthesis and other pathways in *H. mediterranei*. The principle of this gene knock-out system is based on homologous recombination through a double crossover process (Fig. **10**); thus, it can be easily adapted to knock-in desired genes into the genome when this gene is provided between the two homologous arms. Therefore, with knowledge of metabolism mechanism of *H. mediterranei* and the available genetic tools, bioengineering of this haloarchaeon for high-level or tailor-made production of PHBHV becomes more feasible than before.

In addition to the production of PHBHV, *H. mediterranei* is also a producer of exopolysaccharide (EPS) [5]. Previous research has found that EPS could cause increased viscosity of culture broth in liquid media and would also compete for the available exogenous carbon source; thus, EPS production might negatively impact the yield of PHBHV [5, 37, 48, 49]. Based on the structure of this EPS [50], we have successfully predicted the gene cluster involved in EPS biosynthesis in *H. mediterranei*. To overcome the problems caused by EPS in PHBHV production, we deleted the major region (HFX_2145 to HFX_2148) of this gene cluster responsible for the EPS biosynthesis in *H. mediterranei* DF50 [49] with the above gene knock-out system (Fig. **10**).

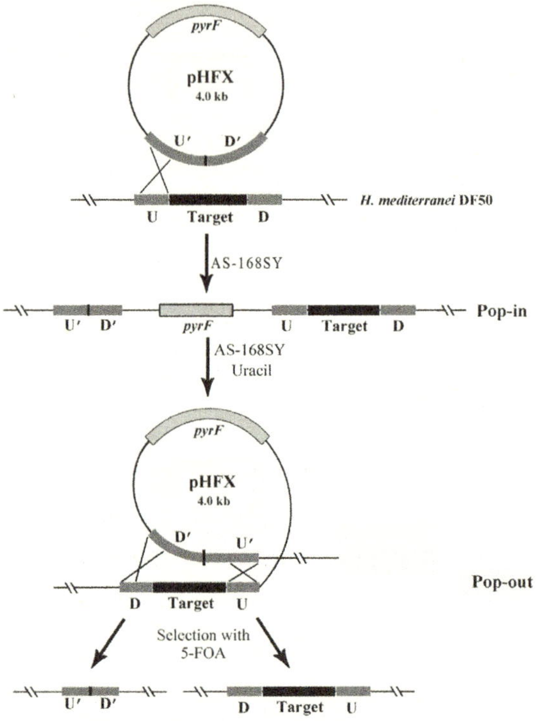

Fig. (10). The *pyrF*-based gene knockout system for *H. mediterranei*. For in-frame deletion of a target gene, the suicide plasmid pHFX carrying the flanking region of the target gene is transferred into *H. mediterranei* DF50. After positive selection of the single-crossover integration recombinants (pop-in) on AS-168SY medium without uracil, and counter-selection of the double-crossover *pyrF*-excised recombinants (pop-out) with 5-fluoroorotic acid (5-FOA), the target gene knockout mutants are confirmed by PCR or Southern blotting [16].

Deletion of this region of the EPS gene cluster completely blocked EPS biosynthesis in the engineered strain [49]. To make this engineered strain more applicable in the future large-scale production of PHBHV, we have also knocked-in the *pyrF* gene into the genome, generating a uracil-independent and EPS biosynthetic defective strain named *H. mediterranei* ES1 [49]. The colonies of *H. mediterranei* ES1 are rough and dry, while the wild-type strain remains smooth and wet on the agar plate, and the deficiency in EPS biosynthesis of the strain ES1 also remarkably decreased the viscosity of culture broth, which increased dissolved oxygen content and decreased the foaming propensity (Fig. **11**). Significantly, compared with the wild-type strain *H. mediterranei* ATCC 33500, the PHBHV productivity of *H. mediterranei* ES1, was enhanced by approximately 20% under a similar culture condition. These results indicate that the carbon sources used for synthesizing EPS are shifted to PHBHV production; thus, the *H. mediterranei* ES1 would be more favorable for future industrial production of PHBHV [49].

Fig. (11). Cultivation of the wide-type strain *H. mediterranei* ATCC 33500 (A) and the EPS-deleted strain *H. mediterranei* ES1 (B) in liquid media (in a 7 L fermenter) or on agar plates (see the inserted photo for each strain).

Production of Tailor-Made PHA

The genetically engineered strain of *H. mediterranei* ES1 was also used for production of tailor-made PHBHV. *H. mediterranei* has a complete β-oxidation cycle for short chain fatty acids and high efficiency in incorporation of 3HV-CoA into the PHBHV; thus, *H. mediterranei* ES1 was used to synthesize higher-order PHBHV (O-PHBHV) by co-feeding valeric acid with glucose [51]. ^{13}C NMR studies showed that the O-PHBHV1 (with 33.3 mol-% 3HV) copolymer consisted of short segments of PHB and PHV covalently linked together with random PHBHV segments. Consistently, two T_g were observed in the DSC curves of O-PHBHV1. In addition, the "blocky" feature and the high 3HV fraction of O-PHBHV1 have endowed this novel copolymer with some good or special material properties, including enhanced crystallinity percentages and improved Young's modulus (Fig. **12a**).

The O-PHBHV1 film showed unique foveolar cluster-like surface morphology with the highest roughness. Compared to the random copolymer PHBHV samples, it also exhibited increased platelet adhesion and accelerated blood clotting (Fig. **12b**). O-PHBHV1 with a novel microstructure and excellent hemostatic properties may endow this type copolymer with great potential in wound healing [51].

Notably, with γ-butyrolactone as the 4-hydroxybutyrate (4HB) precursor, *H. mediterranei* can biosynthesize a poly[(*R*)-3HB-*co*-(*R*)-3HV-*co*-4HB] (PHBHV4HB) terpolyester [40, 42]. In 2007, Koller *et al.* first tested PHBHV 4HB production by *H. mediterranei* from whey lactose by addition of valerate (3HV precursor) and γ-butyrolactone (4HB precursor) and obtained a novel terpolyester P-(3HB-*co*-21.8%-3HV-*co*-5.1%-4HB), providing a viable strategy to obtain different tailor-made PHAs from *H. mediterranei* [40]. With the crude glycerol phase from biodiesel production as the main carbon source for production of biomass, 3HB and 3HV, with γ-butyrolactone as the precursor for 4HB, a terpolyester poly[(*R*)-3HB-*co*-10%-(*R*)-3HV-*co*-5%-4HB) was also produced by *H. mediterranei.* The terpolyester displayed reduced melting temperature T_m (melting endotherms at 122 and 137 °C) and glass transition temperature T_g (2.5 °C). Molar mass increased (391 kDa) compared with the random PHBHV produced from either pure glycerol or crude glycerol phase [42]. Future work may be expected to test the application of these tailor-made PHAs, especially in the medical and pharmaceutical field, and improve the materials features if needed [40, 46].

A

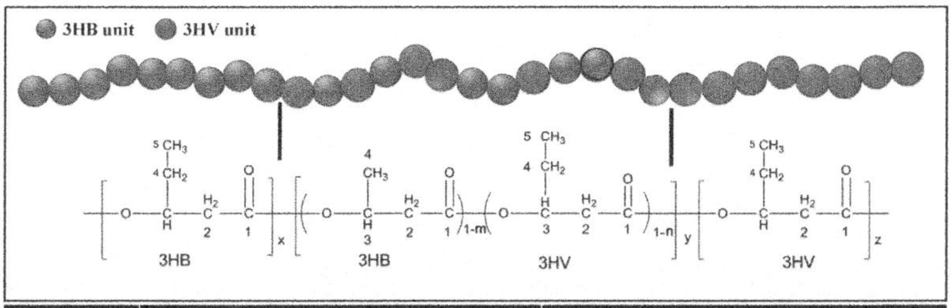

3HV	Molecular weights			Mechanical properties		
(mol%)	M_n 10^6	M_w 10^6	PDI (Mw/Mn)	δ_y (MPa)	E (Mpa)	ε_b (%)
33.3	1.23	2	1.63	12.6	1926.5	125.2
Thermal properties						
T_g (°C)	T_{cc} (°C)	T_m (°C)	W_{PHB} (wt%)	ΔH_m (J/g)	X_{PHB} (%)	T_d (°C)
-8.91, -0.17	64.65	141.3	63.24	36.51	24.91	265

B

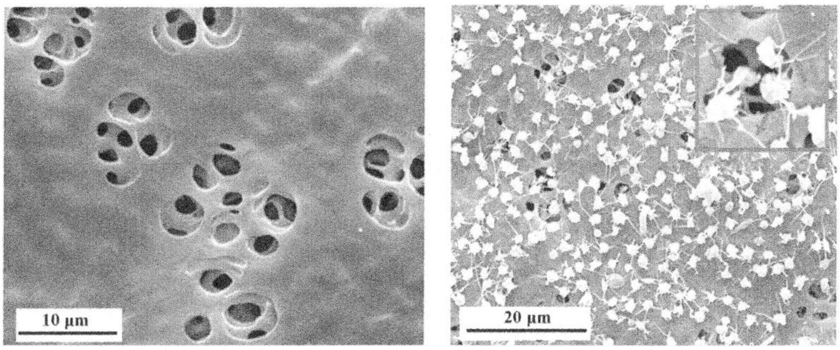

Fig. (12). (**A**) Illustration of the chemical structure of a higher-order copolymer PHBHV (O-PHBHV1) with its characterized material properties; (**B**) The microstructure of the O-PHBHV1 film (left) and the morphology of platelets adhering to the O-PHBHV1 films (right).

CONCLUDING REMARKS

We suggest that *H. mediterranei* is among the most important PHBHV producers, which holds promise for economical production of high-quality PHAs. Identifications of the key genes and pathways, as well as the regulation mechanism of PHBHV biosynthesis in *H. mediterranei*, have provided novel strategies in improving PHA productivity and/or for tailor-made production of higher-value PHAs. A comprehensive use of this available knowledge and other fermentation techniques, including pathway engineering and the optimization of the high-density fermentation with low-cost carbon sources, may result in the industrial-scale production of high-quality PHAs by *H. mediterranei* in the near future.

NOTES

[1]During the publication process of this chapter, this work has been published in Sci Rep (2016). See Liu G, Cai S, Hou J, Zhao D, Han J, Zhou J, Xiang H. Enoyl-CoA hydratase mediates polyhydroxyalkanoate mobilization in *Haloferax mediterranei*. Sci Rep 2016, 6: 24015. [hrrp://dx.doi.org/10.1038/srep24015][PMID: 27052994]

CONFLICT OF INTEREST

The author confirms that author has no conflict of interest to declare for this publication.

ACKNOWLEDGEMENTS

I would like to acknowledge Dr. Jing Han, Dahe Zhao, and Junyu Chen for their help in preparing the photos, proofreading the manuscript, and their discussions. I also appreciate the financial support from the National Natural Science Foundation of China (Grant No. 31330001) and the Hundred Talents Program of the Chinese Academy of Sciences (to H.X.).

REFERENCES

[1] Rodriguez-Valera F, Ruizberraquero F, Ramoscormenzana A. Isolation of extremely halophilic bacteria able to grow in defined inorganic media with single carbon-sources. J Gen Microbiol 1980; 119: 535-8.

[2] Rodriguez-Valera F, Juez G, Kushner DJ. *Halobacterium mediterranei* spec, nov., a New Carbohydrate-Utilizing Extreme Halophile. Syst Appl Microbiol 1983; 4(3): 369-81. [http://dx.doi.org/10.1016/S0723-2020(83)80021-6] [PMID: 23194735]

[3] Torreblanca M, Rodriguez-Valera F, Juez G, Ventosa A, Kamekura M, Kates M. Classification of non-alkaliphilic halobacteria based on numerical taxonomy and polar lipid-composition, and description of *Haloarcula* gen.nov. and *Haloferax* gen.nov. Syst Appl Microbiol 1986; 8(1-2): 89-99. [http://dx.doi.org/10.1016/S0723-2020(86)80155-2]

[4] Gupta RS, Naushad S, Baker S. Phylogenomic analyses and molecular signatures for the class *Halobacteria* and its two major clades: a proposal for division of the class *Halobacteria* into an emended order *Halobacteriales* and two new orders, *Haloferacales* ord. nov. and *Natrialbales* ord. nov., containing the novel families *Haloferacaceae* fam. nov. and *Natrialbaceae* fam. nov. Int J Syst Evol Microbiol 2015; 65(Pt 3): 1050-69.

[http://dx.doi.org/10.1099/ijs.0.070136-0] [PMID: 25428416]

[5] Antón J, Meseguer I, Rodríguez-Valera F. Production of an extracellular polysaccharide by *Haloferax mediterranei.* Appl Environ Microbiol 1988; 54(10): 2381-6. [PMID: 16347749]

[6] Mancinelli RL, Hochstein LI. The occurrence of denitrification in extremely halophilic bacteria. FEMS Microbiol Lett 1986; 35: 55-8. [http://dx.doi.org/10.1111/j.1574-6968.1986.tb01498.x] [PMID: 11542048]

[7] Rajagopalan R, Altekar W. Products of non-reductive CO_2 assimilation in the halophilic archaebacterium *Haloferax mediterranei.* Indian J Biochem Biophys 1991; 28(1): 65-7. [PMID: 1676015]

[8] Rajagopalan R, Altekar W. Characterisation and purification of ribulose-bisphosphate carboxylase from heterotrophically grown halophilic archaebacterium, *Haloferax mediterranei.* Eur J Biochem 1994; 221(2): 863-9. [http://dx.doi.org/10.1111/j.1432-1033.1994.tb18801.x] [PMID: 8174567]

[9] Han J, Hou J, Zhang F, *et al.* Multiple propionyl coenzyme A-supplying pathways for production of the bioplastic poly(3-hydroxybutyrate-co-3-hydroxyvalerate) in *Haloferax mediterranei.* Appl Environ Microbiol 2013; 79(9): 2922-31. [http://dx.doi.org/10.1128/AEM.03915-12] [PMID: 23435886]

[10] Rawal N, Kelkar SM, Altekar W. Ribulose 1,5-bisphosphate dependent CO_2 fixation in the halophilic archaebacterium, *Halobacterium mediterranei.* Biochem Biophys Res Commun 1988; 156(1): 451-6. [http://dx.doi.org/10.1016/S0006-291X(88)80862-3] [PMID: 3140815]

[11] Cheung J, Danna KJ, O'Connor EM, Price LB, Shand RF. Isolation, sequence, and expression of the gene encoding halocin H4, a bacteriocin from the halophilic archaeon *Haloferax mediterranei* R4. J Bacteriol 1997; 179(2): 548-51. [PMID: 8990311]

[12] Englert C, Horne M, Pfeifer F. Expression of the major gas vesicle protein gene in the halophilic archaebacterium *Haloferax mediterranei* is modulated by salt. Mol Gen Genet 1990; 222(2-3): 225-32. [http://dx.doi.org/10.1007/BF00633822] [PMID: 1703266]

[13] Englert C, Pfeifer F. Analysis of gas vesicle gene expression in *Haloferax mediterranei* reveals that GvpA and GvpC are both gas vesicle structural proteins. J Biol Chem 1993; 268(13): 9329-36. [PMID: 7683649]

[14] Fernandez-Castillo R, Rodriguez-Valera F, Gonzalez-Ramos J, Ruiz-Berraquero F. Accumulation of poly (beta-hydroxybutyrate) by Halobacteria. Appl Environ Microbiol 1986; 51(1): 214-6. [PMID: 16346972]

[15] Don TM, Chen CW, Chan TH. Preparation and characterization of poly(hydroxyalkanoate) from the fermentation of *Haloferax mediterranei.* J Biomater Sci Polym Ed 2006; 17(12): 1425-38. [http://dx.doi.org/10.1163/156856206778937208] [PMID: 17260512]

[16] Liu H, Han J, Liu X, Zhou J, Xiang H. Development of pyrF-based gene knockout systems for genome-wide manipulation of the archaea *Haloferax mediterranei* and *Haloarcula hispanica.* J Genet Genomics 2011; 38(6): 261-9.

[http://dx.doi.org/10.1016/j.jgg.2011.05.003] [PMID: 21703550]

[17] Han J, Zhang F, Hou J, *et al.* Complete genome sequence of the metabolically versatile halophilic archaeon *Haloferax mediterranei*, a poly(3-hydroxybutyrate-co-3-hydroxyvalerate) producer. J Bacteriol 2012; 194(16): 4463-4.
[http://dx.doi.org/10.1128/JB.00880-12] [PMID: 22843593]

[18] Yang H, Wu Z, Liu J, *et al.* Activation of a dormant replication origin is essential for *Haloferax mediterranei* lacking the primary origins. Nat Commun 2015; 6: 8321.
[http://dx.doi.org/10.1038/ncomms9321] [PMID: 26374389]

[19] Liu X, Miao D, Zhang F, Wu Z, Liu J, Xiang H. Characterization of the minimal replicon of pHM300 and independent copy number control of major and minor chromosomes of *Haloferax mediterranei.* FEMS Microbiol Lett 2013; 339(1): 66-74.
[http://dx.doi.org/10.1111/1574-6968.12052] [PMID: 23173581]

[20] Rehm BH. Polyester synthases: natural catalysts for plastics. Biochem J 2003; 376(Pt 1): 15-33.
[http://dx.doi.org/10.1042/bj20031254] [PMID: 12954080]

[21] Kirk RG, Ginzburg M. Ultrastructure of two species of *Halobacterium.* J Ultrastruct Res 1972; 41(1): 80-94.
[http://dx.doi.org/10.1016/S0022-5320(72)90040-8] [PMID: 4561659]

[22] Baliga NS, Bonneau R, Facciotti MT, *et al.* Genome sequence of *Haloarcula marismortui*: a halophilic archaeon from the Dead Sea. Genome Res 2004; 14(11): 2221-34.
[http://dx.doi.org/10.1101/gr.2700304] [PMID: 15520287]

[23] Han J, Lu Q, Zhou L, Zhou J, Xiang H. Molecular characterization of the $phaEC_{Hm}$ genes, required for biosynthesis of poly(3-hydroxybutyrate) in the extremely halophilic archaeon *Haloarcula marismortui.* Appl Environ Microbiol 2007; 73(19): 6058-65.
[http://dx.doi.org/10.1128/AEM.00953-07] [PMID: 17675423]

[24] Lu Q, Han J, Zhou L, Zhou J, Xiang H. Genetic and biochemical characterization of the poly(3-hydroxybutyrate-co-3-hydroxyvalerate) synthase in *Haloferax mediterranei.* J Bacteriol 2008; 190(12): 4173-80.
[http://dx.doi.org/10.1128/JB.00134-08] [PMID: 18408025]

[25] Han J, Hou J, Liu H, *et al.* Wide distribution among halophilic archaea of a novel polyhydroxyalkanoate synthase subtype with homology to bacterial type III synthases. Appl Environ Microbiol 2010; 76(23): 7811-9.
[http://dx.doi.org/10.1128/AEM.01117-10] [PMID: 20889776]

[26] Hou J, Feng B, Han J, *et al.* Haloarchaeal-type β-ketothiolases involved in Poly(3-hydroxybutyrate-co-3-hydroxyvalerate) synthesis in *Haloferax mediterranei.* Appl Environ Microbiol 2013; 79(17): 5104-11.
[http://dx.doi.org/10.1128/AEM.01370-13] [PMID: 23793631]

[27] Han J, Lu Q, Zhou L, Liu H, Xiang H. Identification of the polyhydroxyalkanoate (PHA)-specific acetoacetyl coenzyme A reductase among multiple FabG paralogs in *Haloarcula hispanica* and reconstruction of the PHA biosynthetic pathway in *Haloferax volcanii.* Appl Environ Microbiol 2009; 75(19): 6168-75.
[http://dx.doi.org/10.1128/AEM.00938-09] [PMID: 19648370]

[28] Feng B, Cai S, Han J, Liu H, Zhou J, Xiang H. [Identification of the phaB genes and analysis of the PHBV precursor supplying pathway in *Haloferax mediterranei*]. Wei Sheng Wu Xue Bao 2010; 50(10): 1305-12.
[PMID: 21141463]

[29] Cai S, Cai L, Liu H, *et al.* Identification of the haloarchaeal phasin (PhaP) that functions in polyhydroxyalkanoate accumulation and granule formation in *Haloferax mediterranei.* Appl Environ Microbiol 2012; 78(6): 1946-52.
[http://dx.doi.org/10.1128/AEM.07114-11] [PMID: 22247127]

[30] Cai S, Cai L, Zhao D, *et al.* A novel DNA-binding protein, PhaR, plays a central role in the regulation of polyhydroxyalkanoate accumulation and granule formation in the haloarchaeon *Haloferax mediterranei.* Appl Environ Microbiol 2015; 81(1): 373-85.
[http://dx.doi.org/10.1128/AEM.02878-14] [PMID: 25344243]

[31] Pötter M, Madkour MH, Mayer F, Steinbüchel A. Regulation of phasin expression and polyhydroxyalkanoate (PHA) granule formation in *Ralstonia eutropha* H16. Microbiology 2002; 148(Pt 8): 2413-26.
[http://dx.doi.org/10.1099/00221287-148-8-2413] [PMID: 12177335]

[32] Berg IA, Kockelkorn D, Buckel W, Fuchs G. A 3-hydroxypropionate/4-hydroxybutyrate autotrophic carbon dioxide assimilation pathway in Archaea. Science 2007; 318(5857): 1782-6.
[http://dx.doi.org/10.1126/science.1149976] [PMID: 18079405]

[33] Liu G, Hou J, Cai S, *et al.* A patatin-like protein associated with the polyhydroxyalkanoate (PHA) granules of *Haloferax mediterranei* acts as an efficient depolymerase in the degradation of native PHA. Appl Environ Microbiol 2015; 81(9): 3029-38.
[http://dx.doi.org/10.1128/AEM.04269-14] [PMID: 25710370]

[34] Andrews DL, Beames B, Summers MD, Park WD. Characterization of the lipid acyl hydrolase activity of the major potato (*Solanum tuberosum*) tuber protein, patatin, by cloning and abundant expression in a baculovirus vector. Biochem J 1988; 252(1): 199-206.
[http://dx.doi.org/10.1042/bj2520199] [PMID: 3048246]

[35] Hou J, Xiang H, Han J. Propionyl coenzyme A (propionyl-CoA) carboxylase in *Haloferax mediterranei*: Indispensability for propionyl-CoA assimilation and impacts on global metabolism. Appl Environ Microbiol 2015; 81(2): 794-804.
[http://dx.doi.org/10.1128/AEM.03167-14] [PMID: 25398867]

[36] Borjian F, Han J, Hou J, Xiang H, Berg IA. The methylaspartate cycle in haloarchaea and its possible role in carbon metabolism. ISME J 2016; 10(3): 546-7.
[http://dx.doi.org/10.1038/ismej.2015.132] [PMID: 26241502]

[37] Lillo JG, Rodriguez-Valera F. Effects of culture conditions on poly(beta-hydroxybutyric acid) production by *Haloferax mediterranei.* Appl Environ Microbiol 1990; 56(8): 2517-21.
[PMID: 16348261]

[38] Huang TY, Duan KJ, Huang SY, Chen CW. Production of polyhydroxyalkanoates from inexpensive extruded rice bran and starch by *Haloferax mediterranei.* J Ind Microbiol Biotechnol 2006; 33(8): 701-6.
[http://dx.doi.org/10.1007/s10295-006-0098-z] [PMID: 16491353]

[39] Koller M, Hesse P, Bona R, Kutschera C, Atlić A, Braunegg G. Potential of various archae- and eubacterial strains as industrial polyhydroxyalkanoate producers from whey. Macromol Biosci 2007; 7(2): 218-26.
[http://dx.doi.org/10.1002/mabi.200600211] [PMID: 17295410]

[40] Koller M, Hesse P, Bona R, Kutschera C, Atlic A, Braunegg G. Blosynthesis of high quality Polyhydroxyalkanoate co- and terpolyesters for potential medical application by the archaeon *Haloferax mediterranei.* Macromol Symp 2007; 253: 33-9.
[http://dx.doi.org/10.1002/masy.200750704]

[41] Pais J, Serafim LS, Freitas F, Reis MA. Conversion of cheese whey into poly(3-hydroxybutyrate-co-3-hydroxyvalerate) by *Haloferax mediterranei.* N Biotechnol 2016; 33(1): 224-30.
[http://dx.doi.org/10.1016/j.nbt.2015.06.001] [PMID: 26134839]

[42] Koller M, Salerno A, Muhr A, *et al.* Whey lactose as a raw material for microbial production of biodegradable polyester. In: Hosam El-Din MS, Ed. Polyester. InTech, China 2012; pp. 19-60.
[http://dx.doi.org/10.5772/48737]

[43] Bhattacharyya A, Pramanik A, Maji SK, Haldar S, Mukhopadhyay UK, Mukherjee J. Utilization of vinasse for production of poly-3-(hydroxybutyrate-co-hydroxyvalerate) by *Haloferax mediterranei.* AMB Express 2012; 2(1): 34.
[http://dx.doi.org/10.1186/2191-0855-2-34] [PMID: 22776040]

[44] Bhattacharyya A, Jana K, Haldar S, *et al.* Integration of poly-3-(hydroxybutyrate-co-hydroxyvalerate) production by *Haloferax mediterranei* through utilization of stillage from rice-based ethanol manufacture in India and its techno-economic analysis. World J Microbiol Biotechnol 2015; 31(5): 717-27.
[http://dx.doi.org/10.1007/s11274-015-1823-4] [PMID: 25690843]

[45] Bhattacharyya A, Saha J, Haldar S, Bhowmic A, Mukhopadhyay UK, Mukherjee J. Production of poly-3-(hydroxybutyrate-co-hydroxyvalerate) by *Haloferax mediterranei* using rice-based ethanol stillage with simultaneous recovery and re-use of medium salts. Extremophiles 2014; 18(2): 463-70.
[http://dx.doi.org/10.1007/s00792-013-0622-9] [PMID: 24442255]

[46] Hermann-Krauss C, Koller M, Muhr A, Fasl H, Stelzer F, Braunegg G. Archaeal production of polyhydroxyalkanoate (PHA) co- and terpolyesters from biodiesel industry-derived by-products. Archaea 2013.

[47] Hou J, Han J, Cai L, *et al.* Characterization of genes for chitin catabolism in *Haloferax mediterranei.* Appl Microbiol Biotechnol 2014; 98(3): 1185-94.
[http://dx.doi.org/10.1007/s00253-013-4969-8] [PMID: 23674154]

[48] Koller M, Chiellini E, Braunegg G. Study on the production and re-use of poly (3-hydroxybutyrate-*co*-3-hydroxyvalerate) and extracellular polysaccharide by the archaeon *Haloferax mediterranei* strain DSM 1411. Chem Biochem Eng Q 2015; 29(2): 87-98.
[http://dx.doi.org/10.15255/CABEQ.2014.2058]

[49] Zhao D, Cai L, Wu J, *et al.* Improving polyhydroxyalkanoate production by knocking out the genes involved in exopolysaccharide biosynthesis in *Haloferax mediterranei.* Appl Microbiol Biotechnol 2013; 97(7): 3027-36.
[http://dx.doi.org/10.1007/s00253-012-4415-3] [PMID: 23015099]

[50] Parolis H, Parolis LA, Boán IF, *et al.* The structure of the exopolysaccharide produced by the halophilic Archaeon *Haloferax mediterranei* strain R4 (ATCC 33500). Carbohydr Res 1996; 295: 147-56.
[http://dx.doi.org/10.1016/S0008-6215(96)90134-2] [PMID: 9002190]

[51] Han J, Wu LP, Hou J, Zhao D, Xiang H. Biosynthesis, characterization, and hemostasis potential of tailor-made poly(3-hydroxybutyrate-*co*-3-hydroxyvalerate) produced by *Haloferax mediterranei.* Biomacromolecules 2015; 16(2): 578-88.
[http://dx.doi.org/10.1021/bm5016267] [PMID: 25559462]

CHAPTER 7

Bacterial Genetic Modifications for Improving Polyhydroxyalkanoates Production from Inexpensive Carbon Sources

Sergio Casella*, **Lorenzo Favaro** and **Marina Basaglia**

Department of Agronomy Food Natural resources Animals and Environment, University of Padova, AGRIPOLIS, Italy

Abstract: There are several phases of the PHA production chain where genetic tools may help, starting from the microbial strain isolation and characterization, through all the production steps, until the downstream processes.

Leaving aside the huge number of genetic modifications accomplished on many bacterial species to clarify basic biochemical, genetic and metabolic aspects of PHA metabolism, several interesting attempts have been reported aimed at improving the performance of the microorganisms showing potentiality for a possible production process.

Most of the available information deals with the use of genetic tools for (i) characterization and identification of new isolates, (ii) testing new isolates for the presence of relevant genes involved in PHA production/degradation, (iii) metagenomic approach in complex environments for searching relevant genes (and strains), (iv) assessment of PHA genes expression and control of depolymerase genes, (v) metabolic engineering of producing and not producing bacterial strains, (vi) improving downstream processes, (vii) making PHA producing strains able to utilize different carbon sources (simple sugars, fats, starch-lignocellulose, other complex substrates), (viii) cloning PHA genes in other recipient bacteria.

Due to the huge amount of results published in each of the above fields, after a general overview on the methods used to search for new PHA producing microbial strains, this

* **Corresponding author Sergio Casella:** Department of Agronomy Food Natural resources Animals and Environment, Università degli Studi di Padova, Legnaro (Padova), Italy; Tel: +39 049 827 2922; Email: sergio.casella@unipd.it.

Martin Koller (Ed.)

chapter will try to summarize the most relevant results obtained through genetic engineering tools for the production of PHAs from cheap carbon sources in view of possible industrial applications.

Keywords: Bacteria, Cloning PHA genes, Enzymes, *Escherichia coli*, Genes, Genetic engineering, Genetic modification, Inexpensive carbon sources, Inexpensive substrates, Organic by-products, PHA production, Polyhydroxyalkanoates (PHA), Unrelated carbon sources.

TRADITIONAL AND DNA-BASED METHODS TO LOOK FOR PHA PRODUCERS

The accumulation of polyhydroxyalkanoates (PHAs) as energy and carbon storage is a widespread trait among several dozens of Eubacteria and Archaea [1]. PHA can be obtained starting from various in-excess carbon sources; *Cupriavidus necator* (formerly known as *Wautersia eutropha, Ralstonia eutropha* and *Alcaligenes eutrophus,* see below) is the most widely studied microorganism for the production of PHA but other strains, belonging to species such as *Bacillus cereus* and *B. megaterium, Sinorhizobium meliloti, Azotobacter chroococcum*, *Pseudomonas putida* or *Haloferax mediterranei*, were also evaluated for PHA production from various and low-cost carbon sources such as sugars (glucose, maltose, lactose, xylose *etc.*), alkanes (*e.g.* exane, octane *etc.*), alkanoic acids (*e.g.*, acetic, propionic, butyric, valeric, oleic acids *etc.*), alcohols (*e.g.*, ethanol, methanol, glycerol *etc.*) or gases (carbon dioxide and methane) [2 - 5].

Most of the PHA producing strains have been deposited in culture collections such as ATCC or DSMZ, making them commercially available. An exhaustive review of PHA producing microorganisms in culture collections have been recently proposed [6]. Usually genetic and biochemistry of deposited strains in terms of carbon utilization and PHA accumulation, are well known and documented in scientific papers. These knowledges greatly facilitated a quick start-up of research and the following industrial production.

Unfortunately, although the ability to synthesize PHAs is a widespread trait among microorganisms, it is true that not all of these bacteria are suitable for an industrial production due to their lacking of robustness and inability to survive to the highly stressing fermentation conditions [7]. Although to solve these problems a series of optimization strategies have been attempted, such as developing better bacterial strains or supplying renewable low-cost carbon substrates, it is undoubting that the isolation and identification of suitable PHAs producing microorganisms from the environment are key factors for PHA-production processes. Therefore, intensive research was conducted aimed to discover, characterize and improve newly isolated bacterial strains, able to produce huge amount of PHAs from different and unrelated carbon sources, to survive in environments extreme for temperature, pH-value or salinity, and to convert residual and low-cost carbon sources.

Testing New Isolates for the Ability to Produce PHA

Traditional Methods

The most common traditional technique available for analysis of PHAs in bacterial cells is gas chromatography (GC) that has been proposed by Braunegg and colleagues 40 years ago [8] and later refined [9]. The GC method involves hydrolysis and subsequent methanolysis or propanolysis of the PHAs in whole cells in the presence of sulfuric acid and chloroform. Although this method is time-consuming, involves extensive use of solvents and does not enable to test large number of strains in short time, GC analyses still remain the final confirmation technique but cannot be considered a screening methods.

The utilization of phase-contrast light microscope allows to appreciate highly refractive PHA granules directly inside bacteria in a fresh culture, even if it is often difficult to distinguish PHA from other intracellular inclusions such as spores or glycogen. The application of lipophilic dyes such as Sudan black B, Nile blue A, and Nile red is today the "first line screening" of PHA containing bacteria for its easiness: these colorants show in fact a high affinity for lipophilic materials and therefore can stain hydrophobic PHA [10 - 14].

The oxazine dye Nile blue A and its fluorescent oxazone form, Nile red, can be directly included in the medium and used as a simple, specific and highly sensitive staining method to detect poly(3-hydroxybutyric acid) and other polyhydroxyalkanoic acids in growing bacterial colonies, into the culture broth or in smears using a confocal laser scanning or epifluorescence microscopy. Other inclusions, such as glycogen or poly-P do not stain with this dye [15].

Staining techniques were not only applied to bacteria such as *A. vinelandii, Escherichia coli, P. putida* or *R. eutropha* to distinguish producing from non-producing strains [16] but they also allow the i) estimation of the presence of PHAs in viable colonies or cells at any time during the growth, ii) discrimination between PHA-negative and PHA-positive strains and even iii) investigation of cultural conditions useful to enhance PHA production [17].

The fluorescent staining, coupled with flow cytometric determination, also permits PHA quantification in culture [18] if fluorescence intensity of Nile Red stained cells is measured using a spectrofluorometer or a laser flow cytometer. More specifically, high correlation between fluorescence intensity and PHA concentration obtained with the "traditional" GC-method developed by Braunegg has been reported.

Poly-3-hydroxybutyrate, PHB contained in bacterial cells can be also appraised by the infrared spectra, demonstrating the potential of FTIR spectroscopy as a rapid and alternative method [19].

DNA-Based Methods

The phenotypic detection systems, shortly described above, need cultivation of microorganisms on selected carbon sources to raise PHA production. Unfortunately, although PHAs can be occasionally found in some bacteria in normal growth conditions, they are usually produced in specific unbalanced nutritional requirements [20].

For this reason, the phenotypic reported methods are prone to errors: for instance, other non-PHA lipid storage compounds such as triacylglycerol, can accumulate inside the cells giving false positive signals, while false negatives can be caused by unfavorable PHA accumulation conditions.

Molecular biology revolution, starting during 1970s, provided new tools for biological research and the development of molecular techniques to investigate the presence/absence of specific genes. In this way, PHA synthases, key enzymes in PHA synthesis encoded by *phaC* genes, immediately attracted the attention of scientific community because the potentiality to accumulate PHA can be highlighted simply by unveiling *phaC* possessing microorganisms even in unknown growing conditions.

PHA synthases catalyze the polymerization of activated HAs and are grouped in four main classes (I–IV), with different substrate specificity (activated *scl*-PHA or *mcl*-HAs building blocks, respectively) [21]. The type of PHA produced and accumulated by bacteria relies on the metabolic pathway of the microorganism.

The first proposed molecular technique to select microbial potential is Southern Blot Hybridization (SBH) that has been successfully applied as identification method by Timms and coworkers [22]. These authors designed a specific general probe for *pha* genes, based on the conserved regions of PHA synthases in *Wauthersia eutropha* (*C. necator*) and *P. oleovorans*. This probe revealed efficient in detecting class I and class II *pha* genes from various microorganisms such as *P. mendocina* and *P. citronellonis*, but it was not able to disclose classes III and IV *pha* genes.

Unfortunately, SBH is laborious and time consuming. After developing Polymerase chain reaction (PCR) technique in the last 25 years, PCR-based molecular methods became highly sensitive and precise tools for detecting and amplifying specific genes. Nowadays, PHA producing microorganisms can be quickly revealed by amplification of *phaC* genes, coding for PHA synthases typically present in PHA-accumulating, but lacking in non-PHA-accumulating microorganisms. With this purpose, specific primers have been developed allowing to screen a broad range of both short-chain-length (*scl*)- and medium-chain-length (*mcl*-) PHA producing microorganisms [23].

PCR-based PHA synthase detection methods, are powerful first-line screening to identify potential PHA producing organisms allowing to screen large number of newly isolated strains or to unveil potential PHA producing microorganisms in which polymer accumulation is under unknown regulation conditions, and thus staining methods are not applicable for a rapid evaluation.

In 1997 the use of PCR to identify *phaC*-containing organisms in aquatic environment was firstly proposed [24]. Unfortunately, due to the fact that the nondegenerate primers used in the cited work derived solely from the *W. eutropha* (*C. necator*) *pha* gene sequence, the Authors obtained a nonspecific PCR amplification mixture constraining to confirm the *phaC* gene presence by a following Southern blot hybridization.

Starting from *C. necator* class I and *P. putida* class II synthases gene sequences and combining the direct colony PCR and semi-nested PCR techniques, Sheu and colleagues obtained a rapid, broad-specificity and reliable protocol suitable to screen large numbers of environmental isolates belonging to both Gram-positive and Gram-negative groups, with no further culturing or chromosomal DNA extraction procedures [25]. These results were partially confirmed by visualizing PHA granules by Nile Red staining in most of the *pha*-positive isolates.

In the same year, Solaiman *et al.* [26] developed a rapid PCR-based assay for the specific detection of genes coding for type II *mcl*-PHA synthases. The primers used in this research were developed on the basis of the highly conserved sequences in the coding regions of *Pseudomonas phaC1* and *phaC2*. Revelo Romo and colleagues designed three degenerate primers to identify fragments from both class I (*phaC*) and class II (*phaC1* and *phaC2*) synthase genes and selected potentially PHA-accumulating Gram-negative bacteria of different genera by simple colony polymerase chain reaction [27]. Class III PHA synthase genes were also detected using primers for highly conserved regions of *phaC* synthase genes of sulfate-reducing bacteria such as *Desulfococcus multivorans* [28], while Class IV synthase presence was elucidated by McCool *et al.* [29]. These authors discovered that the novel PHA synthase from *B. megaterium* requires PhaC and PhaR for its activity both *in vivo* and *in vitro*. This class IV PhaC showed greatest similarity to the PhaC of class III in both size and sequence.

Molecular and Phenotypic Classification: The Case of C. necator

Historically, identification and classification of eubacteria were based on phenotypic characteristics. Although phenotypic identification of bacteria by biochemical tests (fermentation of sugars, substrate utilization *etc.*) is still relevant, molecular techniques, mainly based on DNA-DNA hybridization and 16S rDNA sequencing, completely transformed bacterial taxonomy. *Cupriavidus necator,* a non-obligate bacterial predator of bacteria in soil and one of the most studied PHA producers, can be considered as a case-study in this sense. Indeed, in the first half of the 20th century, this bacterium was firstly isolated and classified on the basis of some phenotypic assays as *Hydrogenomonas eutrophus.* Studying cell morphology, metabolism and GC content, it was later renamed *A. eutrophus* [30] and on the basis of lipid composition, fatty acid composition and 16S rRNA analysis, renamed again as *R. eutropha* [31]. Upon further study a new genus *Wautersia* was created and *R. eutropha* was renamed *W. eutropha* [32].

DNA-DNA hybridization experiments, extensive biochemical characterization, a deeper evaluation of other phenotypic characteristics, DNA base ratios and 16S rRNA gene sequences demonstrated that *W. eutropha*, the type species of the genus *Wautersia*, is synonym of *Cupriavidus necator* [33].

This story demonstrates that the advent of molecular biology with the development and application of prompt, accurate and convenient methods for identification of microorganisms brought a complete revolution in their classification. New tools for studying macromolecules allowed biologist to figure out relationships between organisms, based not only on their appearance, but also on the sequences of specific subunits of proteins and nucleic acids.

The sequencing of the 16S rRNA and some other genes considered to be phylogenetic markers, is a universal and reliable approach for taxonomic characterization of microorganisms and identification of new isolates. This is particularly important for microorganisms that might be used in large scale fermentative process as PHA production where unequivocal species/strain identification is essential in avoiding the use of possible pathogens.

The Metagenomic Perspective

The "molecular approach" deeply revolutionized the classical method of discovering enzymes endowed with new properties or molecules with potential biotechnological applications.

The development of molecular techniques to investigate the ecology of microbial communities has provided the scientists with a vast array of new tools to deeply investigate the microflora of not-yet-explored environments. The "traditional approach" involves the isolation, the culturing and the growing of microorganisms; unfortunately, the great majority of the microbial diversity still remains unexploited, since the microorganisms found in natural environments have peculiar growth requirements and often cannot be cultured under laboratory conditions for many reasons, including incorrect nutrients or combinations of temperature, pressure, or atmospheric gas composition, accumulation of toxic waste products or intrinsically slow growth rate [34].

In a metagenomic approach, DNA is directly isolated from the environment, cloned into a suitable vector and transferred into a host bacterium, thus obtaining a gene library. The resulting clones can be screened for phylogenetic markers, such as 16S rRNA and *recA*, or for other conserved genes, by hybridization or PCR, or for the expression of specific traits, such as enzyme activity or antibiotic production [35].

Metagenomic libraries has been fruitful in identifying novel genes/enzymes [36, 37] like lipases and amylases [38] with enhanced features and industrial potential, compared to the enzymes from the culturable microorganisms, or in discovering medically active compounds [39].

Metagenomics coupled with a PCR based detection methods has been recently proposed to detect PHA synthase genes in a fosmid metagenomic library derived from an oil-contaminated soil [40]. The Authors successfully isolated a *phaC* gene which enabled, once expressed in a *phaC*-deficient *P. putida*, the production of copolymers with up to 21% in *mcl*-fraction. Moreover, they hypothesized that 3.4% of microorganisms in the soil sample contained *phaC* synthase genes.

This data also demonstrated that the metagenomics and PCR sequence-based approach is a useful tool for the isolation of novel *phaC* genes from the environment: although the primers were designed on the basis of the sequences of known *phaC* synthases, it was possible to identify five *phaC* genes present in bacteria that have not been characterized so far.

Metagenomics, when combined with phenotypic screening, also allowed to isolate novel PHA synthesis genes that would not be predicted through sequencing [41].

Evaluation of mixed microbial cultures (MMCs). Molecular approach coupled with PCR amplification methods are used throughout operating period when population dynamic needs to be evaluated. This is the case of mixed microbial cultures (MMCs) that gained attention as a cost-reducing strategy, when compared to pure cultures [42]. Actually the commercial production of PHB currently employs genetically modified or wild type bacteria in pure culture that require sterile growth conditions and well-defined nutrient media. Unluckily, the pure culture production of PHB has the disadvantages of the high costs for both the pure substrates and the need of sterilization for production process in axenic conditions.

For these reasons, the production of PHA by the use of MMCs, such as activated sludge from wastewater treatment plants and the simultaneous treatment of waste streams, was recently investigated by many authors [43 - 45].

PHA production using MMCs take place in non-sterile bioreactors and natural selection of the microbial consortia for PHA accumulation is realized by alternating periods with substrate (feast phase) and without substrate (famine phase) in an aerobic sequencing batch reactor. Using such natural microbial consortia under no sterile conditions, it was shown that up to 80% of the dry biomass formed in the process can be recovered as PHA [43].

The characterization and quantification of species involved in the microbial consortia working in the bioreactors and their dynamics together with PHA determination, is an important aspect in fine tuning of the process. Among the possible methods to qualify the microbial consortia, the molecular ones are of particular importance since the material (samples) for analysis can be directly extracted from the microorganisms and they do not require specific culturing. As an example, starting from an aerobic activated sludge alternating feast and famine periods, a PHA producing mixed culture was obtained [46]. They used Denaturing Gradient Gel Electrophoresis (DGGE) analysis, 16S rDNA clone library screening and Fluorescent *In Situ* Hybridization (FISH) to study the microbial diversity of a culture-enriched bioreactor, and found one dominating organism, member of the Gammaproteobacteria. The presence of PHA inside the cells was confirmed by the lipophilic stain Nile Blue A. Moreover, they isolated the dominant organism from the mixed culture for further characterization and applications.

MAKING PHA PRODUCING STRAINS ABLE TO UTILIZE DIFFERENT CARBON SOURCES

Successful industrial scale production of PHAs is dependent on several crucial factors, which include the cost of substrate, microbial PHA content and yield, downstream processing, and process development [47]. Most importantly, as the process is scaled-up, the carbon source could account for up to 50% of the total cost [48]. Therefore, the economic feasibility of bulk PHA production is intrinsically coupled to developing efficient biotechnological processes from inexpensive carbon sources. Additionally, the use of industrial by-products as carbon sources takes the advantage of simultaneously enabling a decrease in disposal costs and the production of value-added products.

A broad range of waste and by-product streams associated with agriculture at different stages of the production/utilization can be identified as suitable feedstock for the biotechnological production of PHAs [1, 49]. To be considered as proper substrate, the residue has to support good growth of the aerobic PHA accumulating microorganisms.

In addition, the availability of the waste, both in terms of quantity and timeframe, has to be carefully assessed as well as the possible integration of the PHA producing plant at the site where the by-product is generated.

Other essential aspects towards the large scale and cost-effective production of PHA include nutrient management and recovery (nitrogen and phosphorus sources), as well as exploitation of cellulosic and other complex substrates that are not usual carbon source by the most known PHA producing microbes. All together, these challenges called for research approaches aimed to achieve high titer PHA production in microbial strains from cheap residual materials.

In principle, such microbes could be achieved or developed mainly by three strategies: 1) the isolation, selection and characterization of new natural microbial strains capable of efficiently converting the selected waste into PHAs or 2) the engineering of microorganisms that exhibit high PHAs product yields and titer for the ability to use waste substrates or 3) the engineering of bacteria which can grow on various carbon sources, simple and/or complex, but are unable to produce PHAs.

Although the first approach is fascinating, no wild type strain has been so far reported for high PHA yields and production from low-cost waste streams meanwhile the other two research options, though challenging and time consuming as well, are considered to be the most suitable to achieve the efficient and large scale conversion of abundant and variable by-products into bio-polymers (Fig. **1**).

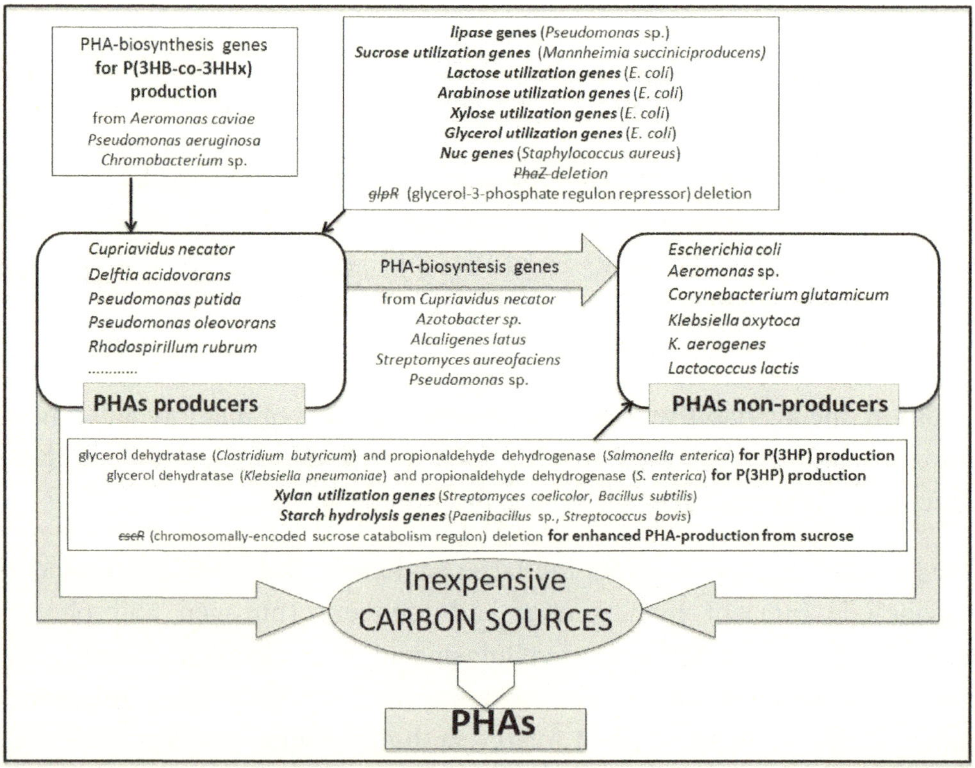

Fig. (1). Overview of the main strategies adopted to improve PHAs production from inexpensive carbon sources by gene transfer from and to PHAs producing and non producing bacteria.

This section gives an overview of the most recent and valuable researches tailored to the PHA production from organic by-products focusing mainly on the second strategy where genetic engineering approaches were undertaken to increase the overall process yield. The following section, entitled "Cloning PHA genes in non-producing bacteria", will address the most notable outcomes on the third research approach.

PHAs from Whey

Many authors have investigated the use of whey as a carbon source for PHA production. It is an attractive substrate as the main by-product from the dairy and cheese industries where about 90% of the volume of handled milk is regarded as a waste. The amount of whey globally produced is nearly 120 million tons per year of which 66% is generated in EU and 25% in North America [50, 51]. About half of the whey is converted into useful products such as human and animal feed while the rest is treated as waste. This disposal is a waste management problem as whey is an environmental hazard due to its high biochemical oxygen demand (almost 60,000 ppm) and chemical oxygen demand (COD: up to 80,000 ppm) [52].

The main components in weight percent of fresh whey are: 94% water, 4.5% lactose, <1% protein, <1% ash, and <1% fat. Lactose accounts for the most of biochemical oxygen demand, so, many studies have aimed at recovering this sugar and finding suitable uses for it [53, 54]. Although the higher-grade lactose obtained from whey is used in infant formulas and pharmaceutical products, the amount of purified lactose required for these applications is only 5-10% of the available whey [55].

This material is a promising feedstock for the biotechnological production of biofuels, added-value compounds as well as PHAs [56 - 59]. Nevertheless, only a rather restricted number of prokaryotic wild-type microorganisms have been described so far for the direct convert of lactose into PHAs. Therefore, one of the most used approach is the hydrolysis of lactose into glucose and galactose, which are readily and efficiently metabolized and converted to PHA. The glycosidic bond can be enzymatically hydrolyzed by β-galactosidase or acid hydrolysis.

The principal possibility of direct conversion of whey lactose towards PHA using different wild type bacteria was first investigated on shaking flask scale. Yellore and Desai isolated a *Methylobacterium* sp. strain from a local pond in India. This organism grew on pure lactose and, on shaking flask scale, produced PHB at a final concentration of 3.1 g/L polymer, corresponding to a PHA content in biomass of about 59%. Using whey as substrate and optimizing the nitrogen supply, PHB levels reached 2.6 g/L, with 44% cell dry mass (CDM) [60]. These findings were extended later on bioreactor scale from cheese whey: in batch mode, final PHB concentrations of 2.1 g/L with 67% of PHB in biomass were reported [61].

Povolo and Casella described the production of PHA from lactose by *Paracoccus denitrificans*, *Sinorhizobium meliloti* and *Hydrogenophaga pseudoflava*. The latter two strains were also able to produce the polymer directly from cheese whey permeate and especially *H. pseudoflava* turned out to be a promising candidate for PHA production from whey [62].

The potential of various microorganisms to convert whey lactose to PHAs has been studied also with the halophilic archaeon *H. mediterranei* as well as two eubacterial strains (*P. hydrogenovora* and *H. pseudoflava*). *H. mediterranei* accumulated 50% of dry cell mass of poly-3-(hydroxybutyrate-*co*-hydroxyvalerate (PHBHV) from hydrolyzed whey, while *P. hydrogenovora* accumulated PHB to 12%. With *H. pseudoflava*, it was possible to reach 40% PHBHV in CDM using non-hydrolyzed whey lactose and valeric acid as 3HV precursor [63, 64]. Recently, Gram-positive *B. megaterium* was also described for the conversion of cheese whey to PHB achieving 1.5 g/L of polymer [65].

Engineering of PHA producing strains was pursued in order to increase both total PHA concentration and yield from whey. Mainly, *C. necator*, a well-known PHA producer unable to split lactose into glucose and galactose, has been interested.

Firstly, in 1990, Pries and colleagues obtained recombinant lactose- and galactose-utilizing strains of *A. eutrophus* (*C. necator*) and *P. saccharophila,* capable of accumulating low amounts of PHB [66]. More recently, a highly proficient *C. necator* strain, harboring the *lacZ*, *lacI* and *lacO* genes of *E. coli*, was developed for lactose conversion [51]. The engineered *E. coli* was able to produce the polymer directly from lactose and from whey permeate. In addition, the insertion of the *lac* operon within *phaZ* gene reduced the amount of PHA depolymerized by the cell, thus improving the final polymer yield. As a result, the engineered strain demonstrated the one-step conversion of cheese whey to biopolymer, producing, on shaking flask scale, 1.4 g/L of PHB with 22% PHB in CDM.

PHAs from Molasses

Molasses, the most valuable by-product from the sugar industry, is residual syrup from which no more crystalline sucrose can be acquired by simple techniques. Its composition varies depending on the source (cane or beet) and type of refining (sulfured or unsulfured). The sugar content is about 50% (mainly sucrose with less amount of glucose and fructose) meanwhile proteins account for 4%, with low concentrations of trace elements (calcium, magnesium, potassium, and iron) and vitamins [67]. It is estimated that from 100 tons of processed cane, 10 tons of sucrose, and 4 tons of molasses is extracted. Molasses is considered as a low-value product that is used in specialized yeast fermentations (production of citric acid, glutamate, and acetone, as soil fertilizer, cattle feed supplement, flavoring agent in some foods, or as a feedstock for bioethanol [68, 69].

Global production of molasses reaches almost 50 million tons per year. Although commercially important, only about 15% of total molasses produced is traded internationally both due to considerable price fluctuations and the fact that it is a difficult product to store and transport. In 2007, 80% of the molasses sold was captured by the bioethanol industry [70]. Despite this, excess of molasses as a residue from sugar processing is still occurring resulting in waste management costs. Thus, the use of such by-product as carbon feedstock for PHA production is a good alternative to its disposal and many efforts have been employed for its conversion to PHA [71 - 73].

The early investigation of molasses for PHA was undertaken using *A. vinelandii* with a production of 2.5 g/L PHB on a shaking flask scale [74]. In a 10-L bioreactor, the production of PHB from green syrup (another intermediate from sucrose production) and sugar beet molasses was reported by two strains of *A. latus* with nearly 3.6 g/L PHB from both substrates [75]. Solaiman and colleagues [76] successfully utilized soy molasses for the production of *mcl*-PHA using *P. corrugata.* From 5% (wt./vol.) soy molasses, the Authors reported the production of poly(3-hydroxydodecanoate-*co*-3-hydroxyoctanoate-*co*-3-hydroxytetradeceno-ate), P(3HDD-3HO-3HTDE), with the highest concentration found to be 0.57 g/L with 17% PHA in CDM.

Cane molasses was found to be an excellent substrate for the growth of *Bacillus* sp. and for subsequent PHB production [77]. A total biomass and PHB content of 6 g/L and 55%, respectively, were achieved using cane molasses as the main carbon source in the production medium in batch fermentation [77]. Omar and colleagues have also observed successful production of PHB by *B. megaterium* on a range of substrates such as date syrup, beet molasses, and the corresponding pure carbohydrates in defined media [78]. A cell concentration of 3 g/L and PHB biomass content of 50% were attained when 5% (wt./vol.) beet molasses was supplied to the microbe.

B. subtilis was recently applied to obtain the maximal production of PHB of 2.5 g/L from 10% sugarcane molasses [79]. Wu *et al.* [80] isolated a *Bacillus* sp. strain from molasses contaminated soil for PHB production using sugarcane molasses. This strain grew at elevated temperatures up to 47°C and was osmotolerant. In a fed-batch fermentation, around 70 g/L dry cell biomass was produced containing 25 to 35% PHB. In other study, using a *B. megaterium* strain, almost 30.5 g/L of PHB was achieved [81]. This is the uppermost concentration of PHA described so far from molasses by wild type microbes.

Considerably high PHA amounts were achieved by genetically engineered sucrose-utilizing microbes such as *E. coli*, *Klebsiella oxytoca*, and *K. aerogenes* and will be reported in the next section. On the contrary, to the knowledge of the authors, only a paper has been published on the engineering of PHA-producer for the sucrose uptake and metabolism. A sucrose utilization pathway was established in *R. eutropha* (*C. necator*) by introducing the *Mannheimia succiniciproducens* MBEL55E *sacC* gene that encodes β-fructofuranosidase [82].

The recombinant *R. eutropha* (*C. necator*) strain expressing the *M. succiniciproducens sacC* could produce PHB to 1.9 g/L with a PHB content of 73% of CDM in a chemically defined medium containing 20 g/L sucrose. Another recombinant *R. eutropha* (*C. necator*), that expresses the *Pseudomonas* sp. 6-19 *phaC1437* and the *Clostridium propionicum pct540* genes [83], once engineered for the expression of both *M. succiniciproducens sacC* and the *E. coli ldhA* genes, was able to produce poly(3-hydroxybutyrate-*co*-lactate), P(3HB-*co*-LA), to a PHA content of 19.5 wt%, using sucrose as the sole carbon source.

PHAs from Waste Lipids

Lipid-rich waste is produced in considerable amounts from a variety of activities: the food processing industry, slaughterhouses, the edible oil processing industry, the dairy products industry, and oil mills. In addition, huge quantities of waste cooking oils and animal fats are globally available [84]. Management of such lipid-rich by-products is challenging because of their disposal problems and possible contamination of the water and land resources. Even though some of this waste cooking oil is used for soap production, a major part of it is discharged into the environment [85]. The Energy Information Administration in the United States estimated that some 11 billion liters of waste cooking oil is produced per year in USA, mainly from industrial deep fryers in potato processing plants, snack food factories, and fast food restaurants. In the EU countries, the total waste cooking oil production was 1 billion liters per year. Large volumes of waste cooking oils are illegally dumped, causing environmental pollution through deoxygenation of water, infiltration into soil sediments and aquifer contamination [86].

Since the theoretical PHA yield from oils and fatty acids (>0.65 g PHA/g carbon source) is greater than from glucose (0.32-0.48 g PHA/g glucose) [87], several lipids of different origin could be applied as substrates for biotechnological PHA production.

Mcl-PHA from oily waste such as residual cooking oil and other lipid waste was accumulated by *P. aeruginosa* up to 66% PHA of the cell biomass, producing almost 2 g/L polymer [88]. *Mcl*-PHA was synthesized by *W. eutropha* (*C. nec-*

ator) when canola oil was used as substrate. In a 5-L fermenter, a final concentration of 18.3 g/L PHA was obtained with 90% of cell dry matter [89]. PHB and its copolymer of PHBHV was produced by *Erwinia* sp. USMI-20 from the crude palm oil with the production of almost 1.7 g/L PHB with 46% CDM [90].

Lipid waste streams have also been a focus for PHA production with *R. eutropha* (*C. necator*). Efficient PHB accumulation using waste frying oil as the sole carbon source in flask cultures of wild-type *R. eutropha* was shown from waste sesame oil (PHB content 63% of CDM) [91] and waste rapeseed oil: PHB content of 62% of CDM [92]. Waste frying oil was also effectively converted to PHB by this species [93].

High production of PHBHV from waste rapeseed oil, using propanol as the 3HV precursor, was obtained [94]. *C. necator* produced a final copolymer concentration of 105 g/L with a 3HV content of 8 mol-%, demonstrating the high yield of 0.83 g PHA/g oil.

Tallow from the slaughtering and rendering industries is one of the cheapest fats available in large amounts. The production of PHA from tallow was demonstrated using *P. resinovorans*, with 0.14 g/L PHA accumulation (15% of CDM), but the process was considered inefficient [95]. Taniguchi and colleagues showed that waste tallow was successfully converted to PHA with relatively high yield by *R. eutropha* (*C. necator*) [91]. Copolyesters containing small amounts of 3HV were found in the cases where lard or tallow was provided as a substrate, most probably caused by the availability of propionyl-CoA as the final product of the β-oxidation of odd-numbered fatty acids found in animal lipids [91].

Although there are some good PHA-producers from waste lipids, on average the polymer accumulations are still low. Hence, more research on the development of superior oily substrates-utilizing strains and improved fermentation technology are needed to further enhance growth and PHA accumulation using these substrates. One avenue to address this issue is the pre-treatment of the lipid waste either through acidogenic fermentation or through transesterification [96]. Alternatively, a number of genetic engineering experiences has been reported on this challenging task in order to achieve cost effective production of PHA from waste or low-cost lipids.

Using random chemical mutagenesis on a wild-type *C. necator* strain, a mutant was isolated for the ability to produce 88% PHB of CDM from waste frying rapeseed oil [92]. The mutant exhibited significantly enhanced specific activities of enzymes involved in oxidative stress response such as malic enzyme and NADP-dependent isocitrate dehydrogenase. The activation of such enzymes was considered to be the main cause of an increase of NADPH/NADP$^+$ ratio which likely supported, in the mutant strain, a significantly higher accumulation of PHB, up to 16 g/L in a bioreactor.

Sequential mutation, using UV and N-methyl-N′-nitro-N-nitrosoguanidine (NTG), was found to be efficient also in selecting mutants of *B. licheniformis* with enhanced ability to produce PHA (19.5 g/L, 69% of CDM) from palm oil mill effluent. Interestingly, the mutant strain could synthesize novel PHA copolymers such as 3-hydroxyvalerate (3HB) and 3-hydroxyhexanoate (3HHx), which were not produced by the wild type [97].

Few engineering research experiences have been exploited to obtain, from oily substrates and by-products, PHAs with higher value relying on PHA-producers as host strain. It is the case of copolymers, such as the poly(3-hydroxybutyrate-*co*--hydroxyhexanoate), P(3HB-*co*-3HHx), which has thermal as well as mechanical properties comparable to petroleum-based plastic materials [98] and polymers with high content in 4HB fractions.

Fukui and Doi investigated the production of PHAs from olive oil, corn oil and palm oil in shake flasks, using a wild-type *C. necator* strain, as well as a PHA-negative *C. necator* mutant, which was transformed with the *Aeromonas caviae* PHA-synthase gene [99]. In all three plant oils, the wild-type strain produced PHB at a high polymer content (80%), but at low concentrations (2.9-3.4 g/L) meanwhile the recombinant strain produced equivalent levels of polymer, however the PHA obtained was the heteropolymer P(3HB-*co*-3HHx), which had 4-5 mol-% 3HHx, independently of the structure of the triglyceride. The same two *C. necator* strains were investigated for PHA production from soybean oil in a 10-

L fed-batch fermenter [100]. In the case of the wild-type *C. necator* strain, PHB was produced at a concentration of 85-95 g/L and a content of 72-76%. The use of the recombinant strain provided P(3HB-*co*-3HHx) containing 5 mol-% 3-hydroxyhexanoate (3HHx), at a concentration of 91-102 g/L and 71-74% PHA in CDM. In both cases, the yields were rather high (0.72-0.76 g PHA/g oil), indicating that PHA production from soybean oil is cost competitive (less than 5 US-$/kg PHA).

Similarly, Loo and coworkers studied the suitability of palm kernel oil, crude palm oil, and palm acid oil as substrates for *scl*-PHA synthesis by a *W. eutropha* (*C. necator*) harboring the PHA synthase gene of *A. caviae*. The copolymer, poly(3-hydroxybutyrate-*co*-3-hydroxyhexanoate), P(3HB-*co*-3HHx), was detected at levels ranging from 1.5 to 6.9 g/L, containing about 5% of (*R*)-3-hydroxyhexanoate in the polyester [101, 102].

Efficient P(3HB-*co*-HHx) accumulation using recombinant *R. eutropha* (*C. necator*) strains, between 72 and 87% CDM, from various plant oils (soybean oil, olive oil, corn oil, jatropha oil, palm oil, and palm kernel oil) was described by other groups (extensively reviewed in [87]). The PHA synthases used in these studies have broad substrate specificity and can incorporate both *scl*- and *mcl*-monomers into the final polymer. However, the 3HHx concentration of the stored polymer was low, only 2-5 mol-%. By inserting a *phaJ* from *A. caviae* into the PHA biosynthesis operon, Mifune *et al.* [103] were able to produce 3.8 g/L of this copolymer containing a higher 3HHx level (up to 9.9 mol-%) together with a high PHA content of 79% of CDM from soybean oil.

A noteworthy attempt to develop recombinant *R. eutropha* (*C. necator*) that can accumulate high levels of P(3HB-*co*-HHx) with significant HHx content directly from palm oil, has been recently realized [104]. The strains expressed a newly characterized PHA synthase gene from the bacterium *Rhodococcus aetherivorans*. Moreover, the expression of an enoyl-CoA hydratase gene (*phaJ*) from *P. aeruginosa* was described to increase PHA accumulation. The authors finally selected two engineered strains for production of P(3HB-*co*-HHx) from palm oil. In palm oil fermentations, one strain accumulated 71% CDM as PHA with 17mol-% HHx, while the other accumulated 66% CDM with 30 mol-% 3HHx.

Riedel and colleagues described the highest concentration of 3HHx in PHAs produced by a recombinant strain of *R. eutropha* (*C. necator*) from waste animal fats (namely the by-products of industrially produced protein hydrolysates) and residual frying oil [105]. The engineered strain was enabled to produce P(3HB-*co*-HHx), and the highest PHA accumulation in shaking flask experiments was about 3.3 g/L with almost 72% CDM and a 3HHx content of 20 mol-%. The recombinant was then grown on waste animal fat of the lowest quality available at lab fermenter scale, resulting in 27 g/L PHA, with the final HHx content of 19 mol-%. This is the first known demonstration of PHA production directly from low quality waste animal fats as carbon feedstock in fermentation scale using recombinant *R. eutropha* (*C. necator*).

Finally, mutagenesis and engineering of *C. necator* was explored towards the controlled and efficient synthesis of PHA terpolymers from triglycerides and fatty acids. The production of P(3HB-*co*-3HV-*co*-3HHx) terpolymer with a wide range of 3HV monomer compositions from mixtures of crude palm kernel oil and 3HV precursors using a mutant *C. necator* harboring the PHA synthase gene (*phaC*) of a locally isolated *Chromobacterium* sp. strain was achieved [106]. The PHA synthase of *Chromobacterium* has an unusually high affinity towards 3HV monomer. About 6.0 g/L of P(3HB-*co*-3HV-*co*-3HHx) terpolymers with 3HV monomer composition ranging from 2 to 91 mol-% was produced at 72% of CDM [106].

The direct conversion of lipids-rich materials and residues to PHA has received little attention. In 2001, two PHA producers, *P. putida* and *P. oleovorans*, which are able to accumulate *mcl*-polyhydroxyalkanoate, were genetically engineered to metabolize triacylglycerols (TAGs), by carrying *Pseudomonas* lipase genes [107]. The recombinants, specially *P. putida,* produced up to 1.6 g/L *mcl*-PHA with lard as substrate. Later on, the processing of slaughterhouse by-products into PHA with 4HB fractions has been proposed through the engineering of *Delftia acidovorans* for the expression of efficient lipase genes [108]. *D. acidovorans* is a well-known producer of PHAs with high molar fractions of 4-hydroxybutyrate (4HB), but unable to grow on fatty substrates. The *lipC* and *lipH* genes of a proficient lipolytic *P. stutzeri* isolate [109], were successfully co-expressed into *D. acidovorans* and the resulting recombinant displayed remarkable extracellular enzymatic activity on corn oil. The PHAs production from corn oil achieved high levels (26% of cell dry mass, with about 7% of 4HB). Surprisingly, the recombinant strain produced greater values directly from slaughterhouse residues such as udder and lard (43 and 39%, respectively, with almost 7% of 4HB). The PHA production levels of the engineered microbe were achieved in shake flasks and are likely to be improved upon at bioreactor scale.

PHAs from Agricultural Residues

Agricultural crop residues and waste represent a vast biomass that could also provide an invaluable feedstock for bioproducts conversion. These cellulose-, starch-, and other carbohydrate-rich materials include field by-products (stalks, stems, leaves, and seed pods) and process residues (peels, husks, seeds, bagasse, brans). To date, the largest quantities of agricultural residues and waste are from the major commodity crops. Crop residues can be considered as an important regional source of platform chemicals for the biobased economy. Processing of such residues may further provide for additional opportunities for nutrient recovery and being resupplied to serve agricultural needs. These agriculture residues or, for example, fermentation products (*i.e.*, volatile fatty acids, VFA) are feedstocks for PHA production in mixed and pure culture bioprocesses.

Starchy residues. Starch, a polymer of glucose, is a form of energy storage of all green plants and it is the most prevalent carbohydrate in the human diet. Globally, the most common resources of starch are the seeds of various types of grain (maize, wheat, and rice), meanwhile in Europe and Asia, significant amounts are also obtained by processing tubers and roots (potatoes and cassava). The worldwide starch production continues to grow because of the increasing demand in food and nonfood industries. About 54% of starch is processed for food applications and 46% for nonfood applications [110].

In 2011, the total consumption of starch and starch derivatives was about 70 million tons [111]. The largest applications of starch in the EU (30%) are the paper, cardboard, and corrugating industries. Other important fields of starch application are biofuels, cosmetics, pharmaceuticals, dyes, textiles and biomaterials [112 - 115]. Starch has also been utilized as a biobased polymer for various types of bioplastics [112] as it is the biggest marketed biopolymer with major operations in Europe (*e.g.*, Novamont S.p.A., BASF SE) and the USA (*e.g.*, National Starch, Cargill). The advantage of starch as a carbon source is that its price is lower than that of glucose. Choi and Lee projected that, with a production scale of 100,000 tons of PHB per year, production costs would decrease from 4.91 to 3.72 US-$ per kg in the case of using hydrolyzed corn starch (0.22 US-$ per kg) instead of glucose (0.49 US-$ per kg) [48].

Starch as well as starch-rich waste generated during crop processing could be used as carbon sources for the PHA production. In a limited number of studies, raw starch was directly converted to PHB using bacteria specifically selected and/or isolated for their ability to use the polysaccharide as carbon source: *A. chroococcum* [116], *B. cereus* [117] and *Saccharophagus degradans* [118]. The production of PHBHV by *Caldimonas taiwanensis* from non-hydrolyzed food-grade starches made from cassava, corn, potato, sweet potato, and wheat, supplemented with valerate was also investigated [25]. More recently, a strain of *Massilia* sp. from *Ulva* sp., a green algae containing starch, was reported to be capable of producing PHA from starch, maltotriose, or maltose as a sole carbon source. The PHA content under the cultivation condition using a 10-L bioreactor was 45% (w/w), with a final titer PHB of 3.8 g/L [119].

However, due to its complex structure and poor solubility, the use of starch as a carbon feedstock in PHA production usually requires hydrolysis pre-treatment to produce glucose [120] or acidogenesis of starchy substrates to form VFAs [121]. Both bacteria and archea have been studied for their ability to produce PHA from starch hydrolyzates [122, 123]. While the majority of studies demonstrated growth and PHB accumulation in shaken flasks, the example of study of *R. eutropha* (*C. necator*) conversion of saccharified waste potato starch in high-cell density fed-batch fermentation to PHB clearly demonstrates a great potential of the starch-based residues. A total biomass of 179 g/L and PHB concentration of 94 g/L containing 55% CDM of PHB were achieved [120].

Lignocellulosic residues. Increasing efforts have been dedicated worldwide to implement biorefinery plants for the conversion of lignocellulosic and cellulosic waste to starting materials for the biotechnological production of biofuels, biopolymers, and a range of fine chemicals [124 - 126]. Lignocellulosic materials, consisting of lignin, cellulose and hemicellulose fibers, represent the most abundant renewable resource [127].

The major obstacles to PHA production from lignocellulosic by-products are their recalcitrance to pre-treatment needed to obtain fermentable hexoses and pentoses. Moreover, during the process of sugar release, toxic by-products are also generated, which can inhibit microbial growth and PHA production. Many methods to detoxify lignocellulosic hydrolysates have been employed with the final aim of obtaining suitable feedstocks for microbial fermentations [128]. In general, detailed studies on the toxicity are needed to determine toxicity limits of strain/hydrolysate combinations. As an example, a surface response methodology was developed for a *B. cepacia* strain on sugar maple hydrolysate to determine the highest concentrations of each inhibitor which did not affect microbial growth and fermentation [129]. Such an approach leads to more optimal feed-conditions, as it can predict the highest concentration of the hydrolysate, which can be supplemented. This directly enhances the PHA productivity by decreasing the dilution factor caused by the hydrolysate feed.

Cellulose hydrolysates have been investigated as carbon sources for PHA production [130, 131]. For instance, 55 bacterial strains isolated from soil were screened for their ability to produce PHAs from xylose and sugarcane bagasse hydrolysate. Two *Burkholderia* sp. strains showed the most interesting performances: when cultivated in bagasse hydrolysate, *B. sacchari* and *B. cepacia* achieved PHB contents of 62% and 53%, respectively, both attaining a cell dry mass of 4.4 g/L. Once the strains were grown in a fed-batch reactor with mixture of xylose and glucose, the PHB content remained similar, but the maximum biomass concentration increased to about 60 g/L.

The strategy used by Yu and Stahl consisted in both adapting *C. necator* cells to hydrolyzed bagasse and using concentrated inocula. PHAs could be accumulated up to 5 g/L with a polymer content of 65% CDM [132].

Hydrolyzates of different woody biomass origins have been employed for PHA production by *E. coli* recombinant strains (see next paragraph), meanwhile no engineered PHA-producer has been reported so far for this application. However, extensive research is currently ongoing to obtain *C. necator* strains capable of using pentoses (such as xylose and arabinose) as PHA carbon sources.

C. necator is unable to utilize xylose or arabinose probably because it lacks enzymes for both the uptake and the catabolism of these sugars [133]. In 2013, a strain of *C. necator* was enabled to metabolize arabinose *via* heterologous expression of a set of genes for L-arabinose uptake and metabolism from *E. coli* [134]. A year later, the engineering of a xylose transporter (*xylE*), *xylA* and *xylB* from *E. coli* was sufficient to establish xylose utilization in *C. necator* [135]. In both cases the recombinant strains produced limited PHB amounts from arabinose or xylose. Nevertheless, these recent research achievements are noteworthy and pose the basis for future implementations towards the development of *C. necator* strains with high PHA producing ability from pentose sugars greatly available in lignocellulosic hydrolyzates.

PHAs from waste glycerol. Due to its widespread presence in nature, a large number of microorganisms utilize glycerol as sole carbon source and energy both aerobically and anaerobically. Such glycerol-utilizing bacteria include *K. pneumoniae*, *Enterobacter aerogenes*, *C. butyricum* and *Lactobacillus reuteri*. Microbes as *Pseudomonas* sp. and *Cupriavidus* sp. can metabolize glycerol aerobically. A broad number of these bacterial genera are also able to accumulate PHA and thus have been the subject of research for glycerol conversion to PHA.

Glycerol was a traditional by-product from fat hydrolysis in soap manufacturing but the prevalence of detergents in the market has resulted in a decline in soap industry and thus glycerol waste production [136]. Nevertheless, waste glycerol is nowadays generated in a large quantity as a major byproduct from biodiesel manufacturing. Biodiesel is traditionally produced from vegetable oils or animal fats through transesterification with methanol or ethanol. This process converts triacylglycerol and methanol into glycerol and fatty acid methyl esters (namely biodiesel) using alkali or acid catalysts. The amount of crude glycerol produced from transesterification reaction accounts for approximately 10% of the final weight of biodiesel. Since global biodiesel production has dramatically increased from 10 billion liters in 1999 to about 26 billion liters in 2014, crude glycerol generated from biodiesel has also been produced proportionally in large amounts. As a result, the world market price of glycerol dropped, making this side-stream an interesting source of microbial PHAs production.

Many researchers have used pure glycerol to study PHA accumulation while others have reported microbial strains that utilize raw glycerol without purification such as degreasing or the removal of methanol [137].

In 2010, Ibrahim and Steinbuchel achieved high polymer production (54.5 g/L PHB with 67% CDM) resulting in the uppermost PHB concentration from glycerol (54 g /L) using a newly isolated strain of *Zobellella denitrificans* [138]. Although PHA productivities from waste glycerol are currently lower than that of pure glycerol, some interesting advances have been made using unrefined glycerol-waste from biodiesel production. Ashby *et al.* [139] used crude glycerol, derived from a soy-based biodiesel production site, as carbon source for the PHA accumulation by *P. oleovorans* and *P. corrugata* strains. *P. oleovorans* grew to 1.3 g/L CDM with 13-27% PHA, while *P. corrugata* grew to 2.1 g/L CDM with 42% PHA. Lately, a *Halomonas* sp. strain was reported for the accumulation of almost 1.4 g/L PHB with 39% CDM from 3% crude glycerol [140].

The suitability of the different crude glycerol streams was evaluated by using *C. necator* and *P. denitrificans* [141]. The authors reported the growth of both microbes up to 50 g/L cell dry mass with a content of 70% PHB, corresponding to nearly 30 g/L of biopolymer. When using crude glycerol containing 5.5% NaCl, a reduced PHB content of 48% was observed. Cavalheiro *et al.* reported PHB production from crude glycerol by using *C. necator* producing up to 31.4 g/L PHB with 38% of CDM [142].

While the ability to utilize unrefined waste glycerol streams greatly reduces the costs of PHA manufacture and improves the economics of biodiesel production, the productivity of cell biomass and PHA need to be improved. The level and type of contaminants present in the by-product were reported to affect both growth and polymer properties [137, 143, 144].

Technological improvements are in progress to handle glycerol contaminations meanwhile few metabolic engineered microbes have been recently developed to obtain higher productivity and/or more valuable polymers.

In addition to PHB, synthesis of copolymers PHBHV and P(3HB-*co*-3-hydroxyvalerate-*co*-4-hydroxybutyrate) (PHBHV4HB) has been established from by-products of biodiesel production [145, 146]. However, the optimization of growth conditions was not sufficient. A metabolic network of intracellular processes during growth on glycerol was analyzed with elementary flux modes, and putative links between central metabolism and PHB synthesis were marked [147]. A linkage between Gly3-DH's cofactor and NAD(H)/NADP(H) transhydrogenases was indicated.

Therefore, *R. eutropha* (*C. necator*) was transformed with the genes (aquaglyceroporin, glycerol kinase, and FAD-dependent glycerol 3-phosphate dehydrogenase), involved in the aerobic metabolism of glycerol found in *E. coli*. In a nitrogen limited shaking flask experiment supplemented with glycerol, both fast growth and PHB accumulation were observed [148].

Overexpressing the glycerol kinase *(glp)* genes implicated in glycerol metabolism was useful to enhance PHA accumulation on glycerol [148]. Alternatively, the transcription repressor gene *glpR* has been deleted in *P. putida*. As a result, lag-phase of mutants was greatly reduced, and *mcl*-PHA accumulation enhanced. The authors also found that the long lag-phase can be avoided by co-feeding the wild type strain with small amounts of octanoate, which, on the other hand, induces the *glp* genes [149]. Both *C. necator* and *P. putida* research experiences ended up with promising, but low, PHA values which have to be further improved and enhanced before opening their industrial exploitations.

A potential use of glycerol in biotechnology is the synthesis of poly(3-hydroxypropionate), P(3HP), a biopolymer with promising properties which is not synthesized by any known wild-type organism. The material properties of P(3HP) are comparable to those of other short-chain polymers, such as PHB. The incorporation of 3HP monomers into other PHAs has valuable effects, like lower crystallinity and fragility of the resulting copolymer [150]. Hence, the production of P(3HP) from low-cost materials, such as waste glycerol, is highly desirable and potentially profitable. The main metabolic engineering efforts on P(3HP) accumulation from pure and waste glycerol have been spent on non-producers, such as *E. coli*, and will be reported in the next section of this chapter.

CLONING PHA GENES IN NON PRODUCING BACTERIA

Natural microorganism can synthesize PHAs from structurally-related precursors which, unfortunately, are usually highly expensive, not always easily available, often unsuitable because toxic to cells, and water insoluble. Therefore, economic production of PHAs can only be attained from cheap carbon sources, namely renewable resources and/or residual materials. The production of PHAs from unrelated carbon sources is an old idea, but the best known and proficient PHA producing bacterial strains are unable to directly utilize the abundant compounds deriving from a range of industries that usually generate these materials as by products or waste (*e.g.*, cellulose, starch, molasses). That's why a number of researches focussed on pre-treatments of these potential substrates to obtain simple sugars for bacterial feeding, with the aim of PHA production. As reported on the above section, nowadays molecular biology could provide suitable tools for engineering PHA producing bacteria, thus making them able to utilize different carbon sources for PHA synthesis (Fig. **1**).

On the other hand, several bacterial species unable to accumulate PHAs under natural conditions, can efficiently hydrolyze complex substrates to obtain simple sugars. Some of those may represent a suitable host for the production of PHAs by taking advantage of available metabolic engineering tools (Fig. **1**). The heterologous expression of an active PHA biosynthesis pathway can take place in recombinant organisms only if a PHA synthase structural gene was successfully transferred to a suitable host, PHA synthase protein is enzymatically active and appropriate pathways feed the enzyme with the right nature and amount of substrates. Substrate specificity and subunits were used to divide PHA synthases into four classes [21]. Class I synthases, utilizing short chain length hydroxyalkanoate monomers, are typical of *C. necator* and a number of alpha- and beta-proteobacteria [151]; class II synthases are characteristic of pseudomonads that polymerize medium chain length PHAs [22]; cyanobacteria, together with *Allochromatium vinosum*, are representative of class III synthases which use short and medium length chain monomers [152], while class IV synthases use only short chain length monomers and was found in some *Bacillus* species [150]. So far, several dozen PHA synthase genes (*phaC*) have been cloned and sequenced from bacteria, and an additional number of homologous synthase sequences have been found in genome databanks. Sequence information of PHA synthase operons will enable the identification or even the construction of new PHA synthases for the production of novel PHAs. The specificity of a PHA synthase (PhaC) is the most important factor controlling PHA monomers composition. As reviewed in [153], at least 14 PHA formation pathways have been reported. Pathways I to IV are natural PHA synthesis, while V to XIV are engineered pathways leading to unconventional PHAs.

The most studied bacterial recipient for PHA gene cloning was assumed to be *E. coli*. It is unable to naturally accumulate the polymer, but it represents a suitable candidate for the production of PHA because (i) it lacks an intracellular depolymerization system, (ii) proper metabolic engineering tools are available and heterologous expression of foreign genes could be easily obtained [154], (iii) high cell density cultivation strategies are well established for these bacteria [155], (iv) the isolation and purification of the biopolymer is simplified by the fragility of the cells that accumulate large amounts of PHA.

The first PHB was heterologously synthesized in *E. coli* in 1988 by cloning PHA genes from *A. eutrophus* (*C. necator*) [156, 157] and later, with the aim to produce PHA from unrelated carbon sources, many other works reported on new metabolic pathways by combining genes from a number of different organisms such as *P. aeruginosa* [158], *A. latus* [159], *C. acetobutylicum* and *Thiocapsa pfennigii* [160], *Streptomyces aureofaciens* [161].

In several studies, homopolymers were obtained by recombinant *E. coli*, including PHB, poly(4-hydroxybutyrate) P(4HB), poly(3-hydroxyvalerate) P(3HV), P(3HP), poly(3-mercaptoalkanoate) PTEs, or polythioesters [157, 162 - 166]. Other Authors reported on the production of random copolymers [167, 168], and copolymers with fully controllable structures, from glycerol without addition of any precursor [169]. Block copolymers were also obtained by recombinant *E. coli* [170, 171]. A summary of PHAs produced and the known metabolic pathways constructed for their biosynthesis in recombinant *E. coli* from unrelated carbon source was recently reported [172]. In addition, a pathway for the simultaneous co-production of extracellular succinate and intracellular PHB from glucose was designed and engineered in *E. coli* [173]. These Authors observed that, in batch fermentation, PHB accumulation (4.95 g/L) increased the production of succinate up to 24.6 g/L, and at the same time reduced pyruvate and acetate secretion.

Moreover, new tools specifically tailored for delivering genes into the chromosome of Gram-negative bacteria were recently proposed [174]. These Authors used a series of new broad-host-range mini Tn5-vectors to insert the entire PHA synthesis pathway from *C. necator* in the chromosome of an *E. coli* mutant strain.

The synthesis of PHAs through recombinant *E. coli* would aim, as one of the main objectives, at the reduction of the whole production chain cost, but so far, in both laboratory and industrial scale, the efficiency of the system remains still too low. Therefore, the use of inexpensive and renewable carbon sources may help to make it feasible, since the substrate may account for 30 to 50% of the total production cost. As reported in the above section, many efforts have been done to obtain PHAs from a variety of waste materials from agricultural and food industries producing huge amounts of feedstocks and co-products mainly deriving from surplus or low-cost processing streams [175].

A representative example is milk whey, deriving from the dairy industry in amounts higher than 150 million tons per year, which contains almost 5% (sweet whey) or more than 20% (whey permeate) lactose. *E. coli*, known to directly convert lactose into glucose and galactose, was originally equipped with PHB-biosynthesis genes of *R. eutropha* (*C. necator*) [176, 177]. The highest PHB concentration (5.2 g/L and 81% of CDM) reported by Lee and coworkers [176] was later improved up to 69 g/L and 87% CDM in a pH-stat fed-batch fermentation using concentrated whey solution containing 210 g/L lactose as nutrient feed [177]. Lately, the production of the polymer by fed-batch culture of recombinant *E. coli*, expressing the *A. latus* PHB biosynthetic genes, was achieved [178]. In this work a concentrated whey solution containing 280 g/L of lactose equivalent was used and the final cell and PHB concentrations reached 119.5 and 96.2 g/L, respectively.

The yields and productivity of PHB was improved by a cell-recycle fed-batch culture [179] and a pilot plant-scale fermentation was also assessed using engineered *E. coli* expressing *A. latus* PHA biosynthesis genes [180]. A recombinant *E. coli* strain containing *pha* biosynthetic genes from *Azotobacter* sp., was specifically designed for the direct production of PHB from milk whey [181]. The resulting strain did not produce the lactose repressor, thus ensuring constitutive expression of genes involved in lactose transport and utilization in *E. coli* and obtaining almost 73% of cell dry mass as polymer in 70 g/L biomass.

More recently, Pais *et al.* [182] engineered *E. coli* through the chromosomal insertion of *C. necator* PHB-synthesis genes, fused to a lactose-inducible promoter. The mutant strain obtained by means of proton suicide method was able to use cheese whey as the sole carbon source for PHB with a considerably higher amount of polymer production (about 19 g/L).

Molasses represent another interesting agro-industrial waste that has been considered for cheaper production of PHAs. It is a sugar-rich co-product deriving mainly from sugar beet and sugar cane refining plants, containing high amounts of sucrose. When provided with PHAs genes, *E. coli* and members of *Klebsiella* genus are likely candidates biopolymer producers from molasses because they can grow well on sucrose and rapidly to high cell density.

In 1998, Liu and colleagues [183] constructed a recombinant *E. coli* strain carrying *A. eutrophus* (*C. necator*) PHB biosynthesis genes with the aim of producing PHA using molasses as carbon source. The recombinant strain was used in fed-batch strategies at different scales. In shake flasks operated in fed-batch mode, the best results were a PHB concentration of 11.2 g/L at a polymer content of 80%. In a fed-batch bioreactor, PHB production was increased to 31.6 g/L at 80% polymer concentration, indicating the relevance of the pH-value and dissolved oxygen control available in the bioreactor.

Three engineered strains of *E. coli*, *K. oxytoca* and *K. aerogenes* were obtained cloning the PHB synthesis pathway from *A. eutrophus* (*C. necator*) [184]. The authors first assessed their ability to convert sugarcane molasses to PHA in shake-flask studies. Of the three bacterial strains, *K. aerogenes* exhibited superior PHB production, accumulating up to 3 g/L of PHB, which represented the 50% of CDM. *K. aerogenes* was then grown in a 10-L bioreactor with molasses as the sole carbon source. The fed-batch studies indicated that this recombinant microbe has excellent potential for production of large amounts of polymer: 24 g/L of PHB, with a final content of 70% per CDM.

Later, a recombinant *E. coli* harbouring PHB genes from *S. aureofaciens* accumulated PHB from glycerol, glucose, palm oil and ethanol, but not when sucrose and molasses were used as carbon source [161]. More recently, an engineered *E.coli* expressing *phaC1* from *Pseudomonas* sp. was used in batch culture for the production of PHAs from molasses [185]: higher PHA yield was obtained from molasses (3.06 g/L and 75.5% CDM) as compared to sucrose (2.5 g/L and 65.1% CDM).

In principle, the promising performances of the above reported recombinant *E. coli* strains could be further improvable playing with native genes limiting the *E. coli* sucrose uptake when the sugar starts to be depleted. *E. coli* can utilize sucrose using the *csc* regulon but, at low sucrose concentrations (<2 g/L), the *cscR* transcription factor represses transcription of the regulon and no growth is observed [186]. This is problematic for fed-batch cultivation where the sugar concentration is maintained at low levels between feeding pulses. To improve sucrose utilization in *E. coli*, Arifin *et al.* [187] developed a *cscR* knock-out mutant. The *C. necator phbABC* operon was then introduced in the final recombinant strain which showed PHB production (47.7 g/L) 1.6-fold that achieved by the *E. coli* strain harboring only the *phbABC* operon. This noteworthy and pioneering approach has been further detailed by the Authors [188] and paves the way for an increased yield of PHB from molasses.

Most of co-products derived from surplus or low-cost processing streams are mixture of sugars and/or fatty acid. Generally, *E. coli* uses to select the carbon sources supporting the highest growth rate through the inhibition of other carbon sources transport capacity, related enzyme activities, and gene expression (carbon catabolite repression) [189]. Due to these phenomena, in the presence of a substrate mixture *E. coli* exhibits sequential sugar utilization and, for instance, fatty acids are not utilized in the presence of sugars, thus missing the possible construction of medium-chain-length PHAs [190].

By the way, fatty acids, which are the main components of triacylglycerols (TAG) deriving from vegetable oils and animal fats, are produced by agricultural and food industries at a rate of about 100 million tons per year. Therefore, as reported in the above section, triacylglycerols and their derivative fatty acids and glycerol may represent an interesting cheap substrate for PHA production.

The main metabolic engineering efforts on pure and waste glycerol have been spent on *E. coli* platform. The biotechnical synthesis of the unusual homopolymer P(3HP) from crude glycerol was first shown in 2010 [162], by introducing the genes for glycerol dehydratase of *C. butyricum*, propionaldehyde dehydrogenase of *Salmonella enterica* serovar Typhimurium LT2, and PHA synthase of *R. eutropha* (*C. necator*) into a recombinant *E. coli* strain. P(3HP) was accumulated up to 1.42 g/L in fed-batch fermentation.

More recently, another recombinant *E. coli* strain was reported for the highest titer of P(3HP) (10.1 g/L) from waste glycerol and glucose as co-substrates [171]. In this study, the genes for glycerol dehydratase and its reactivating factor (*dhaB123* and *gdrAB*, from *K. pneumoniae*), propionaldehyde dehydrogenase (*pduP*, from *Salmonella enterica* serovar Typhimurium), and polyhydroxyalkanoate synthase (*phaC1*, from *C. necator*) were cloned and expressed in *E. coli.* The resulting engineered strain was able to efficiently convert glycerol with high yield.

The problem related to sequential sugar utilization by *E. coli* could be also present in other potential substrates to be adopted for PHA production. Lignocellulose hydrolyzates represent a fitting example of low-cost mixture of different sugars (*e.g.*, glucose and xylose) that generally cause bacterial diauxic growth, leading to long operating time and decrease in productivity. To face this problem, a PTS (phosphotransferase system) mutant of *E. coli* was enabled to simultaneously utilize glucose and xylose in a defined substrate, with increased production of PHA [191]. However, to apply this result in the PHB production, a stable and non-IPTG induced *E. coli* strain would be required.

Dealing with hydrolysates of different woody biomass origins several recombinant strains were developed. Lee investigated PHB production from xylose and cellulose hydrolysates by four recombinant *E. coli* strains harboring *C. necator* PHA biosynthesis genes. The highest accumulation of PHB was 1.7 g/L and 35.8% CDM in a chemically defined medium containing 20 g/L of xylose. By further supplementing the same medium with xylose and 10 g/L of soybean hydrolysate, PHB was obtained at a PHA concentration of 4.4 g/L with a polymer content of 73.9% [192].

Another recombinant *E. coli* strain, found to be resistant to 5-hydroxymethylfurfural (5-HMF), which is a major inhibitory by-product generated in the cellulose pre-treatment, accumulated up to 3.3 g/L PHB with 59% CDM when grown on cellulose hydrolyzate [193].

The first proof of concept on the one step conversion of cellulosic materials into PHA has been recently shared using *E. coli* as host strain. *E. coli* was genetically engineered to express an endoxylanase (*XylB*) from *S. coelicolor* and a β-xylosidase (*XynB*) from *B. subtilis* in order to utilize xylan [194]. This strain was additionally designed to express the PHA synthase from *Pseudomonas* sp. harboring a Ser325Th/Gln481Lys mutation, a propionyl-CoA transferase (*PCT*) from *Megasphaera elsdenii*, and a β-ketothiolase (*PhaA*) and NADPH-dependent acetoacetyl-CoA reductase (*PhaB*) from *R. eutropha* (*C. necator*) under its constitutive promoter [195]. The co-expression of the xylanase system and the P(3HB-*co*-LA) system in *E. coli* resulted in the production of xylan-derived P(3HB-*co*-LA). Although the PHA yields from xylan, with or without arabinose and xylose supplementation, were low, the Authors demonstrate that the Consolidated Bioprocessing of biomasses, recently proposed mainly for biofuels applications [196, 197], could be applied also to biopolymers production.

Only few engineering research activities have been reported so far with the aim of developing PHA-producers from starchy substrates. *E. coli* strains, containing the PHB synthase genes from *C. necator*, produced low PHB levels (about 1 g/L, 50% dry cell) from hydrolyzed corn starch [198]. The direct utilization of starch to produce PHA was recently investigated. An *E. coli* strain, that expresses a functional amylase and accumulates PHB, was constructed using different plasmids containing the amylase gene of *Paenibacillus* sp. and PHB synthesis genes from *R. eutropha* (*C. necator*). This engineered strain can utilize starch as the sole carbon source. The maximum PHB production (1.24 g/L) was obtained with 2% (wt./vol.) starch with an intracellular concentration of 57.4% of CDM [199]. The one step conversion of starch to PHA was even investigated by inserting the operon of *C. necator* (*pha*CAB) in a proficient amylolytic *Aeromonas* sp. isolate. The resulting engineered strain achieved a limited PHA accumulation, about 1.5 g/L, when grown at a bioreactor scale in a minimal medium using starch as sole carbon source [200].

A further cheap feedstock, recently rising biotechnological interest for microbial fermentation purposes, is syngas [201], a mixture of hydrogen, carbon monoxide and carbon dioxide deriving from the gasification of carbonaceous waste material. *Rhodospirillum rubrum*, known to synthesize PHA, is able to use syngas as carbon and energy source and PHA biosynthesis was recently achieved through syngas fermentation [202]. The production costs were found to be lower than sugar fermentation by recombinant *E. coli* strains and H_2, which is of great industrial interest, was also produced. More recently, two *phaC* deletion mutants of *R. rubrum* were constructed, and a number of PHA synthase genes were heterologously expressed in the obtained mutants for PHA biosynthesis. The amount of accumulated PHB was increased by up to 25% (g PHA per g CDM) as compared to the wild type strain [203].

The production of PHAs from recombinant *E. coli* and other Gram negative bacteria, although intensively studied as reported above, may present a problem for medical applications due to the presence of lipopolysaccharide (LPS) endotoxins, which require extensive purification processes to be removed [204]. Therefore, Mifune and colleagues, with the aim to generate functional PHA granules for biomedical applications, tried to utilize the Gram-positive *Lactococcus lactis*, generally regarded as safe, food-grade bacterium [205]. After proper codon optimization, *phaCAB* genes from *R. eutropha* (*C. necator*) were cloned into a plasmid vector as an operon under the control of the *nisA* promoter, and finally transformed into *L. lactis*. Induction with nisin in glucose–supplemented media resulted in PHB granule formation, even if the beads thus obtained were smaller in size and PHB production was less than reported for recombinant *E. coli*.

On the other hand, *Corynebacterium glutamicum* was also used as host strain because of being an endotoxin-free gram-positive microbe, thus ensuring the absence of potentially harmful substances linked with the application of *E. coli* for PHA production. The resulting engineered *C. glutamicum* strain expressed PHB biosynthetic genes (*phbCAB* operon derived from *C. necator*) [206]. The same recombinant strain displayed α-amylase from *Streptococcus bovis* (AmyA) on its cell surface and was successfully utilized on soluble starch, thus accumulating up to 6.4% of PHB of dry cells which stand for 0.39 g/L of PHA.

CONCLUDING REMARKS

Additional approaches related to the various steps of PHA production chain have been also reported. For instance, a possible strategy to help increase the final PHA yield in a bacterial strain already possessing both high polymer production ability and acceptable hydrolytic activities, could originate from the relevant findings regarding poly(3-hydroxybutyrate) depolymerase enzymes, known as PhaZ, and their role in PHB utilization and intracellular granule formation [207 - 209].

The main depolymerase, PhaZ1, was reported to play a significant role in PHB mobilization [209]. Its active role in PHB consumption was indicated by analysing PHB content in *phaZ1* and *phaZ2* deletion strains, and more recently, further elucidations of their role in *C. necator* have been reported [210, 211].

Several studies have indicated that PHB synthesis and PHB degradation can happen simultaneously in *R. eutropha* (*C. necator*) [212 - 214] and that there is constitutive expression of PHB synthase and PHB depolymerase [215]. Therefore, limiting intracellular depolymerisation could lead to enhancement of the final amount of polymer accumulation [51]. In a recent work, by knocking out the relative depolymerase gene, the construction of a *C. necator* recombinant strain impaired in PhaZ1 was achieved [216]. One of the intracellular PHA-depolymerases, was inactivated by chromosomal integration of a cassette containing a kanamycin resistance gene. Since the above studies indicated that production and mobilization of PHB occur simultaneously [212, 213] and that *phaA*, *phaB*, *phaC*, and *phaZ1a* were transcribed throughout cell growth, PHB production, and PHB degradation [215], these Authors expected to find a measurable increase in cell polymer content in the PhaZ1 deficient strain, even from the beginning of the accumulation phase. Indeed, Uchino *et al.* [214] demonstrated that a *phaZ1* mutant of *R. eutropha* (*C. necator*) secretes significantly less amounts of monomeric 3HB into the culture medium, and shows significantly more and bigger PHB granules than the wild type strain during the first 12-18 hours. Therefore, Povolo and colleagues tested the effect of *phaZ1* inactivation on the reduction of PHA depolymerisation in bacterial cultures previously allowed to accumulate the polymer [216]. Under carbon starvation, while the wild type reduced the intracellular PHA content to 30% of the initial content, the *phaZ1*-knockout strain maintained its PHA content at 85% even after 96 hours of incubation. Preserving the biopolymer from degradation at the end of the accumulation phase, thus offering to downstream process a wider interval of time, is a promising strategy for improving the production of PHAs by bacterial strains selected for industrial purposes.

Additionally, the production of enantiomerically pure (*R*)-(-)-hydroxycarboxylic acids was obtained in recombinant *E. coli* expressing PHA biosynthesis operon along with intracellular PHA depolymerase gene of *R. eutropha* (*C. necator*) [217]. While PHA biosynthesis genes were integrated into the chromosome, PHA depolymerase gene was introduced through a high-copy-number plasmid. In this way R3HB could be produced with a yield of 85.6% of the maximum theoretical yield from glucose.

Another possible improvement of the production chain may originate from the reduction of the viscosity of the solution during the purification steps of PHA granules. Integrating a nuclease-encoding gene of *Staphylococcus aureus* into the genomes of several PHA producers was successfully achieved [218]. In such a way, the viscosity of the lysate due to the release of nucleic acids into the medium was reduced in the downstream process to a level similar to that obtained by treating the wild type strain with commercial nuclease. More recently, this approach became object of a patent [219] having the main objective to reduce the overall fermentation process costs by increasing the recovery of the product from the cell biomass.

Recently, growing interest has been devoted to the understanding of phasins, PHA granule-associated proteins (PGAPs), detected in many PHB-accumulating bacteria (see [220] for an overview) with the aim to understand the initiation and formation of PHA granule complex.

In conclusion, the feasibility of commercial-scale production of PHAs is still suffering from a series of glitches, and strongly depends upon a wide variety of scientific attainments covering the entire production chain, ranging from the nature of the substrate to be utilized and focussing on the multiform aspects of the microbial metabolism.

Most of natural microorganisms can synthesize PHAs with varied monomers only at lab scale from structurally-related precursors, which are frequently expensive, not miscible in water, toxic even at low concentrations and high oxygen demanding. That is why researchers have addressed their attention to inexpensive and renewable unrelated carbon sources. Moreover, the exploitation of the metabolic potential of bacteria for the production of PHAs with tailor-made monomer composition from unrelated carbon sources rapidly became one of the main objectives. At the same time, the improvement of yield has to be faced not only with the modification of growth conditions, but possibly by re-engineering metabolic flux to drive carbon utilization toward PHA biosynthesis pathway. The wide range of bacteria capable of naturally producing PHA or those obtained through recombinant DNA techniques are not limited to the actual achievements, but clearly indicate that many other organisms with desirable properties could be exploited in the future for producing recombinant PHAs.

CONFLICT OF INTEREST

The authors confirm that they have no conflict of interest to declare for this publication.

AKNOWLEDGEMENTS

This work was partially financed by the following research projects i) Progetto di Ateneo CPDA137517/13, from University of Padova, ii) ex-60% 60A08-5810/13, 60A08-0721/14 and 60A08-0924/15 from University of Padova, iii) "BioRivaluta" funded by Regione Veneto (PSR 2007-2013, Misura 124, n.

2307660).

REFERENCES

[1] Koller M, Atlić A, Dias M, Reiterer A, Braunegg G. Microbial PHA production from waste raw materials. In: Chen G-Q, Ed. Plastics from bacteria. Springer 2010; pp. 85-119. [http://dx.doi.org/10.1007/978-3-642-03287-5_5]

[2] Kunasundari B, Sudesh K. Isolation and recovery of microbial polyhydroxyalkanoates. Express Polym Lett 2011; 5(7): 620-34. [http://dx.doi.org/10.3144/expresspolymlett.2011.60]

[3] Dhangdhariya JH, Dubey S, Trivedi HB, *et al.* Polyhydroxyalkanoate from marine *Bacillus megaterium* using CSMCRI's Dry Sea Mix as a novel growth medium. Int J Biol Macromol 2015; 76: 254-61. [http://dx.doi.org/10.1016/j.ijbiomac.2015.02.009] [PMID: 25697675]

[4] Cerrone F, Davis R, Kenny ST, *et al.* Use of a mannitol rich ensiled grass press juice (EGPJ) as a sole carbon source for polyhydroxyalkanoates (PHAs) production through high cell density cultivation. Bioresour Technol 2015; 191: 45-52. [http://dx.doi.org/10.1016/j.biortech.2015.04.128] [PMID: 25978856]

[5] Liu CC, Zhang LL, An J, Chen B, Yang H. Recent strategies for efficient production of polyhydroxyalkanoates by micro-organisms. Lett Appl Microbiol 2016; 62(1): 9-15. [PMID: 26482840]

[6] Tan G-Y, Chen C-L, Li L, *et al.* Start a research on biopolymer polyhydroxyalkanoate (PHA): a review. Polymers (Basel) 2014; 6(3): 706-54. [http://dx.doi.org/10.3390/polym6030706]

[7] Chen G-Q. Plastics completely synthesized by bacteria: polyhydroxyalkanoates. In: Chen G-Q, Ed. Plastics from bacteria. Springer 2010; pp. 17-37. [http://dx.doi.org/10.1007/978-3-642-03287-5_2]

[8] Braunegg G. Sonnleitner By, Lafferty R. A rapid gas chromatographic method for the determination of poly-β-hydroxybutyric acid in microbial biomass. Europ J Appl Microbiol Biotechnol 1978; 6(1): 29-37. [http://dx.doi.org/10.1007/BF00500854]

[9] Huijberts GN, van der Wal H, Wilkinson C, Eggink G. Gas-chromatographic analysis of poly (3-hydroxyalkanoates) in bacteria. Biotechnol Tech 1994; 8(3): 187-92. [http://dx.doi.org/10.1007/BF00161588]

[10] Burdon KL. Fatty material in bacteria and fungi revealed by staining dried, fixed slide preparations. J Bacteriol 1946; 52(6): 665-78. [PMID: 16561232]

[11] Bartholomew J W. Stains for micro-organisms in smears. In: Clark G, Ed Staining procedures. Williams and Wilkins 1980; pp. 380-3.

[12] Kranz RG, Gabbert KK, Madigan MT. Positive selection systems for discovery of novel polyester biosynthesis genes based on fatty acid detoxification. Appl Environ Microbiol 1997; 63(8): 3010-3.

[PMID: 9251190]

[13] Ostle AG, Holt JG. Nile blue A as a fluorescent stain for poly-beta-hydroxybutyrate. Appl Environ Microbiol 1982; 44(1): 238-41.
[PMID: 6181737]

[14] Schlegel H-G, Lafferty R, Krauss I. The isolation of mutants not accumulating poly-β-hydroxybutyric acid. Arch Mikrobiol 1970; 71(3): 283-94.
[http://dx.doi.org/10.1007/BF00410161] [PMID: 4097070]

[15] Serafim LS, Lemos PC, Levantesi C, Tandoi V, Santos H, Reis MA. Methods for detection and visualization of intracellular polymers stored by polyphosphate-accumulating microorganisms. J Microbiol Methods 2002; 51(1): 1-18.
[http://dx.doi.org/10.1016/S0167-7012(02)00056-8] [PMID: 12069885]

[16] Spiekermann P, Rehm BH, Kalscheuer R, Baumeister D, Steinbüchel A. A sensitive, viable-colony staining method using Nile red for direct screening of bacteria that accumulate polyhydroxyalkanoic acids and other lipid storage compounds. Arch Microbiol 1999; 171(2): 73-80.
[http://dx.doi.org/10.1007/s002030050681] [PMID: 9914303]

[17] Elain A, Le Fellic M, Corre Y-M, *et al.* Rapid and qualitative fluorescence-based method for the assessment of PHA production in marine bacteria during batch culture. World J Microbiol Biotechnol 2015; 31(10): 1555-63.
[http://dx.doi.org/10.1007/s11274-015-1904-4] [PMID: 26187125]

[18] Degelau A, Scheper T, Bailey J, Guske C. Fluorometric measurement of poly-β hydroxybutyrate in *Alcaligenes eutrophus* by flow cytometry and spectrofluorometry. Appl Microbiol Biotechnol 1995; 42(5): 653-7.
[http://dx.doi.org/10.1007/BF00171939]

[19] Hong K, Sun S, Tian W, Chen G, Huang W. A rapid method for detecting bacterial polyhydroxyalkanoates in intact cells by Fourier transform infrared spectroscopy. Appl Microbiol Biotechnol 1999; 51(4): 523-6.
[http://dx.doi.org/10.1007/s002530051427]

[20] Madison LL, Huisman GW. Metabolic engineering of poly(3-hydroxyalkanoates): from DNA to plastic. Microbiol Mol Biol Rev 1999; 63(1): 21-53.
[PMID: 10066830]

[21] Rehm BH, Steinbüchel A. Biochemical and genetic analysis of PHA synthases and other proteins required for PHA synthesis. Int J Biol Macromol 1999; 25(1-3): 3-19.
[http://dx.doi.org/10.1016/S0141-8130(99)00010-0] [PMID: 10416645]

[22] Timm A, Wiese S, Steinbüchel A. A general method for identification of polyhydroxyalkanoic acid synthase genes from pseudomonads belonging to the rRNA homology group I. Appl Microbiol Biotechnol 1994; 40(5): 669-75.
[http://dx.doi.org/10.1007/BF00173327] [PMID: 7508720]

[23] Solaiman DK, Ashby RD. Rapid genetic characterization of poly(hydroxyalkanoate) synthase and its applications. Biomacromolecules 2005; 6(2): 532-7.
[http://dx.doi.org/10.1021/bm0493640] [PMID: 15762609]

[24] López NI, Pettinari MJ, Méndez BS. Detection of reserve polymer synthesis genes in natural bacterial populations. FEMS Microbiol Ecol 1997; 22(2): 129-36. [http://dx.doi.org/10.1016/S0168-6496(96)00083-9]

[25] Sheu D-S, Wang Y-T, Lee C-Y. Rapid detection of polyhydroxyalkanoate-accumulating bacteria isolated from the environment by colony PCR. Microbiology 2000; 146(Pt 8): 2019-25. [http://dx.doi.org/10.1099/00221287-146-8-2019] [PMID: 10931906]

[26] Solaiman DK, Ashby RD, Foglia TA. Rapid and specific identification of medium-chain-length polyhydroxyalkanoate synthase gene by polymerase chain reaction. Appl Microbiol Biotechnol 2000; 53(6): 690-4. [http://dx.doi.org/10.1007/s002530000332] [PMID: 10919328]

[27] Revelo Romo DM, Grosso MV, Moreno Solano NC, Montoya Castaño D. A most effective method for selecting a broad range of short and medium-chain-length polyhydroxyalkanoate producing microorganisms. Electron J Biotechnol 2007; 10(3): 348-57. [http://dx.doi.org/10.2225/vol10-issue3-fulltext-13]

[28] Hai T, Lange D, Rabus R, Steinbüchel A. Polyhydroxyalkanoate (PHA) accumulation in sulfate-reducing bacteria and identification of a class III PHA synthase (PhaEC) in *Desulfococcus multivorans.* Appl Environ Microbiol 2004; 70(8): 4440-8. [http://dx.doi.org/10.1128/AEM.70.8.4440-4448.2004] [PMID: 15294771]

[29] McCool GJ, Cannon MC. PhaC and PhaR are required for polyhydroxyalkanoic acid synthase activity in *Bacillus megaterium.* J Bacteriol 2001; 183(14): 4235-43. [http://dx.doi.org/10.1128/JB.183.14.4235-4243.2001] [PMID: 11418564]

[30] Bowien B, Schlegel HG. Physiology and biochemistry of aerobic hydrogen-oxidizing bacteria. Annu Rev Microbiol 1981; 35(1): 405-52. [http://dx.doi.org/10.1146/annurev.mi.35.100181.002201] [PMID: 6271040]

[31] Yabuuchi E, Kosako Y, Yano I, Hotta H, Nishiuchi Y. Transfer of two *Burkholderia* and an *Alcaligenes* species to *Ralstonia* gen. Nov.: Proposal of *Ralstonia pickettii* (Ralston, Palleroni and Doudoroff 1973) comb. Nov., *Ralstonia solanacearum* (Smith 1896) comb. Nov. and *Ralstonia eutropha* (Davis 1969) comb. Nov. Microbiol Immunol 1995; 39(11): 897-904. [http://dx.doi.org/10.1111/j.1348-0421.1995.tb03275.x] [PMID: 8657018]

[32] Vaneechoutte M, Kämpfer P, De Baere T, Falsen E, Verschraegen G. *Wautersia* gen. nov., a novel genus accommodating the phylogenetic lineage including *Ralstonia eutropha* and related species, and proposal of *Ralstonia* [*Pseudomonas*] *syzygii* (Roberts *et al.* 1990) comb. nov. Int J Syst Evol Microbiol 2004; 54(Pt 2): 317-27. [http://dx.doi.org/10.1099/ijs.0.02754-0] [PMID: 15023939]

[33] Vandamme P, Coenye T. Taxonomy of the genus *Cupriavidus*: a tale of lost and found. Int J Syst Evol Microbiol 2004; 54(Pt 6): 2285-9. [http://dx.doi.org/10.1099/ijs.0.63247-0] [PMID: 15545472]

[34] Simu K, Hagström A. Oligotrophic bacterioplankton with a novel single-cell life strategy. Appl Environ Microbiol 2004; 70(4): 2445-51. [http://dx.doi.org/10.1128/AEM.70.4.2445-2451.2004] [PMID: 15066843]

[35] Handelsman J. Metagenomics: application of genomics to uncultured microorganisms. Microbiol Mol Biol Rev 2004; 68(4): 669-85.
[http://dx.doi.org/10.1128/MMBR.68.4.669-685.2004] [PMID: 15590779]

[36] Rondon MR, August PR, Bettermann AD, *et al.* Cloning the soil metagenome: a strategy for accessing the genetic and functional diversity of uncultured microorganisms. Appl Environ Microbiol 2000; 66(6): 2541-7.
[http://dx.doi.org/10.1128/AEM.66.6.2541-2547.2000] [PMID: 10831436]

[37] Voget S, Leggewie C, Uesbeck A, Raasch C, Jaeger K-E, Streit WR. Prospecting for novel biocatalysts in a soil metagenome. Appl Environ Microbiol 2003; 69(10): 6235-42.
[http://dx.doi.org/10.1128/AEM.69.10.6235-6242.2003] [PMID: 14532085]

[38] Yun J, Kang S, Park S, *et al.* Characterization of a novel amylolytic enzyme encoded by a gene from a soil-derived metagenomic library. Appl Environ Microbiol 2004; 70(12): 7229-35.
[http://dx.doi.org/10.1128/AEM.70.12.7229-7235.2004] [PMID: 15574921]

[39] Gillespie DE, Brady SF, Bettermann AD, *et al.* Isolation of antibiotics turbomycin A and B from a metagenomic library of soil microbial DNA. Appl Environ Microbiol 2002; 68(9): 4301-6.
[http://dx.doi.org/10.1128/AEM.68.9.4301-4306.2002] [PMID: 12200279]

[40] Cheema S, Bassas-Galia M, Sarma PM, Lal B, Arias S. Exploiting metagenomic diversity for novel polyhydroxyalkanoate synthases: production of a terpolymer poly(3-hydroxybutyrate-c--3-hydroxyhexanoate-co-3-hydroxyoctanoate) with a recombinant *Pseudomonas putida* strain. Bioresour Technol 2012; 103(1): 322-8.
[http://dx.doi.org/10.1016/j.biortech.2011.09.098] [PMID: 22071242]

[41] Schallmey M, Ly A, Wang C, *et al.* Harvesting of novel polyhydroxyalkanaote (PHA) synthase encoding genes from a soil metagenome library using phenotypic screening. FEMS Microbiol Lett 2011; 321(2): 150-6.
[http://dx.doi.org/10.1111/j.1574-6968.2011.02324.x] [PMID: 21631577]

[42] Sabra W, Dietz D, Tjahjasari D, Zeng AP. Biosystems analysis and engineering of microbial consortia for industrial biotechnology. Eng Life Sci 2010; 10(5): 407-21.
[http://dx.doi.org/10.1002/elsc.201000111]

[43] Kleerebezem R, van Loosdrecht MC. Mixed culture biotechnology for bioenergy production. Curr Opin Biotechnol 2007; 18(3): 207-12.
[http://dx.doi.org/10.1016/j.copbio.2007.05.001] [PMID: 17509864]

[44] Reis MA, Serafim LS, Lemos PC, Ramos AM, Aguiar FR, Van Loosdrecht MC. Production of polyhydroxyalkanoates by mixed microbial cultures. Bioprocess Biosyst Eng 2003; 25(6): 377-85.
[http://dx.doi.org/10.1007/s00449-003-0322-4] [PMID: 13680343]

[45] Valentino F, Beccari M, Fraraccio S, Zanaroli G, Majone M. Feed frequency in a sequencing batch reactor strongly affects the production of polyhydroxyalkanoates (PHAs) from volatile fatty acids. N Biotechnol 2014; 31(4): 264-75.
[http://dx.doi.org/10.1016/j.nbt.2013.10.006] [PMID: 24184912]

[46] Johnson K, Jiang Y, Kleerebezem R, Muyzer G, van Loosdrecht MC. Enrichment of a mixed bacterial culture with a high polyhydroxyalkanoate storage capacity. Biomacromolecules 2009; 10(4): 670-6.

[http://dx.doi.org/10.1021/bm8013796] [PMID: 19193058]

[47] Kaur G, Roy I. Strategies for large-scale production of polyhydroxyalkanoates. Chem Biochem Eng Q 2015; 29(2): 157-72.
[http://dx.doi.org/10.15255/CABEQ.2014.2255]

[48] Choi J, Lee SY. Factors affecting the economics of polyhydroxyalkanoate production by bacterial fermentation. Appl Microbiol Biotechnol 1999; 51(1): 13-21.
[http://dx.doi.org/10.1007/s002530051357]

[49] Du C, Sabirova J, Soetaert W, Ki Carol Lin S. Polyhydroxyalkanoates production from low-cost sustainable raw materials. Curr Chem Biol 2012; 6(1): 14-25.

[50] Koller M, Salerno A, Dias M, Reiterer A, Braunegg G. Modern biotechnological polymer synthesis: a review. Food Technol Biotechnol 2010; 48(3): 255-69.

[51] Povolo S, Toffano P, Basaglia M, Casella S. Polyhydroxyalkanoates production by engineered *Cupriavidus necator* from waste material containing lactose. Bioresour Technol 2010; 101(20): 7902-7.
[http://dx.doi.org/10.1016/j.biortech.2010.05.029] [PMID: 20537531]

[52] Prazeres AR, Carvalho F, Rivas J. Cheese whey management: a review. J Environ Manage 2012; 110: 48-68.
[http://dx.doi.org/10.1016/j.jenvman.2012.05.018] [PMID: 22721610]

[53] Dutra Rosolen M, Gennari A, Volpato G, Volken de Souza CF. Lactose hydrolysis in milk and dairy whey using microbial β-galactosidases. Enz Res 2015.

[54] Smithers GW. Whey and whey proteins—from 'gutter-to-gold'. Int Dairy J 2008; 18(7): 695-704.
[http://dx.doi.org/10.1016/j.idairyj.2008.03.008]

[55] Yang S-T. Bioprocessing for value-added products from renewable resources: new technologies and applications. Elsevier 2011.

[56] Cota-Navarro CB, Carrillo-Reyes J, Davila-Vazquez G, Alatriste-Mondragón F, Razo-Flores E. Continuous hydrogen and methane production in a two-stage cheese whey fermentation system. Water Sci Technol 2011; 64(2): 367-74.
[http://dx.doi.org/10.2166/wst.2011.631] [PMID: 22097009]

[57] Dragone G, Mussatto SI, Silva JB, Teixeira JA. Optimal fermentation conditions for maximizing the ethanol production by *Kluyveromyces fragilis* from cheese whey powder. Biomass Bioenergy 2011; 35(5): 1977-82.
[http://dx.doi.org/10.1016/j.biombioe.2011.01.045]

[58] Schirru S, Favaro L, Mangia NP, *et al.* Comparison of bacteriocins production from *Enterococcus faecium* strains in cheese whey and optimised commercial MRS medium. Ann Microbiol 2014; 64(1): 321-31.
[http://dx.doi.org/10.1007/s13213-013-0667-0]

[59] Koller M, Muhr A, Reiterer A, *et al.* Whey lactose as a raw material for microbial production of biodegradable polyesters. INTECH Open Access Publisher 2012.
[http://dx.doi.org/10.5772/48737]

[60] Yellore V, Desai A. Production of poly-3-hydroxybutyrate from lactose and whey by *Methylobacterium* sp. ZP24. Lett Appl Microbiol 1998; 26(6): 391-4. [http://dx.doi.org/10.1046/j.1472-765X.1998.00362.x] [PMID: 9717306]

[61] Nath A, Dixit M, Bandiya A, Chavda S, Desai AJ. Enhanced PHB production and scale up studies using cheese whey in fed batch culture of *Methylobacterium* sp. ZP24. Bioresour Technol 2008; 99(13): 5749-55. [http://dx.doi.org/10.1016/j.biortech.2007.10.017] [PMID: 18032031]

[62] Povolo S, Casella S. Bacterial production of PHA from lactose and cheese whey permeate. Macromol Symp 2003; 197(1): 1-10. [http://dx.doi.org/10.1002/masy.200350701]

[63] Koller M, Atlić A, Gonzalez-Garcia Y, Kutschera C, Braunegg G. Polyhydroxyalkanoate (PHA) biosynthesis from whey lactose. Macromol Symp 2008; 272(1): 87-92. [http://dx.doi.org/10.1002/masy.200851212]

[64] Koller M, Hesse P, Bona R, Kutschera C, Atlić A, Braunegg G. Potential of various archae- and eubacterial strains as industrial polyhydroxyalkanoate producers from whey. Macromol Biosci 2007; 7(2): 218-26. [http://dx.doi.org/10.1002/mabi.200600211] [PMID: 17295410]

[65] Obruca S, Marova I, Melusova S, Mravcova L. Production of polyhydroxyalkanoates from cheese whey employing *Bacillus megaterium* CCM 2037. Ann Microbiol 2011; 61(4): 947-53. [http://dx.doi.org/10.1007/s13213-011-0218-5]

[66] Pries A, Steinbüchel A, Schlegel HG. Lactose-and galactose-utilizing strains of poly (hydroxyalkanoic acid)-accumulating *Alcaligenes eutrophus* and *Pseudomonas saccharophila* obtained by recombinant DNA technology. Appl Microbiol Biotechnol 1990; 33(4): 410-7. [http://dx.doi.org/10.1007/BF00176656]

[67] Yadav R, Solomon S. Potential of developing sugarcane by-product based industries in India. Sugar Tech 2006; 8(2): 104-11. [http://dx.doi.org/10.1007/BF02943642]

[68] Gopal AR, Kammen DM. Molasses for ethanol: the economic and environmental impacts of a new pathway for the lifecycle greenhouse gas analysis of sugarcane ethanol. Environ Res Lett 2009; 4(4): 044005. [http://dx.doi.org/10.1088/1748-9326/4/4/044005]

[69] Saxena R, Adhikari D, Goyal H. Biomass-based energy fuel through biochemical routes: a review. Renew Sustain Energy Rev 2009; 13(1): 167-78. [http://dx.doi.org/10.1016/j.rser.2007.07.011]

[70] Solomon S. Sugarcane by-products based industries in India. Sugar Tech 2011; 13(4): 408-16. [http://dx.doi.org/10.1007/s12355-011-0114-0]

[71] Albuquerque MG, Eiroa M, Torres C, Nunes BR, Reis MA. Strategies for the development of a side stream process for polyhydroxyalkanoate (PHA) production from sugar cane molasses. J Biotechnol 2007; 130(4): 411-21. [http://dx.doi.org/10.1016/j.jbiotec.2007.05.011] [PMID: 17602776]

[72] Albuquerque MG, Torres CA, Reis MA. Polyhydroxyalkanoate (PHA) production by a mixed microbial culture using sugar molasses: effect of the influent substrate concentration on culture selection. Water Res 2010; 44(11): 3419-33. [http://dx.doi.org/10.1016/j.watres.2010.03.021] [PMID: 20427069]

[73] Gumel A, Annuar M, Chisti Y. Recent advances in the production, recovery and applications of polyhydroxyalkanoates. J Polym Environ 2013; 21(2): 580-605. [http://dx.doi.org/10.1007/s10924-012-0527-1]

[74] Page WJ. Suitability of commercial beet molasses fractions as substrates fro polyhydroxyalkanoate production by *Azotobacter vinelandii* UWD. Biotechnol Lett 1992; 14(5): 385-90. [http://dx.doi.org/10.1007/BF01021252]

[75] Braunegg G, Genser K, Bona R, Haage G, Schellauf F, Winkler E. Production of PHAs from agricultural waste material. Macromol Symp 1999; 144(1): 375-83. [http://dx.doi.org/10.1002/masy.19991440135]

[76] Solaiman DK, Ashby RD, Hotchkiss AT Jr, Foglia TA. Biosynthesis of medium-chain-length poly(hydroxyalkanoates) from soy molasses. Biotechnol Lett 2006; 28(3): 157-62. [http://dx.doi.org/10.1007/s10529-005-5329-2] [PMID: 16489492]

[77] Santimano M, Prabhu NN, Garg S. PHA production using low-cost agro-industrial wastes by *Bacillus* sp. strain COLI/Afi. Res J Microbiol 2009; 4(3): 89-96. [http://dx.doi.org/10.3923/jm.2009.89.96]

[78] Omar S, Rayes A, Eqaab A, Voß I, Steinbüchel A. Optimization of cell growth and poly (3-hydroxybutyrate) accumulation on date syrup by a *Bacillus megaterium* strain. Biotechnol Lett 2001; 23(14): 1119-23. [http://dx.doi.org/10.1023/A:1010559800535]

[79] Anjali M, Sukumar C, Kanakalakshmi A, Shanthi K. Enhancement of growth and production of polyhydroxyalkanoates by *Bacillus subtilis* from agro-industrial waste as carbon substrates. Compos Interfaces 2014; 21(2): 111-9. [http://dx.doi.org/10.1080/15685543.2013.834200]

[80] Wu Q, Huang H, Hu G, Chen J, Ho K, Chen G-Q. Production of poly-3-hydroxybutyrate by *Bacillus* sp. JMa5 cultivated in molasses media. A Van Leeuw 2001; 80(2): 111-8.

[81] Kanjanachumpol P, Kulpreecha S, Tolieng V, Thongchul N. Enhancing polyhydroxybutyrate production from high cell density fed-batch fermentation of *Bacillus megaterium* BA-019. Bioprocess Biosyst Eng 2013; 36(10): 1463-74. [http://dx.doi.org/10.1007/s00449-013-0885-7] [PMID: 23340799]

[82] Park SJ, Jang YA, Noh W, *et al.* Metabolic engineering of *Ralstonia eutropha* for the production of polyhydroxyalkanoates from sucrose. Biotechnol Bioeng 2015; 112(3): 638-43. [http://dx.doi.org/10.1002/bit.25469] [PMID: 25258020]

[83] Park SJ, Jang Y-A, Lee H, *et al.* Metabolic engineering of *Ralstonia eutropha* for the biosynthesis of 2-hydroxyacid-containing polyhydroxyalkanoates. Metab Eng 2013; 20: 20-8. [http://dx.doi.org/10.1016/j.ymben.2013.08.002] [PMID: 23973656]

[84] Balat M, Balat H. Progress in biodiesel processing. Appl Energy 2010; 87(6): 1815-35.

[http://dx.doi.org/10.1016/j.apenergy.2010.01.012]

[85] Williams PT. Waste treatment and disposal. U.K: John Wiley & Sons 2013.

[86] Talebian-Kiakalaieh A, Amin NA, Mazaheri H. A review on novel processes of biodiesel production from waste cooking oil. Appl Energy 2013; 104: 683-710.
[http://dx.doi.org/10.1016/j.apenergy.2012.11.061]

[87] Riedel SL, Lu J, Stahl U, Brigham CJ. Lipid and fatty acid metabolism in *Ralstonia eutropha*: relevance for the biotechnological production of value-added products. Appl Microbiol Biotechnol 2014; 98(4): 1469-83.
[http://dx.doi.org/10.1007/s00253-013-5430-8] [PMID: 24343766]

[88] Fernández D, Rodríguez E, Bassas M, *et al.* Agro-industrial oily wastes as substrates for PHA production by the new strain *Pseudomonas aeruginosa* NCIB 40045: effect of culture conditions. Biochem Eng J 2005; 26(2): 159-67.
[http://dx.doi.org/10.1016/j.bej.2005.04.022]

[89] López-Cuellar MR, Alba-Flores J, Rodríguez JN, Pérez-Guevara F. Production of polyhydroxyalkanoates (PHAs) with canola oil as carbon source. Int J Biol Macromol 2011; 48(1): 74-80.
[http://dx.doi.org/10.1016/j.ijbiomac.2010.09.016] [PMID: 20933541]

[90] Majid MI, Akmal DH, Few LL, *et al.* Production of poly(3-hydroxybutyrate) and its copolymer poly(3-hydroxybutyrate-co-3-hydroxyvalerate) by *Erwinia* sp. USMI-20. Int J Biol Macromol 1999; 25(1-3): 95-104.
[http://dx.doi.org/10.1016/S0141-8130(99)00020-3] [PMID: 10416655]

[91] Taniguchi I, Kagotani K, Kimura Y. Microbial production of poly (hydroxyalkanoate) s from waste edible oils. Green Chem 2003; 5(5): 545-8.
[http://dx.doi.org/10.1039/B304800B]

[92] Obruca S, Snajdar O, Svoboda Z, Marova I. Application of random mutagenesis to enhance the production of polyhydroxyalkanoates by *Cupriavidus necator* H16 on waste frying oil. World J Microbiol Biotechnol 2013; 29(12): 2417-28.
[http://dx.doi.org/10.1007/s11274-013-1410-5] [PMID: 23801326]

[93] Verlinden RA, Hill DJ, Kenward MA, Williams CD, Piotrowska-Seget Z, Radecka IK. Production of polyhydroxyalkanoates from waste frying oil by *Cupriavidus necator*. AMB Express 2011; 1(1): 11.
[http://dx.doi.org/10.1186/2191-0855-1-11] [PMID: 21906352]

[94] Obruca S, Marova I, Snajdar O, Mravcova L, Svoboda Z. Production of poly(3-hydroxybutyrate-c--3-hydroxyvalerate) by *Cupriavidus necator* from waste rapeseed oil using propanol as a precursor of 3-hydroxyvalerate. Biotechnol Lett 2010; 32(12): 1925-32.
[http://dx.doi.org/10.1007/s10529-010-0376-8] [PMID: 20814716]

[95] Cromwick A-M, Foglia T, Lenz R. The microbial production of poly (hydroxyalkanoates) from tallow. Appl Microbiol Biotechnol 1996; 46(5-6): 464-9.
[http://dx.doi.org/10.1007/s002530050845]

[96] Titz M, Kettl K-H, Shahzad K, Koller M, Schnitzer H, Narodoslawsky M. Process optimization for efficient biomediated PHA production from animal-based waste streams. Clean Technol Environ

Policy 2012; 14(3): 495-503.
[http://dx.doi.org/10.1007/s10098-012-0464-7]

[97] Sangkharak K, Prasertsan P. The production of polyhydroxyalkanoate by *Bacillus licheniformis* using sequential mutagenesis and optimization. Biotechnol Bioproc Eng 2013; 18(2): 272-9.
[http://dx.doi.org/10.1007/s12257-012-0615-z]

[98] Yu L, Dean K, Li L. Polymer blends and composites from renewable resources. Prog Polym Sci 2006; 31(6): 576-602.
[http://dx.doi.org/10.1016/j.progpolymsci.2006.03.002]

[99] Fukui T, Doi Y. Efficient production of polyhydroxyalkanoates from plant oils by *Alcaligenes eutrophus* and its recombinant strain. Appl Microbiol Biotechnol 1998; 49(3): 333-6.
[http://dx.doi.org/10.1007/s002530051178] [PMID: 9581296]

[100] Kahar P, Tsuge T, Taguchi K, Doi Y. High yield production of polyhydroxyalkanoates from soybean oil by *Ralstonia eutropha* and its recombinant strain. Polym Degrad Stabil 2004; 83(1): 79-86.
[http://dx.doi.org/10.1016/S0141-3910(03)00227-1]

[101] Bhubalan K, Lee W-H, Loo C-Y, *et al.* Controlled biosynthesis and characterization of poly (3-hydroxybutyrate-co-3-hydroxyvalerate-co-3-hydroxyhexanoate) from mixtures of palm kernel oil and 3HV-precursors. Polym Degrad Stabil 2008; 93(1): 17-23.
[http://dx.doi.org/10.1016/j.polymdegradstab.2007.11.004]

[102] Loo C-Y, Lee W-H, Tsuge T, Doi Y, Sudesh K. Biosynthesis and characterization of poly(3-hydroxybutyrate-co-3- hydroxyhexanoate) from palm oil products in a *Wautersia eutropha* mutant. Biotechnol Lett 2005; 27(18): 1405-10.
[http://dx.doi.org/10.1007/s10529-005-0690-8] [PMID: 16215858]

[103] Mifune J, Nakamura S, Fukui T. Engineering of *pha* operon on *Cupriavidus necator* chromosome for efficient biosynthesis of poly (3-hydroxybutyrate-co-3-hydroxyhexanoate) from vegetable oil. Polym Degrad Stabil 2010; 95(8): 1305-12.
[http://dx.doi.org/10.1016/j.polymdegradstab.2010.02.026]

[104] Budde CF, Riedel SL, Willis LB, Rha C, Sinskey AJ. Production of poly(3-hydroxybutyrate-c--3-hydroxyhexanoate) from plant oil by engineered *Ralstonia eutropha* strains. Appl Environ Microbiol 2011; 77(9): 2847-54.
[http://dx.doi.org/10.1128/AEM.02429-10] [PMID: 21398488]

[105] Riedel SL, Jahns S, Koenig S, *et al.* Polyhydroxyalkanoates production with *Ralstonia eutropha* from low quality waste animal fats. J Biotechnol 2015; 214: 119-27.
[http://dx.doi.org/10.1016/j.jbiotec.2015.09.002] [PMID: 26428087]

[106] Bhubalan K, Rathi D-N, Abe H, Iwata T, Sudesh K. Improved synthesis of P (3HB-co-3HV-co-3HHx) terpolymers by mutant *Cupriavidus necator* using the PHA synthase gene of *Chromobacterium* sp. USM2 with high affinity towards 3HV. Polym Degrad Stabil 2010; 95(8): 1436-42.
[http://dx.doi.org/10.1016/j.polymdegradstab.2009.12.018]

[107] Solaiman DK, Ashby RD, Foglia TA. Production of polyhydroxyalkanoates from intact triacylglycerols by genetically engineered *Pseudomonas.* Appl Microbiol Biotechnol 2001; 56(5-6): 664-9.
[http://dx.doi.org/10.1007/s002530100692] [PMID: 11601611]

[108] Romanelli MG, Povolo S, Favaro L, Fontana F, Basaglia M, Casella S. Engineering *Delftia acidovorans* DSM39 to produce polyhydroxyalkanoates from slaughterhouse waste. Int J Biol Macromol 2014; 71: 21-7.
[http://dx.doi.org/10.1016/j.ijbiomac.2014.03.049] [PMID: 24704165]

[109] Povolo S, Romanelli MG, Fontana F, Basaglia M, Casella S. Production of polyhydroxyalkanoates from fatty wastes. J Polym Environ 2012; 20(4): 944-9.
[http://dx.doi.org/10.1007/s10924-012-0485-7]

[110] Kamm B, Gruber PR, Kamm M. Biorefineries–industrial processes and products. Wiley Online Library 2007.

[111] Morrison B, Golden JS. An empirical analysis of the industrial bioeconomy: implications for renewable resources and the environment. BioResources 2015; 10(3): 4411-40.
[http://dx.doi.org/10.15376/biores.10.3.4411-4440]

[112] Bergthaller W. Starch world markets and isolation of starch. Chem Funct Prop Food Sacch 2004; pp. 1-20.

[113] Eliasson A-C. Starch in food: Structure, function and applications. New York: CRC Press 2004.

[114] Favaro L, Jooste T, Basaglia M, *et al.* Codon-optimized glucoamylase sGAI of *Aspergillus awamori* improves starch utilization in an industrial yeast. Appl Microbiol Biotechnol 2012; 95(4): 957-68.
[http://dx.doi.org/10.1007/s00253-012-4001-8] [PMID: 22450569]

[115] Favaro L, Jooste T, Basaglia M, *et al.* Designing industrial yeasts for the consolidated bioprocessing of starchy biomass to ethanol. Bioengineered 2013; 4(2): 97-102.
[http://dx.doi.org/10.4161/bioe.22268] [PMID: 22989992]

[116] Kim BS, Chang HN. Production of poly (3-hydroxybutyrate) from starch by *Azotobacter chroococcum.* Biotechnol Lett 1998; 20(2): 109-12.
[http://dx.doi.org/10.1023/A:1005307903684]

[117] Halami PM. Production of polyhydroxyalkanoate from starch by the native isolate *Bacillus cereus* CFR06. World J Microbiol Biotechnol 2008; 24(6): 805-12.
[http://dx.doi.org/10.1007/s11274-007-9543-z]

[118] González-García Y, Rosales MA, González-Reynoso O, Sanjuán-Dueñas R, Córdova J. Polyhydroxybutyrate production by *Saccharophagus degradans* using raw starch as carbon source. Eng Life Sci 2011; 11(1): 59-64.
[http://dx.doi.org/10.1002/elsc.201000118]

[119] Han X, Satoh Y, Kuriki Y, *et al.* Polyhydroxyalkanoate production by a novel bacterium *Massilia* sp. UMI-21 isolated from seaweed, and molecular cloning of its polyhydroxyalkanoate synthase gene. J Biosci Bioeng 2014; 118(5): 514-9.
[http://dx.doi.org/10.1016/j.jbiosc.2014.04.022] [PMID: 24932969]

[120] Haas R, Jin B, Zepf FT. Production of Poly(3-hydroxybutyrate) from waste potato starch. Biosci Biotechnol Biochem 2008; 72(1): 253-6.
[http://dx.doi.org/10.1271/bbb.70503] [PMID: 18175895]

[121] Castilho LR, Mitchell DA, Freire DM. Production of polyhydroxyalkanoates (PHAs) from waste materials and by-products by submerged and solid-state fermentation. Bioresour Technol 2009;

100(23): 5996-6009.
[http://dx.doi.org/10.1016/j.biortech.2009.03.088] [PMID: 19581084]

[122] Chen CW, Don T-M, Yen H-F. Enzymatic extruded starch as a carbon source for the production of poly (3-hydroxybutyrate-co-3-hydroxyvalerate) by *Haloferax mediterranei.* Process Biochem 2006; 41(11): 2289-96.
[http://dx.doi.org/10.1016/j.procbio.2006.05.026]

[123] Van-Thuoc D, Quillaguamán J, Mamo G, Mattiasson B. Utilization of agricultural residues for poly(3-hydroxybutyrate) production by *Halomonas boliviensis* LC1. J Appl Microbiol 2008; 104(2): 420-8. [PMID: 17887984]

[124] Favaro L, Alibardi L, Lavagnolo MC, Casella S, Basaglia M. Effects of inoculum and indigenous microflora on hydrogen production from the organic fraction of municipal solid waste. Int J Hyd Energ 2013; 38(27): 11774-9.
[http://dx.doi.org/10.1016/j.ijhydene.2013.06.137]

[125] Favaro L, Basaglia M, Trento A, *et al.* Exploring grape marc as trove for new thermotolerant and inhibitor-tolerant *Saccharomyces cerevisiae* strains for second-generation bioethanol production. Biotechnol Biofuels 2013; 6(1): 168.
[http://dx.doi.org/10.1186/1754-6834-6-168] [PMID: 24286305]

[126] Liu S, Abrahamson LP, Scott GM. Biorefinery: Ensuring biomass as a sustainable renewable source of chemicals, materials, and energy. Biomass Bioenergy 2012; 39: 1-4.
[http://dx.doi.org/10.1016/j.biombioe.2010.12.042]

[127] Clark JH, Deswarte F. Introduction to chemicals from biomass. John Wiley & Sons 2015.
[http://dx.doi.org/10.1002/9781118714478]

[128] Jönsson LJ, Alriksson B, Nilvebrant N-O. Bioconversion of lignocellulose: inhibitors and detoxification. Biotechnol Biofuels 2013; 6(1): 16.
[http://dx.doi.org/10.1186/1754-6834-6-16] [PMID: 23356676]

[129] Pan W, Perrotta JA, Stipanovic AJ, Nomura CT, Nakas JP. Production of polyhydroxyalkanoates by *Burkholderia cepacia* ATCC 17759 using a detoxified sugar maple hemicellulosic hydrolysate. J Ind Microbiol Biotechnol 2012; 39(3): 459-69.
[http://dx.doi.org/10.1007/s10295-011-1040-6] [PMID: 21953365]

[130] Obruca S, Benesova P, Marsalek L, Marova I. Use of lignocellulosic materials for PHA production. Chem Biochem Eng Q 2015; 29(2): 135-44.
[http://dx.doi.org/10.15255/CABEQ.2014.2253]

[131] Cesário MT, de Almeida MC. Lignocellulosic hydrolysates for the production of polyhydroxyalkanoates. In: Kamm B, Ed. Microorganisms in Biorefineries. Berlin Heidelberg, Springer 2015; pp. 79-104.

[132] Yu J, Stahl H. Microbial utilization and biopolyester synthesis of bagasse hydrolysates. Bioresour Technol 2008; 99(17): 8042-8.
[http://dx.doi.org/10.1016/j.biortech.2008.03.071] [PMID: 18474421]

[133] Pohlmann A, Fricke WF, Reinecke F, *et al.* Genome sequence of the bioplastic-producing "Knallgas" bacterium *Ralstonia eutropha* H16. Nat Biotechnol 2006; 24(10): 1257-62.

[http://dx.doi.org/10.1038/nbt1244] [PMID: 16964242]

[134] Lu X, Liu G, Wang Y, Ding J, Weng W. Engineering of an L-arabinose metabolic pathway in *Ralstonia eutropha* W50. Wei Sheng Wu Xue Bao 2013; 53(12): 1267-75. [PMID: 24697099]

[135] Liu K, Liu G, Zhang Y, Ding J, Weng W. Engineering of a D-xylose metabolic pathway in *Ralstonia eutropha* W50. Acta Microbiol Sin 2014; 54(1): 42-52.

[136] da Silva GP, Mack M, Contiero J. Glycerol: a promising and abundant carbon source for industrial microbiology. Biotechnol Adv 2009; 27(1): 30-9. [http://dx.doi.org/10.1016/j.biotechadv.2008.07.006] [PMID: 18775486]

[137] Chatzifragkou A, Papanikolaou S. Effect of impurities in biodiesel-derived waste glycerol on the performance and feasibility of biotechnological processes. Appl Microbiol Biotechnol 2012; 95(1): 13-27. [http://dx.doi.org/10.1007/s00253-012-4111-3] [PMID: 22581036]

[138] Ibrahim MH, Steinbüchel A. *Zobellella denitrificans* strain MW1, a newly isolated bacterium suitable for poly(3-hydroxybutyrate) production from glycerol. J Appl Microbiol 2010; 108(1): 214-25. [http://dx.doi.org/10.1111/j.1365-2672.2009.04413.x] [PMID: 19566718]

[139] Ashby RD, Solaiman DK, Foglia TA. Bacterial poly (hydroxyalkanoate) polymer production from the biodiesel co-product stream. J Polym Environ 2004; 12(3): 105-12. [http://dx.doi.org/10.1023/B:JOOE.0000038541.54263.d9]

[140] Kawata Y, Aiba S. Poly(3-hydroxybutyrate) production by isolated *Halomonas* sp. KM-1 using waste glycerol. Biosci Biotechnol Biochem 2010; 74(1): 175-7. [http://dx.doi.org/10.1271/bbb.90459] [PMID: 20057130]

[141] Mothes G, Schnorpfeil C, Ackermann JU. Production of PHB from crude glycerol. Eng Life Sci 2007; 7(5): 475-9. [http://dx.doi.org/10.1002/elsc.200620210]

[142] Cavalheiro JM, de Almeida MC, Grandfils C, Da Fonseca M. Poly (3-hydroxybutyrate) production by *Cupriavidus necator* using waste glycerol. Process Biochem 2009; 44(5): 509-15. [http://dx.doi.org/10.1016/j.procbio.2009.01.008]

[143] Tanadchangsaeng N, Yu J. Microbial synthesis of polyhydroxybutyrate from glycerol: gluconeogenesis, molecular weight and material properties of biopolyester. Biotechnol Bioeng 2012; 109(11): 2808-18. [http://dx.doi.org/10.1002/bit.24546] [PMID: 22566160]

[144] Yang F, Hanna MA, Sun R. Value-added uses for crude glycerol-a byproduct of biodiesel production. Biotechnol Biofuels 2012; 5(13): 1-10. [PMID: 22212435]

[145] Cavalheiro JM, Raposo RS, de Almeida MC, *et al.* Effect of cultivation parameters on the production of poly(3-hydroxybutyrate-co-4-hydroxybutyrate) and poly(3-hydroxybutyrate-4-hydroxybutyra-e-3-hydroxyvalerate) by *Cupriavidus necator* using waste glycerol. Bioresour Technol 2012; 111: 391-7. [http://dx.doi.org/10.1016/j.biortech.2012.01.176] [PMID: 22382294]

[146] Volodina E, Raberg M, Steinbüchel A. Engineering the heterotrophic carbon sources utilization range of *Ralstonia eutropha* H16 for applications in biotechnology. Crit Rev Biotechnol 2015; 2: 1-14. [http://dx.doi.org/10.3109/07388551.2015.1079698] [PMID: 26329669]

[147] Lopar M, Špoljarić IV, Cepanec N, Koller M, Braunegg G, Horvat P. Study of metabolic network of *Cupriavidus necator* DSM 545 growing on glycerol by applying elementary flux modes and yield space analysis. J Ind Microbiol Biotechnol 2014; 41(6): 913-30. [http://dx.doi.org/10.1007/s10295-014-1439-y] [PMID: 24715530]

[148] Fukui T, Mukoyama M, Orita I, Nakamura S. Enhancement of glycerol utilization ability of *Ralstonia eutropha* H16 for production of polyhydroxyalkanoates. Appl Microbiol Biotechnol 2014; 98(17): 7559-68. [http://dx.doi.org/10.1007/s00253-014-5831-3] [PMID: 24878751]

[149] Escapa IF, del Cerro C, García JL, Prieto MA. The role of GlpR repressor in *Pseudomonas putida* KT2440 growth and PHA production from glycerol. Environ Microbiol 2013; 15(1): 93-110. [http://dx.doi.org/10.1111/j.1462-2920.2012.02790.x] [PMID: 22646161]

[150] Park SJ, Kim TW, Kim MK, Lee SY, Lim S-C. Advanced bacterial polyhydroxyalkanoates: towards a versatile and sustainable platform for unnatural tailor-made polyesters. Biotechnol Adv 2012; 30(6): 1196-206. [http://dx.doi.org/10.1016/j.biotechadv.2011.11.007] [PMID: 22137963]

[151] Kozhevnikov IV, Volova TG, Hai T, Steinbüchel A. [Cloning and molecular organization of the polyhydroxyalkanoic acid synthase gene (*phaC*) of *Ralstonia eutropha* strain B5786]. Prikl Biokhim Mikrobiol 2010; 46(2): 153-60. [PMID: 20391757]

[152] Liebergesell M, Schmidt B, Steinbüchel A. Isolation and identification of granule-associated proteins relevant for poly(3-hydroxyalkanoic acid) biosynthesis in *Chromatium vinosum* D. FEMS Microbiol Lett 1992; 78(2-3): 227-32. [http://dx.doi.org/10.1111/j.1574-6968.1992.tb05572.x] [PMID: 1490603]

[153] Meng D-C, Shen R, Yao H, Chen J-C, Wu Q, Chen G-Q. Engineering the diversity of polyesters. Curr Opin Biotechnol 2014; 29: 24-33. [http://dx.doi.org/10.1016/j.copbio.2014.02.013] [PMID: 24632193]

[154] Aldor IS, Keasling JD. Process design for microbial plastic factories: metabolic engineering of polyhydroxyalkanoates. Curr Opin Biotechnol 2003; 14(5): 475-83. [http://dx.doi.org/10.1016/j.copbio.2003.09.002] [PMID: 14580576]

[155] Shiloach J, Fass R. Growing *E. coli* to high cell "density - a" historical perspective on method development. Biotechnol Adv 2005; 23(5): 345-57. [http://dx.doi.org/10.1016/j.biotechadv.2005.04.004] [PMID: 15899573]

[156] Schubert P, Steinbüchel A, Schlegel HG. Cloning of the *Alcaligenes eutrophus* genes for synthesis of poly-beta-hydroxybutyric acid (PHB) and synthesis of PHB in *Escherichia coli*. J Bacteriol 1988; 170(12): 5837-47. [PMID: 2848014]

[157] Slater SC, Voige WH, Dennis DE. Cloning and expression in *Escherichia coli* of the *Alcaligenes*

eutrophus H16 poly-beta-hydroxybutyrate biosynthetic pathway. J Bacteriol 1988; 170(10): 4431-6. [PMID: 3049530]

[158] Langenbach S, Rehm BH, Steinbüchel A. Functional expression of the PHA synthase gene *phaC1* from *Pseudomonas aeruginosa* in *Escherichia coli* results in poly(3-hydroxyalkanoate) synthesis. FEMS Microbiol Lett 1997; 150(2): 303-9. [http://dx.doi.org/10.1016/S0378-1097(97)00142-0] [PMID: 9170275]

[159] Choi JI, Lee SY, Han K. Cloning of the *Alcaligenes latus* polyhydroxyalkanoate biosynthesis genes and use of these genes for enhanced production of Poly(3-hydroxybutyrate) in *Escherichia coli.* Appl Environ Microbiol 1998; 64(12): 4897-903. [PMID: 9835580]

[160] Liu S-J, Steinbüchel A. A novel genetically engineered pathway for synthesis of poly(hydroxyalkanoic acids) in *Escherichia coli.* Appl Environ Microbiol 2000; 66(2): 739-43. [http://dx.doi.org/10.1128/AEM.66.2.739-743.2000] [PMID: 10653745]

[161] Mahishi LH, Tripathi G, Rawal SK. Poly(3-hydroxybutyrate) (PHB) synthesis by recombinant *Escherichia coli* harbouring *Streptomyces aureofaciens* PHB biosynthesis genes: effect of various carbon and nitrogen sources. Microbiol Res 2003; 158(1): 19-27. [http://dx.doi.org/10.1078/0944-5013-00161] [PMID: 12608576]

[162] Andreessen B, Lange AB, Robenek H, Steinbüchel A. Conversion of glycerol to poly(3-hydroxypropionate) in recombinant *Escherichia coli.* Appl Environ Microbiol 2010; 76(2): 622-6. [http://dx.doi.org/10.1128/AEM.02097-09] [PMID: 19933347]

[163] Kawada J, Lütke-Eversloh T, Steinbüchel A, Marchessault RH. Physical properties of microbial polythioesters: characterization of poly(3-mercaptoalkanoates) synthesized by engineered *Escherichia coli.* Biomacromolecules 2003; 4(6): 1698-702. [http://dx.doi.org/10.1021/bm0341327] [PMID: 14606898]

[164] Thakor N, Lütke-Eversloh T, Steinbüchel A. Application of the BPEC pathway for large-scale biotechnological production of poly(3-mercaptopropionate) by recombinant *Escherichia coli*, including a novel *in situ* isolation method. Appl Environ Microbiol 2005; 71(2): 835-41. [http://dx.doi.org/10.1128/AEM.71.2.835-841.2005] [PMID: 15691938]

[165] Zhou Q, Shi Z-Y, Meng D-C, Wu Q, Chen J-C, Chen G-Q. Production of 3-hydroxypropionate homopolymer and poly(3-hydroxypropionate-co-4-hydroxybutyrate) copolymer by recombinant *Escherichia coli.* Metab Eng 2011; 13(6): 777-85. [http://dx.doi.org/10.1016/j.ymben.2011.10.002] [PMID: 22024131]

[166] Zhou X-Y, Yuan X-X, Shi Z-Y, *et al.* Hyperproduction of poly(4-hydroxybutyrate) from glucose by recombinant *Escherichia coli.* Microb Cell Fact 2012; 11(1): 54. [http://dx.doi.org/10.1186/1475-2859-11-54] [PMID: 22550959]

[167] Li Z-J, Shi Z-Y, Jian J, Guo Y-Y, Wu Q, Chen G-Q. Production of poly(3-hydroxybutyrate-c--4-hydroxybutyrate) from unrelated carbon sources by metabolically engineered *Escherichia coli.* Metab Eng 2010; 12(4): 352-9. [http://dx.doi.org/10.1016/j.ymben.2010.03.003] [PMID: 20304089]

[168] Yang JE, Choi YJ, Lee SJ, *et al.* Metabolic engineering of *Escherichia coli* for biosynthesis of poly(3-hydroxybutyrate-co-3-hydroxyvalerate) from glucose. Appl Microbiol Biotechnol 2014; 98(1): 95-

104.
[http://dx.doi.org/10.1007/s00253-013-5285-z] [PMID: 24113828]

[169] Meng D-C, Shi Z-Y, Wu L-P, *et al.* Production and characterization of poly(3-hydroxypropionate--o-4-hydroxybutyrate) with fully controllable structures by recombinant *Escherichia coli* containing an engineered pathway. Metab Eng 2012; 14(4): 317-24.
[http://dx.doi.org/10.1016/j.ymben.2012.04.003] [PMID: 22561235]

[170] Tripathi L, Wu L-P, Meng D, Chen J, Chen G-Q. Biosynthesis and characterization of diblock copolymer of p(3-hydroxypropionate)-block-p(4-hydroxybutyrate) from recombinant *Escherichia coli.* Biomacromolecules 2013; 14(3): 862-70.
[http://dx.doi.org/10.1021/bm3019517] [PMID: 23351169]

[171] Wang Q, Yang P, Xian M, *et al.* Production of block copolymer poly (3-hydroxybutyrate)-block-poly (3-hydroxypropionate) with adjustable structure from an inexpensive carbon source. ACS Macro Lett 2013; 2(11): 996-1000.
[http://dx.doi.org/10.1021/mz400446g]

[172] Wang Q, Zhuang Q, Liang Q, Qi Q. Polyhydroxyalkanoic acids from structurally-unrelated carbon sources in *Escherichia coli.* Appl Microbiol Biotechnol 2013; 97(8): 3301-7.
[http://dx.doi.org/10.1007/s00253-013-4809-x] [PMID: 23474618]

[173] Kang Z, Gao C, Wang Q, Liu H, Qi Q. A novel strategy for succinate and polyhydroxybutyrate co-production in *Escherichia coli.* Bioresour Technol 2010; 101(19): 7675-8.
[http://dx.doi.org/10.1016/j.biortech.2010.04.084] [PMID: 20472425]

[174] Martínez-García E, Aparicio T, de Lorenzo V, Nikel PI. New transposon tools tailored for metabolic engineering of gram-negative microbial cell factories. Front Bioeng Biotechnol 2014; 2: 46.
[PMID: 25389526]

[175] Solaiman DK, Ashby RD, Foglia TA, Marmer WN. Conversion of agricultural feedstock and coproducts into poly(hydroxyalkanoates). Appl Microbiol Biotechnol 2006; 71(6): 783-9.
[http://dx.doi.org/10.1007/s00253-006-0451-1] [PMID: 16708192]

[176] Lee S, Middelberg A, Lee Y. Poly (3-hydroxybutyrate) production from whey using recombinant *Escherichia coli.* Biotechnol Lett 1997; 19(10): 1033-5.
[http://dx.doi.org/10.1023/A:1018411820580]

[177] Wong HH, Lee SY. Poly-(3-hydroxybutyrate) production from whey by high-density cultivation of recombinant *Escherichia coli.* Appl Microbiol Biotechnol 1998; 50(1): 30-3.
[http://dx.doi.org/10.1007/s002530051252] [PMID: 9720197]

[178] Ahn WS, Park SJ, Lee SY. Production of Poly(3-hydroxybutyrate) by fed-batch culture of recombinant *Escherichia coli* with a highly concentrated whey solution. Appl Environ Microbiol 2000; 66(8): 3624-7.
[http://dx.doi.org/10.1128/AEM.66.8.3624-3627.2000] [PMID: 10919830]

[179] Ahn WS, Park SJ, Lee SY. Production of poly (3-hydroxybutyrate) from whey by cell recycle fed-batch culture of recombinant *Escherichia coli.* Biotechnol Lett 2001; 23(3): 235-40.
[http://dx.doi.org/10.1023/A:1005633418161]

[180] Park SJ, Park JP, Lee SY. Production of poly (3-hydroxybutyrate) from whey by fed-batch culture of

recombinant *Escherichia coli* in a pilot-scale fermenter. Biotechnol Lett 2002; 24(3): 185-9. [http://dx.doi.org/10.1023/A:1014196906095]

[181] Nikel PI, Pettinari MJ, Galvagno MA, Méndez BS. Poly(3-hydroxybutyrate) synthesis by recombinant *Escherichia coli* arcA mutants in microaerobiosis. Appl Environ Microbiol 2006; 72(4): 2614-20. [http://dx.doi.org/10.1128/AEM.72.4.2614-2620.2006] [PMID: 16597965]

[182] Pais J, Farinha I, Freitas F, *et al.* Improvement on the yield of polyhydroxyalkanotes production from cheese whey by a recombinant *Escherichia coli* strain using the proton suicide methodology. Enzyme Microb Technol 2014; 55: 151-8. [http://dx.doi.org/10.1016/j.enzmictec.2013.11.004] [PMID: 24411458]

[183] Liu F, Li W, Ridgway D, Gu T, Shen Z. Production of poly-β-hydroxybutyrate on molasses by recombinant *Escherichia coli.* Biotechnol Lett 1998; 20(4): 345-8. [http://dx.doi.org/10.1023/A:1005367011378]

[184] Zhang H, Obias V, Gonyer K, Dennis D. Production of polyhydroxyalkanoates in sucrose-utilizing recombinant *Escherichia coli* and *Klebsiella* strains. Appl Environ Microbiol 1994; 60(4): 1198-205. [PMID: 8017916]

[185] Saranya V, Shenbagarathai R. Production and characterization of PHA from recombinant *E. coli* harbouring *phaC1* gene of indigenous *Pseudomonas* sp. LDC-5 using molasses. Braz J Microbiol 2011; 42(3): 1109-18. [http://dx.doi.org/10.1590/S1517-83822011000300032] [PMID: 24031729]

[186] Jahreis K, Bentler L, Bockmann J, *et al.* Adaptation of sucrose metabolism in the *Escherichia coli* wild-type strain EC3132. J Bacteriol 2002; 184(19): 5307-16. [http://dx.doi.org/10.1128/JB.184.19.5307-5316.2002] [PMID: 12218016]

[187] Arifin Y, Sabri S, Sugiarto H, Krömer JO, Vickers CE, Nielsen LK. Deletion of *cscR* in *Escherichia coli* W improves growth and poly-3-hydroxybutyrate (PHB) production from sucrose in fed batch culture. J Biotechnol 2011; 156(4): 275-8. [http://dx.doi.org/10.1016/j.jbiotec.2011.07.003] [PMID: 21782859]

[188] Sabri S, Nielsen LK, Vickers CE. Molecular control of sucrose utilization in *Escherichia coli* W, an efficient sucrose-utilizing strain. Appl Environ Microbiol 2013; 79(2): 478-87. [http://dx.doi.org/10.1128/AEM.02544-12] [PMID: 23124236]

[189] Brückner R, Titgemeyer F. Carbon catabolite repression in bacteria: choice of the carbon source and autoregulatory limitation of sugar utilization. FEMS Microbiol Lett 2002; 209(2): 141-8. [http://dx.doi.org/10.1016/S0378-1097(02)00559-1] [PMID: 12007797]

[190] DiRusso CC, Metzger AK, Heimert TL. Regulation of transcription of genes required for fatty acid transport and unsaturated fatty acid biosynthesis in *Escherichia coli* by FadR. Mol Microbiol 1993; 7(2): 311-22. [http://dx.doi.org/10.1111/j.1365-2958.1993.tb01122.x] [PMID: 8446033]

[191] Li R, Chen Q, Wang PG, Qi Q. A novel-designed *Escherichia coli* for the production of various polyhydroxyalkanoates from inexpensive substrate mixture. Appl Microbiol Biotechnol 2007; 75(5): 1103-9. [http://dx.doi.org/10.1007/s00253-007-0903-2] [PMID: 17334755]

[192] Lee SY. Poly (3-hydroxybutyrate) production from xylose by recombinant *Escherichia coli.* Bioprocess Eng 1998; 18(5): 397-9.
[http://dx.doi.org/10.1007/s004490050462]

[193] Nduko JM, Suzuki W, Matsumoto K, *et al.* Polyhydroxyalkanoates production from cellulose hydrolysate in *Escherichia coli* LS5218 with superior resistance to 5-hydroxymethylfurfural. J Biosci Bioeng 2012; 113(1): 70-2.
[http://dx.doi.org/10.1016/j.jbiosc.2011.08.021] [PMID: 21993429]

[194] Salamanca-Cardona L, Ashe CS, Stipanovic AJ, Nomura CT. Enhanced production of polyhydroxyalkanoates (PHAs) from beechwood xylan by recombinant *Escherichia coli.* Appl Microbiol Biotechnol 2014; 98(2): 831-42.
[http://dx.doi.org/10.1007/s00253-013-5398-4] [PMID: 24287934]

[195] Taguchi S, Yamada M, Matsumoto K, *et al.* A microbial factory for lactate-based polyesters using a lactate-polymerizing enzyme. Proc Natl Acad Sci USA 2008; 105(45): 17323-7.
[http://dx.doi.org/10.1073/pnas.0805653105] [PMID: 18978031]

[196] Favaro L, Viktor MJ, Rose SH, *et al.* Consolidated bioprocessing of starchy substrates into ethanol by industrial *Saccharomyces cerevisiae* strains secreting fungal amylases. Biotechnol Bioeng 2015; 112(9): 1751-60.
[http://dx.doi.org/10.1002/bit.25591] [PMID: 25786804]

[197] Lynd LR, van Zyl WH, McBride JE, Laser M. Consolidated bioprocessing of cellulosic biomass: an update. Curr Opin Biotechnol 2005; 16(5): 577-83.
[http://dx.doi.org/10.1016/j.copbio.2005.08.009] [PMID: 16154338]

[198] Fonseca GG, Fonseca GG, de Arruda-Caulkins JC, Vasconcellos Antonio R. Production and characterization of poly-(3-hydroxybutyrate) from recombinant *Escherichia coli* grown on cheap renewable carbon substrates. Waste Manag Res 2008; 26(6): 546-52.
[http://dx.doi.org/10.1177/0734242X07088443] [PMID: 19039071]

[199] Bhatia SK, Shim Y-H, Jeon J-M, *et al.* Starch based polyhydroxybutyrate production in engineered *Escherichia coli.* Bioprocess Biosyst Eng 2015; 38(8): 1479-84.
[http://dx.doi.org/10.1007/s00449-015-1390-y] [PMID: 25820819]

[200] Chien CC, Ho LY. Polyhydroxyalkanoates production from carbohydrates by a genetic recombinant *Aeromonas* sp. Lett Appl Microbiol 2008; 47(6): 587-93.
[http://dx.doi.org/10.1111/j.1472-765X.2008.02471.x] [PMID: 19120931]

[201] Do YS, Smeenk J, Broer KM, *et al.* Growth of *Rhodospirillum rubrum* on synthesis gas: conversion of CO to H2 and poly-β-hydroxyalkanoate. Biotechnol Bioeng 2007; 97(2): 279-86.
[http://dx.doi.org/10.1002/bit.21226] [PMID: 17054121]

[202] Choi D, Chipman DC, Bents SC, Brown RC. A techno-economic analysis of polyhydroxyalkanoate and hydrogen production from syngas fermentation of gasified biomass. Appl Biochem Biotechnol 2010; 160(4): 1032-46.
[http://dx.doi.org/10.1007/s12010-009-8560-9] [PMID: 19247588]

[203] Klask C, Raberg M, Heinrich D, Steinbüchel A. Heterologous expression of various PHA synthase genes in *Rhodospirillum rubrum.* Chem Biochem Eng Q 2015; 29(2): 75-85.

[http://dx.doi.org/10.15255/CABEQ.2014.2249]

[204] Draper JL, Rehm BH. Engineering bacteria to manufacture functionalized polyester beads. Bioengineered 2012; 3(4): 203-8. [http://dx.doi.org/10.4161/bioe.19567] [PMID: 22705844]

[205] Mifune J, Grage K, Rehm BH. Production of functionalized biopolyester granules by recombinant *Lactococcus lactis*. Appl Environ Microbiol 2009; 75(14): 4668-75. [http://dx.doi.org/10.1128/AEM.00487-09] [PMID: 19465535]

[206] Song Y, Matsumoto K, Tanaka T, Kondo A, Taguchi S. Single-step production of polyhydroxybutyrate from starch by using α-amylase cell-surface displaying system of *Corynebacterium glutamicum*. J Biosci Bioeng 2013; 115(1): 12-4. [http://dx.doi.org/10.1016/j.jbiosc.2012.08.004] [PMID: 22959444]

[207] Baei MS, Najafpour G, Younesi H, Tabandeh F, Eisazadeh H. Poly (3-hydroxybutyrate) synthesis by *Cupriavidus necator* DSMZ 545 utilizing various carbon sources. World Appl Sci J 2009; 7(2): 157-61.

[208] Reinecke F, Steinbüchel A. *Ralstonia eutropha* strain H16 as model organism for PHA metabolism and for biotechnological production of technically interesting biopolymers. J Mol Microbiol Biotechnol 2009; 16(1-2): 91-108. [http://dx.doi.org/10.1159/000142897] [PMID: 18957865]

[209] York GM, Lupberger J, Tian J, Lawrence AG, Stubbe J, Sinskey AJ. *Ralstonia eutropha* H16 encodes two and possibly three intracellular Poly[D-(-)-3-hydroxybutyrate] depolymerase genes. J Bacteriol 2003; 185(13): 3788-94. [http://dx.doi.org/10.1128/JB.185.13.3788-3794.2003] [PMID: 12813072]

[210] Brigham CJ, Reimer EN, Rha C, Sinskey AJ. Examination of PHB depolymerases in *Ralstonia eutropha*: further elucidation of the roles of enzymes in PHB homeostasis. AMB Express 2012; 2(1): 26. [http://dx.doi.org/10.1186/2191-0855-2-26] [PMID: 22537946]

[211] Jendrossek D. Polyhydroxyalkanoate granules are complex subcellular organelles (carbonosomes). J Bacteriol 2009; 191(10): 3195-202. [http://dx.doi.org/10.1128/JB.01723-08] [PMID: 19270094]

[212] Doi Y, Segawa A, Kawaguchi Y, Kunioka M. Cyclic nature of poly(3-hydroxyalkanoate) metabolism in *Alcaligenes eutrophus*. FEMS Microbiol Lett 1990; 55(1-2): 165-9. [http://dx.doi.org/10.1111/j.1574-6968.1990.tb13856.x] [PMID: 2328906]

[213] Taidi B, Mansfield DA, Anderson AJ. Turnover of poly (3-hydroxybutyrate)(PHB) and its influence on the molecular mass of the polymer accumulated by *Alcaligenes eutrophus* during batch culture. FEMS Microbiol Lett 1995; 129(2-3): 201-5. [PMID: 7649441]

[214] Uchino K, Saito T, Gebauer B, Jendrossek D. Isolated poly(3-hydroxybutyrate) (PHB) granules are complex bacterial organelles catalyzing formation of PHB from acetyl coenzyme A (CoA) and degradation of PHB to acetyl-CoA. J Bacteriol 2007; 189(22): 8250-6. [http://dx.doi.org/10.1128/JB.00752-07] [PMID: 17720797]

[215] Lawrence AG, Schoenheit J, He A, *et al.* Transcriptional analysis of *Ralstonia eutropha* genes related to poly-(R)-3-hydroxybutyrate homeostasis during batch fermentation. Appl Microbiol Biotechnol 2005; 68(5): 663-72.
[http://dx.doi.org/10.1007/s00253-005-1969-3] [PMID: 15924243]

[216] Povolo S, Basaglia M, Fontana F, Morelli A, Casella S. Poly (hydroxyalkanoate) production by *Cupriavidus necator* from fatty waste can be enhanced by *phaZ1* inactivation. Chem Biochem Eng Q 2015; 29(2): 67-74.
[http://dx.doi.org/10.15255/CABEQ.2014.2248]

[217] Lee SY, Lee Y. Metabolic engineering of *Escherichia coli* for production of enantiomerically pure (R)-(--)-hydroxycarboxylic acids. Appl Environ Microbiol 2003; 69(6): 3421-6.
[http://dx.doi.org/10.1128/AEM.69.6.3421-3426.2003] [PMID: 12788745]

[218] Boynton ZL, Koon JJ, Brennan EM, *et al.* Reduction of cell lysate viscosity during processing of poly(3-hydroxyalkanoates) by chromosomal integration of the staphylococcal nuclease gene in *Pseudomonas putida.* Appl Environ Microbiol 1999; 65(4): 1524-9.
[PMID: 10103246]

[219] Huisman GW, Luo LZ, Peoples OP. Microbial strains and processes for the manufacture of biomaterials. Google Patents 2014.

[220] Jendrossek D, Pfeiffer D. New insights in the formation of polyhydroxyalkanoate granules (carbonosomes) and novel functions of poly(3-hydroxybutyrate). Environ Microbiol 2014; 16(8): 2357-73.
[http://dx.doi.org/10.1111/1462-2920.12356] [PMID: 24329995]

CHAPTER 8

Medium-Chain-Length Poly-3-hydroxyalkanoates

Bruce A. Ramsay*

Polyferm Canada Inc. 4530 CamdenE-Portland Bdry Rd RR1 Harrowsmith, ON K0H1V0, Canada, and Department of Chemical Engineering, Queen's University, Kingston, ON K7L 3N6, Canada

Abstract: Medium-chain-length poly-3-hydroxyalkanoates (*mcl*-PHAs) were discovered in the 1980s, 60 years after their more famous biopolyester ancestor poly--hydroxybutyrate. Accumulated only by a group of related *Pseudomonas* species, these polyesters are remarkable in their structural diversity and applications. *mcl*-PHAs are highly water resistant, yet readily susceptible to enzymatic degradation. They can be used in packaging, paints, toners, adhesives and in many biomedical applications but have yet to be produced commercially. *mcl*-PHAs are most efficiently produced from carboxylic (fatty) acids such as octanoic acid but can also be produced from many non-related substrates such as glucose. Recent advances in fermentation and separation processes, as well as strain development are improving polymer quality and reducing production cost, which should soon lead to the adoption of *mcl*-PHAs by industry.

Keywords: Applications, Biosynthesis, *mcl*-PHA, Medium-chain-length PHA (*mcl*-PHA), Poly-3-hydroxyalkanoate, Production, Recovery, Thermoelastomer.

INTRODUCTION

This chapter is focused on different types of polyesters including some that are biologically produced. Although they all contain subunits linked by an ester bond, polyester properties vary greatly. For example, both poly-3-hydroxyalkanaotes (PHAs) and polylactides (PLAs) are polyesters that degrade in the natural environment, but PLAs are polymerized in an *ex vivo* chemical process and

* **Corresponding author Bruce A. Ramsay:** Polyferm Canada Inc. 4530 CamdenE-Portland Bdry Rd RR1 Harrowsmith, ON K0H1V0, Canada, and Department of Chemical Engineering, Queen's University, Kingston, ON K7L 3N6, Canada; Tel: 613-372-2421; Email: bruce.ramsay@queensu.ca.

Martin Koller (Ed.)

degrade *via* chemical hydrolysis while PHA degradation occurs *via* extracellular enzymes. Among the PHAs (Fig. **1**), the short-chain-length (scl)-PHAs have less than three carbon atoms in the R group while medium-chain-length (*mcl*)-PHAs have three or more carbon atoms in the R group. *mcl*-PHA films may absorb water, but the rate of hydrolysis near neutral pH-values is very slow [1]. *mcl*-PHAs are unique in many ways. In contrast to most *scl*-PHAs, they are thermoelastomers and are produced by only certain species of the genus *Pseudomonas*. *mcl*- and *scl*-PHAs are degraded by different types of depolymerase enzymes. Both *mcl*- and *scl*-PHAs can be produced from glucose but the biosynthetic pathways are distinct and the thermomechanical properties of the resulting materials (*e.g.*, poly-3-hydroxydecanoate (PHD) and poly--hydroxybutyrate (PHB)) are strikingly different. This chapter will examine the production and possible applications of *mcl*-PHAs focusing on their relationship to other bioplastics, especially *scl*-PHAs.

DISCOVERY AND POTENTIAL APPLICATIONS

While studying *Bacillus megaterium* in the 1920s, Maurice Lemoigne identified and began the characterization of the first and most ubiquitous PHA, PHB [2]. He proved that it was a high molecular weight polyester long before the pioneering work of Carothers [3] on polyesters produced *via* chemical synthesis and only shortly after Hermann Staudinger proposed the existence of covalently bonded macromolecules.

Baptist examined thc thcrmomechanical properties of PHB and developed applications as a thermoplastic during the 1960s, especially as sutures [4], but found that it suffered from limitations as a thermoplastic, particularily thermal instability. As a homopolymer, its physical properties are limited. For use as a thermoplastic, PHB needs to be melted, but degrades at its melting temperature T_m and loss of molecular weight further deteriorates its mechanical properties. The breakthough in application of PHAs as thermoplastics came from the realization that the *scl*-PHA polymerases could polymerize other monomers than 3-hydroxybutyrate. Most naturally occurring *scl*-PHAs are PHB homopolymers since many, if not most, environmentally available carbon sources such as glucose are metabolized to acetate, the building block for 3-hydroxybutyrate synthesis. However, it was found that feeding propionic or valeric acid to *Alcaligenes eutrophus* (now known as *Cupriavidus necator* and previously named *Wausteria eutropha* and also *Hydrogenomonas eutrophus*) produced 3-hydroxyvalerate which is also a good substrate for *scl*-PHA polymerases. Incorporation of subunits other than HB reduces crystallinity which lowers the melting point and increases flexibility. However, even with this discovery, the thermomechanical properties of these *scl*-PHAs are limited.

Although they were also reported as bacterial inclusions, until the early 1980s *mcl*-PHAs existed unidentified by the bacteriologists studying these organisms. Indeed, Stanier in his massive treatise on Pseudomonad taxonomy [5], and adopted by Bergy's Manual of Systematic Bacteriology to classify pseudomonads, described these *Pseudomonas* species as "do not accumulate poly-β - hydroxybutyrate as a cellular reserve material, and are unable to attack this polymer by means of extracellular enzymes". *mcl*-PHA emerged as distinct group of PHA polyesters only three decades ago. Perhaps in a similar manner to Lemoigne, researchers in the Witholt lab were working on medium-chain-length alkane (*e.g.*, octane and 1-octene) utilization by *P. oleovorans* and wondered at the composition of the inclusion bodies [6]. They isolated the granules, expecting PHB, but instead found another 3-hydroxy polyester composed primarily of 3-hydroxyoctanoate. It was eventually named *mcl*-PHA by the discoverers. This was a wise decision as its synthesis requires enzymes able to utilize medium-chai- -length carboxylic acids and since even longer chain PHAs have been discovered [7]. It was soon understood that most, if not all ribosomal RNA type 2 *Pseudomonas* species can accumulate *mcl*-PHA, but not *scl*-PHAs. PHA composition from specific carbon substrates depends mostly on the nature of the synthase. The synthase responsible for *mcl*-PHA production in pseudomonads is now classified as type 2.

There are now well over 100 different known subunits, all with structures fitting the formula of (Fig. **1**). They can be incorporated into *mcl*-PHAs due to the broad substrate specificity of the polymerases used in their synthesis. Some of the side-chains of these subunits are chemically reactive, so further diversity may be obtained through chemical modification, including grafting to other biomaterials

[8, 9]. This diversity is translated into a wide range of potential thermal and mechanical properties.

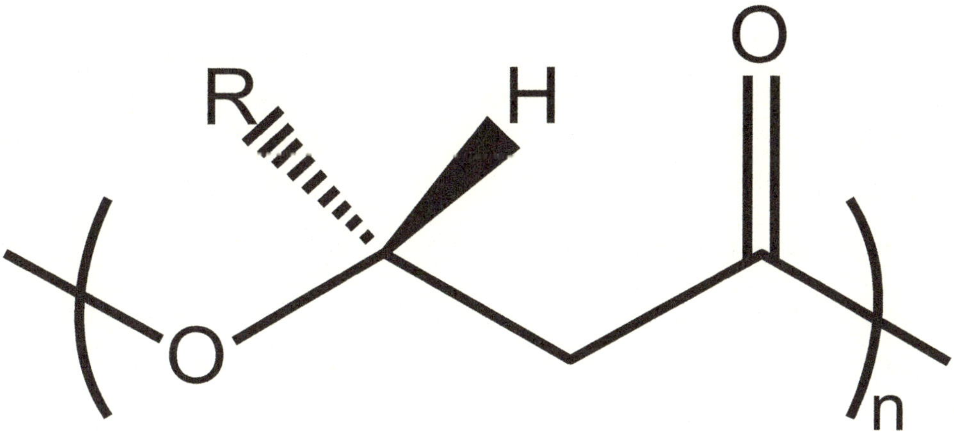

Fig. (1). PHA structure. R is less than three carbon atoms in *scl*-PHA; R is equal to or greater than three carbon atoms in *mcl*-PHA.

Due to their unique properties, many commercial applications have been suggested for these materials. These include paint formulations, adhesives, coatings, controlled release compounds and in tissue engineering scaffolds [10 - 13]. They can be depolymerized to chiral monomers for use in the synthesis of antibiotics and other medical applications [14], vitamins [15], and various bioactive compounds [16]. Under specific physiological conditions, these polymers can account for well over half of a bacterium's dry weight, so they must perform an important ecological function. Suggested benefits to the cells include storage of carbon and energy for future needs and control of redox conditions in the cell. However, due to the tantalizing possibilities for industrial application of these materials, the present document will focus on *mcl*-PHAs as thermoelastomers. Not only do many opportunities exist, but the majority will allow for commercial exploitation without much of the environmental harm often associated with the use of polymers based on petroleum-derived chemicals.

METABOLIC ROUTES TO *MCL*-PHA

Differences in *mcl*-PHA production, properties and the peculiarities of production strains cannot be discussed without a firm understanding of the biochemical pathways leading to its synthesis. Most work has concentrated on the use of *mcl*-carboxylic acids as substrates but it was observed that many *mcl*-PHA accumulators could produce PHA from glucose, albeit less efficiently, and that this material invariably contained primarily 3-hydroxydecanoate subunits. It is now understood that most *mcl*-PHA accumulators can use either β-oxidation or *de novo* fatty acid synthesis to produce the 3-hydroxyalkanoate-CoA substrates required by the synthase.

Structurally Related Carbon Sources

Once activated by coenzyme A (CoA), many organic compounds can be incorporated into *mcl*-PHA virtually intact. These "structurally related carbon sources" are compounds that are processed by enzymes of the β-oxidation pathway. β-oxidation is the main pathway to *mcl*-PHA when fatty acids are available. During β-oxidation, one molecule of acetate (*i.e.*, two carbons) is lost from each molecule of ketoacyl-CoA in every cycle but a considerable amount of energy is captured from the reduction of $NADP^+$ (Fig. **2**). In addition, the acetate can enter the TCA cycle to generate more energy, while the carbons are lost as CO_2. However, if conditions are correct, 2-trans-enoyl-CoA can be directed to PHA synthesis with little loss of acetate.

The main enzyme linking PHA synthesis to β-oxidation is (*R*)-specific enoyl-CoA hydratase (*phaJ*). Two other enzymes have been suggested that may link *mcl*-PHA synthesis to β-oxidation. 3-Ketoacyl-CoA reductase (*fabG*) has been shown to effectively provide precursors for *mcl*-PHA synthases [17]. However, its overexpression in *P. putida* was much less effective than overexpression of *phaJ* [18]. The other putitive link is 3-hydroxyacyl-CoA epimerase but as yet, there is no evidence that this enzyme links β-oxidation to PHA synthesis. Further evidence for the importance of phaJ can be found in data generated using mutants defective in fadA and/or fadB. Without fadB, the substrates for 3-ketoacyl-CoA reductase and 3-hydroxyacyl-CoA epimerase cannot be produced, yet mutants where these genes have been completely knocked out still accumulate substantial amounts of *mcl*-PHA from fatty acids. In fact, such mutations have been found to increase *mcl*-PHA production [19].

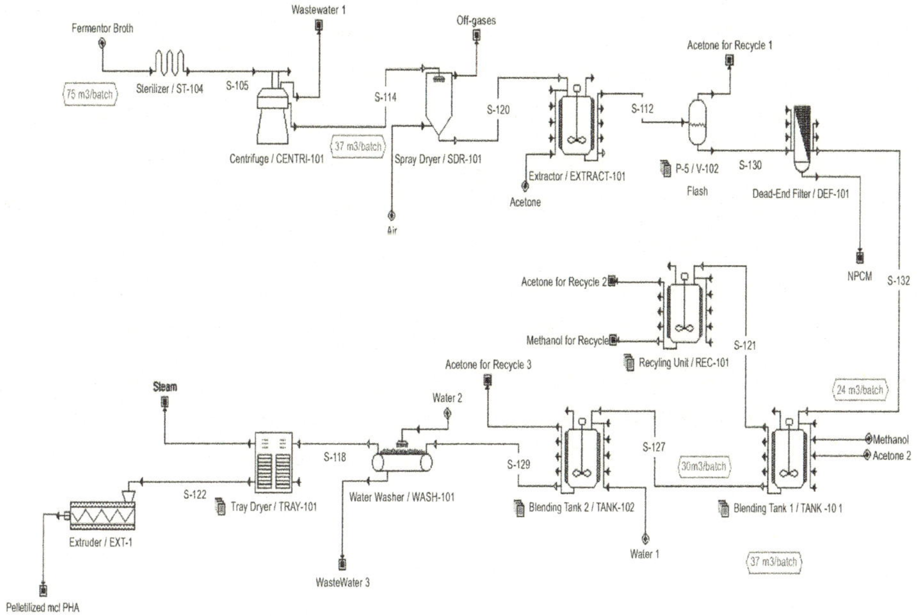

Fig. (2). The two major routes to synthesis of *mcl*-PHAs are β-oxidation of carboxylic acids and *de novo* fatty acid synthesis of carboxylic acids from unrelated carbon sources. Genes expressing key enzymes are in red while cofactors are in blue. *phaJ* codes for (*R*)-specific enoyl-CoA hydratase which links *mcl*-PHA synthesis to β-oxidation. *phaG* codes for 3-hydroxyacylCoA transferase which links *de novo* fatty acid synthesis to *mcl*-PHA synthesis. *phaC* codes for *mcl*-PHA synthase, the enzyme responsible for the *mcl*-PHA polymerization. *fadA,B,D* and *E* code for β-oxidation enzymes while *fabG* plays a key role in fatty acid (*de novo*) biosynthesis. Other key enzymes include acyl-CoA synthetase (*fadD*) which is required for conversion of free fatty acids into acyl-CoA thioesters, the first step in fatty acid degradation and in *mcl*-PHA synthesis from carboxylic acids. The highest productivities of *mcl*-PHA production occur from *mcl*-carboxylic acids. These must be activated by linkage to coenzyme A, which is the function of *fadD* and requires ATP with Mg^{2+} as cofactor.

Structurally Unrelated Compounds

Most organic compounds have structures very different from *mcl*-PHA subunits and cannot be processed *via* β-oxidation. However, these compounds can usually be converted, at least in part, to acetate. They include most common carbon sources such as gluconate, acetate or organic waste such as cheese whey (which contains lactose). Upon activation to form acetyl-CoA, this carbon compound can be used in fatty acid *de novo* biosynthesis. Reports of *mcl*-PHAs produced from "unrelated' carbon sources and composed primarily of C10 subunits emerged in 1990 [20, 21]. It was speculated that their mechanism of synthesis must involve the same β-addition process used in fatty acid synthesis from acetate [22], but there was a missing link. The fatty acid is carried by acyl carrier protein (ACP) in *de novo* fatty acid synthesis, while in β-oxidation and in *mcl*-PHA synthesis, the substrate must be linked to Coenzyme A (CoA) (Fig. **2**). This was particularly a problem for genetic engineers such as those wishing to produce *mcl*-PHA in the oil-producing plastids of plants. They required a gene that coded for the missing enzyme(s). An examination of mutants not able to accumulate *mcl*-PHA from gluconate enabled the identification of the gene responsible, which was named *phaG* [23]. It was demonstrated that *phaG* could restore activity in the mutant strains and in wild type *P. putida* GPo1, which normally cannot produce *mcl*-PHA from gluconate.

Chain Elongation Reactions, in which Acetyl-Coa Moieties are condensed to 3-Hydroxyacyl-CoA

It was noticed that two or sometimes multiples of two carbons were added to some subunits when certain related substrates were supplied. This was particularly the case when hexanoic acid was used for polymer synthesis. Since octanoic acid is preferred to hexanoate for *mcl*-PHA synthesis, its specific rate of uptake should be higher. If chain elongation occurs, octanoate is quickly used for PHA synthesis. While this is a minor mechanism of supplying PHA subunits, it can cause problems. For example, if one wishes to produce homopolymeric poly--hydroxyhexanoate (PHHx) from hexanoic acid, the formation of C8 subunits will be undesirable.

mcl-PHA Synthesis

The vast majority of *mcl*-PHA synthesis occurs on/in inclusion bodies commonly referred to as granules. *mcl*-PHA granules have been much less studied than *scl*-PHA granules but appear to function in a similar manner. Their formation is still the subject of inquiry but at least two models (budding and autoformation) exist. Once a granule has formed, the appropriate substrate ((*R*)-3-hydroxyacl-CoA) is available and the synthase (*phaC*) is produced, polymer synthesis may take place. Micrographs of *mcl*-PHA granules, the putative granule structure and the organization of the *mcl*-PHA operon for *P. putida* KT2442 are shown in Fig. (**3**). The (*R*)-3-hydroxyl-CoA substrate of *phaC* is already activated by CoA, so no additional energy is required for chain elongation. It would seem that production of large amounts of *mcl*-PHA depends entirely on the presence of enough active synthase and provisioning of sufficient ((*R*)-3-hydroxyacyl-CoA). *P. putida* GPo1 has two acyl-CoA synthases (*fadE*) [24]. One binds to the granule membrane and the other to the cell membrane. While granule-bound *fadE* is probably involved in intracellular depolymerization by *phaZ*, it also plays a key role in providing 2-trans-enoyl-CoA for *phaJ* (Fig. **2**).

There are likely mechanisms that control the molecular weight of the polymer chains as well as granule size. All PHA produced by wild type bacteria and isolated by non-destructive means has a low polydispersity index. In other words, the polymer chain lengths are very similar, so there must be a mechanism that terminates a chain after it reaches a certain length. These properties may be controlled by small proteins called phasins, which sit in the phospholipid monolayer at the surface of the granule, and have very similar N-terminal domains that serve as binding sites (Fig. **3**).

Mcl-PHA operons typically consist of promotor sites, two type 2 synthases, a depolymerase, the regulatory protein *phaD*, as well as phasins *phaF* and *phaI* which are transcribed in the opposite direction (Fig. **3a**). Interpretation of experimental data from strains with modified genes in the operon can be challenging as any modification may affect expression of other genes in the operon. For example, deletion of *phaZ* may result in less transcription of *phaC1* with a concomitant up-regulation of *phaC2* [26]. It is of great interest that there are invariably two synthases separated by the depolymerase gene, *phaZ*. Reasons for this could be that synthesis and degradation are linked in some way. Both *mcl*-PHA synthesis [27] and degradation [28] enzymes show similarities to lipases and that both may therefore have a genetic relationship to lipases. Although the two synthases catalyze a similar range of substrates, there appear to be differences in their activity [29].

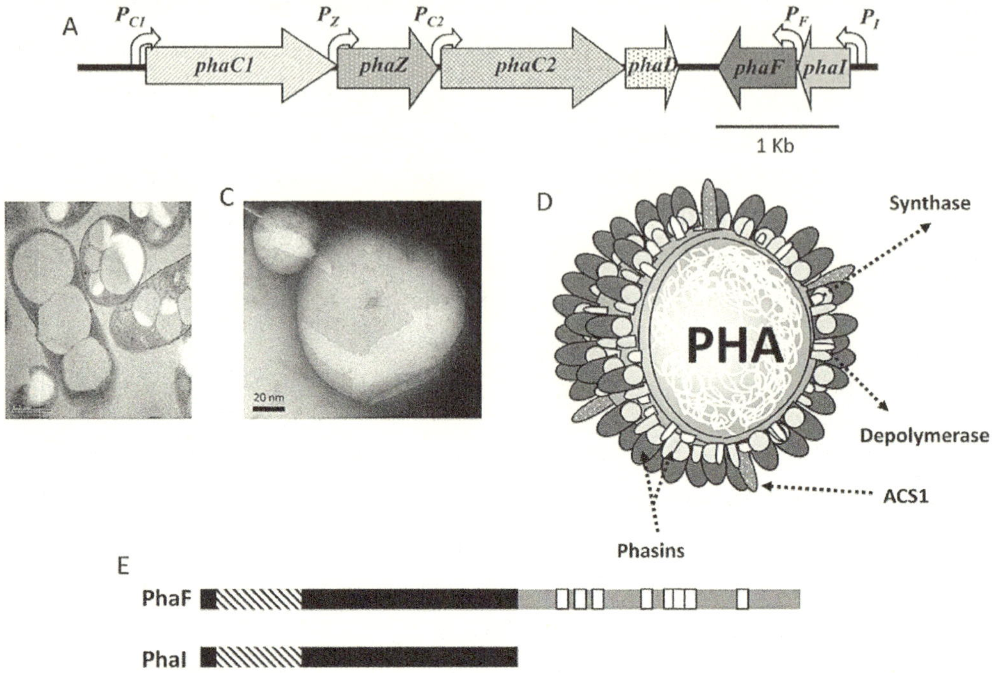

Fig. (3). From Galán *et al.* 2011 [25]. *pha* gene cluster organization, granule structure and scheme of phasins domains in *P. putida* KT2442. A. Genetic organization of *pha* cluster, in *P. putida* KT2442. The open arrows indicate the directions of gene transcription. The *phaC1* and *phaC2* genes encode two synthases and are separated by the *phaZ* gene that encodes an intracellular depolymerase. *phaD* gene encodes a transcriptional regulator. *phaF* and *phaI* genes code for phasins. Promoter regions are indicated as curved white arrows. B. TEM image of *mcl*-PHA-producing cells of *P. putida* KT2442. C. TEM image of a PHA granule from *P. putida* KT2442. D. Model of PHA granule structure. Granules are composed of PHA enclosed by a monolayer of phospholipids and granule associated proteins (GAPs). The most abundant GAPs are phasins. Polymerases (or synthases), depolymerases and acyl CoA synthetase (ACS1 coded by *fadD*) are also associated with the granule. E. Domain architecture of the phasins *PhaF* and *PhaI*. The N-terminal domain of *PhaF* is similar to *PhaI* protein. Hatched and white boxes indicate the putative granule binding sites and the AAKP repeated motifs respectively.

PhaD is a *TetR*-like transcriptional regulator that can bind onto the P_{C1} and P_I promotors in the pha cluster (Fig. **3**) and appears to be a carbon source dependent activator [30]. It can regulate its own synthesis as well as the expression of *phaI* and *phaF*. Of the two phasins, *phaF* appears to be the most important. It has been demonstrated in *P putida* U that *phaC1* and *phaF* are the only genes in the PHA operon required for “normal” production [31].

Key Enzymes/Genes

The exact composition of the PHA produced depends on the substrates supplied, the rate of substrate uptake as well as the quantity, specificity and activity of enzymes leading to the final product. However, it is generally believed that the key enzyme is the synthase. Studies evaluating the substrate specificities of PHA synthases have produced varied results depending on how they were conducted. Typically studies have found that *phaC1* has higher activity than *phaC2* and substrate specificity may also be different [32, 33].

The preferred substrate of *phaC1* of *P. aeruginosa* is 3-hydroxydecanoyl-CoA [29] but the conversion of decanoic acid to 3-hydroxydecanoyl-CoA involves transporting this poorly water-soluble substrate into the cells and a series of enzymatic reactions, any one of which could limit flux. For this reason and others, the choice of production strain is crucial to the productivity of any *mcl*-PHA fermentation process.

NATURALLY OCCURRING *MCL*-PHA PRODUCING MICROORGANISMS

Scl-PHAs of various sorts can be produced in high yield in recombinant *Escherichia coli* strains. However, although interesting advances have been made [34], this is not yet the case for *mcl*-PHA. Thus, high *mcl*-PHA yielding strains are all rRNA type 2 *Pseudomonas* species. Several stand out as being the most effective producers. Choice of production strain is critical to producing high purity *mcl*-PHA of defined structure. These qualities may be affected, not only by the PHA synthase, but the affinities, number and activities of several key enzymes in the pathway from substrate to product. Organisms able to accumulate high concentrations of the desired *mcl*-PHA are critical to eventual product quality as a higher degree of purification may be more easily achieved.

One consideration in the choice of strain regards the propensity for alginate production which reduces productivity and increases broth viscosity which in turn reduces oxygen transfer. Many *mcl*-PHA producers including *P. aeruginosa*, *P. fluorescens*, *P. mendocina*, *P. stutzeri* [35] and *P. resinovorans* produce alginate especially when glucose is used. Rhamnolipids are another extracellular product that may "waste" carbon. These glycolipids can serve as biodegradable surfactants and simultaneous production of rhamnolipids and *mcl*-PHA has been proposed and studied [36].

Formerly known as *P. oleovorans* and the first organism known to accumulate *mcl*-PHAs, *Pseudomonas putida* GPo1 is unusual in several ways. It is an effective user of *mcl*-alkanes since it possesses the OCT plasmid containing the *alk* genes for the oxidation of *mcl*-alkanes to carboxylic acids [37]. These genes allow this strain to accumulate certain monomers that other strains cannot. Biorefinery carboxylic acids are often produced in the form of methyl esters by transesterification [38, 39]. For this reason carboxylic methyl esters are typically considerably cheaper than their acid form. Since *P. putida* GPo1 has the ability to carboxylate alkanes, methyl esters of carboxylic acids are converted to monomethyl esters of 3-hydroxycarboxylic acids. These are suitable substrates for *mcl*-PHA synthesis and result in the production of monomers whose side-chains have terminal ester groups [40]. There have been few reports of their use in *mcl*-PHA synthesis [41]. The thermal and mechanical characteristics of such PHAs have not been evaluated. *P. putida* GPo1 has also been shown to accumulate a variety of other subunits of potential interest such as those with methyl branched [42] or aromatic groups [43] on the side-chains.

Pseudomonas putida GPo1 is also the only well known producer of *mcl*-PHAs that cannot produce PHA from unrelated carbon sources, since, although its genome contains a putative *phaG,* it is not expressed in a functional form [44]. Another interesting feature of this bacterium is that it can, apparently, grow on naphthenic acids, as reported in the original publication describing its isolation from machine shop cutting fluids [45].

P putida strain mt-2 was isolated from a soil enrichment culture using *m*-toluene in Japan in 1960 [46]. It was found to degrade a range of aromatic compounds due to its TOL plasmid. After being cured of its plasmid (in Germany) mt-2 became KT2440 where "kt" is derived from Kenneth Timmis. KT2442 is a spontaneous rifampicin resistant mutant of KT2440, so it should be almost identical. However, resistance to rifampicin in *P. putida* is likely the result of mutation of the *rpoB* gene encoding for β subunit of RNA polymerase [47]. Although this was found not to alter transcription profiles when grown under succinic acid limitation [48], significant differences in growth and PHA production rates on specific carbon sources have been identified [49].

Pseudomonas sp. 61-3 has a type II synthase (*phaC*) but can incorporate substantial amounts of *scl*-PHA subunits and this affinity has been greatly amplified by site-specific mutagenesis [50]. *P. putida* U (isolated in the US on m-cresol) and *P. putida* BM01 (isolated in Korea on octanoate [51]) can, like *P. putida* GPo1, accumulate PHA with phenyl reactive groups in the side-chains. Another aromatic degrader, *P. putida* CA-3 was isolated *via* growth on styrene and can accumulate substantial amounts of *mcl*-PHA. *P. citronellolis*, *P. resinovorans*, *P. mendocina*, *P. aeruginosa* and *P. fluorescens* are other notable *mcl*-PHA producers and each has its own peculiar characteristics.

MCL-PHA MATERIAL PROPERTIES

Many applications of *mcl*-PHAs, such as in films or food packaging may require specific thermomechanical properties. In this domain, *mcl*-PHAs must compete commercially with other "green" themoelastomers such as Pebax® Rnew (polyether block amide), Versaflex® Bio, Pearlthane® ECO (polyurethane based) or Hytrel® RS (a block copolymer of polybutylene terephthalate and a polyether glycol). Although none of these materials are fabricated totally from renewable resources and BASF's Ecoflex® (a copolymer of 1,4-butanediol, adipic acid and terephthalic acid) is not biodegradable, *mcl*-PHAs must still compete in terms of price and properties. These competing thermoelastomers have similar properties to some *mcl*-PHAs. All have low glass transition temperatures like *mcl*-PHA and all have melting temperatures above 145 °C.

Tensile, Thermal and Crystallization Properties

As with other PHAs, the thermomechanical properties of *mcl*-PHA are highly dependent on crystallinity. This, in turn, is dependent on monomer composition and method of preparation. *mcl*-PHAs that are allowed to crystallize at higher temperatures also have higher crystallinity and melting temperature (Fig. **4**), presumably due to greater chain mobility during the crystallization process, so thermal history may be of great importance to properties. *mcl*-PHA homopolymers typically have higher crystallinity and melting temperature than copolymers. Crystallinity and melting temperature increase as the amount of the dominant subunit increases (Table **1**).

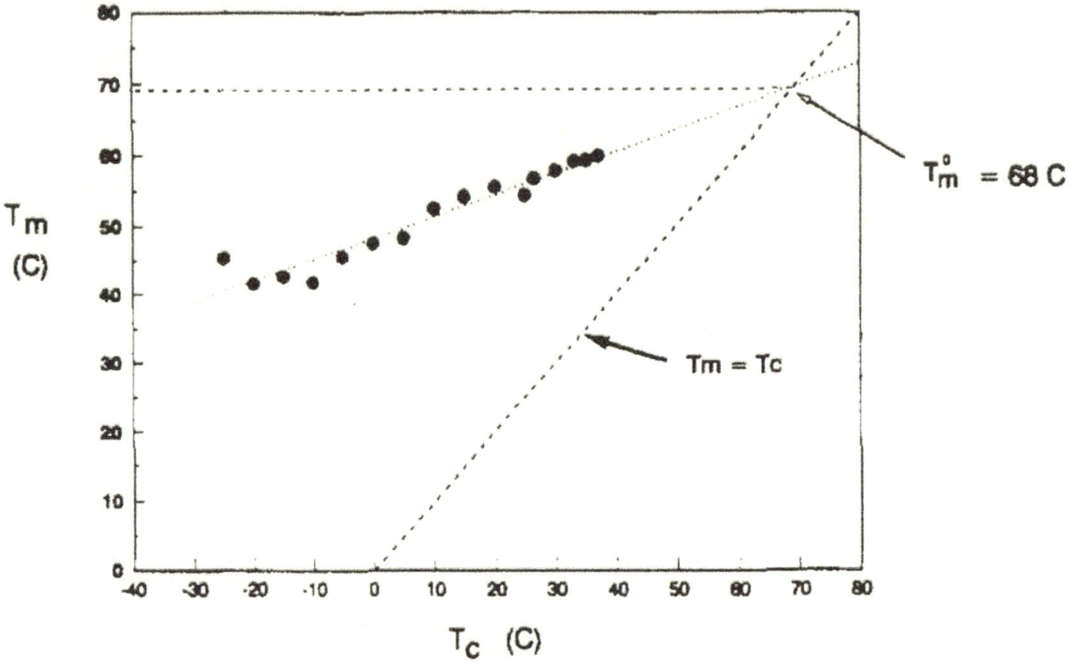

Fig. (4). Melting temperature (T_m) as a function of the crystallization temperature, T_C, for poly--hydroxyoctanoate (PHO) copolymer with an unspecified quantity of other subunits crystallized for 24 h from the melt. From Gagnon *et al.* 1992 [52].

Block Copolymers

Block copolymers of *scl-* and *mcl*-PHAs (*b*-PHA) can be effectively produced in fed-batch fermentations in *scl-mcl*-PHA accumulating bacteria. *b*-PHA have well known advantages over random copolymers [54, 55]. There have been several reports of PHA block copolymers (reviewed in [56]) and improved properties have been shown [48]. Probably the greatest success so far used *P. putida* with an *A. caviae phaC* and a weakened β-oxidation pathway [57].

Differential solvent extraction can be used to determine whether the material is a mixture of copolymers of significantly different composition. Differential scanning calorimetry studies of extracted polymer can detect the presence of two

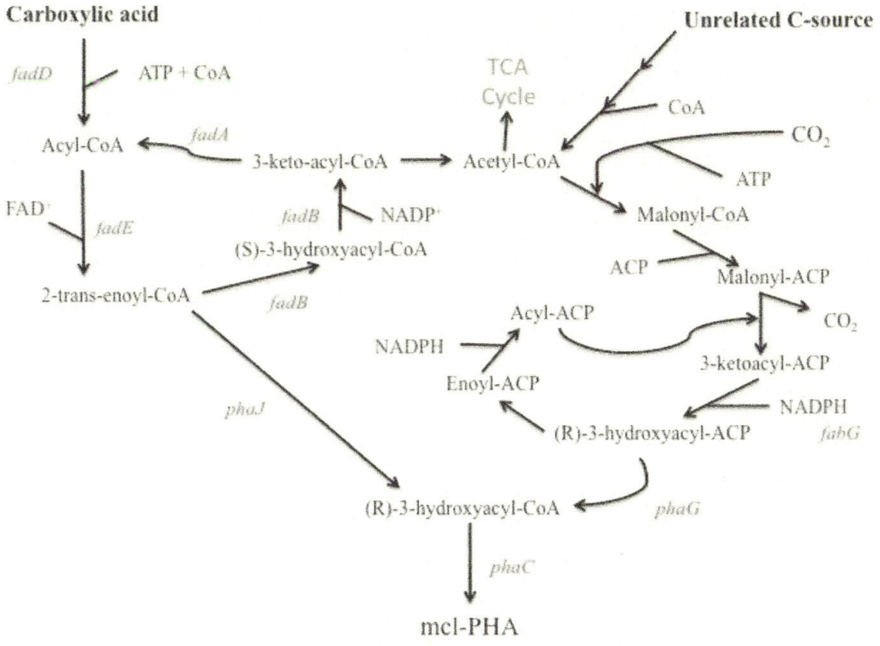

Fig. (2). The two major routes to synthesis of *mcl*-PHAs are β-oxidation of carboxylic acids and *de novo* fatty acid synthesis of carboxylic acids from unrelated carbon sources. Genes expressing key enzymes are in red while cofactors are in blue. *phaJ* codes for (*R*)-specific enoyl-CoA hydratase which links *mcl*-PHA synthesis to β-oxidation. *phaG* codes for 3-hydroxyacylCoA transferase which links *de novo* fatty acid synthesis to *mcl*-PHA synthesis. *phaC* codes for *mcl*-PHA synthase, the enzyme responsible for the *mcl*-PHA polymerization. *fadA,B,D* and *E* code for β-oxidation enzymes while *fabG* plays a key role in fatty acid (*de novo*) biosynthesis. Other key enzymes include acyl-CoA synthetase (*fadD*) which is required for conversion of free fatty acids into acyl-CoA thioesters, the first step in fatty acid degradation and in *mcl*-PHA synthesis from carboxylic acids. The highest productivities of *mcl*-PHA production occur from *mcl*-carboxylic acids. These must be activated by linkage to coenzyme A, which is the function of *fadD* and requires ATP with Mg^{2+} as cofactor.

Structurally Unrelated Compounds

Most organic compounds have structures very different from *mcl*-PHA subunits and cannot be processed *via* β-oxidation. However, these compounds can usually be converted, at least in part, to acetate. They include most common carbon sources such as gluconate, acetate or organic waste such as cheese whey (which contains lactose). Upon activation to form acetyl-CoA, this carbon compound can process, the two polymers must be mixed above the melting temperature of the higher melting material. For example, *mcl*-PHAs must be heated above 180 °C (the melting point of PHB) for impact modification of PHB. Viscosity mismatches result in poor mixing, creating large domain (*i.e.* droplet) sizes of the minor blend component. Large domain size results in poor mechanical properties. For these reasons, methods have been developed to modify polymer melt viscosity.

Peroxide treatment can significantly increase the complex viscosity (Fig. **5a**) and storage modulus (Fig. **5b**) of *mcl*-PHAs such as PHO, decreasing the loss tangent, tan δ, by orders of magnitude (Fig. **5c**), and causing a transformation from viscoelastic liquid to a viscoelastic solid [59]. The viscosity and elasticity of *mcl*-PHA can be increased substantially by reactive compounding with chain extenders and cross-linking agents such as dicumyl peroxide and triallyl trimesate [60]. This approach can also result in cross-linking of different phases in polyester and other blends. Consideration must always be given to the end-use of these materials, which may be marketed as "healthy" and "biodegradable". Some chain extenders such as diisocyanate are highly toxic, but have still been proposed for use in reactive blending of PLA with *mcl*-PHAs [61]. It is often claimed that covalently cross-linked materials are biodegradable, because they appear to biodegrade in certain standard tests. However, such cross-linking is almost certain to impede enzymatic degradation, possibly producing tiny undegradable particles during composting. The hazards of such materials are unknown, but commercially produced microparticles for cosmetic use are banned in many jurisdictions due to their effect on wildlife. *mcl*-PHAs generally draw their tensile strength from crystalline cross-links. Chemical cross-linking reduces crystallinity and properties such as tensile strength and melting temperature associated with crystallinity to produce a gel phase, which will no longer dissolve in solvents. Thus, minor modifications can greatly improve processability but may also produce undesired effects.

PRODUCTION METHODS AND ECONOMIC CONSIDERATIONS

Fermentation Methods

While shake-flask culture can quickly provide initial data, fermentors are required to control physiological conditions such as pH-value, dissolved oxygen and key nutrient concentrations to allow more meaningful information. Chemostat culture is an especially useful research tool as steady-state conditions can be achieved at a controlled specific growth rate while fed-batch culture is currently required for commercial production.

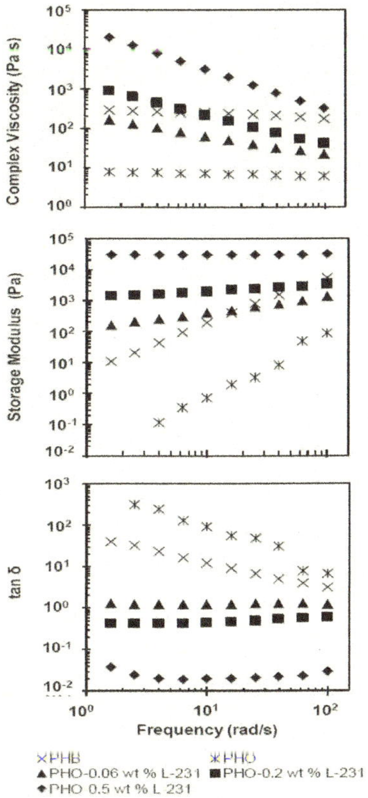

Fig. (5). Rheological properties of PHB and PHO and effect of lauroyl peroxide (L-231) cross-linking on a) complex viscosity, b) storage modulus and c) loss tangent, tan δ, measured at 190°C (From Nerkar *et al.* 2014 [59]).

Chemostat Studies

The carboxylic acids favoured for *mcl*-PHA production inhibit growth at relatively low concentrations, so production of dense cultures is not possible in batch mode. To achieve high biomass density these fatty acids and other inhibitory substrates must be fed at a controlled rate to keep a low concentration in the fermentor. In carbon-limited chemostat, culture liquid medium is fed into the bioreactor at a constant rate while fermentation broth is removed at that same rate. This maintains a constant liquid volume in the bioreactor. High production rates have been achieved in single-stage chemostats. For example, *P. putida* GPo1 biomass with a PHA content of 29.8% of cell dry mass (CDM) and a volumetric productivity of 0.74 g/(Lh) has been produced using n-octane as the sole source of carbon and energy [62]. *P. putida* KT2440 has been grown on mixtures of nonanoic acid and glucose to produce biomass containing CDM with 63% PHA content at a productivity of 1.52 g/(Lh) while using acrylic acid to inhibit β-oxidation [63]. All bacterial strains shown in Table **2** are variants of *P. putida* with the recombinant GPo1 POMC1 showing the lowest volumetric productivity.

Several studies have used 2-stage chemostat to increase PHA content in the final biomass. One study [64] employed dual (carbon and nitrogen) limitation and noted simultaneous PHA accumulation and degradation inside the cell. The most successful of these had a PHA content of 38% of CDM in the first reactor, 63% in the second and an overall productivity of 1.06 g/(Lh) using N-limited *P. putida* GPo1 on octane [65]. One of the advantages of the use of chemostat culture is that high biomass production can be achieved with lower peak oxygen demand than batch culture. Indeed, none of these studies required the use of pure O_2, and most if not all could be optimized for very substantial gains in productivity. It is interesting to note that although the chemostat data in Table **1** are from scientific studies and was not designed for high productivity, the productivities achieved were quite substantial.

Fed-Batch Studies

In contrast to chemostat studies, most fed-batch reports concentrated on maximum productivity. The highest productivities have been obtained with oleic acid Table **3** which is much less toxic than the lower molecular weight carboxylic acids such as octanoic and nonanoic. The high productivity resulted from the very high final biomass obtained, although the maximum PHA attained was only 51.4% of the biomass and contained a wide range of subunits. A nitrogen-limited, fed-batch study using octane as the carbon source produced similar biomass levels (112 g/L) but the PHA content and hence productivity were low. An earlier study demonstrated that higher PHA content (33% of CDM) can be obtained from octane but this is still much lower than typically achieved from carboxylic acids. The other studies that achieved high productivity both fed a mixture of glucose and either octanoic or nonanoic acid.

Table 2. Selected *mcl*-PHA chemostat production processes.

Conditions	**Organism**	**Subunits**	**Biomass [g/L]**	**PHA [wt.-%]**	**Productivity [g((Lh)]**	**Ref.**
Single stage; D=0.30 1/h; nonanoic acid and glucose co-feeding; C-limited; acrylic acid	*P. putida* KT2440	C9 C7	8.0	63.1	1.52	[63]
Two stage; D_1=0.30 1/h; D_2=0.25 1/h; nonanoic acid and glucose co-feeding; C-limited; acrylic acid	*P. putida* KT2440	C9 C7	7.3 (R1) 9.2 (R2)	51.7 (R1) 58.5 (R2)	1.14 (R1) 0.40 (R2) 0.76 (Overall)	[63]
Single stage; D=0.25 1/h; octanoic acid and glucose and nonanoic acid and glucose co-feeding; C-limited; acrylic acid	*P. putida* KT2440	C8 C6 (octanoic acid) C9 C7 (nonanoic acid)	5.0 (octanoic acid) 5.8 (nonanoic acid)	48 (octanoic acid) 50 (nonanoic acid)	0.60 (octanoic acid) 0.73 (nonanoic acid)	[53]
Two stage; D_1=0.21 1/h, D_2=0.16 1/h; *n*-octane; N-limited	*P. putida* GPo1	C8 C6	R_1=10.5 R_2=18.0	R_1=38.0 R_2=63.0	R_1=0.84 R_2=1.18 Overall=1.06	[65]
Two stage; D_1=0.10 1/h, D_2=0.10 1/h; octanoic acid in R1; 10-undecenoic acid in R2; dual nutrient limited growth regime	*P. putida* GPo1	C11:1 C9:1 C8 C7:1 C6	R_1=1.13 R_2=1.53	R_1=24.2 R_2=52.4	R_1=0.027 R_2=0.053 Overall=0.04	[64]
Single stage; D=0.20 1/h; *n*-octane; N-limited	*P. putida* GPo1	C8 C6	12.4	29.8	0.74	[62]
Single stage; D=0.10 1/h; oleic acid; O_2-limited	*P. putida* KT2442	C14:1 C12:1 C12 C10 C8 C6	30	23	0.67	[66]
Single stage; D=0.20 1/h; octanoic acid; IPTG inducer	*P. putida* GPo1 POMC1 (recombinant)	C8 C6	1.64	53.2	0.17	[67]

* Productivity is defined as volumetric productivity (PHA concentration X volumetric flow rate/ liquid reactor volume).

Table 3. Selected *mcl*-PHA fed-batch production processes.

Conditions	Organism	Subunits	Biomass [g/L]	PHA [wt.-%]	Productivity [g((Lh)]	Ref.
Exponential feeding at μ=0.15 1/h; nonanoic acid and glucose; C-limited; acrylic acid; pure O_2	*P. putida* KT2440	C9 C7	71.4	75.5	1.8	[69]
DO-stat feeding; oleic acid; O_2-limited; air only	*P. putida* KT2442	C14:1 C12 C10 C8 C6	92	45	1.6	[70]
Exponential feeding (μ=0.10 1/h); Na octanoate; N-limited; pure O_2	*P. putida* KT2442	C8 C6	51.5	34	0.40	[71]
DO-stat and pH-stat feeding; oleic acid; P-limited; pure O_2	*P. putida* KT2442	C14:1 C12 C10 C8 C6	141	51.4	1.91	[72]
Staged feeding; glucose and Na octanoate; N-limited, O_2-limited; air	*P. putida* BM01	C8 C6	55	65	0.92	[73]
Fed-batch; batch mode followed by constant rate feeding; *n*-octane; N-limited; air	*P. putida* GPo1	C8 C6	37.1	33	0.25	[74]
Exponential feeding (μ=0.19 1/h) followed by constant rate; octanoic acid and ammonium octanoate; N-limited; air	*P. putida* GPo1	C8 C6	47	55	0.53	[75]
Exponential feeding (μ=0.05 1/h); *n*-octane; N-limited; pure O_2	*P. putida* GPo1	C8 C6	112	4	0.07	[76]
pH-stat ; octanoic acid; N-limited; air	*P. putida* GPo1	C8 C6	63	62	1	[77]
Constant feed rate; palmitic acid; inducer; air only	*E. coli* 193MC1 (recombinant)	C10 C8 C6	3.55	12	0.011	[67]
Constant rate feeding; glucose and fructose; P-limited; air only	*P. putida* IPT 046	C10 C8	50	63	0.75	[68]

* Productivity is defined as overall volumetric productivity (total PHA produced per unit volume/total fermentation time).

A variety of methods are used to control feeding of the substrates. The aim is to supply enough substrate for rapid growth and production while not overfeeding which may produce toxic levels of substrate in the bioreactor. The exponential feeding method increases that feed rate in a manner designed to maintain a constant specific growth rate in the fermentor. DO-stats and pH-stats rely on oxygen and pH probe data respectively to sense when growth substrate limitation occurs at which point the feed rate is automatically increased. Approaches for the control of both fed-batch and continuous production of *mcl*-PHA have been reviewed [68].

Oxygen Supply

As a fed-batch fermentation progresses, the oxygen uptake rate (OUR) of the microorganisms in the bioreactor typically increases greatly as their concentration increases. The bioreactor oxygen transfer rate (OTR) must be sufficient to match this demand. The OTR depends on the volumetric O_2 transfer coefficient (k_La) of the reactor under specified conditions of agitation and aeration. It also depends on the equilibrium dissolved O_2 concentration (C^*) such that $OTR = k_La(C^* - C)$ where C is the dissolved O_2 concentration. Once the maximum k_La has been achieved, the only way to increase OTR is to increase C^*. This is typically done by replacing air with pure O_2 achieving a 4.75 times higher OTR. However, pure O_2 substantially increases production cost [79] or may be dangerous [80]. Another way to increase OTR is to pressurize the bioreactor [81] and has been tested in PHA production [82], although this increases the cost of pumping air and needs specially designed equipment. O_2 carriers such as hexadecane [83] can be employed but tend to be expensive and cause problems in PHA production. *Vitreoscilla* hemoglobin (VHb) is an O_2-binding protein that allows a bacterium to grow aerobically even under very low O_2 conditions. The gene responsible for VHb production has been cloned into *E. coli* for ethanol [84] and PHB production [85, 86]. However, since PHA is already produced at very low dissolved O_2 concentration [78], OTR is already maximized. Thus use of VHb cannot increase the OTR.

A less studied but a potentially highly effective method of increasing O_2 transfer is to use peroxides, particularly H_2O_2, to provide O_2. Many fermentation technologists believe that H_2O_2 will impair productivity since O_2 radicals are known to damage DNA [87]. However, like any toxic substrate, if handled properly it may be of use. Schlegel [88] was the first to demonstrate this and later studies have also been successful [89 - 91]. To date the most economically attractive practical approach to increasing OTR and thus productivity is to use pressure swing adsorption (PSA) to produce a high oxygen concentration in the inlet gas and recycle the off-gas (which still contains substantial amounts of oxygen) through the PSA unit. This has been applied to the production of *scl*-PHA where it has been shown to result in substantial cost reduction [92] but has not yet been evaluated for *mcl*-PHA.

SEPARATION METHODS

There are two major routes to the separation of *mcl*-PHA from bacterial biomass with *scl*-PHAs. Both routes can produce material in the form of pellets or as a latex (Fig. **6**). When ready for harvest, bacterial cells should contain greater than 70% *mcl*-PHA in CDM in the form of granules coated with a monolayer of phospholipid, which contains various proteins associated with granule formation. One can dissolve the biomass other than PHA with chemicals and/or enzymes to release the granules, thus forming an aqueous suspension of PHA. However, the solubility in different solvents as well as the density of the native granules is quite different for *mcl*-PHAs compared to *scl*-PHAs. These differences greatly affect how separation processes may be conducted.

Solvent Extraction

Biomass is first concentrated, usually by centrifugation; then washed to further remove impurities. Water is normally removed to improve the efficiency of solvent extraction. This may be accomplished by lyophilization (lab scale) or by spray drying (large scale). *mcl*-PHA is then extracted into a suitable solvent of which there are many. The most effective extraction methods are Soxhlet (lab scale) and countercurrent (large scale) since these methods maintain the maximum concentration differential between the solvent and the material to be extracted. The solvent containing *mcl*-PHA is then added to a non-solvent while mixing at a cold temperature to facilitate precipitation. The solvent may or may not be soluble in the nonsolvent. In either case the liquid phase will contain contaminants, so this phase must be removed, by gravity, centrifugation or filtration. If water is used as the non-solvent, then centrifugation may result in the *mcl*-PHA forming a pellet at the bottom of the tube, a film at the surface of the water or, as often occurs, both. This occurs because the density of *mcl*-PHA is close to that of water and because *mcl*-PHA is often not completely homogeneous, with individual polymer chains containing different subunit ratios.

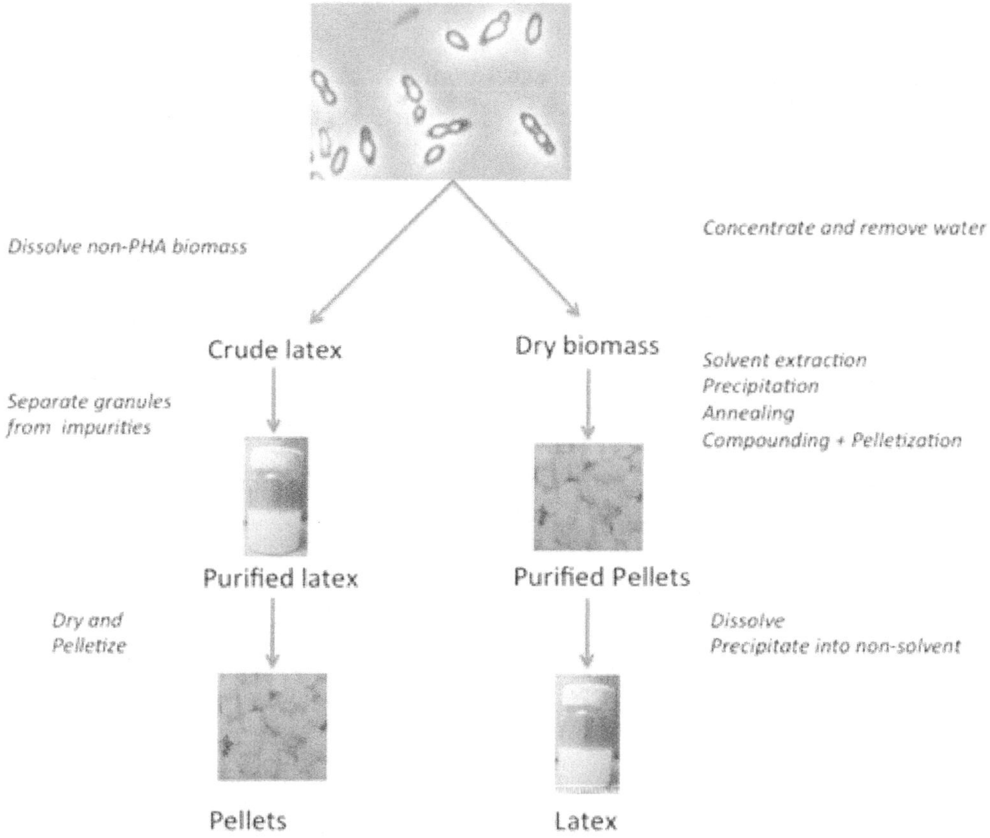

Fig. (6). Two general approaches to separate PHA from microbial biomass. Either method can result in production of a latex or solid PHA. These are applicable to any type of PHA but must be modified accordingly.

Solubility change at different temperatures is another approach to the separation of PHA from biomass. The most notable example of this is the use of ethylene carbonate or 1,2-propylene carbonate for PHB [93]. However, since to be effective, some contaminants must remain in the solvent. This reduces its effectiveness after every cycle unless the solvent is purified.

A typical solvent extraction process is detailed in Fig. (**7**) where acetone is the solvent and methanol is the non-solvent. Ideally the solvent should be good for extraction of *mcl*-PHA, but poor at the extraction of any other biomass components. The non-solvent should be poor for *mcl*-PHA extraction and good for any contaminants in the solvent (acetone in this case). It is hoped that most of the contaminants remain with the biomass solids (*i.e.* unextracted) or dissolved in the solvents since PHA will precipitate. Often the precipitated PHA is re-extracted in the same or a similar process. Each cycle should result in improved purity. All solvents should be safe to handle and easy to recycle.

Pilot scale production of *mcl*-PHA has been reported in 650 L bioreactors with a production of up to 55 g/L *P. putida* GPo1 biomass containing 60% *mcl*-PHA [94]. The bacterial cells were separated from the spent medium by centrifugation and freeze-dried before extraction of the polymer into acetone, filtration, then evaporation (with solvent recovery) to concentrate the polymer. This concentrated solution was then added to a methanol/ethanol/water mixture causing the PHA to readily precipitate before drying. Use of an alcohol/water mixture appears to produce higher purity PHA than using either alone.

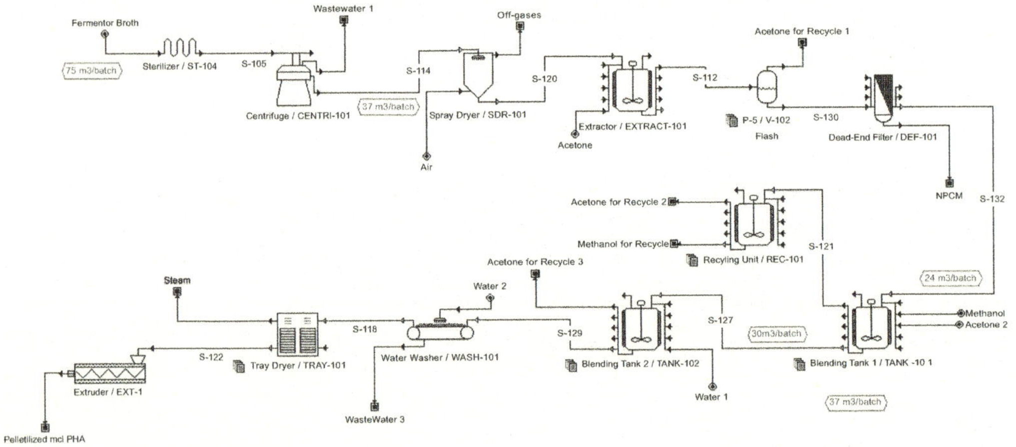

Fig. (7). Solvent extraction of *mcl*-PHA [95].

Differential Digestion of Biomass and PHA

Many of the most interesting applications of PHAs and especially *mcl*-PHAs are as films, paint, adhesives, toner particles or nanoparticles, particularly in drug release applications. In most instances this requires that the material be in the form of a stable latex. Bacterial cells can be broken by shear force using high-pressure homogenization, by oxidation methods such as ozonolysis or hypochlorite treatment, or by a cocktail of enzymes. Depending on the process, the resulting suspension contains all cytoplasmic components such as PHA granules, enzymes, RNA and DNA, as well as cell envelope debris. The challenge is to selectively remove the granules, which are much larger than the other remaining cell components. Sedimentation or centrifugation do not usually help, as the density of *mcl*-PHA is close to that of water. Addition of salts that increase the density of water will cause all of the PHA to float to the surface but further contaminates the material. DNA must be hydrolyzed to reduce viscosity either *via* enzymatic or chemical hydrolysis.

The first reported "pilot" scale separation process reported for a *mcl*-PHA separation utilized *P. putida* kt2442 that had been grown on sodium octanoate up to 200 L [96]. The recovery conditions were similar to those previously employed to digest *C. necator* biomass for *scl*-PHA separation [97], but tailored to *P. putida*. The fermentor content was heated to 121°C to kill cells and to denature nucleic acid polymers (for viscosity reduction) and then rapidly cooled to 55°C. The pH-value was adjusted to 8.5 and excess EDTA, sodium dodecyl sulphate (SDS), and a commercially available proteases (Alcalase) added. EDTA chelates Mg^{2+} and Ca^{2+} which destabilizes the outer membrane. SDS is an anionic surfactant that dissolves lipids and proteins while the proteases digest proteins, thereby increasing their solubility. SDS is known to sorb to *mcl*-PHA granules which may affect their colloidal stability. After half an hour of treatment, a sterile filter recirculation loop was connected and diafiltration started. Water was added to the filtrate to maintain a constant volume and regular back flushing was applied using pressurized air. Initially, the filtrate was yellow with high absorbance at 350 nm. When the filtrate optical density was sufficiently reduced, the water was disconnected and the process was converted to common filtration, producing a 300 g/L retentate. Choice of filtration unit is critical. This report used a 0.1 µm ceramic filter since it could be backflushed and heated and is solvent resistant which allows it to be readily regenerated. Indeed, in our experience, flux through membrane or ceramic filters decreases quickly due to filter blockage. However, if successful this process produces a stable concentrated latex. The stability at this high concentration is likely due to residual SDS. Contaminants include non-PHA biomass that has not been sufficiently digested to pass thorough the filter as well as anything that binds to the granule surface.

One of the major concerns in using the aforementioned process in a commercial process is the high cost of proteases. A less expensive alternative is the replacement of enzymes with a strong base such as NaOH to dissolve proteins. Such a process is illustrated in (Fig. **8**). Unlike enzymes, strong alkalinity may decrease *mcl*-PHA molecular weight, so care must be taken to limit exposure. This process also produces a latex with impurities in the aqueous phase. Changing the pH-value of a *mcl*-PHA suspension changes the amount of ionized carboxylic groups (pKa of about 3.9), which affects the surface charge and thus the stability of the particles in suspension. Separation of the granules from the aqueous phase again usually involves a filtration process. Filtration must be rapid to avoid polymer degradation as the pH-value cannot be adjusted to a neutral value until the dissolved impurities are removed. Otherwise they may fall out of solution and cannot be separated by filtration.

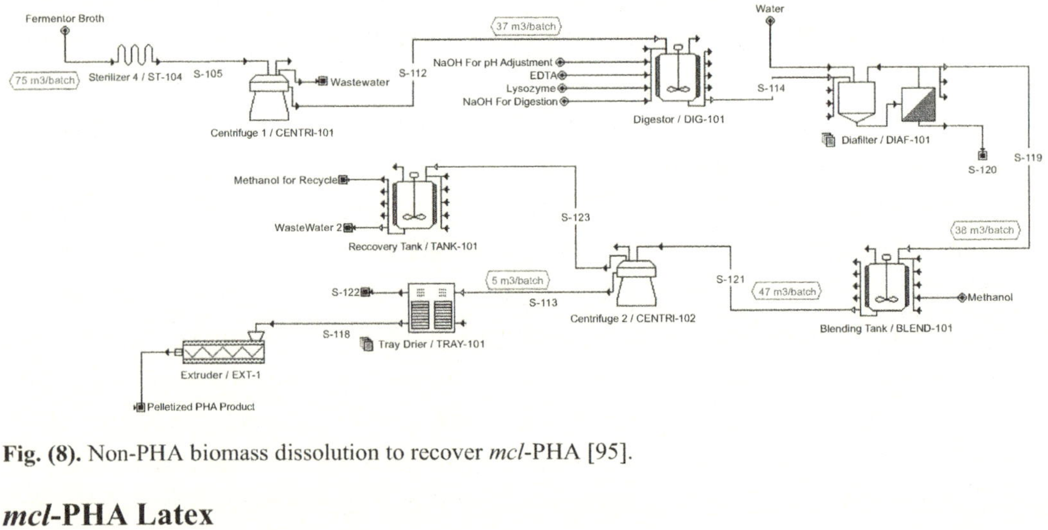

Fig. (8). Non-PHA biomass dissolution to recover *mcl*-PHA [95].

mcl-PHA Latex

Synthesis of PHA granules has been compared to emulsion polymerization, the process commonly used for commercial production of polystyrene [98]. As shown in Fig. (**6**), a latex may be produced in either approach to *mcl*-PHA purification. Dissolution of non-PHA biomass produces "native" granules, which retain their internal morphology and can possibly be used to produce materials with superior properties due to the arrangement of polymer chains as they are produced in the granules. But "artificial" granules may also be produced. PHA granules, *in vivo*, are amorphous (probably stabilized by bound water) and covered with a monolayer of phospholipid containing proteins. The monolayer is often disturbed during purification especially if surfactants or oxidizing agents are employed. Hypochlorite has been used to dissolve non-PHA biomass to produce stable *mcl*-PHA latex used for film formation [75].

Like most polymeric materials in aqueous suspension, PHB artificial granule suspensions can be stabilized with surfactants. However, removal of the surfactant coating destabilizes PHB suspensions [99]. In contrast, *mcl*-PHA artificial granules have been found to be self-stabilizing, apparently due to steric stabilization due to the longer side-chains [100]. A PHO suspension was observed to be stable over a period of three years [100], but contained only 0.4% solids. We have found that, unless stabilizers are added, particle agglomeration always occurs above 2% (wt/vol). Mixing causes agglomeration to occur more quickly by increasing contact between particles. *mcl*-PHA particles have a negative surface charge. Increasing the cation concentration neutralizes the surface charge on the particles, decreasing the repulsive force between them.

The stability of a latex suspension can be explained by DLVO-theory [101]. The two main forces involved in DLVO are the attractive van der Waals London force and the repulsive electric double layer. The van der Waals London force is a result of random fluctuations in atomic charge distribution and its strength is highly dependent on the distance between particles [102]. Knowledge of the surface properties of the latex particles is required to understand how the repulsive double layer can be controlled to influence latex stability. The number of hydrophobic side-chains per carboxyl end group is very high indicating that hydrophobic interactions would dominate colloidal properties. However, while uncharged, the polar carbonyl group can also influence properties. For example, both PHB [103] and mcl-PHA [104] can form Langmuir-Blodgett monolayers at air-water interfaces with monolayers maintained after crystallization. Thus, PHAs are somewhat amphiphilic. We found that a 0.40% (wt./vol.) PHN suspension always agglomerated below pH-value 4.0 and a corresponding to a zeta potential of -30 mV. Above pH-value 6.0, the zeta potential for PHN was below -30 mV.

THE FUTURE

It is fortunate that *P. putida* KT2440 and related strains are the most productive *mcl*-PHA accumulators because KT2440 was one of the first bacteria to have its genome sequenced and putative genes identified [105, 106], greatly aiding in the construction of more complete metabolic models. The current era of "omics" should provide information that will allow a much deeper understanding of how bacteria function based on the knowledge of how the transcriptome, proteome and metabolome respond to changes in physiological conditions. Such data, especially when combined with ^{13}C-metabolic flux analysis are leading to construction of more useful models that will allow much more accurate predictions concerning the modification of genes or physiological conditions. The effects of N-limitation on the proteome of *P. putida* CA-3 have been examined [107] as well as the effects on the transcriptome and proteome of *P. putida* LS46 from waste glycerol and carboxylic acids under several physiological conditions [108]. A genome-scale constraint-based model of *P. putida* KT2440 metabolism has been generated [109]. *Pseudomonas* species, especially those related to *P. putida* have the genetic machinery to produce a wide variety of useful materials [110]. Since this "machinery" is already in place, it makes sense to use such organisms even if genetic manipulation is required to achieve adequate productivity. Although not nearly as many tools are currently available as for *E. coli*, many approaches to site-specific gene insertion and deletion are being developed for these bacteria [111].

mcl-PHAs have many potential applications but cannot yet compete with the polymers currently used due to its high cost of production. *Scl*-PHAs are also expensive to produce due to fermentation and separation costs, but substrate (mostly glucose) cost is much less than the carboxylic acids typically used for production of *mcl*-PHAs. In *scl*-PHA production, PHA content is typically well over 80% of CDM, as opposed to only 70% for *mcl*-PHA production. However, since the specific gravity of *mcl*-PHA is typically about 1.0, while that of *scl*-PHA may exceed 1.2, volumetric polymer content is about equal. Thus, if *mcl*-PHA could be produced in high yield from glucose, it could be produced for about the same cost as PHB per unit volume. In N- or P-limited fermentations on glucose, wild type *P. putida* KT2440 can produce about 33% of CDM as PHA. Based on *in silico* optimization using a mathematical model, KT2440 was genetically modified to produce PHD with PHA accumulation of 42% [112]. Presumably production from glucose can be increased further, but even at this yield the carboxylic acid requirement is greatly reduced. To be economically competitive in the near future, commercial *mcl*-PHA production will likely rely on some *de novo* synthesis from carbohydrates, supplemented with the appropriate carboxylic acids.

Since PHAs, along with single-cell protein, are among the few aerobically produced commodity fermentation products, their production may require advances in fermentation technology. ICI developed their well-known 1,500,000 L pressure cycle reactor to meet such demands of high productivity combined with efficient aeration and heat removal. It can be easily demonstrated that multistage continuous or plug-flow modes are the most economical methods of producing PHA [113]. These have been avoided since industrial scale aerobic plug flow bioreactors do not yet exist and since continuous fermentations have not been used commercially due to concerns with contamination. However, the favourable economics of running such systems may lead to the development of more reliable (in terms of avoiding contamination) fermenters or the use of production strains under conditions that do not require media sterilization. The concept of multistage chemostat *scl*-PHA production has already been demonstrated at the laboratory scale [114] while methods of producing PHA from waste organics with no media sterilization are also being developed.

CONFLICT OF INTEREST

The authors confirm that they have no conflict of interest to declare for this publication.

ACKNOWLEDGEMENTS

Declared none.

REFERENCES

[1] Marois Y, Zhang Z, Vert M, Deng X, Lenz R, Guidoin R. Mechanism and rate of degradation of polyhydroxyoctanoate films in aqueous media: A long-term in vitro study. J Biomed Mater Res 2000; 49(2): 216-24. [http://dx.doi.org/10.1002/(SICI)1097-4636(200002)49:2<216::AID-JBM9>3.0.CO;2-X] [PMID: 10571908]

[2] Lemoigne M. Produit de déshydratation et de polymérisation de l'acide β-oxybutyrique. Bull Soc Chim Biol (Paris) 1926; 8: 770-82.

[3] Mark H, Whitby GS. Collected Papers of Wallace H Carothers on Polymerization. New York: Interscience 1940.

[4] Baptist JN, Ziegler JB. Method of making absorbable surgical sutures from poly beta hydroxy acids. US Patent 3,225,766 (A) 1965.

[5] Stanier RY, Palleroni NJ, Doudoroff M. The aerobic pseudomonads: a taxonomic study. J Gen Microbiol 1966; 43(2): 159-271. [http://dx.doi.org/10.1099/00221287-43-2-159] [PMID: 5963505]

[6] de Smet MJ, Eggink G, Witholt B, Kingma J, Wynberg H. Characterization of intracellular inclusions formed by *Pseudomonas oleovorans* during growth on octane. J Bacteriol 1983; 154(2): 870-8. [PMID: 6841319]

[7] Singh AK, Mallick N. Enhanced production of SCL-LCL-PHA co-polymer by sludge-isolated *Pseudomonas aeruginosa* MTCC 7925. Lett Appl Microbiol 2008; 46(3): 350-7. [http://dx.doi.org/10.1111/j.1472-765X.2008.02323.x] [PMID: 18221276]

[8] Hazer B, Steinbüchel A. Increased diversification of polyhydroxyalkanoates by modification reactions for industrial and medical applications. Appl Microbiol Biotechnol 2007; 74(1): 1-12. [http://dx.doi.org/10.1007/s00253-006-0732-8] [PMID: 17146652]

[9] Kurth N, Renard E, Brachet F, Robic D, Guerin P, Bourbouze R. Poly(3-hydroxyoctanoate) containing pendant carboxylic groups for the preparation of nanoparticles aimed at drug transport and release. Polymer (Guildf) 2002; 43(4): 1095-101. [http://dx.doi.org/10.1016/S0032-3861(01)00692-9]

[10] Sodian R, Sperling JS, Martin DP, *et al.* Fabrication of a trileaflet heart valve scaffold from a polyhydroxyalkanoate biopolyester for use in tissue engineering. Tissue Eng 2000; 6(2): 183-8. [http://dx.doi.org/10.1089/107632700320793] [PMID: 10941212]

[11] Ashby RD, Foglia TA. Poly(hydroxyalkanoate) biosynthesis from triglyceride substrates. Appl Microbiol Biotechnol 1998; 49: 431-7. [http://dx.doi.org/10.1007/s002530051194]

[12] van der Walle GA, de Koning GJ, Weusthuis RA, Eggink G. Properties, modifications and applications of biopolyesters. Adv Biochem Eng Biotechnol 2001; 71: 263-91. [http://dx.doi.org/10.1007/3-540-40021-4_9] [PMID: 11217415]

[13] Anderson AJ, Haywood GW, Williams DR, Dawes EA. The production of polyhydroxyalkanoates from unrelated carbon sources. Novel biodegradable microbial polymers. Springer Netherlands 1990; pp. 119-29.

[http://dx.doi.org/10.1007/978-94-009-2129-0_11]

[14] Schnurrenberger P, Hungerbuhler E, Seebach D. Total synthesis of (+)-colletodiol from (S,S)-tartrate and (R)-3-hydroxybutanoate. Liebigs Ann Chem 1987; 9: 733-44. [http://dx.doi.org/10.1002/jlac.198719870821]

[15] Chiba T, Nakai T. A synthetic approach to (+)-thienamycin from methyl (R)-3-hydroxybutanoate. A new entry to (3R, 4R)-3-((*R*)-1-hydroxyethyl)-4-acetoxy-2-azetidinone. Chem Lett 1985; (5): 651-4. [http://dx.doi.org/10.1246/cl.1985.651]

[16] Seebach D, Zuger M. On the depolymerization of poly-(R)-3-hydroxy-butanoate (PHB). Helv Chim Acta 1982; 65: 495 503. [http://dx.doi.org/10.1002/hlca.19820650208]

[17] Ren Q, Sierro N, Witholt B, Kessler B. FabG, an NADPH-dependent 3-ketoacyl reductase of *Pseudomonas aeruginosa*, provides precursors for medium-chain-length poly-3-hydroxyalkanoate biosynthesis in *Escherichia coli.* J Bacteriol 2000; 182(10): 2978-81. [http://dx.doi.org/10.1128/JB.182.10.2978-2981.2000] [PMID: 10781572]

[18] Vo MT, Lee KW, Jung YM, Lee YH. Comparative effect of overexpressed phaJ and fabG genes supplementing (R)-3-hydroxyalkanoate monomer units on biosynthesis of mcl-polyhydroxyalkanoate in *Pseudomonas putida* KCTC1639. J Biosci Bioeng 2008; 106(1): 95-8. [http://dx.doi.org/10.1263/jbb.106.95] [PMID: 18691538]

[19] Vo MT, Lee KW, Kim TK, Lee YH. Utilization of fadA knockout mutant *Pseudomonas putida* for overproduction of medium chain-length-polyhydroxyalkanoate. Biotechnol Lett 2007; 29(12): 1915-20. [http://dx.doi.org/10.1007/s10529-007-9476-5] [PMID: 17653511]

[20] Haywood GW, Anderson AJ, Ewing DF, Dawes EA. Accumulation of a polyhydroxyalkanoate containing primarily 3-hydroxydecanoate from simple carbohydrate substrates by Pseudomonas sp. strain NCIMB 40135. Appl Environ Microbiol 1990; 56(11): 3354-9. [PMID: 16348341]

[21] Timm A, Steinbüchel A. Formation of polyesters consisting of medium-chain-length 3-hydroxyalkanoic acids from gluconate by *Pseudomonas aeruginosa* and other fluorescent pseudomonads. Appl Environ Microbiol 1990; 56(11): 3360-7. [PMID: 2125185]

[22] Eggink G, Waard P, Huijberts GN. The role of fatty acid biosynthesis and degradation in the supply of substrates for poly (3-hydroxyalkanoate) formation in *Pseudomonas putida.* FEMS Microbiol Lett 1992; 103(2-4): 159-63. [http://dx.doi.org/10.1111/j.1574-6968.1992.tb05833.x]

[23] Rehm BH, Krüger N, Steinbüchel A. A new metabolic link between fatty acid de novo synthesis and polyhydroxyalkanoic acid synthesis. The PHAG gene from *Pseudomonas putida* KT2440 encodes a 3-hydroxyacyl-acyl carrier protein-coenzyme a transferase. J Biol Chem 1998; 273(37): 24044-51. [http://dx.doi.org/10.1074/jbc.273.37.24044] [PMID: 9727022]

[24] Ruth K, de Roo G, Egli T, Ren Q. Identification of two acyl-CoA synthetases from *Pseudomonas putida* GPo1: one is located at the surface of polyhydroxyalkanoates granules. Biomacromolecules 2008; 9(6): 1652-9.

[http://dx.doi.org/10.1021/bm8001655] [PMID: 18471011]

[25] Galán B, Dinjaski N, Maestro B, *et al.* Nucleoid-associated PhaF phasin drives intracellular location and segregation of polyhydroxyalkanoate granules in *Pseudomonas putida* KT2442. Mol Microbiol 2011; 79(2): 402-18.
[http://dx.doi.org/10.1111/j.1365-2958.2010.07450.x] [PMID: 21219460]

[26] Arias S, Sandoval A, Arcos M, *et al.* Poly-3-hydroxyalkanoate synthases from *Pseudomonas putida* U: substrate specificity and ultrastructural studies. Microb Biotechnol 2008; 1(2): 170-6.
[http://dx.doi.org/10.1111/j.1751-7915.2007.00016.x] [PMID: 21261834]

[27] Cai L, Yuan MQ, Liu F, Jian J, Chen GQ. Enhanced production of medium-chain-length polyhydroxyalkanoates (PHA) by PHA depolymerase knockout mutant of *Pseudomonas putida* KT2442. Bioresour Technol 2009; 100(7): 2265-70.
[http://dx.doi.org/10.1016/j.biortech.2008.11.020] [PMID: 19103481]

[28] Jendrossek D, Handrick R. Microbial degradation of polyhydroxyalkanoates. Annu Rev Microbiol 2002; 56(1): 403-32.
[http://dx.doi.org/10.1146/annurev.micro.56.012302.160838] [PMID: 12213937]

[29] Qi Q, Rehm BH, Steinbüchel A. Synthesis of poly(3-hydroxyalkanoates) in Escherichia coli expressing the PHA synthase gene phaC2 from Pseudomonas aeruginosa: comparison of PhaC1 and PhaC2. FEMS Microbiol Lett 1997; 157(1): 155-62.
[http://dx.doi.org/10.1111/j.1574-6968.1997.tb12767.x] [PMID: 9418250]

[30] de Eugenio LI, Galán B, Escapa IF, *et al.* The PhaD regulator controls the simultaneous expression of the pha genes involved in polyhydroxyalkanoate metabolism and turnover in *Pseudomonas putida* KT2442. Environ Microbiol 2010; 12(6): 1591-603.
[PMID: 20406286]

[31] Sandoval A, Arias-Barrau E, Arcos M, Naharro G, Olivera ER, Luengo JM. Genetic and ultrastructural analysis of different mutants of *Pseudomonas putida* affected in the poly-3-hydrox--n-alkanoate gene cluster. Environ Microbiol 2007; 9(3): 737-51.
[http://dx.doi.org/10.1111/j.1462-2920.2006.01196.x] [PMID: 17298373]

[32] Ren Q, de Roo G, Witholt B, Zinn M, Thöny-Meyer L. Overexpression and characterization of medium-chain-length polyhydroxyalkanoate granule bound polymerases from *Pseudomonas putida* GPo1. Microb Cell Fact 2009; 8: 60.
[http://dx.doi.org/10.1186/1475-2859-8-60] [PMID: 19925642]

[33] Kim TK, Jung YM, Vo MT, Shioya S, Lee YH. Metabolic engineering and characterization of phaC1 and phaC2 genes from *Pseudomonas putida* KCTC1639 for overproduction of medium-chain-length polyhydroxyalkanoate. Biotechnol Prog 2006; 22(6): 1541-6.
[http://dx.doi.org/10.1002/bp0601746] [PMID: 17137299]

[34] Wang Q, Tappel RC, Zhu C, Nomura CT. Development of a new strategy for production of medium-chain-length polyhydroxyalkanoates by recombinant *Escherichia coli* via inexpensive non-fatty acid feedstocks. Appl Environ Microbiol 2012; 78(2): 519-27.
[http://dx.doi.org/10.1128/AEM.07020-11] [PMID: 22101037]

[35] Fialho AM, Zielinski NA, Fett WF, Chakrabarty AM, Berry A. Distribution of alginate gene sequences in the Pseudomonas rRNA homology group I-Azomonas-Azotobacter lineage of superfamily B

procaryotes. Appl Environ Microbiol 1990; 56(2): 436-43.
[PMID: 1689562]

[36] Hori K, Marsudi S, Unno H. Simultaneous production of polyhydroxyalkanoates and rhamnolipids by *Pseudomonas aeruginosa.* Biotechnol Bioeng 2002; 78(6): 699-707.
[http://dx.doi.org/10.1002/bit.10248] [PMID: 11992535]

[37] Chakrabarty AM, Chou G, Gunsalus IC. Genetic regulation of octane dissimilation plasmid in Pseudomonas. Proc Natl Acad Sci USA 1973; 70(4): 1137-40.
[http://dx.doi.org/10.1073/pnas.70.4.1137] [PMID: 4515610]

[38] Chew TL, Bhatia S. Catalytic processes towards the production of biofuels in a palm oil and oil palm biomass-based biorefinery. Bioresour Technol 2008; 99(17): 7911-22.
[http://dx.doi.org/10.1016/j.biortech.2008.03.009] [PMID: 18434141]

[39] Naik SN, Goud VV, Rout PK, Dalai AK. Production of first and second generation biofuels: a comprehensive review. Renew Sustain Energy Rev 2010; 14: 578-97.
[http://dx.doi.org/10.1016/j.rser.2009.10.003]

[40] Scholz C, Fuller RC, Lenz RW. Production of poly (β-hydroxyalkanoates) with β-substituents containing terminal ester groups by *Pseudomonas oleovorans.* Macromol Chem Phys 1994; 195(4): 1405-21.
[http://dx.doi.org/10.1002/macp.1994.021950424]

[41] Scholz C, Lenz RW, Fuller RC. Growth and polymer incorporation of *Pseudomonas oleovorans* on alkyl esters of heptanoic acid. Macromol 1994; 27: 2886-9.
[http://dx.doi.org/10.1021/ma00088a033]

[42] Hazer B, Lenz RW, Fuller RC. Biosynthesis of methyl branched poly(β-hydroxyalkanoate)s with *Pseudomonas oleovorans.* Macromol 1994; 27: 45-9.
[http://dx.doi.org/10.1021/ma00079a008]

[43] Curley J, Hazer B, Lenz RW, Fuller RC. Production of poly (3-hydroxyalkanoates) containing aromatic substituents by *Pseudomonas oleovorans.* Macromol 1996; 29: 1762-6.
[http://dx.doi.org/10.1021/ma951185a]

[44] Hoffmann N, Steinbüchel A, Rehm BH. Homologous functional expression of cryptic phaG from *Pseudomonas oleovorans* establishes the transacylase-mediated polyhydroxyalkanoate biosynthetic pathway. Appl Microbiol Biotechnol 2000; 54(5): 665-70.
[http://dx.doi.org/10.1007/s002530000441] [PMID: 11131392]

[45] Lee M, Chandler AC. A study of the nature, growth and control of bacteria in cutting compounds. J Bacteriol 1941; 41(3): 373-86.
[PMID: 16560407]

[46] Nakazawa T. Travels of a Pseudomonas, from Japan around the world. Environ Microbiol 2002; 4(12): 782-6.
[http://dx.doi.org/10.1046/j.1462-2920.2002.00310.x] [PMID: 12534461]

[47] Jatsenko T, Tover A, Tegova R, Kivisaar M. Molecular characterization of Rif(r) mutations in *Pseudomonas aeruginosa* and *Pseudomonas putida.* Mutat Res 2010; 683(1-2): 106-14.
[http://dx.doi.org/10.1016/j.mrfmmm.2009.10.015] [PMID: 19887074]

[48] Hervás AB, Canosa I, Santero E. Transcriptome analysis of *Pseudomonas putida* in response to nitrogen availability. J Bacteriol 2008; 190(1): 416-20. [http://dx.doi.org/10.1128/JB.01230-07] [PMID: 17965157]

[49] Follonier S, Panke S, Zinn M. A reduction in growth rate of *Pseudomonas putida* KT2442 counteracts productivity advances in medium-chain-length polyhydroxyalkanoate production from gluconate. Microb Cell Fact 2011; 10(1): 25. [http://dx.doi.org/10.1186/1475-2859-10-25] [PMID: 21513516]

[50] Takase K, Taguchi S, Doi Y. Enhanced synthesis of poly(3-hydroxybutyrate) in recombinant *Escherichia coli* by means of error-prone PCR mutagenesis, saturation mutagenesis, and *in vitro* recombination of the type II polyhydroxyalkanoate synthase gene. J Biochem 2003; 133(1): 139-45. [http://dx.doi.org/10.1093/jb/mvg015] [PMID: 12761209]

[51] Yoon SC, Song JJ, Kim TU. Isolation and characterization of *Pseudomonas putida* BM01 accumulating high amount of PHA MCL. J Microbiol Biotechnol 1994; 4: 126-33.

[52] Gagnon KD, Lenz RW, Farris RJ, Fuller RC. Crystallization behavior and its influence on the mechanical properties of a thermoplastic elastomer produced by *Pseudomonas oleovorans*. Macromolecules 1992; 25(14): 3723-8. [http://dx.doi.org/10.1021/ma00040a018]

[53] Jiang X, Sun Z, Marchessault RH, Ramsay JA, Ramsay BA. Biosynthesis and properties of medium-chain-length polyhydroxyalkanoates with enriched content of the dominant monomer. Biomacromolecules 2012; 13(9): 2926-32. [http://dx.doi.org/10.1021/bm3009507] [PMID: 22871146]

[54] Schacher FH, Rupar PA, Manners I. Functional Block Copolymers: Nanostructured Materials with Emerging. Applications Chem Int Ed 2012; 51: 7898-921.

[55] Ruzette AV, Leibler L. Block copolymers in tomorrow's plastics. Nat Mater 2005; 4(1): 19-31. [http://dx.doi.org/10.1038/nmat1295] [PMID: 15689991]

[56] Gao X, Chen JC, Wu Q, Chen GQ. Polyhydroxyalkanoates as a source of chemicals, polymers, and biofuels. Curr Opin Biotechnol 2011; 22(6): 768-74. [http://dx.doi.org/10.1016/j.copbio.2011.06.005] [PMID: 21705209]

[57] Li SY, Dong CL, Wang SY, Ye HM, Chen GQ. Microbial production of polyhydroxyalkanoate block copolymer by recombinant *Pseudomonas putida.* Appl Microbiol Biotechnol 2011; 90(2): 659-69. [http://dx.doi.org/10.1007/s00253-010-3069-2] [PMID: 21181145]

[58] Kamiya N, Yamamoto Y, Inoue Y, Chujo R, Doi Y. Microstructure of bacterially synthesized poly(3-hydroxybutyrate-co-3-hydroxyvalerate). Macromol 1989; 22: 1676-82. [http://dx.doi.org/10.1021/ma00194a030]

[59] Nerkar M, Ramsay JA, Ramsay BA, Kontopoulou M. Melt compounded blends of short and medium chain length poly-3-hydroxyalkanoates. J Polym Environm 2014; 22: 236-43. [http://dx.doi.org/10.1007/s10924-013-0635-6]

[60] Nerkar M, Ramsay JA, Ramsay BA, Kontopoulou M. Improvements in the extensional viscosity, crystallization kinetics and morphology of poly(lactic acid)/ poly-3-hydroxyoctanoate blends through reactive modification. Polymer (Guildf) 2015; 64: 51-61.

[http://dx.doi.org/10.1016/j.polymer.2015.03.015]

[61] Lee J, McCarthy S. Biodegradable poly(lactic acid) blends with chemically modified polyhydroxyoctanoate through chain extension. J Polym Environ 2009; 17(4): 240-7. [http://dx.doi.org/10.1007/s10924-009-0144-9]

[62] Hazenberg W, Witholt B. Efficient production of medium chain-length poly(3-hydroxyalkanoates) from octane by Pseudomonas oleovorans: economic considerations. Appl Microbiol Biotechnol 1997; 48: 588-96.
[http://dx.doi.org/10.1007/s002530051100]

[63] Gillis J. Production of medium-chain-length poly(3-hydroxyalkanoates) using *Pseudomonas citronellolis* DSM50332 and *P. putida* KT2440 in continuous reactor systems. Masters thesis 2011.

[64] Hartmann R, Hany R, Witholt B, Zinn M. Simultaneous biosynthesis of two copolymers in *Pseudomonas putida* GPo1 using a two-stage continuous culture system. Biomacromolecules 2010; 11(6): 1488-93.
[http://dx.doi.org/10.1021/bm100118t] [PMID: 20459087]

[65] Jung K, Hazenberg W, Prieto M, Witholt B. Two-stage continuous process development for the production of medium-chain-length poly(3-hydroxyalkanoates). Biotechnol Bioeng 2001; 72(1): 19-24.
[http://dx.doi.org/10.1002/1097-0290(20010105)72:1<19::AID-BIT3>3.0.CO;2-B] [PMID: 11084589]

[66] Huijberts GN, Eggink G. Production of poly(3-hydroxyalkanoates) by *Pseudomonas putida* KT2442 in continuous cultures. Appl Microbiol Biotechnol 1996; 46: 233-9.
[http://dx.doi.org/10.1007/s002530050810]

[67] Prieto MA, Kellerhals MB, Bozzato GB, Radnovic D, Witholt B, Kessler B. Engineering of stable recombinant bacteria for production of chiral medium-chain-length poly-3-hydroxyalkanoates. Appl Environ Microbiol 1999; 65(8): 3265-71.
[PMID: 10427005]

[68] Sun Z, Ramsay JA, Guay M, Ramsay BA. Fermentation process development for the production of medium-chain-length poly-3-hyroxyalkanoates. Appl Microbiol Biotechnol 2007; 75(3): 475-85.
[http://dx.doi.org/10.1007/s00253-007-0857-4] [PMID: 17453197]

[69] Jiang X, Sun Z, Ramsay J, Ramsay B. Fed-batch Production of Medium-Chain-Length Polyhydroxyalkanoates with Elevated 3-hydroxynonanoate content. AMB Express 2013; 3: 50.
[http://dx.doi.org/10.1186/2191-0855-3-50] [PMID: 23987136]

[70] Weusthuis RA, Huijberts GN, Eggink G. Production of mcl-poly(hydroxyalkanoates) (review). In: Eggink G, Steinbüchel A, Poirer Y, Witholt B, Eds. International symposium on bacterial polyhydroxyalkanoates. Ottawa: NRC Research Press 1996.

[71] Kellerhals MB, Kessler B, Witholt B. Closed-loop control of bacterial high-cell-density fed-batch cultures: production of mcl-PHAs by *Pseudomonas putida* KT2442 under single-substrate and cofeeding conditions. Biotechnol Bioeng 1999; 65(3): 306-15.
[http://dx.doi.org/10.1002/(SICI)1097-0290(19991105)65:3<306::AID-BIT8>3.0.CO;2-0] [PMID: 10486129]

[72] Lee SY, Wong HH, Choi Ji, Lee SH, Lee SC, Han CS. Production of medium-chain-length

polyhydroxyalkanoates by high-cell-density cultivation of *Pseudomonas putida* under phosphorus limitation. Biotechnol Bioeng 2000; 68(4): 466-70.
[http://dx.doi.org/10.1002/(SICI)1097-0290(20000520)68:4<466::AID-BIT12>3.0.CO;2-T] [PMID: 10745215]

[73] Kim GJ, Lee IY, Yoon SC, Shin YC, Park YH. Enhanced yield and a high production of medium-chain-length poly(3-hydroxyalkanoates) in a two-step fed-batch cultivation of *Pseudomonas putida* by combined use of glucose and octanoate. Enzyme Microb Technol 1997; 20: 500-5.
[http://dx.doi.org/10.1016/S0141-0229(96)00179-2]

[74] Preusting H, Hazenberg W, Witholt B. Continuous production of poly(3-hydroxyalkanoates) by *Pseudomonas oleovorans* in a high-cell-density, 2-liquid-phase chemostat. Enzyme Microb Technol 1993; 15: 311-6.
[http://dx.doi.org/10.1016/0141-0229(93)90156-V]

[75] Dufresne A, Samain E. Preparation and characterization of a poly(beta-hydroxyoctanoate) latex produced by *Pseudomonas oleovorans.* Macromol 1998; 31: 6426-33.
[http://dx.doi.org/10.1021/ma980508a]

[76] Kellerhals MB, Hazenberg W, Witholt B. High cell density fermentations of *Pseudomonas oleovorans* for the production of mcl-PHAs in two-liquid phase media. Enzyme Microb Technol 1999; 24: 111-6. a
[http://dx.doi.org/10.1016/S0141-0229(98)00113-6]

[77] Kim BS. Production of medium chain length polyhydroxyalkanoates by fed-batch culture of *Pseudomonas oleovorans.* Biotechnol Lett 2002; 24: 125-30.
[http://dx.doi.org/10.1023/A:1013898504895]

[78] Diniz SC, Taciro MK, Gomez JG, da Cruz Pradella JG. High-cell-density cultivation of *Pseudomonas putida* IPT 046 and medium-chain-length polyhydroxyalkanoate production from sugarcane carbohydrates. Appl Biochem Biotechnol 2004; 119(1): 51-70.
[http://dx.doi.org/10.1385/ABAB:119:1:51] [PMID: 15496728]

[79] Choi J, Lee SY. Factors affecting the economics of polyhydroxyalkanoate production by bacterial fermentation. Appl Microbiol Biotechnol 1999; 51(1): 13-21.
[http://dx.doi.org/10.1007/s002530051357]

[80] Shim H, Hwang B, Lee SS, Kong SH. Kinetics of BTEX biodegradation by a coculture of *Pseudomonas putida* and *Pseudomonas fluorescens* under hypoxic conditions. Biodegradation 2005; 16(4): 319-27.
[http://dx.doi.org/10.1007/s10532-004-1842-6] [PMID: 15865337]

[81] Knoll A, Maier B, Tscherrig H, Büchs J. The oxygen mass transfer, carbon dioxide inhibition, heat removal, and the energy and cost efficiencies of high pressure fermentation. Technology Transfer in Biotechnology. Springer Berlin Heidelberg 2005; pp. 77-99.
[http://dx.doi.org/10.1007/b98918]

[82] Follonier S, Henes B, Panke S, Zinn M. Putting cells under pressure: a simple and efficient way to enhance the productivity of medium-chain-length polyhydroxyalkanoate in processes with *Pseudomonas putida* KT2440. Biotechnol Bioeng 2012; 109(2): 451-61.
[http://dx.doi.org/10.1002/bit.23312] [PMID: 21858788]

[83] Liu YS, Wu JY. Use of n-hexadecane as an oxygen vector to improve *Phaffia rhodozyma* growth and carotenoid production in shake-flask cultures. J Appl Microbiol 2006; 101(5): 1033-8. [http://dx.doi.org/10.1111/j.1365-2672.2006.03009.x] [PMID: 17040227]

[84] Sanny T, Arnaldos M, Kunkel SA, Pagilla KR, Stark BC. Engineering of ethanolic *E. coli* with the Vitreoscilla hemoglobin gene enhances ethanol production from both glucose and xylose. Appl Microbiol Biotechnol 2010; 88(5): 1103-12. [http://dx.doi.org/10.1007/s00253-010-2817-7] [PMID: 20717665]

[85] Yu H, Shi Y, Zhang Y, Yang S, Shen Z. Effect of Vitreoscilla hemoglobin biosynthesis in *Escherichia coli* on production of poly(β-hydroxybutyrate) and fermentative parameters. FEMS Microbiol Lett 2002; 214(2): 223-7. [PMID: 12351235]

[86] Horng YT, Chang KC, Chien CC, Wei YH, Sun YM, Soo PC. Enhanced polyhydroxybutyrate (PHB) production via the coexpressed phaCAB and vgb genes controlled by arabinose P promoter in Escherichia coli. Lett Appl Microbiol 2010; 50(2): 158-67. [http://dx.doi.org/10.1111/j.1472-765X.2009.02772.x] [PMID: 19943886]

[87] Imlay JA, Linn S. DNA damage and oxygen radical toxicity. Science 1988; 240(4857): 1302-9. [http://dx.doi.org/10.1126/science.3287616] [PMID: 3287616]

[88] Schlegel HG. Aeration without air: oxygen supply by hydrogen peroxide. Biotechnol Bioeng 1977; 19(3): 413-24. [http://dx.doi.org/10.1002/bit.260190311] [PMID: 321047]

[89] Ibrahim M, Schlegel HG. Oxygen supply to bacterial suspensions of high cell densities by hydrogen peroxide. Biotechnol Bioeng 1980; 22(9): 1877-94. [http://dx.doi.org/10.1002/bit.260220908] [PMID: 7407339]

[90] Nies D, Schlegel HG. Use of catalase from *Escherichia coli* in model experiments for oxygen supply of microorganisms with hydrogen peroxide. Biotechnol Bioeng 1984; 26(7): 737-41. [http://dx.doi.org/10.1002/bit.260260716] [PMID: 18553439]

[91] Sonnleitner B, Hahnemann U. Robust oxygen supply by controlled addition of hydrogen peroxide to microbial cultures. Bioprocess Eng 1997; 17(4): 215-9. [http://dx.doi.org/10.1007/s004490050377]

[92] Chang HN, Kim MI, Fei Q, *et al.* Economic evaluation of off-gas recycle pressure swing adsorption (PSA) in industrial scale poly(3-hydroxybutyrate) fermentation. Biotechnol Bioprocess Eng 2010; 15(6): 905-10. [http://dx.doi.org/10.1007/s12257-010-0114-z]

[93] Lafferty RM, Heinzle E. Cyclic carbonic acid esters as solvents for poly-(β-hydroxybutyric acid. US Patent No 4,101,533 1978.

[94] Elbahloul Y, Steinbüchel A. Large-scale production of poly(3-hydroxyoctanoic acid) by Pseudomonas putida GPo1 and a simplified downstream process. Appl Environ Microbiol 2009; 75(3): 643-51. [http://dx.doi.org/10.1128/AEM.01869-08] [PMID: 19047387]

[95] Douglas W. Economic evaluation of scaling up the recovery of polyhydroxyalkanoates 4th year thesis project, Chemical Engineering. Queen's University 2013.

[96] De Koning GJ, Kellerhals M, Van Meurs C, Witholt B. A process for the recovery of poly(hydroxyalkanoates) from Pseudomonads Part 2: Process development and economic evaluation. Bioprocess Eng 1997; 17(1): 15-21. [http://dx.doi.org/10.1007/s004490050346]

[97] Holmes PA, Lim GB. Separation process. US Patent 4,910,145 1990.

[98] De Koning GJ, Maxwell IA. Biosynthesis of poly-(R)-3-hydroxyalkanoate: an emulsion polymerization. J Environ Polym Degrad 1993; 1(3): 223-6. [http://dx.doi.org/10.1007/BF01458030]

[99] Horowitz DM, Sanders JK. Amorphous, biomimetic granules of polyhydroxybutyrate: preparation, characterization, and biological implications. J Am Chem Soc 1994; 116(7): 2695-702. [http://dx.doi.org/10.1021/ja00086a001]

[100] Marchessault RH, Morin FG, Wong S, Saracovan I. Artificial granule suspensions of long side chain poly (3-hydroxyalkanoate). Can J Microbiol 1995; 41(13): 138-42. [http://dx.doi.org/10.1139/m95-180]

[101] Verwey E, Overbeek J. Theory of the stability of lyophobic colloids. Amsterdam: Elsevier Publishing 1948.

[102] Karaman M, Pashley R. Applied colloids and surface chemistry. West Sussex: John Wiley and Sons Ltd 2004.

[103] Lambeek G, Vorenkamp EJ, Schouten AJ. Structural Study of Langmuir-Blodgett Mono-and Multilayers of Poly(beta-hydroxybutyrate). Macromol 1995; 28(6): 2023-32. [http://dx.doi.org/10.1021/ma00110a041]

[104] Jagoda A, Zinn M, Bieler E, Meier W, Kita-Tokarczyk K. Biodegradable polymer–lipid monolayers as templates for calcium phosphate mineralization. J Mater Chem B Mater Biol Med 2013; 1(3): 368-78. [http://dx.doi.org/10.1039/C2TB00083K]

[105] Nelson KE, Weinel C, Paulsen IT, *et al.* Complete genome sequence and comparative analysis of the metabolically versatile *Pseudomonas putida* KT2440. Environ Microbiol 2002; 4(12): 799-808. [http://dx.doi.org/10.1046/j.1462-2920.2002.00366.x] [PMID: 12534463]

[106] Weinel C, Nelson KE, Tümmler B. Global features of the *Pseudomonas putida* KT2440 genome sequence. Environ Microbiol 2002; 4(12): 809-18. [http://dx.doi.org/10.1046/j.1462-2920.2002.00331.x] [PMID: 12534464]

[107] Nikodinovic-Runic J, Flanagan M, Hume AR, Cagney G, O'Connor KE. Analysis of the Pseudomonas putida CA-3 proteome during growth on styrene under nitrogen-limiting and non-limiting conditions. Microbiology 2009; 155(Pt 10): 3348-61. [http://dx.doi.org/10.1099/mic.0.031153-0] [PMID: 19608612]

[108] Fu J, Sharma P, Spicer V, *et al.* Quantitative 'Omics Analyses of Medium Chain Length Polyhydroxyalkanaote Metabolism in *Pseudomonas putida* LS46 Cultured with Waste Glycerol and Waste Fatty Acids. PLoS One 2015; 10(11): e0142322. [http://dx.doi.org/10.1371/journal.pone.0142322] [PMID: 26544181]

[109] Puchałka J, Oberhardt MA, Godinho M, *et al.* Genome-scale reconstruction and analysis of the *Pseudomonas putida* KT2440 metabolic network facilitates applications in biotechnology. PLOS

Comput Biol 2008; 4(10): e1000210.
[http://dx.doi.org/10.1371/journal.pcbi.1000210] [PMID: 18974823]

[110] Loeschcke A, Thies S. *Pseudomonas putida*-a versatile host for the production of natural products. Appl Microbiol Biotechnol 2015; 99(15): 6197-214.
[http://dx.doi.org/10.1007/s00253-015-6745-4] [PMID: 26099332]

[111] Nikel PI, Martínez-García E, de Lorenzo V. Biotechnological domestication of pseudomonads using synthetic biology. Nat Rev Microbiol 2014; 12(5): 368-79.
[http://dx.doi.org/10.1038/nrmicro3253] [PMID: 24736795]

[112] Borrero-de Acuña JM, Bielecka A, Häussler S, *et al.* Production of medium chain length polyhydroxyalkanoate in metabolic flux optimized *Pseudomonas putida.* Microb Cell Fact 2014; 13: 88.
[http://dx.doi.org/10.1186/1475-2859-13-88] [PMID: 24948031]

[113] Koller M, Braunegg G. Potential and prospects of continuous Polyhydroxyalkanoate (PHA) production. Bioeng 2015; 2(2): 94-121.

[114] Atlić A, Koller M, Scherzer D, *et al.* Continuous production of poly([*R*]-3-hydroxybutyrate) by *Cupriavidus necator* in a multistage bioreactor cascade. Appl Microbiol Biotechnol 2011; 91(2): 295-304.
[http://dx.doi.org/10.1007/s00253-011-3260-0] [PMID: 21503760]

SUBJECT INDEX

A

A. caviae 289, 290, 291, 292, 398
Accumulation, polyester 180, 290
Acetate 77, 124, 125, 178, 179, 180, 182, 183, 187, 193, 195, 198, 201, 231, 232, 233, 238, 243, 245, 246, 247, 248, 250, 252, 266, 287, 288, 318, 437, 440, 441, 442
Acetate and propionate 245, 246, 247, 248, 252, 287
  mixture of 245, 248, 252
Acetate supplementation 198, 199, 200, 202
Acetoacetate 189, 190
Acetoacetyl-CoA 134, 178, 181, 182, 183, 186, 189, 190, 202, 232, 233, 270, 286, 289, 354, 355
Acetoacetyl-CoA reductase 7, 12, 36, 122, 181, 199, 209, 232, 273, 291, 349, 356, 412
Acetyl-CoA 12, 13, 134, 181, 182, 189, 190, 200, 202, 209, 232, 233, 271, 286, 287, 308, 309, 310, 313, 315, 316, 348, 349, 354, 355, 356, 361, 362, 365
  carboxylase (ACC) 209
  molecules 182, 286, 354, 355
Acid(s) 5, 7, 9, 10, 15, 16, 17, 25, 27, 28, 29, 30, 31, 32, 36, 38, 43, 45, 48, 51, 58, 61, 63, 68, 77, 78, 104, 105, 106, 108, 114, 124, 130, 133, 134, 137, 138, 140, 141, 142, 143, 144, 149, 150, 151, 152, 178, 179, 180, 182, 187, 232, 233, 238, 239, 241, 244, 246, 247, 250, 251, 253, 272, 287, 289, 290, 300, 318, 322, 367, 372, 381, 395, 399, 400, 410, 411, 436, 440, 441, 442, 449, 451, 452, 453, 454
  acetic 28, 30, 32, 38, 106, 137, 180, 244, 247, 251, 253
  acrylic 449, 452, 453, 454
  diluted 25, 27, 38
  eutropha strains 3-mercaptopropionic 300
  fatty5, 7, 9, 10, 15, 16, 17, 63, 77, 78, 143, 178, 180, 182, 187, 232, 233, 272, 287, 289, 290, 318, 322, 372, 395, 399, 410, 411, 440, 441, 442, 451
  lactic 36, 48, 77, 105, 108, 133, 137, 143, 144, 149
  lactobionic 104, 105
  levulinic 28, 30, 61, 68
  nonanoic 452, 453, 454
  octanoic 436, 442, 453, 454
  oleic 381, 452, 453, 454
  pentanoic 114, 138, 140, 141, 142, 150, 151, 152
  plant oils and fatty 289, 290
  short-chain organic (SCOAs) 228, 229, 230, 231, 232, 236, 237, 238, 239, 241, 246, 250, 253
  synthesis, fatty 440, 441, 442
  volatile fatty (VFAs) 77, 78, 124, 137, 142, 143, 144, 400
Acid hydrolysis 27, 28, 68, 391
Acidobacteria 244, 245, 248
Activated carbon 32, 368
Activated sludge 77, 227, 228, 251, 253, 254, 388
Active PhaC homodimer 275
Acyl carrier protein (ACP) 209, 442

ADF conditions 246, 247, 250, 251, 252, 253
ADF process 238, 241, 245
Aerobic anoxygenic phototrophic bacteria (AAPB) 180
Aerobic dynamic feeding (ADF) 227, 229, 237, 240, 241, 245, 246, 253, 254
A. eutrophus 145, 202, 267, 386, 392, 407, 409
AFEX process 39, 40
Agricultural waste 3, 37, 101
Alcaligenes eutrophus 7, 16, 182, 200, 205, 234, 312, 381, 437
Amylases 65, 68, 69, 71, 72, 387
Amylopectin 65, 67, 68, 69
Amylose 65, 67, 68, 69, 254
Anaerobic biotechnological conversion 105
Anaerobic digestion 39, 77, 238
Antibiotics 106, 115, 116, 132, 299, 439
Arabinose 27, 36, 403, 412
Autotrophic growth 178, 268, 269, 284, 301, 302, 304, 306, 321
Azotobacter sp 56, 57, 409

B

Bacillus megaterium 4, 42, 46, 52, 57, 144, 152, 233, 437
Bacillus sp 47, 136, 234, 394
Bacteria 6, 7, 9, 10, 11, 12, 21, 27, 33, 34, 36, 41, 54, 69, 71, 78, 116, 176, 188, 189, 193, 208, 227, 228, 229, 230, 233, 235, 238, 243, 253, 266, 267, 269, 292, 301, 304, 312, 314, 352, 354, 357, 358, 363, 381, 382, 383, 384, 385, 386, 387, 390, 401, 403, 406, 407, 416, 462
  producing 54, 71, 78, 390, 406
Bacterial cells 4, 12, 16, 54, 279, 298, 382, 383, 456, 458, 459
Bacterial counterparts 353, 354, 355, 356
Bacterial strains 12, 13, 15, 21, 22, 23, 36, 42, 47, 55, 56, 63, 71, 76, 382, 402, 409, 414, 415, 452
Bagasse 37, 42, 43, 129, 145, 400, 402
  hydrolysate 42, 43, 402
Batch, coffee ground oil 19
Batch cultivations 296, 297
B. cepacia 61, 62, 135, 402
B. cereus 53, 234, 401
Beet sugar molasses 44
B. funghorum 140, 141, 149
Bicarbonate 200, 302, 303
Biochemical oxygen demand (BOD) 100, 103, 391
Biocompatible bioplastics 176
Biodegradable polymers 24, 298
Biodiesel-glycerol 62
Biodiesel industry 3, 59, 368, 369
Biodiesel production 14, 15, 58, 60, 62, 64, 139, 251, 369, 372, 404, 405
Biomass 20, 21, 24, 35, 36, 37, 38, 39, 41, 42, 43, 47, 48, 50, 52, 56, 57, 60, 61, 62, 63, 70, 72, 73, 74, 75, 79, 118, 119, 132, 135, 136, 138, 140, 143, 144, 145, 180, 190, 203, 207, 230, 242, 254, 268, 294, 296, 297, 372, 392, 400, 409, 412, 452, 453, 454, 456, 457, 458, 460, 461
  active 132, 143, 144

grass 41, 42
highest 36, 37, 73
non-PHA 457, 460, 461
wood 35, 36, 37
Biomass concentration, total 294
Biomass formation 20, 134, 136, 291
Bioplastics 4, 6, 108, 109, 191, 208, 348, 350, 401, 437
Bioreactor 22, 63, 118, 124, 126, 138, 147, 150, 151, 152, 292, 294, 388, 393, 397, 401, 409, 410, 452, 454, 455, 458
laboratory 20, 136, 137, 153
Biosynthesis 4, 7, 21, 112, 149, 190, 265, 271, 273, 289, 293, 297, 300, 355, 407, 436, 441
Biosynthetic genes 57
latus PHA 57
Biotechnological applications 188, 193
Biotechnological conversion 104, 105, 106
Biotechnological process 13, 28, 68, 79, 104, 389
developing efficient 389
Biotechnological production of biofuels 391, 401
Biotechnological production of PHAs 3, 5, 12, 13, 24, 34, 74, 79, 389
Biotechnological production process 51, 119
Block copolymers of scl- and mcl-PHAs 448
Bovine origin 102, 152, 153
B-oxidation 182, 232, 283, 287, 349, 396, 440, 441, 442, 449
links mcl-PHA synthesis to 440, 441
Building blocks 112, 113, 114, 124, 139, 141, 437
Burkholderia cepacia 21, 42, 43, 62, 135
Burkholderia sacchari 37, 39, 42, 43
By-products, agricultural 366

C

Candidatus competibacter 244, 245, 246
Cane molasses 44, 45, 46, 77, 245, 394
fermented sugar 77, 245
Carbon 13, 41, 51, 65, 76, 111, 117, 124, 125, 126, 131, 176, 177, 179, 196, 197, 200, 202, 228, 229, 230, 246, 253, 254, 265, 280, 290, 299, 300, 308, 312, 317, 318, 319, 320, 321, 350, 413, 439, 440, 441, 442, 452
sole 41, 308, 317, 319
Carbon atoms 6, 14, 113, 176, 285, 437, 439
Carbon catabolite repression (CCR) 35, 39, 410
Carbon feedstock 101, 104, 119, 146, 295, 393, 399, 401
Carbon flux 111, 140, 147, 318, 320
Carbonic anhydrase 302
Carbonic anhydrase enzymes 302, 303
Carbon monoxide 306, 413
Carbon reserve 195, 196
Carbon-rich waste 3, 101
Carbon sources 7, 12, 15, 16, 20, 21, 36, 40, 42, 43, 44, 55, 60, 61, 62, 63, 70, 71, 73, 74, 112, 138, 140, 141, 144, 149, 150, 151, 152, 153,

154, 178, 190, 193, 199, 233, 235, 240, 243, 247, 265, 267, 270, 273, 274, 281, 284, 286, 288, 289, 293, 294, 295, 296, 301, 306, 307, 317, 318, 320, 323, 349, 366, 367, 368, 369, 371, 380, 389, 390, 391, 395, 401, 402, 404, 406, 409, 410, 440, 442, 444, 447, 452
Carbon substrate 3, 47, 136, 137, 145, 146, 229, 232, 253, 273, 293, 438
Carboxylic acids 440, 441, 446, 451, 453, 462, 463
Caseins 48, 49, 131, 149
Cassava 66, 71, 73, 400, 401
Cassava starch 67, 71, 73
Cassava starch hydrolysate 71, 72
Cassava starch wastewater 71, 72
Catabolism 134, 189, 190, 314, 315, 316, 403
Catalysis, covalent 8, 9, 184, 185
Catechol 308, 309, 310, 311, 315
CBB cycle 303, 304
CDM concentration 144, 145
Cell concentration 71, 366, 394
Cell dry mass (CDM) 17, 18, 19, 20, 21, 22, 23, 36, 39, 41, 46, 47, 52, 53, 54, 56, 63, 64, 74, 75, 76, 111, 124, 125, 138, 140, 141, 144, 145, 146, 150, 151, 152, 153, 178, 187, 195, 197, 198, 199, 201, 240, 241, 243, 244, 245, 247, 248, 250, 251, 253, 254, 287, 289, 290, 292, 295, 299, 301, 306, 368, 392, 393, 394, 395, 396, 397, 398, 399, 400, 402, 404, 408, 409, 410, 412, 413, 452, 456, 462, 463
Cell growth 41, 63, 187, 229, 240, 288, 290, 414
Cell membrane 28, 29, 279, 443
Cellulose 24, 25, 26, 27, 28, 29, 32, 37, 38, 39, 40, 64, 177, 369, 400, 402, 406, 412
Cellulose hydrolysates 41, 42, 402, 412
Cell wall 24, 25, 30, 56
Cheese production 100, 101, 102
Cheese whey 3, 14, 48, 49, 50, 51, 52, 53, 54, 55, 56, 57, 79, 101, 122, 142, 147, 153, 392, 393, 409, 441
  powdered 49, 56
Chemical oxygen demand (COD) 51, 100, 103, 124, 238, 252, 254, 391
Chromosome 147, 274, 303, 307, 350, 352, 354, 408, 415
Class III PHA synthases 183, 184
Cloning PHA genes 380, 381, 406, 407
CO_2 fixation 175, 176, 208, 304, 321, 348, 362
Coenzyme A (CoA) 178, 182, 189, 190, 281, 286, 354, 355, 440, 441, 442, 443
Coffee grounds 19, 20, 38, 111
Commercial refined glycerol 62, 63
Compounds 28, 30, 33, 34, 116, 176, 179, 198, 205, 266, 307, 406, 440, 441
  inhibitory 28, 30, 33
  toxic 27, 31, 33, 34
Conditions 179, 180, 187, 195, 247, 249, 273, 318, 439, 451, 462
  limiting 187, 195, 247, 249, 273, 318
  nitrogen-fixing 179, 180

operating 249
physiological 439, 451, 462
Copolyesters 54, 62, 101, 115, 126, 138, 139, 141, 150, 151, 152, 349, 396
Copolymers 7, 8, 10, 17, 36, 53, 61, 73, 190, 200, 206, 230, 254, 273, 285, 288, 289, 290, 291, 297, 372, 387, 397, 398, 405, 407, 447, 448
block 407, 447, 448, 449
Corn oil 18, 19, 300, 397, 398, 400
Cornstarch, extruded 71, 73
Crude glycerol 58, 59, 60, 61, 62, 63, 64, 139, 241, 251, 295, 403, 404, 411
Cultivation media 18, 21, 138
Cultivation medium 139, 142, 144, 146
Culture 78, 117, 123, 137, 143, 148, 228, 229, 243, 250, 251, 389
enriched 228, 243, 250, 251
mixed 78, 117, 123, 137, 143, 148, 229, 389
Culture conditions 195, 199, 208, 323, 371
Culture medium 56, 123, 194, 198, 203, 204, 415
Cupriavidus necator 7, 12, 15, 36, 38, 40, 41, 54, 55, 57, 60, 61, 105, 110, 119, 122, 182, 232, 233, 234, 235, 265, 267, 268, 381, 386, 437
bacterial strain 119
eubacterial strain 110
Cupriavidus necator mRePT 153
Cyanobacteria 120, 175, 176, 177, 183, 186, 192, 193, 194, 195, 197, 198, 199, 200, 204, 205, 206, 208, 301, 406
Cyanophycin production 301

D

Dairy industry 49, 101, 134, 367, 408
Dairy waste 101, 144, 152
Degradation, nPHA granule 363
Degradation products 28
Dehydrogenase, propionaldehyde 411
Denaturing gradient gel electrophoresis (DGGE) 242, 245, 246, 248, 249, 252, 388
Depolymerases 121, 191, 192, 280, 281, 282, 363, 443, 444
Deproteinized whey 52, 57
Dinoroseobacter sp 180
Dissolved organic carbon (DOC) 180
Downstream processing 54, 101, 123, 125, 135, 139, 231, 242, 389
Dry Sweet Whey 134

E

Endoamylases 66, 68, 69
Engineered *E. coli* 393, 409
Enoyl-CoA hydratase 209, 290, 362, 364, 374, 440, 441
Enzymatic detoxification 34
Enzyme activities 186, 274, 275, 286, 387, 397, 410, 445
Enzymes 8, 9, 26, 27, 29, 32, 34, 55, 66, 68, 69, 106, 110, 112, 142, 148, 182, 183, 186, 191, 202, 228,

229, 233, 271, 272, 274, 275, 276, 279, 283, 296, 298, 302, 303, 309, 310, 313, 314, 315, 316, 320, 348, 355, 362, 381, 386, 387, 397, 403, 406, 437, 438, 440, 441, 443, 445, 456, 459, 460
β-ketothiolase 182, 272
debranching 66, 68, 69
extracellular 437, 438
starch-converting 68, 69
EPS biosynthesis 370, 371
Ethanol 11, 12, 27, 46, 52, 58, 63, 108, 144, 145, 193, 238, 269, 319, 381, 403, 410, 455
Extruded corn starch (ECS) 366
Extruded rice bran (ERB) 71, 366, 368

F

Fatty acid production 322
Fatty acids, free 14, 15, 110, 441
Fatty acids biosynthesis 232, 233
Fed-batch 19, 20, 35, 61, 147, 265, 287, 290, 291, 292, 293, 295, 296, 297, 320, 394, 410, 411, 448, 455
cultivation 19, 20, 35, 61, 147, 265, 291, 292, 293, 297, 320, 410
cultures 287, 290, 292, 293, 408, 451
fermentations 36, 47, 61, 62, 293, 295, 296, 394, 411, 448, 455
mode of cultivation 20
Fermentable sugars 24, 27, 33, 34, 68
Fermentation 77, 245, 285, 294, 307, 396
acidogenic 77, 245, 396
controlled 285, 294, 307
Fermentation medium 50, 123, 140
Fermented paper mill wastewater 244, 245, 251, 252
Fermentor 18, 19, 23, 52, 57, 60, 61, 73, 450, 451, 455
fed-batch 60, 61, 73
Fermentor PHBHV 18
Flavobacterium 246, 248, 252
Fluorescent *in Situ* Hybridization (FISH) 228, 242, 245, 247, 251, 252, 388
Free fatty acids (FFA) 14, 15, 110, 441
Furfurals 28, 30, 31, 33, 34, 38

G

Galactose 27, 38, 51, 107, 136, 139, 149, 367, 391, 392, 408
Gene cluster 317, 357, 369, 370, 371
Genes 17, 55, 122, 181, 195, 199, 200, 201, 202, 208, 272, 273, 276, 277, 278, 281, 282, 283, 290, 301, 307, 308, 309, 310, 312, 313, 316, 318, 320, 349, 352, 355, 356, 357, 369, 370, 380, 381, 383, 384, 386, 395, 403, 405, 407, 408, 409, 411, 440, 441, 442, 443, 444, 446, 455, 462
phasin 277
target 370
Genes encoding 121, 122, 145, 273, 357, 362
Genetic engineering 32, 120, 121, 122, 135, 140, 199, 235, 323, 369, 381, 391, 396
Genome 6, 183, 201, 308, 312, 313, 316, 317, 349, 350, 351, 369, 371, 415, 446, 462
Gentisate 310, 311, 312

Glucoamylase 65, 69, 71, 72
Gluconate 10, 441, 442
Glucose co-feeding 453
Glycerol 3, 14, 15, 17, 21, 58, 59, 60, 61, 62, 63, 64, 110, 120, 122, 133, 138, 177, 208, 251, 295, 300, 366, 368, 372, 381, 403, 404, 405, 407, 410, 411
  mixture of 63
  pure 59, 62, 63, 372, 404
Glycerol concentration 62, 63
Glycerol dehydratase 411
Glycerol kinase 405
Glycerol phase, crude 369, 372
Glycogen 4, 69, 176, 196, 197, 201, 227, 232, 233, 236, 237, 250, 382, 383
Glycogen-accumulating organisms (GAOs) 228, 232, 236, 237, 241, 243, 244, 245, 246
Granule-associated proteins 188, 277, 356, 358, 363, 416
Granule formation 277, 356, 357, 360, 363, 456
Granules 10, 64, 186, 191, 194, 270, 276, 277, 278, 281, 357, 359, 363, 438, 442, 443, 444, 456, 457, 459, 460, 461
  nPHA 363
Green plastics 107, 108, 109, 110
Growth 15, 41, 73, 76, 195, 196
  bacterial 15, 41, 73, 76
  photoautotrophic 195, 196
Growth conditions 6, 175, 176, 178, 208, 405, 416
Growth phase 10, 22, 150, 195, 293
Growth rate 28, 54, 178, 292, 293, 301, 309, 451, 455

H

Haloarchaea 348, 353, 354, 357, 360, 362, 365, 366, 369
Haloferax mediterranei DSM 152
Halophilic archaea 138, 349
Hardwood sulfite spent liquor (HSSL) 251
Hemicellulose fractions 39, 40
Hemicelluloses 24, 25, 26, 27, 28, 29, 32, 38, 39, 40, 208
Hemicelluloses and cellulose 27, 29, 32, 38
Homopolyester 101, 113, 115, 141, 150, 151, 152, 153
Host strain 397, 412, 414
HoxJ, histidine protein kinase 305
H. pseudoflava 54, 136, 137, 141, 392
Hydraulic retention time (HRT) 239, 240, 246, 248, 250, 251, 252, 253, 254
Hydrogenases 179, 190, 304, 305, 307
Hydrogenomonas eutrophus 386, 437
Hydrolysates 24, 27, 29, 30, 32, 34, 35, 36, 37, 38, 39, 41, 42, 43, 54, 61, 119, 402, 411
  lignocellulosic 30, 34, 39, 40
  wood 35, 36, 42
Hydrolysed wheat straw 40
Hydrolysis 25, 26, 27, 28, 29, 34, 37, 40, 68, 69, 71, 105, 107, 116, 131, 138, 143, 147, 191, 236, 238, 367, 382, 401, 437

Enzymatic 26, 27, 28, 68, 126, 136, 137
Hydroxyl groups 184, 185, 308, 310

I

Implant materials 115
Industrial production 16, 56, 235, 371, 381
Intracellular concentration 8, 413
Isobutanol 265, 319, 320, 321, 322, 323
  production 320
  production 321

J

Jar bioreactor 151, 153

K

K. aerogenes 394, 409, 410
Ketothiolase 209, 233, 273, 356
Key enzymes 195, 199, 302, 303, 310, 352, 356, 363, 364, 383, 441, 445

L

Lac genes 55
Lactobacilli 105, 108, 137, 143
Lactose 48, 49, 51, 52, 53, 55, 56, 101, 103, 105, 120, 122, 131, 135, 136, 137, 141, 142, 143, 144, 147, 149, 150, 151, 153, 367, 381, 391, 392, 393, 408, 441
  hydrolysis of 135, 136, 391
  pure 136, 150, 151, 153, 392
Lactose content 146
Lag phase 34, 195, 276, 280
Lignin 24, 25, 26, 28, 29, 33, 39, 130, 402
Lignocellulose-based agricultural waste for PHA production 37
Lignocelluloses 3, 129
Lignocellulosic biomass 23, 24, 27
Lipase box 9, 183, 191, 192, 275, 276
Lipases 9, 17, 18, 183, 191, 387, 443
Lipid substrates 3, 13, 14, 18, 21, 22
Liquefaction 68, 69, 70, 71
Liquid phase 61, 456

M

Maleylacetate 316
Malonate semialdehyde reductase (MSR) 362
Materials 24, 26, 27, 28, 33, 66, 68, 70, 101, 102, 110, 153, 402
  lignocellulosic 24, 26, 27, 28, 33, 402
  starchy 66, 68, 70, 110
  surplus 101, 102, 153
Maximum PHA content 77, 240, 242, 250
Mcl-PHA 7, 12, 13, 22, 47, 53, 63, 394, 438, 439, 440, 441, 442, 443, 445, 446, 447, 451, 458, 459, 462, 463
  elastomer 63
  extraction 458
  granules 442, 443, 459
  producers 445, 447
  producing microorganisms 445
  production 12, 22, 47, 53, 63, 394, 438, 439, 440, 441, 451, 462

commercial 463
productivity 13
synthases 440, 441
synthesis 7, 22, 441, 442, 443, 446
volumetric productivity 22
Mcl-PHAs 22, 23, 457, 459
biotechnological production of 22
density of 457, 459
flasks 23
production of 13, 62, 462
Mechanical properties 60, 68, 115, 130, 133, 270, 285, 287, 289, 397, 437, 438, 450
Mediterranei 120, 138, 139, 140, 141, 149, 348, 349, 350, 351, 352, 353, 354, 355, 356, 357, 358, 359, 360, 362, 363, 364, 365, 366, 367, 368, 369, 370, 372, 373, 374, 392
cells 362, 363
PHBHV production 368, 369
Membrane-bound hydrogenase (MBH) 304, 305, 306
Methanogenesis 238, 239
Methyl esters 59, 63, 446
Methyl ketone production 322
Methylmalonyl-CoA 182, 233, 361
Methylobacterium sp 52, 57, 136, 151, 392
Microbes 5, 51, 52, 55, 103, 108, 120, 121, 132, 134, 137, 138, 139, 150, 175, 176, 177, 190, 208, 272, 389, 394, 403, 404
photosynthetic 175, 176, 177, 208
Microbial cells 110, 242
Microbial communities 226, 244, 245, 248, 387
Microbial consortia 388
Microbial contamination 53, 120, 138, 142
Microbial growth 28, 31, 119, 121, 228, 402
Microbial populations 123, 230, 231, 246, 250
Microbial storage materials 176
Mixture, glucose/galactose 135, 137, 141, 150
MMC processes 229, 230, 231, 237, 242
Mobilization 196, 265, 269, 281, 282, 414
Molar mass characterization 150, 151, 152
Molasses 3, 43, 44, 45, 46, 47, 48, 51, 66, 110, 122, 300, 393, 394, 406, 409, 410
beet 44, 45, 46, 47, 394
sugarcane 46, 394, 409

N

Necator genes 146, 202
Necator strains 8, 146, 392, 397, 398, 403
Nitrogen deficiency 74, 196
Nitrogen deprivation 196, 201
Nitrogen source 19, 46, 53, 73, 76, 150, 151, 178, 180, 194, 297
Nitrogen starvation 178, 187, 200
Non-hydrolyzed whey permeate 147, 150
Non-PHA cell mass (NPCM) 126
Nutraceuticals 101, 104

O

Octanoate 291, 292, 405, 442, 447, 454
Oils 14, 15, 16, 18, 19, 20, 21, 22, 23, 38, 54, 58, 59, 74, 75, 119, 177, 182, 208, 296, 395, 396, 397, 398
 coconut 18, 22, 23
 olive 18, 21, 22, 397, 398
 seed 21, 22, 59
 sesame 18, 19, 22
 sunflower 14, 22, 23, 59
 waste cooking 59, 395
Open reading frames (ORFs) 311, 312
Operon, phaCAB 318, 320, 322
O-PHBHV1 films 372, 373
Organic by-products 381, 391
Organic loading rate (OLR) 239, 240, 246, 247, 248, 249, 251, 252

P

Packed-bed biofilm reactor (PBBR) 142
Palm oil 14, 17, 18, 19, 22, 23, 59, 273, 291, 292, 297, 397, 398, 410
 saponified waste 22, 23
Parental strain 11, 319, 321
Pathways 181, 271, 308, 310, 311, 312, 362, 407
 3-hydroxypropionate 362
 engineered 181, 271, 407
 gentisate 310, 311, 312
 peripheral 308
Pentoses 27, 28, 34, 35, 402, 403
PE pyrolysis wax 76
Petrochemical plastics 3, 75, 76, 114, 115
PHA accumulation 10, 40, 54, 63, 71, 74, 76, 124, 134, 154, 178, 187, 198, 202, 237, 238, 239, 242, 244, 247, 253, 254, 357, 358, 359, 360, 381, 388, 396, 398, 404, 463
PHA applications 298
PhaB genes 202, 272, 273, 356
PHA biosynthesis 17, 112, 122, 127, 134, 139, 140, 141, 146, 150, 271, 279, 288, 292, 293, 301, 317, 322, 349, 352, 358, 362, 364, 365, 413
PHA biosynthetic pathway 7, 289, 318, 356
PhaC 9, 17, 181, 183, 184, 188, 209, 234, 270, 274, 275, 276, 280, 286, 351, 352, 353, 357, 385, 395, 399, 407, 411, 414, 443, 444, 445, 447
 catalytic subunit 234
 tagged 275
PhaCAc 290
PhaC and phaE 183
PhaC enzymes 271, 276, 279
PhaC genes 352, 353, 384, 385, 387
PHA concentration 71, 72, 146, 150, 294, 366, 383, 412, 453
PHA content 40, 46, 53, 56, 61, 63, 71, 75, 121, 178, 180, 245, 297, 392, 395, 401, 415, 452, 462
PhaC subunits 183, 186, 352
PhaC synthase genes 385, 387
PHA depolymerases 10, 55, 191, 192, 281, 349
PhaEC genes 202, 354, 356
PhaE subunits 183, 185, 186, 188
PHA granule formation 279, 359
PHA granules 10, 123, 125, 188, 189, 204, 228, 277, 278, 279, 349, 356, 357, 358, 359, 360, 362, 365, 415, 416, 444, 459, 460, 461
PHA granule surface 277

PHA homeostasis 271, 272, 276, 277, 280, 284, 325
PHA mobilization 349, 357, 364, 365
PHA on waste glycerol based substrate 59, 60
PHA Produced on Monomeric Level 150, 151, 152, 153
PHA producer Ralstonia eutropha H16 in order 122
PHA-producers 394, 396, 397
PHA Producing Strains 5, 389
PHA production 4, 5, 11, 19, 20, 25, 34, 35, 36, 37, 40, 45, 46, 52, 53, 54, 56, 60, 63, 68, 70, 71, 74, 78, 109, 117, 118, 119, 120, 123, 124, 125, 127, 136, 138, 139, 141, 142, 150, 154, 177, 178, 179, 180, 186, 193, 194, 195, 198, 199, 203, 204, 226, 228, 231, 234, 237, 238, 244, 248, 249, 250, 254, 256, 284, 288, 289, 291, 292, 293, 294, 296, 299, 324, 357, 369, 381, 383, 386, 388, 391, 392, 393, 396, 397, 398, 399, 400, 401, 402, 406, 411, 414, 455
PHA production 4, 54, 56, 58, 118, 119, 121, 123, 126, 127, 131, 139, 144, 146, 154, 193, 199, 226, 229, 230, 231, 235, 238, 240, 249, 251, 256, 290, 380, 395, 414, 447
  biotechnological 131, 395
  by MMC 229, 231, 235, 238, 251, 256
  chain 380, 414
  continuous 119, 127
  costs 226, 229, 230, 249
  cyanobacterial 193
  industrial 4, 290
  on surplus whey 144
  on whey by rec 146
  processes 56, 118, 126
  processes by MMC 240
  processes market-fit 154
  rates 54, 447
  strains 123
  whey-based 121, 139, 146
PHA productivity 12, 64, 121, 123, 145, 146, 368, 402, 404
  higher 121, 368
PHA Properties Available 150, 151, 152, 153
PHA recovery 54, 117, 118, 126, 297
PhaR protein 278
PHAs 4, 15, 35, 36, 45, 53, 63, 70, 72, 74, 76, 79, 124, 112, 113, 135, 141, 142, 143, 176, 179, 183, 186, 194, 201, 208, 251, 231, 232, 285, 290, 298, 318, 321, 348, 362, 365, 374, 381, 384, 392, 393, 401, 406, 409, 411, 413, 416, 452, 454, 463
  accumulated 35, 45, 76, 135, 141, 179, 186
  accumulating 63, 72, 124, 194
  containing 362, 365
  conversion to 70, 393
  higher-value 348, 374
  high-quality 373, 374
  maximum 74, 452
  medium chain length 113, 176
  medium-chain-length 285, 411
  microbially-produced 298
  natural 112, 113

potential 384
produced 4, 53, 142, 232, 290
producing 15, 79, 201, 251, 401, 409, 416, 463
short-chain 231, 232
short chain length 113, 176, 183
synthesize 208, 318, 321, 381, 406, 413, 416
total 35, 36, 142, 454
whey lactose to 143, 392
PHAs accumulation 4, 12, 13, 47
PHAs by bacteria species 40
PHAs contents 19, 35
PHAs family 4, 17
PHAs genes 409
Phasin proteins 188, 277, 281, 360
Phasins 277, 357, 415, 443, 444
roles of 277
Phasins PhaF 443, 444
PHAs process 15
PHAs production 5, 11, 12, 13, 15, 16, 18, 19, 20, 21, 22, 34, 35, 36, 37, 38, 39, 41, 77, 78, 128, 390, 400, 404
cost-efficient 128
microbial 404
PHAs production on lipids 18
PHAs production process 5, 21, 35, 77
PHAs production processes 59
PHA storage 187, 228, 241, 251
PHA-storing bacteria 228, 229, 256
PHA-storing microorganisms 238, 239, 243, 245, 249, 250, 252, 253
PHA-storing organisms 123, 229, 240
PHA synthase activity 202, 275, 360
PHA synthase enzymes 108, 275, 286
PHA synthase genes 6, 17, 21, 121, 194, 233, 290, 291, 353, 387, 398, 399, 407, 413
PHA synthases 9, 11, 36, 121, 183, 184, 185, 189, 209, 232, 233, 243, 274, 280, 289, 349, 352, 354, 356, 357, 383, 384, 398, 406, 407, 411, 412, 445
haloarchacal 352, 354
PHA synthesis 112, 121, 147, 148, 176, 179, 183, 186, 188, 198, 231, 232, 278, 322, 356, 360, 384, 406, 440, 442
PHB 4, 6, 7, 8, 10, 12, 13, 14, 15, 17, 18, 19, 20, 21, 22, 23, 36, 37, 38, 39, 41, 42, 43, 46, 47, 52, 53, 55, 56, 60, 61, 62, 64, 70, 71, 72, 73, 74, 101, 112, 113, 119, 129, 130, 134, 135, 140, 141, 144, 145, 146, 147, 178, 180, 181, 187, 188, 189, 190, 195, 196, 197, 198, 200, 201, 202, 204, 205, 206, 207, 232, 246, 253, 254, 255, 269, 270, 272, 273, 274, 276, 277, 278, 280, 281, 282, 285, 286, 287, 288, 289, 290, 291, 295, 296, 299, 301, 302, 306, 307, 320, 354, 355, 372, 383, 388, 392, 393, 394, 395, 396, 397, 398, 401, 404, 405, 407, 408, 409, 410, 412, 414, 437, 449, 450, 451, 457, 461, 463
flasks 18, 22, 23
intracellular 280, 282, 408
PHB biosynthesis 6, 7, 8, 135, 181, 268, 272, 273, 277, 286, 289, 301

PHB concentrations 21, 55, 136, 144, 287, 295, 296, 401, 408, 409
PHB content 39, 46, 47, 55, 56, 136, 195, 196, 197, 200, 201, 204, 206, 287, 290, 394, 395, 396, 402
PHB content in biomass 20, 21, 144
Phb genes 9, 10
PHB granules 111, 181, 188, 204, 205, 271, 274, 277, 279, 280
PHB homeostasis 271, 273, 277, 278, 283
PHB homopolyester 150, 151, 152, 153
PHBHV 19, 21, 23, 42, 43, 53, 54, 60, 61, 73, 125, 128, 129, 130, 138, 145, 189, 207, 230, 234, 246, 247, 250, 251, 271, 273, 285, 287, 288, 289, 295, 299, 348, 349, 350, 356, 360, 361, 362, 366, 367, 368, 369, 370, 371, 372, 392, 396, 401
PHBHV accumulation 354, 356, 366, 369
PHBHV Biosynthesis 348, 350, 352, 354, 356, 361, 362, 366, 369, 373
PHBHV concentration 54, 366
PHBHV content 248, 366, 367, 368
  highest 247, 248
PHBHV copolyesters 122, 137, 138, 141, 142, 145
PHBHV production 61, 198, 287, 288, 350, 355, 362, 366, 367, 368, 369, 370, 371, 401
PHB production 12, 14, 37, 38, 39, 41, 45, 46, 47, 52, 55, 56, 70, 72, 78, 119, 136, 142, 149, 179, 196, 198, 200, 201, 234, 273, 277, 292, 295, 296, 301, 394, 409, 411, 413, 414, 455
PHB production on whey 53, 144
PHB synthase 6, 7, 8, 9, 272, 274, 414
PHB synthesis 7, 8, 10, 64, 119, 150, 181, 187, 195, 196, 200, 272, 405, 414
Phenolics 28, 29, 31, 36
Phenylacetate 312, 313
Phosphate limitation 19, 197, 294, 295
Photosynthetic bacteria 186, 190, 192, 208
Phototrophic bacteria, aerobic anoxygenic 180
PLA (Poly(Lactic Acid)) 104, 105, 108, 115, 128, 137, 143, 436, 450
Plant oils 14, 18, 21, 122, 289, 290, 397, 398
Polyester granule associated proteins (PGAPs) 188, 356, 357, 416
Polyhydroxyalkanoates 3, 5, 58, 60, 101, 104, 112, 176, 192, 208, 226, 227, 265, 266, 349, 350, 356, 381
Polymerase chain reaction (PCR) 370, 384, 385, 387
Polymer content 36, 40, 136, 187, 242, 402, 409, 412
Polymerization 7, 25, 67, 108, 110, 112, 182, 270, 352, 384
Polymer synthesis 195, 198, 276, 279, 287, 302, 442, 443
Polyphosphate-accumulating organisms (PAOs) 228, 232, 236, 241, 243, 246
Polystyrene 74, 75, 460
Production of PHA on waste glycerol based substrate 59, 60
Production of scl- and mcl-PHA on lipid substrates 22

Propionyl-CoA 17, 182, 232, 233, 271, 287, 288, 349, 354, 355, 356, 361, 362, 365, 396
Protocatechuate 308, 309, 312
Pseudomonas mediterranea 60, 63
Pseudomonas sp 22, 23, 36, 62, 105, 123, 243, 245, 395, 403, 410, 412, 447
Pure culture processes 229, 230, 231
Pure lactose PHB homopolyester 151, 152, 153
Purple photosynthetic bacteria 176
Purple sulfur bacterium (PSB) 179, 180

R

Ralstonia eutropha strain 312
Rastolnia eutropha cassava starch PHB Batch 73
Recombinant *E. coli* 42, 56, 407, 408, 413, 415
Refined glycerol 62, 63, 64
R. eutropha 188, 201, 265, 266, 267, 268, 269, 270, 271, 272, 273, 274, 275, 276, 277, 278, 279, 281, 282, 283, 284, 285, 286, 287, 288, 289, 290, 291, 292, 293, 294, 295, 296, 297, 299, 300, 301, 302, 303, 304, 305, 306, 307, 313, 318, 319, 320, 321, 322, 323, 324, 325, 383, 385, 386, 394, 395, 396, 398, 399, 401, 405, 408, 411, 412, 413, 414, 415
regulatory hydrogenase (rh) 305, 306
R. eutropha biomass 297, 301
R. eutropha cells 269, 276, 278, 279, 280, 281, 282, 286, 288, 301, 303, 304, 320
R. eutropha cells replete 280
R. eutropha cultures 294, 307, 321, 324
R. eutropha genome 272, 282
R. eutropha growth 289, 307
R. eutropha PhaA enzyme 273
R. eutropha PhaC 275, 276
R. eutropha Productions 284
Rhamnolipids 76, 445, 446
Rhodobacter sphaeroides 179, 186, 188
R-3-hydroxyacyl-CoA monomers 181

S

Saccharification 40, 68, 69, 72
Scl- and mcl-PHAs 22, 448
Selected mcl-PHA fed-batch production processes 454
Serine pathway 233, 234
Shaking flasks 47, 150, 151, 152, 153
Shaking flask scale 119, 135, 144, 145, 147, 392, 393
Single cell oil (SCO) 106
Single-cell protein program 284
Sludge retention time (SRT) 239, 240, 244, 246, 247, 249, 250, 251, 252, 253, 254
Sole carbon and energy source 41, 308, 317, 319
Solid state fermentation (SSF) 127, 128
Soluble hydrogenase (SH) 304, 305, 306
Solution of processed whey powder of bovine origin 153
Solution of whey powder of bovine origin 152, 153
Sophorolipids 104, 106
Source, biodegradable carbon 238

Southern Blot Hybridization (SBH) 384, 385
Soybean oil 14, 18, 19, 22, 59, 283, 290, 291, 397, 398
Soy molasses 44, 47, 394
Spent coffee grounds (SCG) 14, 19, 20, 38, 43, 296
Spirulina 196, 206, 207
Starch hydrolysis 68, 69, 72
Streams, low-cost processing 408, 410
Substrate binding domain (SBDs) 192
Substrates16, 23, 42, 64, 65, 70, 73, 175, 179, 208, 236, 237, 238, 311, 315, 316, 397, 401, 412
  lignocellulose-based 23, 42
  oily 16, 397
  organic 175, 179, 208, 236, 237, 238
  starch-based 64, 70, 73
  starchy 65, 401, 412
  starting 311, 315, 316
Succiniciproducens sacC 395
Succinyl-CoA 182, 233, 309, 313, 316, 361
Sugar beets 43, 44, 409
Sugar industry 3, 43, 44, 393
Sugar molasses 43, 44, 249
  cane 44
  fermented 249
Surfactants 104, 106, 461
Sweet whey 49, 134, 408
Synechococcus sp 194, 197, 200, 202
Syngas 122, 306, 307, 413
Synthases 9, 185, 186, 188, 233, 270, 274, 406, 438, 440, 443, 444, 445
Synthesizing PHA, cyanobacterial strains capable of 193

T

Tailor-made PHA 101, 349, 372, 373
Tensile strength 115, 130, 207, 450
Terpolyesters 62, 101, 113, 139, 141, 151, 152, 349, 372
Terpolymers 8, 36, 53, 399
Tetrad-forming organisms (TFOs) 246
Thauera 246, 247, 248, 250, 252
Thermal analysis 150, 151, 152
Thermoelastomer 436, 437, 439, 447
Thermomechanical properties 437, 438, 447
Thermoplastics 113, 254, 437
Three-step PHA production process 239
Transesterification 14, 58, 116, 396, 403, 446
Triacylglycerols 14, 16, 274, 383, 403, 411
Triglycerides 58, 397, 399

U

Unrelated carbon sources 53, 287, 381, 382, 406, 407, 416, 441, 446
Upstream processing 3, 101

V

Vegetable oils 14, 16, 403, 411
Volumetric productivity 5, 20, 22, 118, 119, 127, 138, 230, 291, 366, 367, 368, 369, 452, 453, 454

W

Waste 3, 4, 5, 14, 15, 16, 18, 19, 22, 23, 37, 38, 40, 42, 49, 54, 58, 60, 61, 63, 64, 70, 73, 77, 78, 79, 103, 110, 111, 119, 120, 125, 152, 226, 227, 232, 236, 238, 239, 244, 253, 256, 266, 295, 296, 300, 304, 389, 390, 391, 395, 396, 397, 400, 403, 404, 405, 406, 408, 411, 462
  frying oils 14, 19, 22, 23, 54, 119, 152, 296, 300, 396
  glycerol 60, 61, 63, 64, 403, 404, 405, 406, 411, 462
  lipids 14, 18, 395, 396
  materials 37, 38, 408
  oils 14, 15, 111, 120, 296
  potato starch 70, 73
  products 3, 5, 60, 79
  substrates 4, 16, 40, 78, 390
  treatment 226, 227, 256
  waters 58, 70, 77, 125, 232, 236, 239, 256
    condensate 244
Wheat starch 73
Whey 48, 49, 50, 51, 52, 53, 55, 56, 57, 77, 100, 101, 102, 103, 104, 105, 106, 108, 119, 120, 122, 131, 132, 133, 134, 136, 139, 141, 143, 144, 145, 146, 147, 148, 149, 150, 151, 152, 153, 154, 177, 366, 367, 368, 391, 392
  acidic 48, 49
  acidified 145, 151, 152
  concentrated 57, 131
  conversion of 105, 122
  fat 131
  generated 101, 103
  hydrolyzed 367, 368, 392
  native 148, 149
Whey-Based PHA 101, 135, 136, 137, 138, 139, 140, 144, 145, 146, 148, 154, 155
  -Producers 135, 144, 145
  production 101, 135, 136, 137, 138, 139, 140, 146, 148, 154, 155
Whey hydrolysates 51, 54, 119, 142, 367
Whey lactose 104, 105, 111, 124, 136, 137, 143, 147, 367, 372, 392
  direct conversion of 136, 392
  hydrolysis 51, 143
Whey permeate 51, 53, 55, 57, 120, 131, 133, 134, 137, 138, 146, 147, 148, 149, 151, 152, 153, 393, 408
  cheese 77, 136, 392
  hydrolysed 147, 149
  hydrolyzed 53, 55, 57
Whey powder 102, 140, 146, 152, 153
  of bovine origin 152, 153
Whey proteins 19, 48, 50, 132, 133
Whey retentate 132, 134
Whey solution, concentrated 55, 146, 408
Whey sugars 54, 148
Whey supernatant 151
Whey transportation 50
Wild type strain 122, 282, 390, 405, 413, 415

X

Xylans 25, 36, 412
Xylose 27, 35, 36, 39, 42, 43, 135, 251, 381, 402, 403, 411, 412

Y

Yeast extract 51, 54, 141, 366, 367, 368

Made in the USA
Middletown, DE
04 November 2026

36496628R10291